John Belknap Marcou

Bibliography of publications relating to the collection of fossil invertebrates in the United States National Museum:

Including complete lists of the writings of Fielding B. Meek, Charles A. White and

Charles D. Walcott

John Belknap Marcou

Bibliography of publications relating to the collection of fossil invertebrates in the United States National Museum:
Including complete lists of the writings of Fielding B. Meek, Charles A. White and Charles D. Walcott

ISBN/EAN: 9783337719029

Printed in Europe, USA, Canada, Australia, Japan

Cover: Foto ©ninafisch / pixelio.de

More available books at **www.hansebooks.com**

Department of the Interior:

U. S. NATIONAL MUSEUM.

—— Serial Number 40 ——

BULLETIN

OF THE

UNITED STATES NATIONAL MUSEUM.

No. 30.

PUBLISHED UNDER THE DIRECTION OF THE SMITHSONIAN INSTITUTION.

WASHINGTON:
GOVERNMENT PRINTING OFFICE.
1885.

ADVERTISEMENT.

This work (Bulletin No. 30) is the fortieth of a series of papers intended to illustrate the collections of natural history and ethnology belonging to the United States, and constituting the National Museum, of which the Smithsonian Institution was placed in charge by the act of Congress of August 10, 1846. It constitutes the third of the series of bibliographies illustrating the work of the Museum.

It has been prepared at the request of the Institution, and printed by authority of the honorable Secretary of the Interior.

The publications of the National Museum consist of two series—the Bulletins, of which this is No. 30, in continuous series, and the Proceedings, of which the eighth volume is now in press.

The volumes of Proceedings are printed signature by signature, each issue having its own date, and a small edition of each signature is distributed to libraries promptly after its publication.

From time to time the publications of the Museum which have been issued separately are combined together and issued as volumes of the Miscellaneous Collections. These are struck off from the stereotype plates from which the first edition was printed, and in this form are distributed by the Smithsonian Institution to libraries and scientific societies throughout the world. Volume 13 of these collections includes Bulletins 1 to 10, inclusive; volume 19, volumes 1 and 2 of the Proceedings; volume 22, volumes 3 and 4 of the Proceedings; and volume 23, Bulletins 11 to 15, inclusive.

Full lists of the publications of the Museum may be found in the current catalogues of the publications of the Smithsonian Institution.

SPENCER F. BAIRD,
Secretary of the Smithsonian Institution.

SMITHSONIAN INSTITUTION,
Washington, December 1, 1885.

BIBLIOGRAPHIES OF AMERICAN NATURALISTS.

III.

BIBLIOGRAPHY OF PUBLICATIONS

RELATING TO

THE COLLECTION OF FOSSIL INVERTEBRATES

IN THE

UNITED STATES NATIONAL MUSEUM,

INCLUDING

COMPLETE LISTS OF THE WRITINGS OF FIELDING B. MEEK, CHARLES A. WHITE, AND CHARLES D. WALCOTT.

BY

JOHN BELKNAP MARCOU.

WASHINGTON:
GOVERNMENT PRINTING OFFICE.
1885.

CONTENTS.

	Page.
Introductory note	7
Part I.—The published writings of Fielding Bradford Meek	9
Part II.—The published writings of Charles Abiathar White	113
Part III.—The published writings of Charles Doolittle Walcott	183
Part IV.—Publications based upon the Paleontological collections of the United States Government by—	

	Page.		Page.
Jacob Whitman Bailey	203	Alpheus Hyatt	239
Timothy Abbot Conrad	205	Jules Marcou	241
James Dwight Dana	223	John Strong Newberry	245
Christian Gottfried Ehrenberg	229	David Dale Owen	247
James Hall	231	James Schiel	253
Angelo Heilprin	237	Benjamin F. Shumard	255
	Robert Parr Whitfield	259	

Supplement			273
	Page.		Page.
J. W. Bailey	273	Hiram A. Prout	273
T. N. Nicollet	273	Benjamin F. Shumard	274
Index of genera and species			275
General Index			327

INTRODUCTORY NOTE.

In preparing the following catalogues chronologic order has been followed under the different authors, and in Part IV the fifteen different authors are arranged, first, alphabetically and then chronologically under each author. Correctness of bibliographic form and detail has not been so much sought after as completeness and accuracy in the numerous references. A general alphabetic index of species will be found at the end of the volume. The compiler will be glad to have his attention called to any errors or omissions. Dr. White's bibliography is prepared from data furnished by himself. I am much indebted to Dr. White for his kind advice and help in the preparation of the work.

PART I.

THE PUBLISHED WRITINGS OF FIELDING BRADFORD MEEK.

1.—THE PUBLISHED WRITINGS OF FIELDING BRADFORD MEEK.

BIOGRAPHIC SKETCH OF FIELDING BRADFORD MEEK.*

On December 10, 1817, Fielding Bradford Meek was born in the city of Madison, Ind. His grandparents were Irish Presbyterians, who removed from Armagh County to America about 1768, and who finally settled in Hamilton County, Ohio. His father, together with his family, removed from there to Madison, where he was a lawyer of considerable eminence. The family, including those born in Madison, consisted of the parents, two sons and two daughters besides Fielding, all of whom died several years before him. The father died when the son who was to become so distinguished a paleontologist was only three years old, leaving the family in moderate circumstances. Mr. Meek's early youth was passed in Madison. His education was much impeded by the delicate condition of his health. Upon reaching manhood, by advice of his friends and against his own inclination, for he was of a studious and retiring disposition, he invested his small patrimony in mercantile business, first in his native place, and afterward in Owensborough, Ky. The result was financial failure and loss of all he possessed. After this, while laboring for his support and struggling with ill health and poverty, he continued his studies, general and special, for he began early to devote himself to natural history. His first public work was during the years 1848 and 1849, and was performed, as an assistant of Dr. D. D. Owen, upon the United States geological survey of Iowa, Wisconsin, and Minnesota.

Having closed this work, he returned to his home in Owensborough, but soon after, in the year 1852, went to Albany, N. Y., as assistant to Professor Hall, in the paleontologic work of that State. He remained there until 1858, serving three summers. Two of these summers were spent on the geologic survey of Missouri; the other, that of 1853, was employed in exploring the Bad Lands of Nebraska, together with Dr. F. V. Hayden, both being commissioned by Professor Hall for that work. Three

---
*This sketch is compiled from data taken down at Mr. Meek's dictation by a friend shortly before his death.

years after this exploration, he prepared for publication, in conjunction with Professor Hall, an important memoir on cretaceous fossils from Nebraska. In 1858, Mr. Meek left Albany and took up his residence in Washington, where he continued to live until his death. His home, and the place of his scientific work, except when in the field, was at the Smithsonian Institution, and it was within its walls that the greatest part of his scientific work was accomplished. The association which he formed with Dr. Hayden in 1853 was tacitly continued until Mr. Meek's death. When Dr. Hayden commenced his explorations in the Western Territories, and afterwards organized the Geological Survey of the Rocky Mountain region, Mr. Meek was entrusted with all the invertebrate paleontology, much of which appeared under their joint names. It was the custom of Mr. Meek to publish preliminary descriptions of his new species, and afterward elaborate and illustrate the subjects for final publication. Thoroughness, scrupulous exactness, and nice powers of discrimination are manifested in all his labors; and with such merits his works will shed luster upon his name as long as paleontology is studied. No one in America has done more than he to systematize and advance the science to which he devoted his life. His personal character cannot be too highly eulogized, for it was without a blemish. He was a genial, sincere, pure-minded, honorable man. Gentleness and candor were apparent in every expression of his face, and in every word he uttered; but he was self-reliant and ready at all times to defend what he believed to be right, and with his keen sense of justice, he was seldom mistaken as to what was right. He was never in vigorous health and often ill; but never complaining, always hopeful, always cheerful, always at the work he loved so well, always helpful of others. His hearing began to fail in early manhood, and the affliction increased until he became entirely deaf several years before he died. Even when cut off from conversation with his fellow-men his cheerfulness did not forsake him; but he seemed to derive great pleasure from written communication with his friends. He was never married, and leaves no near relatives; but all with whom he was ever brought in contact will remember him with pleasure, while to those who were permitted to enjoy scientific intercourse or correspondence with him during his life, his memory will be especially dear. He died at Washington, D. C., on the 21st of December, 1876, having only a few days before completed his 59th year. He had been in ill health for several years past, and indeed almost all his life, for his malady was inherited *phthisis pulmonalis*. It had been his custom for several years to spend the winter in Florida, and the summer months in the Alleghany Mountains. He had made preparations to leave Washington for Florida early in December, but was taken with hemorrhage of the lungs on the day before the one set for his departure. He never rallied from this attack, but gradually sank to a peaceful and quiet death.

1.

MEEK, F. B. Mr. Meek's report on Moniteau County. <The first and second annual reports of the geological survey of Missouri, by G. C. Swallow, State geologist. Part II. pp. 96-117. Jefferson City, 1855.

CONTENTS OF THE REPORT.

	Page.
Quaternary deposits	96
Carboniferous system.—Encrinital limestone	98
Chemung group	101
Devonian system.—Upper Helderberg.—Onondaga limestone	103
Lower Silurian strata:	
Saccharoidal sandstone	105
Second magnesian limestone	106
Economical geology:	
Soil	109
Building stones	110
Fire-stones, limestones for lime	110
Sand for making mortar, hydraulic limestone, millstones, materials for the construction of roads, clays for the manufacture of bricks	111
Coal	112
Iron ore	115
Lead	115
High-Point Mine	117

Mr. Meek also drew the figures illustrating the paleontological report in this volume. A geological map of the county is also given.

2.

HALL, JAMES, and MEEK, F. B. Descriptions of new species of fossils from the Cretaceous formations of Nebraska; with observations upon Baculites ovatus and B. compressus, and the progressive development of the septa in Baculites, Ammonites, and Scaphites. <Mem. Am. Acad. Arts and Sci., vol. v, new ser., pp. 379-411, 8 plates. 1856. Cambridge, 1856.

The fossils described in this paper were collected in 1853 by F. B. Meek and F. V. Hayden. The paper was communicated June 27, 1854, but was not published till 1856.

	Page.
Callianassa danai, n. s., H. & M., 1856, pl. i, fig. 1 a, b	379-380
Lingula subspatulata, n. s., H. & M., 1856, pl. i, figs. 2 a, b	380
Caprinella coralloidea, n. s., H. & M., 1856, pl. ii, figs. 3 a-f	380-381
Pecten rigida, n. s., H. & M., 1856, pl. i, figs. 4 a-c	381
Avicula haydeni, n. s., H. & M., 1856, pl. i, figs. 5 a, b	382
Lucina subundata, n. s., H. & M., 1856, pl. i, figs. 6 a, b	382
Cytherea orbiculata, n. s., H. & M., 1856, pl. i, fig. 7	382-383
Cytherea tenuis, n. s., H. & M., 1856, pl. i, figs. 8 a-c	383
Crassatella evansii, n. s., H. & M., 1856, pl. i, figs. 9 a-e	383-384
Pectunculus siouxensis, n. s., H. & M., 1856, pl. i, fig. 12	384
Nucula subnasuta, n. s., H. & M., 1856, pl. i, figs. 10 a-c	384-385
Nucula ventricosa, n. s., H. & M., 1856, pl. i, fig. 11 a, b	385
Capulus occidentalis, n. s., H. & M., 1856, pl. i, fig. 13 a-d	385-386
Inoceramus sublaevis, n. s., H. & M., 1856, pl. ii, fig. 1 a, b	386
Inoceramus convexus, n. s., H. & M., 1856, pl. ii, fig. 2 a, b	386-387
Inoceramus tenuilineatus, n. s., H. & M., 1856, pl. ii, fig. 3 a, b	387
Inoceramus conradi, n. s., H. & M., 1856, pl. ii, fig. 5 a, b	387-388
Inoceramus fragilis, n. s., H. & M., 1856, pl. ii, figs. 6 a, b	388
Natica obliquata, n. s., H. & M., 1856, pl. iii, figs. 1 a, b	389
Natica concinna, n. s., H. & M., 1856, pl. iii, figs. 2 a-d	389
Natica paludinaeformis, n. s., H. & M., 1856, pl. iii, fig. 3 a-c	389-390
Actaeon concinnus, n. s., H. & M., 1856, pl. iii, figs. 4 a-d	390
Buccinum? vinculum, n. s., H. & M., 1856, pl. iii, figs. 5 a-b	390-391
Fusus shumardii, n. s., H. & M., 1856, pl. iii, figs. 6 a-c	391
Fusus constrictus, n. s., H. & M., 1856, pl. iii, figs. 7 a-d	391-392
Fusus? tenuilineata, n. s., H. & M. 1856, pl. iii, figs. 8 a-c and 9 a-c	392
Rostellaria fusiformis, n. s., H. & M., 1856, pl. iii, fig. 10 a, b	393
Dentalium gracilis, n. s., H. & M., 1856, pl. iii, fig. 11 a-c	393

	Page.
H. I. x bridge n. s., H. & M., 1856, pl. iii, fig. 12 *a*, *b*	394
Ammonites complexus, n. s., H. & M., 1856, pl. iv, fig. 1 *a–f*	394–395
Ammonites percarinatus, n. s., H. & M., 1856, pl. iv, fig. 2 *a–e*	396
Hamites mortoni, n. s., H. & M., 1856, pl. iv, figs. 3 *a–e*	396–397
Ancyloceras ? nicolletii, n. s., H & M., 1856, pl. iv, fig. 4	397–398
Baculites ovatus and *B. compressus*, Say	398–399
Baculites ovatus, Say, pl. v, fig 1. *a–e* ; pl. vi, figs. 1–7	399–400
Baculites compressus, Say, pl. v, figs. 2 *a–b* ; pl. vi, figs. 8–9	400–402
Baculites grandis, n. s., H. & M., 1856, pl vii, fig. 1, 2 ; pl. viii, figs. 1, 2 ; pl. vi, fig. 10	402–403

The same paper contains—

"Section of the members of the Cretaceous formation as observed on the Missouri River, and thence westward to the Mauvaises Terres	405
List of fossils heretofore identified and described from the Cretaceous formation of Nebraska	405
List of species common to the Cretaceous formations of Nebraska and New Jersey	406
List of the new species of fossils described in the preceding paper	406
In the additions and corrections	411

Fusus constrictus is referred to the genus *Buccinum*.
Hamites mortoni is regarded as belonging to the genus *Ancyloceras* as defined by Pictet.

3.

MEEK, F. B., *and* HAYDEN, F. V. Descriptions of new species of Gasteropoda, from the Cretaceous formation of Nebraska Territory. <Proc. Acad. Nat. Sci. Phila., vol. viii, pp. 63–69. 1856. Philadelphia, 1857.

	Page.
Scalaria cerethiformis, n. s., M. & H., 1856	63
Acteon subellipticus, n. s., M. & H., 1856	63–64
Avalana subglobosa, n. s., M. & H., 1856	64
Natica ? ambigua, n. s., M. & H., 1856	64
Natica occidentalis, n. s., M. & H., 1856	64
Natica moreauensis, n. s., M. & H., 1856	64
Turbo nebrascensis, n. s., M. & H., 1856	64
Turbo tenuilineatus, n. s., M. & H., 1856	64–65
Rostellaria biangulata, n. s., M. & H., 1856	65
Fusus dakotaensis, n. s., M. & H., 1856	65
Fusus galpinianus, n. s., M. & H., 1856	65
Fusus contortus, n. s., M. & H., 1856	65
Fusus culbertsoni, n. s., M. & H., 1856	66
Fusus dexnocostatus, n. s., M. & H., 1856	66
Fusus newberryi, n. s., M. & H., 1856	66
Pyrula bairdi, n. s., M. & H., 1856	66
Fasciolaria cretacea, n. s., M. & H., 1856	66–67
Fasciolaria buccinoides, n. s., M. & H., 1856	67
Buccinum ? nebrascensis, n. s., M. & H., 1856	67
Capulus fragilis, n. s., M. & H., 1856	68
Helcion Montfort	68
Helcion sexsulcatus, n. s., M. & H., 1856	68
Helcion patelliformis, n. s., M. & H., 1856	68
Helcion alveolus, n. s., M. & H., 1856	68
Helcion subovatus, n. s., M. & H., 1856	68
Helcion carinatus, n. s., M. & H., 1856	68–69
Dentalium fragilis, n. s., M. & H., 1856	69
Bulla ? laria, n. s., M. & H., 1856	69
Bulla minor, n. s., M. & H., 1856	69
Bulla occidentalis, n. s., M. & H., 1856	69

4.

MEEK, F. B., *and* HAYDEN, F. V. Descriptions of new species of Gasteropoda and Cephalopoda, from the Cretaceous formations of Nebraska Territory. <Proc. Acad. Nat. Sci. Phila., vol. viii, pp. 70–72. 1856. Philadelphia, 1857.

	Page.
Turritella convexa, n. s., M. & H., 1857	70
Turritella moreauensis, n. s., M. & H., 1857	70

	Page.
Belemnitella? bulbosa, n. s., M. & H., 1857	70
Ammonites halli, n. s., M. & H., 1857	70–71
Ancyloceras? nobrascensis, n. s., M. & H., 1857	71
Ancyloceras? cheyenensis, n. s., M. & H., 1857	71–72

5,

MEEK, F. B., *and* HAYDEN, F. V. Descriptions of twenty-eight new species of Acephala and one Gasteropod, from the Cretaceous formations of Nebraska Territory. <Proc. Acad. Nat. Sci. Phila., vol. viii, pp. 81–87. 1856. Philadelphia, 1857.

	Page.
Pholadomya undata, n. s., M. & H., 1856	81
Goniomya americana, n. s., M. & H., 1856	81–82
Solen subplicatus, n. s., M. & H., 1856	82
Tellina gracilis, n. s., M. & H., 1856	82
Tellina equilateralis, n. s., M. & H., 1856	82
Tellina? cheyenensis, n. s., M. & H., 1856	82
Tellina scitula, n. s., M. & H., 1856	82
Tellina subelliptica, n. s., M. & H., 1856	83
Tellina prouti, n. s., M. & H., 1856	83
Cytherea deweyi, n. s., M. & H., 1856	83
Cytherea nebrascensis, n. s., M. & H., 1856	83
Corbula ventricosa, n. s., M. & H., 1856	83
Corbula moreauensis, n. s., M. & H., 1856	83–84
Corbula? gregaria, n. s., M. & H., 1856	84
Astarte gregaria, n. s., M. & H., 1856	84
Nucula scitula, n. s., M. & H., 1856	84
Nucula evansi, n. s., M. & H., 1856	84
Nucula equilateralis, n. s., M. & H., 1856	84–85
Nucula subplana, n. s., M. & H., 1856	85
Nucula cancellata, n. s., M. & H., 1856	85
Nucula planomarginata, n. s., M. & H., 1856	85
Pectunculina parvula, n. s., M. & H., 1856	85
Arca (Cucullæa) cordata, n. s., M. & H., 1856	86
Arca (Cucullæa) shumardi, n. s., M. & H., 1856	86
Mytilus attenuatus, n. s., M. & H., 1856	86
Avicula? fibrosa, n. s., M. & H., 1856	86–87
Inoceramus ventricosus, n. s., M. & H., 1856	87
Pecten nebrascensis, n. s., M. & H., 1856	87
Natica suberassa, n. s., M. & H., 1856	87

6.

MEEK, F. B., *and* HAYDEN, F. V. Descriptions of new species of Acephala and Gasteropoda, from the Tertiary formations of Nebraska Territory; with some general remarks on the geology of the country about the sources of the Missouri River. <Proc. Acad. Nat. Sci. Phila., vol. viii, pp. 111–126. 1856. Philadelphia, 1857.

	Page.
Formations immediately beneath the Tertiary in this district	113
Formations at the base of the Cretaceous of this district	114
Cyclas formosa, n. s., M. & H., 1856	115
Cyclas fragilis, n. s., M. & H., 1856	115
Cyclas subellipticus, n. s., M. & H., 1856	115
Cyrena moreauensis, n. s., M. & H., 1856	115–116
Cyrena intermedea, n. s., M. & H., 1856	116
Cyrena occidentalis, n. s., M. & H., 1856	116
Corbula subtrigonalis, n. s., M. & H., 1856	116
Corbula perundata, n. s., M. & H., 1856	116
Corbula mactriformis, n. s., M. & H., 1856	117
Unio priscus, n. s., M. & H., 1856	117
Bulimus? teres, n. s., M. & H., 1856	117–118
Bulimus? vermiculus, n. s., M. & H., 1856	118

	Page.
Bulimus limneaformis, n. s., M. & H., 1856	118
Bulimus nebrascensis, n. s., M. & H., 1856	118
Pupa helicoides, n. s., M. & H., 1856	118–119
Limnæa tenuicosta, n. s., M. & H., 1856	119
Physa longiuscula, n. s., M. & H., 1856	119
Physa rhomboidea, n. s., M. & H., 1856	119
Physa nebrascensis, n. s., M. & H., 1856	119–120
Physa subelongata, n. s., M. & H., 1856	120
Planorbis subumbilicatus, n. s., M. & H., 1856	120
Planorbis convolutus, n. s., M. & H., 1856	120
Valvata (Ancylus) minuta, n. s., M. & H., 1856	120
Paludina multilineata, n. s., M. & H., 1856	120–121
Paludina retusa, n. s., M. & H., 1856	121
Paludina leai, n. s., M. & H., 1856	121–122
Paludina retusa, n. s., M. & H., 1856	122
Paludina conradi, n. s., M. & H., 1856	122
Paludina peculiaris, n. s., M. & H., 1856	122
Paludina trochiformis, n. s., M. & H., 1856	122–123
Paludina leidyi, n. s., M. & H., 1856	123
Valvata parvula, n. s., M. & H., 1856	123
Melania minutula, n. s., M. & H., 1856	123–124
Melania anthonyi, n. s., M. & H., 1856	124
Melania multistriata, n. s., M. & H., 1856	124
Melania nebrascensis, n. s., M. & H., 1856	124–125
Melania convexa, n. s., M. & H., 1856	125
Cerithium nebrascensis, n. s., M. & H., 1856	125

On page 126 *Pyrula bairdi* is changed to Busycon (Bolten), and called *B. bairdi*.

7.

MEEK, F. B., *and* HAYDEN, F. V. Descriptions of new fossil species of Mollusca collected by Dr. F. V. Hayden in Nebraska Territory, together with a complete catalogue of all the remains of Invertebrata hitherto described and identified from the Cretaceous and Tertiary formations of that region. <Proc. Acad. Nat. Sci. Phila., vol. viii, pp. 265–286, 1856. Philadelphia, 1857.

	Page.
Vertical section of the geological formation of Nebraska Territory, so far as determined	269
Natica tuomyana, n. s., M. & H., 1856	270
Bulla subcylindrica, n. s., M. & H., 1856	270
Panopæa occidentalis, n. s., M. & H., 1856	270
Mactra formosa, n. s., M. & H., 1856	271
Mactra warrenana, n. s., M. & H., 1856	271
Mactra alta, n. s., M. & H., 1856	271–272
Tellina subtortuosa, n. s., M. & H., 1856	272
Venus? circularis, n. s., M. & H., 1856	272
Cytherea pellucida, n. s., M. & H., 1856	272–273
Cytherea arenaria, n. s., M. & H., 1856	273
Lucina occidentalis, n. s., M. & H., 1856	273–274
Hettangia americana, n. s., M. & H., 1856	274
Cardium speciosum, n. s., M. & H., 1856	274–275
Nucula obsoletistriata, n. s., M. & H., 1856	275
Cucullæa exigua, n. s., M. & H., 1856	275–276
Mytelus subarenatus, n. s., M. & H., 1856	276
Gervilia subtortuosa, n. s., M. & H., 1856	276
Inoceramus pectenuis, n. s., M. & H., 1856	276–277
Inoceramus incurvus, n. s., M. & H., 1856	277
Ostrea patina, n. s., M. & H., 1856	277
Catalogue of all the invertebrate fossil remains hitherto described and identified, from the Tertiary and Cretaceous formations of Nebraska Territory	278
Tertiary species	278
Cretaceous species	280

8.

MEEK, F. B. Description of new organic remains from the Cretaceous rocks of Vancouver's Island. < Trans. Albany Inst., vol. iv, pp. 37-49. 1857. Albany, 1858-1864.

The fossils here described were collected by Dr. J. S. Newberry, geologist of Lieutenant Williamson's North California and Oregon Exploring Expedition. The author describes twelve new species.

	Page.
Nucula traskana, n. s., Meek, 1857	39
Arca (Cucullœa) equilateralis, n. s., Meek, 1857	39–40
Arca vancouverensis, n. s., Meek, 1857	40
Cardium scitulum, n. s., Meek, 1857	40–41
Ph[o]ladomya (Goniomya) borealis, n. s., Meek, 1857	41–42
Ph[o]ladomya subelongata, n s., Meek, 1857	42
Trigonia evansana, n. s., Meek, 1857	43
Thracia? occidentalis, n. s., Meek, 1857	43–44
Thracia? subtruncata, n. s., Meek, 1857	44
Dentalium nanaimoensis, n. s., Meek, 1857	44–45
Ammonites (Scaphites?) ramosus, n. s., Meek, 1857	45–47
Ammonites newberryanus, n. s., Meek, 1857	47–48
Baculites ovatus? (Say)	48–49

Mr. Meek gives the date of these species as 1857 when he republished them with illustrations in the Bull. U. S. Geol. and Geogr. Surv. Terr., vol. ii, No. 4. Washington, 1876.

9.

MEEK, F. B., *and* HAYDEN, F. V. Descriptions of new species and genera of fossils collected by Dr. F. V. Hayden in Nebraska Territory, under the direction of Lieut. G. K. Warren, U. S. Topographical Engineers; with some remarks on the Tertiary and Cretaceous formations of the Northwest and the parallelism of the latter with those of other portions of the United States and Territories. < Proc. Acad. Nat. Sci. Phila., 1st series, vol. ix, pp. 117-148. 1857. Philadelphia, 1858.

Genera Pseudobuccinum, Corbulamella.

	Page.
Gives a brief account of the results of previous explorations	117–121
Section of Tertiary beds thirteen miles above Fort Clark	122
Section of fresh-water and estuary deposits near the mouth of Judith River	124
Section of the older deposits at the mouth of Judith River, in the descending order	125
Alabama section, from facts communicated by Prof. A. Winchell	126
New Jersey section, compiled from the reports of that State	127
Vertical section of the geological formations of Nebraska Territory, so far as determined	128
Sections of the rocks in Northeastern Kansas (above the Coal Measures). By Major F. Hawn, U. S. deputy surveyor	130
Section of the formations at Pyramid Mountain, New Mexico [by Jules Marcou]	132
Conclusions	133
Ptychoceras mortoni, n. s., M. & H., 1857	134
Serpula? tenuicarinatus, n. s., M. & H., 1857	134
Vitrina obliqua, n. s., M. & H., 1857	134
Helix occidentalis, n. s., M. & H., 1857	135
Helix vitrinoides, n. s., M. & H., 1857	135
Planorbis tenuivolvis, n. s., M. & H., 1857	135
Planorbis amplexus, n. s., M. & H., 1857	135–136
Planorbis fragilis, n. s., M. & H., 1857	136
Melania subtortuosa, n. s., M. & H., 1857	136
Melania omitta, n. s., M. & H., 1857	136
Melania sublaevus, n. s., M. & H., 1857	136–137
Melania invenusta, n. s., M. & H., 1857	137
Melania warrenana, n. s., M. & H., 1857	137
Melania tenuicarinata, n. s., M. & H., 1857	137–138
Melania convexa, M. & H.	138
Fusus vaughani, n. s., M. & H., 1857	138–139
Fusus subturritus, n. s., M. & H., 1857	139

	Page.
Fusus intertextus, n. s., M. & H., 1857	139
Fusus (Pleurotoma?) scarboroughi, n. s., M. & H., 1857	139-140
Pseudobuccinum, n. g., M. & H., 1857	140
Pseudobuccinum nebrascensis, M. & H.	140-141
Xylophaga elegantula, n. s., M. H., 1857	141
Xylophaga stimpsoni, n. s., M. & H., 1857	141-142
Pholadomya subventricosa, n. s., M. & H., 1857	142
Solen? dakotensis, n. s., M. & H., 1857	142
Corbulamella, n. g., M. & H., 1857	142-143
Corbulamella gregarea, M. & H	143
Cyprina arenaria, n. s., M. & H., 1857	143
Cyprina cordata, n. s., M. & H., 1857	143
Cyprina compressa, n. s., M. & H., 1857	144
Cyprina subtumida, n. s., M. & H., 1857	144
Cyprina ovata, n. s., M. & H., 1857	144
Unio danai, n. s., M. & H., 1857	145
Unio deweyanus, n. s., M. & H., 1857	145-146
Unio subspatulata, n. s., M. & H., 1857	146
Pectunculus subimbricatus, n. s., M. & H., 1857	146
Ostrea glabra, n. s., M. & H., 1857	146-147
Ostrea translucida, n. s., M. & H., 1857	147
Ileminster? humphreysanus, n. s., M. & H., 1857	147-148

10.

MEEK, F. B., *and* HAYDEN, F. V. Fossils of Nebraska. Letter from F. B. Meek and F. V. Hayden to G. K. Warren, Lieut. Topog. Eng., dated Washington February 8, 1858; printed in the National Intelligencer of March 16. <Am. Journ. Sci., vol. XXV, 2d ser., pp. 439-442. New Haven, 1858.

This article is mainly geologic, only the genera, which characterize the formations under discussion, being enumerated.

11.

MEEK, F. B., *and* HAYDEN, F. V. Descriptions of new organic remains collected in Nebraska Territory, in the year 1857, by Dr. F. V. Hayden, geologist to the exploring expedition under the command of Lieut. G. K. Warren, Top. Eng., U. S. Army; together with some remarks on the geology of the Black Hills and portions of the surrounding country. <Proc. Acad. Nat. Sci. Phila., vol. x, pp. 41-59. 1858. Philadelphia, 1859.

Jurassic fossils. Afterward republished and illustrated, in 1865, in Paleontology of the Upper Missouri, Smithsonian Contributions to Knowledge, 172.

	Page.
General section of the geological formations seen in and near the Black Hills (descending)	43-44
Carboniferous rocks of the Black Hills	47
Lower Silurian	49
Potsdam sandstone	49
Metamorphic and igneous rocks	49
Pentacrinus astericus, n. s., M. & H., 1858	49-50
Lingula brevirostris, n. s., M. & H., 1858	50
Inoceramus umbonatus, n. s., M. & H., 1858	50
Avicula (Monotis) tenuicostata, n. s., M. & H., 1858	50-51
Mytilus pertenuis, n. s., M. & H., 1858	51
Arca (Cucullæa) inornata, n. s., M. & H., 1858	51
Unio nuculis, n. s., M. & H., 1858	52
Corbula inornata, n. s., M. & H., 1858	52
Panopæa (Myacites) subelliptica, n. s., M. & H., 1858	52-53
Teredo globosa, n. s., M. & H., 1858	53
Pholas cuneata, n. s., M. & H., 1858	53
Acteon (Solidula) attenuata, n. s., M. & H., 1858	54
Helicoceras? tortus, n. s., M. & H., 1858	54-55
Turrilites (Helicoceras) cochleatus, n. s., M. & H., 1858	55-56
Turrilites? umbilicatus, n. s., M. & H., 1858	56

WRITINGS OF F. B. MEEK AND F. V. HAYDEN.   19

	Page.
Helicoceras tenuicostatus, n. s., M. & H., 1858	56
Ancyloceras (Hamites) uncus, n. s., M. & H., 1858	56-57
Ammonites cordiformis, n. s., M. & H., 1858	57
Ammonites henryi, n. s., M. & H., 1858	57-58
Scaphites larvæformis, n. s., M. & H., 1858	58
Belemnites densus, n. s., M. & H., 1858	58-59

12.

MEEK, F. B., *and* HAYDEN, F. V.   Remarks on the Lower Cretaceous beds of Kansas and Nebraska, together with descriptions of Carboniferous fossils from the valley of Kansas River. <Proc. Acad. Nat. Sci. Phila., vol. x, pp. 256–264. 1858. Philadelphia, 1859.

	Page.
Description of new Carboniferous fossils	260
Fusulina cylindrica, Fischer	260-261
F. cylindrica var. *ventricosa*, M. & H., 1858	261
Orthisina crassa, n. s., M. & H., 1858	261
Chonetes mucronata, n. s., M. & H., 1858	262
Axinus (Schizodus) oratus, n. s., M. & H., 1858	262
Allorisma? altirostrata, n. s., M. & H., 1858	263
Allorisma subcuneata, n. s., M. & H., 1858	263
Allorisma? leavenworthensis, n. s., M. & H., 1858	263-264
Allorisma? cooperi, M. & H., (1858)	264
Pleurotomaria subturbinata, n. s., M. & H., 1858	264
Pleurotomaria humerosa, n. s., M. & H., 1858	264

13

MEEK, F. B., *and* HAYDEN, F. V.   Descriptions of new organic remains from Northeastern Kansas, indicating the existence of Permian rocks in that territory. <Trans. Albany Institute, vol. iv, pp. 73–88. 1858. Albany, 1858–1864.

This includes a note in relation to the priority of discovery of these fossils of Permian type.

DESCRIPTIONS OF FOSSILS.

	Page.
Monotis hawni, n. s., M. & H., 1858	76-77
Myalina (Mytelus) perattenuata, n. s., M. & H., 1858	77-78
Bakevellia parva, n. s., M. & H., 1858	78-79
Leda (Nucula) subscitula, n. s., M. & H., 1858	79
Edmondia? calhouni, n. s., M. & H., 1858	80
Pleurophorus? occidentalis, n. s., M. & H., 1858	80-81
Pleurophorus? (Cardinea) subcuneata, n. s., M. & H., 1858	81-82
Lyonsia (Panopœa) concara, n. s., M. & H., 1858	82-83
Panopœa cooperi, n. s., M. & H., 1858	83
Nautilus eccentricus, n. s., M. & H., 1858	83-84

14

MEEK, F. B., *and* HAYDEN, F. V.   Geological explorations in Kansas Territory. <Proc. Acad. Nat. Sci. Phila., vol. xi, pp. 8–30. 1859. Philadelphia, 1860.

	Page.
General section of the rocks of Kansas valley from the Cretaceous down, so as to include portions of the upper Coal measures	16-18
List of the species mentioned in this paper, with some remarks on the synonymy, and references to the works in which they are described	24-30
Foraminifera	24
Fusulina cylindrica, Fischer	24
Fusulina cylindrica var *ventricosa*, M. & H	24
Bryozoa	24
Synocladia biserialis, Swallow	24
Acanthocladia americana, Swallow	24
Echinodermata	24
Cyathocrinus ——?	24
Archæocidaris ——?	25
Archæocidaris ——?	25

	Page.
Brachiopoda	25
Discina tenuilineata, n. s., M. & H., 1859	25
D. manhattanensis, n. s., M. & H., 1859	25
Productus splendens (?) Norwood & Pratten	25
P. norwoodi, Swallow	25
P. rogersi, Norwood & Pratten	26
P. pustulosus (?), Phillips	26
P. prattenianus, Norwood	26
P. calhounianus, Swallow	26
Chonetes vermeuiliana, Norwood & Pratten	26
C. mucronata, M. & H., 1858	26
Orthisina crassa, M. & H., 1858	26
O. umbraculum ?, Schlot sp	26
O. missouriensis, Swallow	26
O. shumardiana, Swallow	26
Terebratula millepunctata, Hall	26–27
Rhynchonella uta, Marcou	27
Retzia mormonii, Marcou	27
Spirifer kentuckensis, Shumard	27
S. cameratus, Morton	27
S. hemiplicata, Hall	28
S. lineatus, Phillips	28
Spirifer, ——	28
S. planoconvexa, Shumard	28
Spirigera subtilita, Shumard	28
Spirigera —— ?	28
Lamellibranchiata	28
Monotis hawni, M. & H., 1858	28
Myalina (Mytilus) perattenuata, M. & H., 1858	28–29
Myalina squamosa (J. de C.) Sowerby	29
M. subquadrata, Shumard	29
Edmondia ? *calhouni*, M. & H., 1858	29
Bakerellia parva, M. & H., 1858	29
Arca carbonaria, Cox	29
Leda subscitula, M. & H., 1858	29
Pleurophorus ? *subcuneata*, M. & H., 1858	29
Axinus (Schizodus) ovatus, M. & H., 1858	29
Axinus rotundatus, Brown	29
Allorisma ? *leavenworthensis*, M. & H., 1858	29
A. subcuneata, M. & H., 1858	29
A. ? *altirostrata*, M. & H., 1858	29
A. ? *cooperi*, M. & H., 1858	29
Leptodomus granosus, Shumard	29
Gasteropoda	30
Pleurotomaria humerosa, M. & H., 1858	30
P. subturbinata, M. & H., 1858	30
Bellerophon —— ?	30
Euomphalus —— ?	30
Cephalopoda	30
Nautilus eccentricus, M. & H., 1858	30
Fishes	30
Xystracanthus arcuatus, Leidy	30
Cladodus occidentalis, Leidy	30
Petalodus alleghaniensis, Leidy	30

15.

MEEK, F. B., *and* HAYDEN, F. V. On a new genus of Patelliform shells from the Cretaceous rocks of Nebraska. <Am. Journ. Sci., vol. xxix, 2d ser., pp. 33–35, plate i. 1860. New Haven, 1860.

Genus Anisomyon,* n. g., M. & H., 1860.

	Page.
Anisomyon patelliformis, M. & H., pl. i, figs. 1–3	34–35

*"Ἄνισος, unequal; μυῶν, muscle; in allusion to the unsymmetrical muscular scar.

The authors refer the following Nebraska species to this genus:

	Page.
Anisomyon borealis (=*Hipponyx borealis*, Morton, 1842, =*Helcium carinatum*, M. & H., 1856)	35
Anisomyon sexsulcatus (=*Helcium sexsulcatum*, M. & H.)	35
Anisomyon alveolus (=*Helcium alveolum*, M. & H.)	35
Anisomyon patelliformis (=*Helcium patelliforme*, M. & H.)	35
Anisomyon subovatus (=*Helcium subovatum*, M. & H.)	35

16.

MEEK, F. B., and HAYDEN, F. V. Descriptions of new organic remains from the Tertiary, Cretaceous, and Jurassic rocks of Nebraska. <Proc. Acad. Nat. Sci. Phila., vol. xii, pp. 175-184. 1860. Philadelphia, 1861.

A description of the Carboniferous species *Myalina aviculoides* is also added; also a corrected list of fossils.

TERTIARY SPECIES.

	Page.
Gasteropoda	175
Helix evansi, n. s., [M.] & H., 1860	175
Planorbis vetulus, n. s., M. & H., 1860	175
Planorbis leidyi, n. s., M. & H., 1860	175
Conchifera	175
Sphærium planum, n. s., M. & H., 1860	175-176
Sphærium recticardinale, n. s., M. & H., 1860	176
Cyrena (Corbicula?) cytheriformis, n. s., M. & H., 1860	176

CRETACEOUS SPECIES.

Cephalopoda	176
Phylloteuthis, n. g., M. & H., 1860	176
Phylloteuthis subovatus, n. s., M. & H., 1860	176
Helicoceras angulatum, n. s., M. & H., 1860	176
Ammonites placenta var. *intercalaris*, M. & H.	177
Ammonites vermilionensis, n. s., M. & H., 1860	177
Scaphites nodosus var. *plenus*, M. & H.	177-178
Gasteropoda	178
Aporrhais parva, n. s., M. & H., 1860	178
Aporrhais sublevis, n. s., M. & H., 1860	178
Dentalium pauperculum, n. s., M. & H., 1860	178
Cylichna scitula, n. s., M. & H., 1860	178
Conchifera	178
Teredo selliformis, n. s., M. & H., 1860	178-179
Mactra siouxensis, n. s., M. & H., 1860	179
Mactra gracilis, n. s., M. & H., 1860	179
Tellina? formosa, n. s., M. & H., 1860	179
Cyprina humilis, n. s., M. & H., 1860	179-180
Avicula subgibbosa, n. s., M. & H., 1860	180
Inoceramus cuneatus, n. s., M. & H., 1860	180
Inoceramus vanuxemi, n. s., M. & H., 1860	180
Inoceramus balchii, n. s., M. & H., 1860	180-181
Inoceramus subcompressus, n. s., M. & H., 1860	181
Inoceramus aviculoid(a)es, n. s., M. & H., 1860	181
Anomia obliqua, n. s., M. & H., 1860	181
Anomia subtrigonalis, n. s., M. & H., 1860	181
Ostrea inornata, n. s., M. & H., 1860	181-182

JURASSIC SPECIES.

Conchifera	182
Pholadomya humilis, n. s., M. & H., 1860	182
Myacites nebrascensis, n. s., M. & H., 1860	182
Thracia? sublevis, n. s., M. & H., 1860	182
Thracia? arcuata, n. s., M. & H., 1860	182
Cardium shumardi, n. s., M. & H., 1860	182-183
Tancredia? æquilateralis, n. s., M. & H., 1860	183
Tancredia warrenana, n. s., M. & H., 1860	183

	Page.
Astarte fragilis, n. s., M. & H., 1860	183
Astarte inornata, n. s., M. & H., 1860	183
Trigonia conradi, n. s., M. & H., 1860	183-184
Pecten extenuatus, n. s., M. & H., 1860	184

17.

MEEK, F. B. Descriptions of new fossil remains collected in Nebraska and Utah by the exploring expeditions under the command of Capt. J. H. Simpson, of the U. S. topographical engineers (extracted from that officer's forthcoming report). <Proc. Acad. Nat. Sci. Phila., vol. xii, pp. 308-315. 1860. Philadelphia, 1861.

Devonian, Carboniferous, Jurassic, Cretaceous, and Tertiary. Republished with illustrations in a quarto volume of Captain Simpson's Reports, in 1876.

DEVONIAN SPECIES.

	Page.
Brachiopoda	308
Spirifera norwoodi, n. s., Meek, 1860	308
Spirifera engelmanni, n. s., F. B. Meek, 1860	308-309
Spirifera macra, n. s., F. B. Meek, 1860	309

CARBONIFEROUS SPECIES.

Brachiopoda	309
Productus semistriatus, n. s., F. B. Meek, 1860	309
Productus multistriatus, n. s., F. B. Meek, 1860	309-310
Spirifera scobina, n. s., F. B. Meek, 1860	310
Spirifera pulchra, n. s., F. B. Meek, 1860	310
Conchifera	310
Pecten utahensis, n. s., F. B. Meek, 1860	310
Cephalopoda	
Orthoceras baculum, n. s., F. B. Meek, 1860	310-311

JURASSIC SPECIES.

Conchifera	311
Ostrea engelmanni, n. s., F. B. Meek, 1860	311
Pecten bellistriata, n. s., F. B. Meek, 1860	311
Gasteropoda	311
Dentalium? subquadratum, n. s., F. B. Meek, 1860	311

CRETACEOUS SPECIES.

Conchifera	311
Anomia concentrica, n. s., F. B. Meek, 1860	311-312
Inoceramus simpsoni, n. s., F. B. Meek, 1860	312

TERTIARY SPECIES.

Conchifera	312
Unio retusus, n. s., Meek, 1860	312
Unio haydeni, n. s., Meek, 1860	312
Corbula (Potamomya?) pyriformis, n. s., Meek, 1860	312-313
Corbula (Potamomya?) concentrica, n. s., Meek, 1860	313
Corbula (Potamomya?) engelmanni, n. s., Meek, 1860	313
Gasteropoda	313
Melania humerosa, n. s., Meek, 1860	313
Melania simpsoni, n. s., Meek, 1860	313-314
Melania arcta, n. s., Meek, 1860	314
Melania? nitidula, n. s., Meek, 1860	314
Planorbis spectabilis, n. s., Meek, 1860	314
Planorbis utahensis, n. s., Meek, 1860	314
Limnæa retusta, n. s., Meek, 1860	314
Limnæa similis, n. s., Meek, 1860	314
Melampus priscus, n. s., Meek, 1860	315

18.

MEEK, F. B., and HAYDEN, F. V. Systematic catalogue with synonymy, &c., of Jurassic, Cretaceous, and Tertiary fossils, collected in Nebraska by the exploring expeditions under the command of Lieut. G. K. Warren, of U. S. Topographical Engineers. <Proc. Acad. Nat. Sci. Phila., vol. xii, pp. 417-432. 1860. Philadelphia, 1861.

Of the 276 species and varieties enumerated in the following catalogue, 25 are from Jurassic rocks, 194 from Cretaceous, and the remaining 57 from Tertiary strata.

	Page.
Jurassic species	417-419
Cretaceous species	419-430
Tertiary species	430-432

19.

MEEK, F. B., and WORTHEN, A. H. Descriptions of new species of Crinoidea and Echinoidea, from the Carboniferous rocks of Illinois and other western States. <Proc. Acad. Nat. Sci. Phila., vol. xii, pp. 379-397. 1860. Philadelphia, 1861.

Afterward republished in the Illinois Geological Reports, vol. ii.

	Page.
Platycrinus, Miller	379
P. prattenanus, n. s., M. & W., 1860	379-380
P. penicillus, n. s., M. & W., 1860	380
P. plenus, n. s., M. & W., 1860	380-381
Dichocrinus, Munster	381
D. constrictus, n. s., M. & W., 1860	381
D. conus, n. s., M. & W., 1860	381-382
D. (Pterotocrinus) crassus, n. s., M. & W., 1860	382-383
D. (Pterotocrinus) chesterensis, n. s., M. & W., 1860	383
Trematocrinus, Hall, 1860	383
T. fascellus, n. s., M. & W., 1860	383-384
Actinocrinus, Miller	384
A. validus, n. s., M. & W., 1860	384-385
A. asteriscus, n. s., M. & W., 1860	385-386
A. speciosus, n. s., M. & W., 1860	386
A. scilutus, n. s., M. & W., 1860	386-387
A. araneolus, n. s., M. & W., 1860	387-388
A. (Amphoracrinus) subturbinatus, n. s., M. & W., 1860	388-389
Forbsiocrinus, De Koninck & Le Hon	389
F. ? norwoodi, n. s., M. & W., 1860	389
F. ? semiovatus, n. s., M. & W., 1860	389-390
Zeacrinus, Troost	390
Z. discus, n. s., M. & W., 1860	390
Z. troostanus, n. s., M. & W., 1860	390-391
Z. planobrachiatus, n. s., M. & W., 1860	391
Cyathocrinus, Miller	391
C. saffordi, n. s., M. & W., 1860	391-392
C. ? sangamonensis, n. s., M. & W., 1860	392
C. ? crassus, n. s., M. & W., 1860	392-393
C. scitulus, n. s., M. & W., 1860	393
C. angulatus, n. s., M. & W., 1860	393-394
Poteriocrinus, Miller	394
P. (Scaphiocrinus) decadactylus, n. s., M. & W., 1860	394
P. swallovi, n. s., M. & W., 1860	394-395
Echinoidea	
Archæocidaris, McCoy	395
A. mucronatus, n. s., M. & W., 1860	395-396
Palæchinus, McCoy	396
P. burlingtonensis, n. s., M. & W., 1860	396
Melonites, Owen & Norwood	396-397
M. danae, n. s., M. & W., 1860	397

20.

MEEK, F. B., and WORTHEN, A. H. Descriptions of new Carboniferous fossils, from Illinois and other Western States. < Proc. Acad. Nat. Sci. Phila., vol. xii, pp. 447–472. 1860. Philadelphia, 1861.

Genera Sphenopoterium, Soleniscus. Afterward republished in the Illinois Geological Reports, vol. ii.

	Page.
Zoophyta.	
Sphenopoterium,* n. g., M. & W., 1860	447–448
S. obtusum, n. s., M. & W., 1860	448
S. compressum, n. s., M. & W., 1860	448
S. enorme, n. s., M. & W., 1860	448
S. cuneatum, n. s., M. & W., 1860	448
Echinodermata	449
Asteridæ	449
Palasterina, McCoy	449
Subgenus Schoenaster	449
Palasterina (Schoenaster) fimbriata, n. s., M. & W., 1860	449
Mollusca	450
Brachiopoda	450
Chonetes, Fischer	450
C. planumbona, n. s., M. & W., 1860	450
Productus, Sowerby	450
P. nanus, n. s., M. & W., 1860	450
P. parvus, n. s., M. & W., 1860	450–451
P. scitulus, n. s., M. & W., 1860	451
Rhynchonella, Fischer	451
R. subtrigona, n. s., M. & W., 1860	451
Athyris, McCoy	451
A. parvirostra, n. s., M. & W., 1860	451–452
Conchifera	452
Pecten, Linnaeus	452
P. tenuilineatus, n. s., M. & W., 1860	452
Aviculopecten, McCoy	452
A. oweni, n. s., M. & W., 1860	452–453
A. coxanus, n. s., M. & W., 1860	453
A. burlingtonensis, n. s., M. & W., 1860	453
A. koninckii, n. s., M. & W., 1860	453–454
A. interlineatus, n. s., M. & W., 1860	454
A. amplus, n. s., M. & W., 1860	454–455
A. pellucidus, n. s., M. & W., 1860	455
Avicula, Klein	455
A. oblonga, n. s., M. & W., 1860	455
Myalina, Koninck	455
M. angulata, n. s., M. & W., 1860	455–456
M. concentrica, n. s., M. & W., 1860	456
M. recurvirostra, n. s., M. & W., 1860	456
Solemya, Lamarck	457
S. radiata, n. s., M. & W., 1860	457
Leda, Schumaker	457
L. (Yoldia?) levistriata, n. s., M. & W., 1860	457
Schizodus, King	457
S. chesterensis, n. s., M. & W., 1860	457
Cardiomorpha, Koninck	458
C. radiata, n. s., M. & W., 1860	458
Gasteropoda	458
Bellerophon, Montfort	458
B. crassus, n. s., M. & W., 1860	458
Pleurotomaria, Defrance	458
P. subconstricta, n. s., M. & W., 1860	458–459
P. granulostriata, n. s., M. & W., 1860	459
P. tenuicincta, n. s., M. & W., 1860	459

* Σφην, a wedge; ποτηριον, a cup.

	Page.
P. pratteni, n. s., M. & W., 1860	459-460
P. subsinuata, n. s., M. & W., 1860	460
P. chesterensis, n. s., M. & W., 1860	460
P. subscalaris, n. s., M. & W., 1860	460-461
P. speciosa, n. s., M. & W., 1860	461
P. turbiniformis, n. s., M. & W., 1860	461
P. scitula, n. s., M. & W., 1860	461-462
P. shumardi, n. s, M. & W., 1860	462
Straparollus, Montfort? =*Euomphalus*, Sowerby	462
Euomphalus planodorsatus, n. s., M. & W., 1860	462
E. umbilicatus, n. s., M. & W., 1860	462-463
Naticopsis, McCoy	463
N. nodosus, n. s., M. & W., 1860	463
N. hollidayi, n. s., M. & W., 1860	463
Platyostoma, Conrad	463
P. nana, n. s., M. & W., 1860	463
P. ? tumida, n. s., M. & W., 1860	463-464
Eunema, Salter	464
E. ? salteri, n. s., M. & W., 1860	464
Loxonema, Phillips	464
L. scitula, n. s., M. & W., 1860	464-465
L. rugosa, n. s., M. & W., 1860	465
L. cerithiformis, n. s., M. & W., 1860	465
L. inornata, n. s., M. & W., 1860	465
L. nitidula, n. s., M. & W., 1860	465-466
Eulima, Risso	466
E. ? peracuta, n. s., M. & W., 1860	466
Macrocheilus, Phillips	466
M. medialis, n. s., M. & W., 1860	466-467
M. intercalaris, n. s., M. & W., 1860	467
M. pulchellus, n. s., M. & W., 1860	467
Soleniscus, n. g., M. & W., 1860	467
S. typicus, n. s., M. & W., 1860	467-468
Cephalopoda	468
Orthoceras, Breynius	468
O. expansum, n. s., M. & W., 1860	468
Cyrtoceras, Goldfuss	468
C. curtum, n. s., M. & W., 1860	468
C. ? dilatatum, n. s., M. & W., 1860	468-469
Nautilus, Breynius	469
N. subglobosus, n. s., M. & W., 1860	469
N. chesterensis, n. s., M. & W., 1860	469
N. spectabilis, n. s., M. & W., 1860	469
N. (Discus) planorbiformis, n. s., M. & W., 1860	469-470
N. (Discus) trisulcatus, n. s., M. & W., 1860	470
N. (Discus) digonus, n. s., M. & W., 1860	470
N. (Discus) sangamonensis, n. s., M. & W., 1860	470-471
Goniatites, De Haan	471
G. globulosus, n. s., M. & W., 1860	471
G. iowensis, n. s., M. & W., 1860	471
G. lyoni, n. s., M. & W., 1860	471-472
Subgenus *Oligoporus*,[*] n. s. g., M. & W., 1860	472

21.

MEEK, F. B., and WORTHEN, A. H. Remarks on the age of the Goniatite limestone at Rockford, Indiana, and its relation to the "Black Slate" of the western States, and to some of the succeeding rocks above the latter. <Am. Journ. Sci., vol. xxxii, 2d ser., pp. 167-177. 1861. New Haven, 1861.

The authors conclude it to be probably of Carboniferous age, and, at any rate, much more recent than the Chemung, and not equivalent to any New York rock.

---

[*] ὀλίγος, few; and πόρος, opening a pore.

22.

MEEK, F. B., *and* WORTHEN, A. H. Descriptions of new Paleozoic fossils from Illinois and Iowa. <Proc. Acad. Nat. Sci. Phila., vol. xiii, pp. 128-148. 1861. Philadelphia. 1862.

Genera *Bursacrinus*, *Cardiopsis*, *Orthonema*. Afterward republished in the Illinois Geological Reports, vol. ii.

	Page.
Echinodermata:	
Crinoidea:	
Platycrinus, Miller, 1821.	
P. oweni, n. s., M. & W., 1861	128-129
P. scobina, n. s., M. & W., 1861	129
P. (Pleurocrinus) asper, n. s., M. & W., 1861	129-130
Forbesiocrinus, Koninck & Le Hon	130
F. monroensis, n. s., M. & W., 1861	130-131
F. agassizi var. *giganteus*, n. s., M. & W., 1861	131
Actinocrinus, Miller, 1821	131
A. dodecadactylus, n. s., M. & W., 1861	131
A. pyriformis var. *rudis*, M. & W., 1861	131-132
A. (Amphoracrinus ?) concavus, n. s., M. & W., 1861	132-133
A. (Pradocrinus ?) amplus, n. s., M. & W., 1861	133-134
A. sillimani, n. s., M. & W., 1861	134-135
Agaricocrinus, Troost	135
A. gracilis, n. s., M. & W., 1861	135
Platycrinus multi-branchiatus, n. s., M. & W., 1861	135-136
Cyathocrinus, Miller, 1821	136
C. wachsmuthi, n. s., M. & W., 1861	136
Bursacrinus, n. g., M. & W., 1861	136
B. wachsmuthi, n. s., M. & W., 1861	137
Poteriocrinus, Miller, 1821	137
P. ? enormis, n. s., M. & W., 1861	137-138
P. sub-impressus, n. s., M. & W., 1861	138
P. tenuibrachiatus, n. s., M. & W., 1861	138-139
P. carinatus, n. s., M. & W., 1861	139-140
Subgenus *Scaphiocrinus*, Hall, 1858	140
P. (Scaphiocrinus ?) carbonarius, n. s., M. & W., 1861	140
P. (Scaphiocrinus) solidus, n. s., M. & W., 1861	140-141
P. (Scaphiocrinus) wachsmuthi, n. s., M. & W., 1861	141
Blastoidea:	
Pentremites, Say, 1820	141
P. cornutus, n. s, M. & W., 1861	141-142
P. melo var. *projectus*, n. s., M. & W., 1861	142
Astereidea:	
Petraster, Billings, 1858	142
P. wilberanus, n. s., M. & W., 1861	142
Mollusca:	
Brachiopoda:	
Productus, Sowerby, 1812	142
P. magnus, n. s., M. & W., 1861	142-143
Spirifera glabra var. *contracta*, n. s., M. & W., 1861	143-144
Lamellibranchiata:	
*Cardiopsis, n. g., M. & W., 1861	144
Leda, Schumacher, 1817	144
L. curta, n. s., M. & W., 1861	144-145
Gasteropoda	
Dentalium, Lin., 1740	145
D. venustum, n. s., M. & W., 1861	145
Straparollus, Montfort, 1810	145
S. similis, n. s., M. & W., 1861	145-146
S. similis var. *planus*, M. & W., 1861	146

* Cardium and ὄψις, from its resemblance to Cardium.

	Page.
† *Orthonema*, n. g., M. & W., 1861	146
Loxonema, Phillips, 1841	146
L. multicostata, n. s., M. & W., 1861	146-147
Cephalopoda:	
Orthoceras, Breynius, 1732	147
O. annulocostatum, n. s., M. & W., 1861	147
Nautilus, Breynius, 1732	147
‡ Subgenus *Trematodiscus*, n. s. g., M. & W., 1861	147

Corrections in regard to a few fossils described in papers of September and October, 1860.

Cyathocrinus scitulus (September, 1860) should be *C. sculptilus*.
Platyostoma nana (October, 1860) should be *Naticopsis*.
Eulima peracuta (October, 1860) should be ranged under *Polyphemopsis*, of Portlock, probably a section of the genus *Loxonema*.
Orthoceras expansum (October, 1860) belongs to the genus or subgenus *Actinoceras*.
Cyrtoceras curtum (October, 1860) should be ranged under the subgenus *Aploceras*.

23.

MEEK, F. B. Descriptions of new Cretaceous fossils collected by the Northwestern Boundary Commission on Vancouver and Sucia Islands. <Proc. Acad. Nat. Sci. Phila., vol. xiii, pp. 314-318. 1861. Philadelphia, 1862.

Lamellibranchiata:	Page.
Inoceramus subundatus, Meek, n. s., 1861	315
Dosinia ? tenuis, n. s., Meek, 1861	315
Mactra gibbsana, n. s., Meek, 1861	315-316

Cephalopoda:	
Baculites inornatus, n. s., Meek, 1861	316
Baculites occidentalis, n. s., Meek, 1861	316-317
Ammonites complexus var. *suciaensis*, n. s., Meek, 1861	317
Ammonites vancouverensis, n. s., Meek, 1861	317-318
Nautilus campbelli, n. s., Meek, 1861	318

24.

MEEK, F. B., *and* HAYDEN, F. V. Descriptions of new Lower Silurian (Primordial), Jurassic, Cretaceous, and Tertiary fossils, collected in Nebraska by the exploring expedition under the command of Capt. Wm. F. Raynolds, U. S. Top. Engrs., with some remarks on the rocks from which they were obtained. <Proc. Acad. Nat. Sci. Phila., vol. xiii, pp. 415-447. 1861. Philadelphia, 1862.

	Page.
Lower Silurian (Primordial) rocks	415-416
Jurassic rocks	416-417
Cretaceous rocks	417-432
Lower series	417-424
General section of the Cretaceous rocks of Nebraska	419
Upper Cretaceous series of Nebraska	424-432
New Jersey section, compiled from the reports of that State	426
Fox Hill beds	427
Relations of the upper Cretaceous series of Nebraska to European divisions	428-432
Tertiary rocks	432-435
General section of the Tertiary rocks of Nebraska	433

SILURIAN (PRIMORDIAL) FOSSILS.

Brachiopoda:	
Obolella, Billings	435
O. nana, n. s., M. & H., 1861	435-436

---

† ορθος, straight; νημα, thread.
‡ τρῆμα and δισκος, in allusion to the perforated umbilicus and the discoid form of the shell in the typical species.

	Page.
Pteropoda ?	436
Theca, Morris	436
Theca (Pugiunculus) gregarea, n. s., M. & H., 1861	436
Trilobites	436
Arionellus, Barrande	436
A. (Crepicephalus) oweni, n. s., M. & H., 1861	436–437

JURASSIC FOSSILS.

Lamellibranchiata	437
Gryphaea, Lamarck	437
G. calceola var. *nebrascensis*, M. & H., 1861	437–439
Modiola, Lamarck	439
M. (Perna) formosa, n. s., M. & H., 1861	439

CRETACEOUS FOSSILS.

Lamellibranchiata	440
Ostrea, Linnaeus	440
O. gabbana, n. s., M. & H., 1861	440
Leda, Schumacher	440
L. bisulcata, n. s., M. & H., 1861	440
Gervillia, Defrance	441
G. recta, n. s., M. & H., 1861	441
Crenella, Brown	441
C. elegantula, n. s., M. & H., 1861	441–442
Cardium, Linnaeus	442
C. (Hemicardium?) curtum, n. s., M. & H., 1861	442
C. pertenue, n. s., M. & H., 1861	442
Callista, Poli	443
C. deweyi, M. & H., 1856	443
Tellina, Linnaeus	443
T. nitidula, n. s., M. & H., 1861	443
Lingula, Bruguière	443
L. nitida, n. s., M. & H., 1861	443–444
Gasteropoda	444
Neritella, Humphrey	444
N. nebrascensis, n. s., M. & H., 1861	444
Melania, Lamarck	444
M. (Potodoma) veterna, n. s., M. & H., 1861	444–445
Cephalopoda	445
Baculites, Lamarck	445
B. baculus, n. s., M. & H., 1861	445

TERTIARY FOSSILS.

Gasteropoda	446
Vivipara, Lamarck	446
V. raynoldsana, n. s., M. & H., 1861	446
Helix, Linnaeus	446
H. spatiosa, n. s., M. & H., 1861	446–447
H. veterna, n. s., M. & H., 1861	447
Correction, notes the omission of the name Arcadæ on p. 428 of the Proc. for 1860	447

25.

MEEK, F. B., *and* HAYDEN, F. V. Descriptions of new Cretaceous fossils from Nebraska Territory, collected by the expedition sent out by the Government under the command of Lieut. John Mullan, U. S. Topographical Engineers, for the location of a wagon road from the sources of the Missouri to the Pacific Ocean. <Proc. Acad. Nat. Sci. Phila., vol. xiv, pp. 21–28. 1862. Philadelphia, 1863.

Cephalopoda:

	Page.
Scaphites, Parkinson	22
S. ventricosus, n. s., M. & H., 1862	22
S. vermiformis, n. s., M. & H., 1862	22–23
Ammonites, Bruguière	23

	Page.
A. mullananus, n. s., M. & H., 1862	23-25
Nautilus, Breynius	25
N. elegans, Sowerby var. nebrascensis, M. & H., 1862	25
Lamellibranchiata:	
Inoceramus, Sowerby	26
I. undabundus, n. s., M. & H., 1862	26
I. exogyroides, n. s., M. & H., 1862	26-27
I. tenuirostratus, n. s., M. & H., 1862	27
Venilia, Morton	27
V. mortoni, n. s., M. & H., 1862	27-28
Pholadomya, Sowerby	28
P. papyracea, n. s., M. & H., 1862	28

26.

MEEK, F. B. Remarks on the family Actæonidæ, with descriptions of some new genera and subgenera. <Am. Journ. Sci., vol. xxxv, 2d ser., pp. 84-94. 1863. New Haven, 1863.

	Page.
Actæoninæ:	
Actæonella, d'Orbigny (as restricted)	89
Trochactæon, Meek, 1863 (Actæonella (part) d'Orb.)	89-90
Subgenus, Spiractæon Meek, 1863	90
Cylindrites (Auct.) Morris & Lycett (as restricted)	90-91
Subgenus Goniocylindrites	91
Actæonina, d'Orbigny (as restricted)	91
Subgenus, Trochactæonina Meek, 1863	91
Euconactæon, Meek, 1863 (Acteonina (part) d'Orb. & Alt.)	91-92
Subgenus ? Conactæon, Meek	92
Ringiculinæ:	
Cinulia, Gray	92
Subgenus Avellana, d'Orbigny	92
Subgenus Euptycha, Meek, 1863	93
Aptycha, Meek, 1863	93-94

27.

MEEK, F. B. Remarks on the family Pterriidæ (=Aviculidæ), with descriptions of some new fossil genera. <Am. Journ. Sci., vol. xxxvii, 2d ser., pp. 212-220. 1864. New Haven, 1864.

	Page.
Pteriidæ (or Aviculidæ)	214-215
Pterininæ (or Pterinia group)	215
Pteriinæ (or Aviculinæ)	215
Melininæ (Perna or Isognomen group)	215
Gryphorhyncus, n. g., Meek, 1864	217-218
Eumicrotis, n. g., Meek, 1864	218-220

28.

MEEK, F. B. Carboniferous and Jurassic fossils. <Paleontology of California (Whitney), vol. i, pp. 1-16, 2 plates, and pp. 39-53, pls. vii and viii. 1864. Published by authority of the legislature of California, 1864.

DESCRIPTION OF CARBONIFEROUS FOSSILS.

	Page.
Foraminifera:	
Fusulina, Fischer	3
F. robusta, n. s., Meek, 1864, pl. i, figs. 3 and 3 a-c	3-4
F. gracilis, n. s., Meek, 1864, pl. ii, fig. 1 and 1 a-c	4
F. cylindrica, Fischer ?, 1837, pl. ii, figs. 2 and 2 a	4

Zoophyta. Page.
Lithostrotion, Fleming ....
L. mamillare (?), Castlenau (sp.), 1843, pl. i, figs. 4 and 4 a, b 5-6
L. ? californiense, n. s., Meek, 1864, pl. i, figs. 2 and 2 a-c 6-7
Lithostrotion ———?, Meek, 1864, pl. i, figs. 3 and 3 a 7
Clisiophyllum, Dana 8
C. gabbi, n. s., Meek, 1864, pl. i, fig. 1 and 1 a, b 8-9

Mollusca.
Brachiopoda 10
Orthis, Dalman 10
Orthis, sp. undt., 1864, pl. ii, figs. 5 and 5 a-c 10-11
Productus, Sowerby 11
P. semireticulatus, Martin (sp.), 1809, pl. ii, figs. 4 and 4 a 11
Rhynchonella, Fischer 12
Rhynchonella, sp. undt., 1864 12
Spirifer, Sowerby 13
Subgenus *Martinia*, McCoy 13
S. (Martinia) lineatus, Martin ? (sp.), 1809, pl. ii, figs. 6 and 6 a-d 13
Spiriferina, Davidson 14
Spiriferina, sp. undt., 1864 14
Retzia, King 14
R. compressa, n. s., Meek, 1864, pl. ii, figs. 7 and 7 a-c 14
Gasteropoda 15
Euomphalus, Sowerby = *Straparollus*, Montfort ? 15
Subgenus *Omphalotrochus*, Meek 15
E. (Omphalotrochus) whitneyi, n. s., Meek, 1864, pl. ii, figs. 8 and 8 a 15-16

DESCRIPTION OF THE JURASSIC FOSSILS.

Brachiopoda 39
Rhynchonellidæ 39
Rhynchonella, Fischer, 1809 39
R. gnathophora, n. s., Meek, 1864, pl. viii, fig. 1 and 1 a-f 39-40
Terebratulidæ 41
Terebratula, Müller, 1776 41
Terebratula ———, 1864, pl. viii, figs. 2 and 2 a, b 41
Lamellibranchiata 42
Ostreidæ 42
Gryphæa, Lamarck, 1801 42
Gryphæa ———, 1864, pl. viii, figs. 4 and 4 a 42
Limidæ 43
Lima, Bruguière, 1792 43
L. ? sinuata, n. s., Meek, 1864, pl. vii, figs. 4 and 4 a 43
L. recticostata, n. s., Meek, 1864, pl. vii, fig. 5 44
L. ? cuneata, n. s., Meek, 1864, pl. vii, figs. 6 and 6 a 44-45
Pectenidæ 46
Pecten, Müller, 1776 46
P. acutiplicatus, n. s., Meek, 1864, pl. viii, fig. 3 46
Pteriidæ 47
Inoceramus, Sowerby, 1814 47
I. ? obliquus, n. s., Meek, 1864, pl. vii, figs. 2 and 2 a 47
I. ? rectangulus, n. s., Meek, 1864, pl. vii, figs. 1 and 1 a 47-48
Trigoniidæ 48
Trigonia, Bruguière, 1789 48
T. pandicosta, n. s., Meek, 1864, pl. viii, fig. 7 48-49
Mytilidæ 49
Mytilus, Linnæus, 1758 49
M. multistriatus, n. s., Meek, 1864, pl. vii, figs. 7 and 7 a 49
Crassatellidæ 50
Astarte, Sowerby, 1816 50
A. ventricosa, n. s., Meek, 1864, pl. viii, figs. 5 and 5 a 50
Lucinidæ 50
Unicardium, d'Orbigny, 1852 50
U. ? gibbosum, n. s., Meek, 1864, pl. viii, figs. 8 and 8 a 50-51
Anatinidæ 51

	Page.
Myacites (Schlot.), Munster, 1840	51
M. depressus, n. s., Meek, 1864, pl. viii, figs. 6 and 6 *a*	51-52
Cephalopoda	53
Belemnitidœ	53
Belemnites, Auct	53
Belemnites, ———, 1864, pl. viii, figs. 9 and 9 *a*	53

29.

MEEK, F. B., *and* HAYDEN, F. V. Palœontology of the Upper Missouri. <Smithsonian contributions to knowledge (172), pp. 1-135, pls. i-v. 1864. Washington, 1865.

Primordial, Carboniferous, Permian, and Jurassic. Genera Camptonectes, Lioplacodes, Eumicrotis, Chœnomya. This work contains descriptions of new fresh-water Jurassic species, the first of that age discovered in North America. The work has additional importance in consequence of the philosophical discussion of important questions.

	Page.
Introduction	VII-IX

SILURIAN AGE. (POTSDAM OR PRIMORDIAL PERIOD.)

Mollusca.
Brachiopoda.

Lingulidæ	1
Lingulepis, Hall, 1863	1-2
L. pinniformis, Owen, 1852, pl. i, figs. 1 *a, b*	2-3
L. prima (Conrad), Hall, 1847, pl. i, figs. 2 *a, b*	3
Obolella, Billings, 1861	3-4
O. nana, M. & H., 1861, pl. i, figs. 3 *a-d*	4

Gasteropoda.
Pteropoda.
Thecosomata

Carolinidæ	4-5
Theca, Sowerby, 1845	5
T. gregaria, M. & H., 1861, figs. *a-d*, p. 5	5-6

Articulata.
Crustacea.
Trilobita.

Paradoxidæ	7
Agraulos, Corda, 1847	7-9
A. oweni, M. & H., 1861, figs. *a-c*, p. 9	9-10
Agraulos ——— ? pl. i, fig. 4	10

CARBONIFEROUS AGE. (CARBONIFEROUS PERIOD.)

Protozoa.
Rhizopoda.
Foraminifera.

Camerinidæ	11-13
Fusulina, Fischer, 1837	13-14
F. cylindrica, Fischer, 1837, pl. i, figs. 6 *a-i*	14-15

Mollusca.
Brachiopoda.

Spiriferidæ	16
Spirifer, Sowerby, 1813	17-20
Trigonotreta, Kœnig, = *Spirifer* of most authors	19
Martinia, McCoy, = *Amboccelia*, Hall	19
S. (*Martinia*) *plano-convexus* Shumard, 1855, figs. *a-e*, p. 21	20-21
Productidæ	21-22
Chonetes, Fischer, 1837	22
C. mucronata, M. & H., 1858, pl. i, fig. 5, *a-e*	22-23
Strophomenidæ	23
Hemipronites, Pauder, 1830	24-26
H. crassus, M. & H., 1858, pl. i, fig. 7 *a-d*	26-27

Lamellibranchiata.   Page.
  Pteriidæ (=*Aviculidæ*) 27–30
    Pteriniinæ (or Pterinia group) 28
    Pteriinæ (or *Aviculinæ*) 28
    Melininæ (Perna or Isognomon group) 28
    Pteriniinæ 30
      Myalina, Koninck, 1842 30–32
      Shell structure of *Myalina angulata*, fig. — 31
      M. perattenuata, M. & H., 1858, pl. i, figs. 12 *a, b* 32
      M. subquadrata, Shumard, figs. *a, b*, p. 33 32–33
    Crassatellidæ 34
      Pleurophorus, King, 1844 34–35
      Subgenus *Oleidophorus*, Hall, 1847 35
      P. occidentalis, M. & H., pl. i, figs. 11 *a, b* 35
    Anatinidæ 36
      Allorisma, King, 1844 36–37
      A. subcuneata, M. & H., 1858, pl. i, figs. 10 *a, b* 37–38
      Sedgwickia, McCoy, 1844 38–40
      S. topekaensis ?, Shumard, 1858, figs. *a, b*. p. 40 40–41
      S.? concava, M. & H., 1858, pl. i, figs. 8 *a, b* 41
      S.? altirostrata, M. & H., 1858, pl. i, fig. 9 41–42
      Chænomya, n. g., Meek, 1865 42–43
      C. leavenworthensis, M. & H., 1858, pl. ii, figs. 1 *a–c* 43–44
      C. cooperi, M. & H., 1858, pl. ii, figs. 2 *a, b* 44
Gasteropoda.
  Prosobranchiata.
    Rhipidoglossata.
      Podophthalma.
        Pleurotomariidæ 44–45
          Pleurotomaria, Defrance, 1826 45–46
          P. humerosa, M. & H., 1858, pl. i, figs. 14 *a, b* 46–47
          P. subturbinata, M. & H., 1858, pl. i, fig. 13 47

CARBONIFEROUS AGE. (PERMIAN PERIOD.)
Mollusca.
  Lamellibranchiata.
    Pectinidæ 48
      Pectininæ 48
      Aviculo-pectininæ 49
    Aviculopectininæ 49
      Aviculopecten, McCoy, 1851 49
      A. amplus, M. & W., fig. —, p. 50 50
      Aviculopecten, ———— *?*, pl. ii, fig. 10 50
      A. macoyi, pl. ii, fig. 9 50–51
    Pteriidæ.
      Pteriniinæ 51
        Myalina, Koninck, 1842 51
        M. aviculoides, M. & H., 1860, pl. ii, figs. 8 *a–d* 51–52
        M. permiana, Swallow, 1858, pl. ii, figs. 7 *a–c* 52
      Pteriinæ.
        Eumicrotis, Meek, 1864 53–54
        E. hawni, M. & H., pl. ii, figs. 5 *a–c*, and fig. 1, p. 54 54–55
        Shell structure of *E. curta*, No. 2, p. — 54
        E. hawni var. *ovata*, pl. ii, figs. 5 *a, b* 55
      Melininæ.
        Bakerellia, King, 1848 57
        B. parva, M. & H., 1858, pl. ii, figs. 12 *a, b* 57
    Trigoniidæ 57–58
      Schizodus, King, 1840 58–59
      S. ovatus, M. & H., 1858, pl. ii, figs. 11 *a, b* 59
    Nuculanidæ 59–60
      Nuculaninæ 60
      Malletinæ 60

*Χαιρω, to open or gape; and *mya*.

	Page.
Nuculaninæ	60
Yoldia, Möller, 1842	60
Y.? subscitula, M. & H., 1858, pl. ii, figs. 4 a, b	60–61
Crassatellidæ	61
Pleurophorus, King, 1844	61
P. ? subcuneatus, M. & H., 1858, pl. ii, fig. 3	61
P.? calhouni, M. & H., 1858, pl. ii, figs. 13 a, b	62
Cephalopoda.	
Tetrabanchiata.	
Nautilidæ	63–64
Nautilus, Linnæus, 1758	64–65
N. eccentricus, M. & H., pl. ii, figs. 14 a, b	65

REPTILIAN AGE. (JURASSIC PERIOD.)

Radiata.
Echinodermata.
 Crinoidea.
  Pentacrinidæ.

Pentacrinites, Miller, 1821	66
Chladocrinus or *Cladocrinus* (Agassiz)	66
Pentacrinus asteriscus, M. & H., 1858, pl. iii, figs. 2 a, b and fig.—. p. 67	67

Mollusca.
 Brachiopoda.

Lingulidæ	68
Lingula, Bruguière, 1792	68–69
L. brevirostris, M. & H., 1858, pl. iii, figs. 3 a, b	69
Rhynchonellidæ	70
Rhynchonella, Fischer, 1809	70–71
Rhynchonella, ——, pl. iii, fig. 4	71–72

Lamellibranchiata.

Ostreidæ	72
Ostrea, Linnæus, 1758	72–73
O. engelmanni, Meek, 1860, figs. A, B, p. 73	73–74
Gryphæa, Lamarck, 1801	74
G. calceola var. *nebrascensis*, M. & H., 1861, pl. iii, figs. 1 a–e and figs. A–E, p. 75	74–76
Pectinidæ	76
Pectininæ	76
Camptonectes, Agassiz, 1865	76–77
C. bellistriatus, Meek, 1860, figs. A–D, p. 77	77–78
C.? extenuatus, M. & H., 1860, pl. iii, fig. 6	78
Pteriidæ	79
Pteriinæ	79
Pteria, Scopoli, 1777	79
Subgenus *Oxytoma*, n. s. g., Meek, 1865	79–80
Pteria (Oxytoma) munsteri, Bronn, 1829, figs. A, B, p. 80	80–81
Eumicrotis, Meek, 1864	81
E. curta, Hall, 1852, pl. iii, figs. 10 a–d	81
Trigoniidæ	82
Trigonia, Bruguière, 1789	82–83
Subdivisions:	
Les Scaphoides, Agassiz. (Jurassic and Cretaceous.)	
Les Clavellees, Agassiz. (Mainly Jurassic.)	
Les Carrees, Agassiz. (Upper Jurassic and Cretaceous.)	
Les Scabres, Agassiz. (Mainly Cretaceous.)	
Les Ondulees, Agassiz. (Jurassic and Cretaceous.)	
Les Costees, Agassiz. (Jurassic and Cretaceous.)	
Les Lissees, Agassiz. (Jurassic and Cretaceous.)	
Les Pectinacees, Agassiz. (Existing seas.)	
Trigonia conradi, M. & H., 1860, pl. iii, fig. 11	83–84
Mytilidæ	84
Volsella, Scopoli, 1777	85–86
V. pertenuis, M. & H., 1858, pl. iii, figs. 5, 5 a	86
V. formosa, M. & H., 1861, figs. A, B, p. 87	86–87
Arcidæ	87–89
Arcinæ	88

34   BULLETIN NO. 30, UNITED STATES NATIONAL MUSEUM.

	Page.
Macrodontinæ	88
Axininæ	88
Macrodontina	89
Grammatodon, M. & H., 1858	89
G. inornatus, M. & H., 1858, pl. iii, figs. 9, 9 a, 9 b	90
Unionidæ	90–91
Unio, Retzius, 1788	92
U. nucalis, M. & H., 1858, pl. iii, figs. 13 a–c	92–93
Crassatellidæ	93
Astarte, Sowerby, 1816	93–94
A. fragilis, M. & H., 1860, pl. iv, fig. 7	94
A. inornata, M. & H., 1860, pl. iii, figs. 12 a, b	94
Tancrediidæ	95
Tancredia, Lycett, 1850	95–96
T. warrenana, M. & H., 1860, pl. iii, fig. 7	96
T. ? æquilateralis, M. & H., 1860, pl. iii, fig. 8	96–97
Cardiidæ	97
Protocardia, Beyrich, 1845	97–98
P. shumardi, M. & H., 1860, figs. A, B on p. 98	98–99
Anatinidæ	99
Myacites (Schlot.), Munster, 1840	99–100
M. nebrascensis, M. & H., 1860, pl. iv, fig. 5	100–101
M. subellipticus, M. & H., 1858, pl. iv, figs. 6 a–c	101
Thracia, Leach, 1819	101–102
T.? subleris, M. & H., 1860, pl. iv, figs. 4, 4 a	102
T.? arcuata, M. & H., 1860, pl. iv, fig. 8	102
Pholadomya, Sowerby, 1823	102–103

Section *I.—Species without a circumscribed cardinal area.*

Multicostatæ, Agassiz. (Jurassic and Cretaceous.)
Trigonatæ, Agassiz. (Cretaceous, Tertiary, and Recent.)
Bucardinæ, Agassiz. (Lias to Tertiary.)

Section *II.—Species with a circumscribed cardinal area.*

Flabellatæ, Agassiz. (Jurassic.)
Ovales, Agassiz. (Jurassic.)
Cardissoides, Agassiz. (Jurassic.)

Pholadomya humilis, M. & H., 1860, pl. iv, figs. 3, a, b	104
Gasteropoda.	
Pulmonifera.	
Inoperculata.	
Limnophila.	
Limnæidæ	105–106
Limnæinæ	105
Physinæ	105
Planorbinæ	105
Planorbis, Müller, 1774	106
Planorbella, Haldeman	106
Helisoma, Swainson	107
Taphius, H. & A. Adams	107
Menetus, H. & A. Adams	107
Anisus, Fitzinger	107
Bathyomphalus, Agassiz (= *Spirorbis*, Swainson, not Lamk.)	107
Gyraulus, Agassiz (= *Nautilina*, Stein)	107
Planorbis veternus, M. & H., 1860, pl. iv, figs. 1 & 1 a, b	107
Prosobranchiata.	
Riphidoglossata.	
Podopthalma.	
Neritidæ	108
Neritella, Humphrey, 1797	108–109
N. nebrascensis, M. & H., 1861, figs. —, on p. 109	109–110
Cyclobranchiata.	
Dentaliidæ	110
Dentalium, Lin., 1740	111
D. subquadratum, Meek, 1860, fig. —, p. 111	111
Ctenobranchiata (= *Pectinibranchiata*).	

Rostrifera.
  Valvatidæ.
    Valvata, Müller, 1774 112
      Valvata (proper) =*Gyrorbis,* Fitzinger, =*Planella,* Schlut. 112
      Tropidina, H. & A. Adams 112–113
      Subgenus *Tropidina,* H. & A. Adams 113
      Valvata ? (Tropidina) scabrida, M. & H., 1860, pl. iv, figs. 2 a, b 113
    Viviparidæ 113–114
      Viviparus, Montfort, 1810 114–115
      V. gilli, pl. v, figs. 3 a, b 115
      Lioplacodes, Meek, 1864 115–116
      L. veternus, M. & H., 1861, figs. —, p. 116 116
Cephalopoda.
  Tetrabranchiata.
    Ammonitidæ 116–118
      Trigonellites, Parkinson, 1811, fig. 1, p. 119, figs. 2–4, p. 120 118–120
      Ammonites, Bruguière, 1789 121–122
      A. cordiformis, M. & H., 1856, pl. v, figs. 2 a–e 122–123
      A. henryi, M. & H., 1858, pl. iv, figs. 9 a–c 123–124
  Dibranchiata.
  Decapoda.
    Belemnitidæ 124
      Belemnites, Auct 124–125
      Acoeli, Bronn 125
      Gastrocoeli, D'Orbingy =*Notosiphites,* Duval 125
      Notocoeli, D'Orbigny = *Gastrosiphites,* Duval 125
      Belemnites densus, M. & H., 1858, pl. iv, figs. 10 a–c, and pl. v, figs. 1, 1 a–h 126–127
Articulata.
  Annulata.
    Tubicola.
      Serpulidæ 137
      Serpula, Linnæus, 1758 127–128
      Serpula, undt. pl. v, fig. 4 128

30.

MEEK, F. B. Description of fossils from the auriferous slates of California. <Geol. surv. California, Geology, vol. i. (Appendix B), pp. 477–482, 1 plate, 1865. Published by authority of the legislature of California, 1865. Philadelphia, 1865.

Jurassic fossils.—This article is in the volume of Geology, and not in either of those devoted to Paleontology exclusively.

Page.
Amussium (Klein), Bolton, 1798 478
Subgenus *Entolium,* Meek 478
A. aurarium, n. s., Meek, 1865, pl. i, figs. 6 and 6 a 478–479
Aucella, Keyserling, 1843 479
A. erringtoni, Gabb, sp., 1864, pl. i, figs. 2, 3, 7, 5, a–d 479–480
A. erringtoni, var. *linguiformis,* Meek, 1865, pl. i, figs. 1 and 1 a 481
Pholadomya, Sowerby, 1823 481
Pholadomya (?) *orbiculata,* Gabb, pl. i, fig. 4 481
Belemnites, Auct 482
Belemnites pacificus, Gabb 482

31.

MEEK, F. B. Remarks on the Carboniferous and Cretaceous Rocks of Eastern Kansas and Nebraska, and their relations to those of the adjacent States, and other localities farther eastward, in connection with a review of a paper recently published on this subject by M. Jules Marcou, in the Bulletin of the Geological Society of France. <Am. Journ. Sci., vol. xxxix, 2d ser., pp. 157–174. 1865. New Haven, 1865.

The paper of Professor Marcou's referred to is the "Reconnaissance geologique du Nebraska," par Jules Marcou. Bull. Geol. Soc. France, xxi, pp. 132–147, January, 1864.

At the close of the paper Meek & Worthen describe: Page.
Erisocrinus, n. g., M. & W., 1865 174
Erisocrinus typus, n. s., M. & W., 1865 174
Erisocrinus nebrascensis, n. s., M. & W., 1865 174

32.

MEEK, F. B., *and* WORTHEN, A. H. Note in relation to a genus of Crinoids from the Coal measures of Illinois and Nebraska, proposed by them on page 174 of this volume of the Journal. <Am. Journ. Sci., vol. xxxix, 2d ser., p. 350. 1865. New Haven, 1865.

> The authors regard their genus Erisocrinus as identical with Philocrinus de Koninck. *Erisocrinus typus* is changed to *Philocrinus pelvis*, M. & W.
> *E. nebrascensis* to *P. nebrascensis*.

33.

MEEK, F. B. Preliminary notice of a small collection of fossils found by Dr. Hays (Hayes) on the west shore of Kennedy channel, at the highest northern localities ever explored. <Am. Journ. Sci., vol. xl, 2d ser., pp. 31–34. 1865. New Haven, 1865.

Upper Silurian.

	Page.
Zaphrentis haysii, n. s., Meek, 1865	32
Syringopora, sp. undt	32
Favosites, sp. undt	32
Strophomena rhomboidalis, Wahlb	33
Strophodonta headleyana, Hall ?	33
Strophodonta beckii, Hall ?	33
Rhynchonella, sp. undt	33
Coelospira concava, Hall	33
Spirifer, sp. undt	33
Loxonema ? kanei, n. s., 1865, Meek	33
Orthoceras, sp. undt	33
Illaenus, sp. undt	33

34.

MEEK, F. B. Note on the genus Gilbertsocrinus Phillips. <Proc. Acad. Nat. Sci. Phila., vol. xvii, pp. 166–167. 1865. Philadelphia, 1865.

> The author takes the ground that the difference between Gilbertsocrinus Phillips, Goniasteroidocrinus Lyon & Casseday, and Trematocrinus Hall, is at most not more than subgeneric.

35.

MEEK, F. B. Observations on the microscopic shell structure of Spirifer cuspidatus, Sowerby, and some similar American forms. <Proc. Acad. Nat. Sci. Phila., vol. xvii, pp. 275–277. 1865. Philadelphia.

36.

MEEK, F. B., *and* WORTHEN, A. H. Notice of some new types of organic remains from the Coal-measures of Illinois. <Proc. Acad. Nat. Sci. Phila., vol. xvii, pp. 41–53. 1865. Philadelphia, 1865.

> Genera Acanthotelson, Palaeocaris, Anthracerpes, Palaeocampa, afterward republished and illustrated in vol. iii of the Illinois Geological Reports.

CRUSTACEA.

Palæocaridæ.
Palæocaris n. g., M. & W., 1865 48–49
P. typus, n. s., M. & W., 1865 49–50
Decapoda,
Macrura.
? Anthrapalæmon, Salter, 1861 50
A. gracilis, n. s., M. & W., 1865 50–51
Myriapoda.
? Anthracerpes, n. g., M. & W., 1865 51
A. typus, n. s., M. & W., 1865 51–52
Insecta.
Lepidoptera.
Palæocampa, n. g., M. & W., 1865 52
P. anthrax, n. s., M. & W., 1865 52–53

37.

MEEK, F. B., *and* WORTHEN, A. H. Remarks on the genus Taxocrinus (Phillips). McCoy, 1844; and its relations to Forbesiocrinus de Koninck and Le Hon, 1854, with descriptions of new species. < Proc. Acad. Nat. Sci. Phila., vol. xvii, pp, 138–143. 1865. Philadelphia, 1865.

Republished in vol. II of the Illinois Geological Reports.

Page.
Taxocrinus (Phillips) McCoy, 1844 138–142
Table showing the gradations of structure from *Taxocrinus* to *Forbesiocrinus* 140
T. gracilis, n. s., M. & W., 1865 142–143

38.

MEEK, F. B., *and* WORTHEN, A. H. Descriptions of new species of Crinoidea, &c., from the Paleozoic rocks of Illinois and some of the adjoining States. < Proc. Acad. Nat. Sci. Phila., vol. xvii, pp. 143–155. 1865. Philadelphia, 1865.

A "note in regard to the name Cincinnati group used in the foregoing paper" is appended on page 155; the descriptions are republished and the views restated in the Illinois Geological Reports, vol. i.
Radiata.
Echinodermata.
Cystidea. Page.
Comarocystites, Billings, 1854 143
C. shumardi, n. s., M. & W., 1865 143–144
C. shumardi var. obconicus, n. s., M. & W., 1865 144–145
Crinoidea.
Porocrinus, Billings, 1856 145
P. crassus, n. s., M. & W., 1865 145–146
P. pentagonius, n. s., M. & W., 1865 146–147
Heterocrinus, Hall, 1847 147
H. crassus, n. s., M. & W., 1865 147–148
H. subcrassus, n. s., M. & W., 1865 .... 148, 149
Subgenus *Anomalocrinus*, M. & W 148–149
H. ? (Anomalocrinus) incurvus, n. s., M. & W., 1865 148–149
Erisocrinus, n. g., M. & W., 1865 149
E. conoideus, n. s., M. & W., 1865 150
E. tuberculatus, n. s., M. & W., 1865 150
Cyathocrinus, Miller, 1821 150
C. quinquelobus, n. s., M. & W., 1865 150–151
C. subtumidus, n. s., M. & W., 1865 151–152
C. enormis, M. & W., 1861 152
Poteriocrinus, Miller, 1821 152
P. (Zeacrinus) carbonarius, M. & W., 1861 152
Actinocrinus, Miller, 1821 152
A. pistillus, M. & W., n. s., 1865 152–154
Subgenus *Sphærocrinus*, M. & W., 1865 154
A. (Sphærocrinus) concavus, M. & W., 1861 154

Mollusca.
Cephalopoda.
    Goniatites compactus, n. s., M. & W., 1865 154–155

39.

MEEK, F. B., *and* WORTHEN, A. H. Descriptions of New Crinoidea, &c., from the Carboniferous rocks of Illinois and some of the adjoining States. <Proc. Acad. Nat. Sci. Phila., vol. xvii, 1865, pp. 155–166. Philadelphia, 1865.

                                                                                                                        Page.

Poteriocrinus, Miller, 1821 155
P. indianensis, n. s., M. & W., 1865 155–156
P. (Scaphiocrinus) tenuidactylus, n. s., M. & W., 1865 156–157
P. (Scaphiocrinus) bayensis, n. s., M. & W., 1865 157–158
P. (Scaphiocrinus) ? norwoodi, n. s., M. & W., 1865 158–159
P. (Scaphiocrinus) subtumidus, n. s., M. & W., 1865 159
Cyathocrinus, Miller, 1821 160
C. arboreus, n. s., M. & W., 1865 160
Platycrinus, Miller, 1821 160–162
    Centrocrinus, Austin 161
    Cupellœcrinus, Troost 161
    Pleurocrinus, Austin 161
Platycrinus niotensis, n. s., M. & W., 1865 162
P. hemisphæricus, n. s., M. & W., 1865 162–163
P. parvulus, n. s., M. & W., 1865 163–164
Actinocrinus, Miller, 1821 164
Subgenus *Alloprosallocrinus,* Lyon & Casseday, 1860 164
A. (Alloprosallocrinus) cuconus, n. s., M. & W., 1865 164–165
Pentremites (Granatocrinus) granulosus, n. s., M. & W., 1865 165

POLYZOA.

Evactinopora, n. g., M. & W., 1865 165
E. radiata, n. s., M. & W., 1865 165–166

40.

MEEK, F. B., *and* WORTHEN, A. H. Contributions to the Paleontology of Illinois and other Western States. <Proc. Acad. Nat. Sci. Phila., vol. xvii, pp. 245–273, 1865. Philadelphia, 1865.

    Silurian, Devonian, Carboniferous. Genus Endolobus. Afterward republished in the Illinois Geological Reports, vol. ii.
Mollusca.
    Lamellibranchiata.   Page.
        Lithophaga, Bolten, 1798 245
        (*Lithodomus,* Cuvier, 1817.) 245
        Lithophaga ? pertenuis, n. s., M. & W., 1865 245
        L. ? lingualis Phillips (?), sp 245
        Modiolopsis, Hall, 1847 246
        M. perovata, n. s., M. & W., 1865 246
        Pleurophorus, King, 1844 246
        P. subcostatus, n. s., M. & W., 1865 246–247
        P. ? angulatus, n. s., M. & W., 1865 247
        P. costatiformis, n. s., M. & W., 1865 247–248
        Grammysia, De Verneuil, 1847 248
        G. ? rhomboidalis, n. s., M. & W., 1865 248–249
        Conocardium, Bronn, 1837 249
        C. obliquum, n. s., M. & W., 1865 249
        Edmondia, De Koninck, 1842 249
        E. ? peroblonga, n. s., M. & W., 1865 249–250
        Chænomya, Meek, 1865 250
        C. ? rhomboidea, n. s., M. & W., 1865 250
        C. ? hybrida, n. s., M. & W., 1865 250–251
        Sedgwickia, McCoy, 1844 251
        S. (Sanguinolites?) subarcuata, n. s., M. & W., 1865 251

Gasteropoda.
 Holopea, Hall, 1847 251
 Subgenus *Isonema*, M. & W 221
 H. (Isonema) depressa, n. s., M. & W., 1865 251–252
 Pleurotomaria, Defrauce, 1825 252
 P. (Murchisonia ?), meta, n. s., M. & W., 1865 252

Pteropoda.
 Conularia, Miller, 1818 252
 C. multicostata, n. s., M. & W., 1865 252–253
 C. subcarbonaria, n. s., M. & W., 1865 253
 C. whitei, n. s., M. & W., 1865 253–254
 Tentaculites, Schlotheim, 1820 254
 T. tenuistriatus, n. s., M. & W., 1865 254
 T. oswegoensis, n. s., M. & W., 1865 254–255
 T. sterlingensis, n. s., M. & W., 1865 255

Cephalopoda.
 Orthoceras, Auct 255
 O. crebristriatum, n. s., M. & W., 1865 255–256
 O. subbaculum, n. s., M. & W., 1865 256
 O. jolietense, n. s., M. & W., 1865 256
 O. nobile, n. s., M. & W., 1865 256–257
 O. winchellii, n. s., M. & W., 1865 257
 Phragmoceras, Broderip, 1834 257
 P. walshii, n. s., M. & W., 1865 257
 Gomphoceras, Sowerby, 1839 258
 G. sacculum, n. s., M. & W., 1865 258
 G. (Apioceras) turbiniforme, n. s., M. & W., 1865 258–259
 Nautilus, Linnæus, 1758 259
 Subgenus *Endolobus*, M. & W., 1865 259
 N. (Endolobus) peramplus, n. s., M. & W., 1865 259
 N. (Temnocheilus) niotensis, n. s., M. & W., 1865 260
 Subgenus *Discites*, McCoy, 1844 260
 N. (Discites) ornatus Hall, var. *amplus*, M. & W., 1865 260–261
 N. (Discites) disciformis, n. s., M. & W., 1865 261
 N. lasallensis, n. s., M. & W., 1865 261–262
 Subgenus *Cryptoceras*, d'Orbigny, 1847 262
 N. (Cryptoceras) capax, n. s., M. & W., 1865 262
 N. (Cryptoceras ?) leidyi, n. s., M. & W., 1865 262–263
 Trochoceras, Barrande, 1847 263
 T. baeri, n. s., M. & W., 1865 263–264

Articulata.
 Crustacea.
 Trilobita.
  Dalmania, Emmerich, 1845 264
  D. danæ, n. s., M. & W., 1865 264–266
  Lichas, Dalman, 1827 266
  L. cucullus, n. s., M. & W., 1865 266–267
  Prœtus Steininger, 1830 267
  P. ellipticus, n. s., M. & W., 1865 267–268
  Phillipsia, Portlock, 1843 268
  Subgenus *Griffithides*, Portlock, 1843 268
  P. (Griffithides) portlockii, n. s., M. & W., 1865 268–270
  P. (Griffithides) scitula, n. s., M. & W., 1865 270–271
  P. (Griffithides ?) sangamonensis, n. s., M. & W., 1865 271–273

41.

MEEK, F. B. Note on the affinities of the Bellerophontidæ. <Proc. Chicago Acad. of Sci., pp. 9–11, 1866. Chicago, 1866.

The author places this family near Fissurellidæ, Haliotidæ, and Pleurotomariidæ.

42.

MEEK, F. B., *and* WORTHEN, A. H. Contributions to the Paleontology of Illinois and other Western States. <Proc. Acad. Nat. Sci. Phila., vol. xviii, pp. 251-275, 1866. Philadelphia, 1866.

Carboniferous fossils. Afterward republished and illustrated in the Illinois Geological Reports.

Radiata.
Echinodermata.
Crinoidea.   Page.
   Belemnocrinus whitii, n. s., M. & W., 1866 251
   Subgenus *Nematocrinus*, M. & W. 1866 251
   Synbathocrinus wachsmuthi, n. s., M. & W., 1866 251-252
   Cyathocrinus farleyi, n. s., M. & W., 1866 252-253
   Actinocrinus calyculus var. *pardinensis*, M. & W 253
   Strotocrinus, n. g., M. & W., 1866 253
   Steganocrinus, n. g., M. & W., 1866 253-254
   Rhodocrinus nanus, n. s., M. & W., 1866 254
   Onychocrinus, Lyon & Casseday 255-256
   Onychocrinus diversus, n. s., M. & W., 1866 256-257
   Granatocrinus shumardi, n. s., M. & W., 1866 257-258
   Granatocrinus norwoodi, O. & S.? 258-259
Asteroidea.
   Schœnaster wachsmuthi, n. s., M. & W., 1866 259

Mollusca.
Lamellibranchiata.
   Pteria (Pterinea?) morganensis, n. s., M. & W., 1866 259-260
   Dolabra sterlingensis, n. s., M. & W., 1866 260-261
   Macrodon micronema, n. s., M. & W., 1866 261
Gasteropoda.
   Platyceras, Conrad, 1840 262-263
   Subgenus *Orthonychia*, Hall 263
   Subgenus *Igoceras*, Hall 263
   P. lævigatum, n. s., M. & W., 1866 263-264
   P. haliotoides, n. s., M. & W., 1866 264
   P. uncum, n. s., M. & W., 1866 264-265
   P. (Orthonychia) chesterense, n. s., M. & W. 1866 265
   P. (Orthonychia) subplicatum, n. s., M. & W., 1866 265-266
   P. (Orthonychia) infundibulum, n. s., M. & W., 1866 266
   Metoptoma, Phillips, 1836 266-267
   Metoptoma (Platyceras?) umbella, n. s., M. & W., 1866 267
   Polyphemopsis chrysallis, n. s., M. & W., 1866 267-268
   Naticopsis littonana var. *genevievensis*, M. & W 268
   Anomphalus, n. g., M. & W., 1866 268
   A. rotulus, n. s., M. & W., 1866 268-269
   Microdoma, n. g., M. & W., 1866 269
   M. conica, n. s., M. & W., 1866 269-270
   Orthonema conica, n. s., M. & W., 1866 270
   Trochita? carbonaria, n. s., M. & W., 1866 270-271
   Platyschisma pelicoides, Sowerby? sp 271
   Pleurotomaria conoides, n. s., M. & W., 1866 271-272
   Pleurotomaria cozana, n. s., M. & W., 1866 272
   Pleurotomaria spironema, n. s., M. & W., 1866 272-273
   Pleurotomaria valvatiformis, n. s., M. & W., 1866 273
   Murchisonia inornata, n. s., M. & W., 1866 274
Cephalopoda.
   Nautilus (Trematodiscus) sulcatus, Sowerby? 274
   Nautilus (Cryptoceras) rockfordensis, n. s., M. & W., 1866 275

In a note the authors consider their previously proposed Evactinopora as equivalent to Conodictyum Münster, and call their species C. radiatum.

43.

MEEK, F. B., and WORTHEN, A. H. Descriptions of Paleozoic fossils from the Silurian, Devonian, and Carboniferous rocks of Illinois and other Western States. <Proc. Chicago Acad. Sci., pp. 11-23. 1866. Chicago, 1866.

Genera Monopteria Megaptera afterward republished in the Illinois Geological Reports, vol. ii.

Amorphozoa.

	Page.
Astylospongia ? carbonaria, n. s., M. & W., 1866	11-12
Astræospongia hamiltonensis, n. s., M. & W., 1866	12

Mollusca.
Lamellibranchiata.

Placunopsis carbonaria, n. s., M. & W., 1866	13
Aviculopecten randolphensis, n. s., M. & W., 1866	14
A. indianensis, n. s., M. & W., 1866	14-15
A. fimbriatus, n. s., M. & W., 1866	15-16
Vanuxemia dixonensis, n. s., M. & W., 1866	16-17
Macrodon tenuistriatus, n. s. g., M. & W., 1866	17
Schizodus curtus, n. s., M. & W., 1866	18
Anthracoptera ? fragilis, n. s., M. & W., 1866	18-19
Myalina meliniformis, n. s., M. & W., 1866	19-20
Monopteria, n. s., g., M. & W., 1866	20
Pterinea (Monopteria) gibbosa, n. s., M. & W., 1866	20-21
Pterinea ? subpapyracea, n. s., M. & W., 1866	21-22
Ambonychia (Megaptera) casei, n. s., M. & W., 1866	22-23

44.

MEEK, F. B., and WORTHEN, A. H. Descriptions of Invertebrates from the Carboniferous System. <Geological Survey of Illinois, vol. ii, pp. 145-411, plates, 14-20 and 23-32. 1866. Published by authority of the legislature of Illinois, Springfield, 1866.

Genera Sphenopoterium, Cardiopsis, Trematodiscus, Strotocrinus, Steganocrinus, Cæloerinus, Oligoporus, Erisocrinus, Syntriclasma, Enmicrotis, Trachydomia, Orthonema, Seleniscus, Acanthotelson, Palæocaris, Anthracerpes, Palæocampa, Shænaster. Volumes ii, iii, v, and vi of these reports all comprise very important works on Invertebrate Paleontology, in which are not only species and genera described, but higher groups are defined, and many important questions are philosophically discussed.

INVERTEBRATE FOSSILS OF THE KINDERHOOK GROUP.

Protozoa.
Spongiæ.

	Page.
* *Sphenopoterium,* M. & W., 1866	145-146
S. enorme, M. & W., 1866, pl. xiv, figs. 1 *a*, 1 *b*	146
S. enorme var. *depressum,* M. & W., 1866, pl. xiv, figs. 2 *a*, 2 *b*	146

Radiata.
Echinodermata.
Crinoidea.

Actinocrinus, Miller	147-149
Batocrinus, Casseday (*Eretmocrinus* Lyon?)	150-151
Subgenus *Batocrinus,* Casseday	151
Sec. C. (*Uperocrinus,* M. & W.)	151
Actinocrinus (Batocrinus) pistilliformis, M. & W., 1861, pl. xiv, fig. 8	151-153

Mollusca.
Brachiopoda.

Rhynchonella, Fischer, 1809	153
R. missouriensis, Shumard, 1855, pl. xiv, figs. 4 *a*, 4 *b*	153-154
Spirifer, Sowerby, 1815	155
Subgenus *Martinia,* McCoy, 1844	155
S. (Martinia) cooperensis, Swallow, 1860, pl. xiv, figs. 5 *a*, 5 *b*	155-156

---

* σφην, a wedge, ποτηριον, a drinking cup.

Lamellibranchiata. Page

Cardiopsis, M. & W., 1861 156–157
C. radiata, M. & W., 1860, pl. xiv, figs. 6 a, 6 b 157–158

Gasteropoda.
Straparollus, Montfort, 1810 158–159
S. lens, Hall, sp., 1860, pl. xiv, figs. 7 a, 7 b 159–160
Bellerophon, Montfort, 1810 160
B. cyrtolites, Hall, 1860, pl. xiv, figs. 8 a, 8 b 160–161

Cephalopoda.
Nautilus, Linnæus, 1758 161
Subgenus *Trematodiscus*, M. & W., 1861 161–162
N. (Trematodiscus) trisulcatus, M. & W., 1860, pl. xiv, figs. 10 a–c 162–163
N. digonus, M. & W., 1860, pl. xiv, figs. 9 a–d 163–164
Goniatites, de Haan, 1825 165
G. lyoni, M. & W., 1860, pl. xiv, figs. 11 a–c 165–166

INVERTEBRATE FOSSILS OF THE BURLINGTON GROUP.

Radiata.
Echinodermata.
Crinoidea.
Dichocrinus, Miller, 1821 167–169
D. conus, M. & W., 1860, pl. xvi, figs. 5 a, b 169–170
Platycrinus, Miller, 1821 170–172
Centrocrinus, Austin 172
Pleurocrinus, Austin 172
Marsupiocrinites, Phillips 172
Subgenus *Pleurocrinus*, Austin 173
Platycrinus (Pleurocrinus) subspinosus, Hall, 1858, pl. xv, fig. 6 173–175
Cyathocrinus, Miller, 1821 175–178
C. sculptilis, Hall, 1860, pl. xv, fig. 2 a, b 178–179
Poteriocrinus, Miller, 1821 179–182
P. swallovi, M. & W., 1860, pl. 16, figs. 4 a, b, and fig. 3, p. 184 183–184
Zeacrinus, Troost 185–186
Z. troostanus, M. & W., 1860, pl. xvi, fig. 2 and fig. 4, p. 187 186–187
Strotocrinus,* M. & W., [1866] 188–192
S. perumbrosus?, Hall, fig. 5, p. 188 188
S. regalis, sp., Hall, 1860, pl. xvi, figs. 6 a, b, fig. 6, p. 191, fig. 7, p. 192, and fig. 8, p. 194 . 191–195
Steganocrinus, n. g., M. & W. [1866] 195
S. pentagonus,† Hall, figs. 9 a, d 196
S. sculptus, Hall, figs. 10, a, b, and d. p. 197 197–198
S. araneolus, M. & W., 1860, pl. xv, figs. 1 a, b 198–200
Actinocrinus, Miller, 1821 200
A. concinnus, Shumard, 1855, pl. xv, figs 9 a, b., fig. 11, p. 202 200–202
A. scitulus, M. & W., 1860, pl. xv, figs. 7 a, b, and fig. 12, p. 204 202–205
A. sillimani, M. & W., 1860, fig. 13, p. 204 204–205
Subgenus *Batocrinus*, Casseday 205
A. (Batocrinus) dodecadactylus, M. & W., 1861, pl. xv, figs. 3 a–c., fig. 14, p. 206 ... 205–207
A. Batocrinus asteriscus, M. & W., 1860, pl. xv, figs. 8 a–c 207–209
Amphoracrinus, Austin, 1848 209–211
A. subturbinatus, M. & W., 1860, pl. xv, figs. 4 a, b., and fig. 15, p. 213 ... 212–213
Coelocrinus, ;M. & W., 1865 214–215
C. concavus, M. & W., 1861, pl. xv, figs. 10 a–c., and fig. 16, p. 215 215–216
Gilbertsocrinus, Phillips = *Ollacrinus*, Cumberland 217–219
Subgenus *Goniasteroidocrinus*, Lyon & Casseday=*Trematocrinus*, Hall 219
Gilbertsocrinus bursa, Phillips, fig. 17, p. 217 217
G. calcaratus, Phillips, fig. 18 a-c, p. 217 217
Goniasteroidocrinus tuberosus, Lyon & Casseday, fig. 19 a-d, p. 220 220
G. (Goniasteroidocrinus) fiscellus, M. & W., 1860, pl. xv, fig. 5, and fig. 20, p. 224 ... 222–225
Echinoidea.
Perischæchinidæ.
Melonites multipora, figs. 21 and 22, pp. 227–228 227–228

---

*στρωτος, spread; χρινον, a lily.
†στεγανος, covered; χρινον, a lily; in allusion to the covered free rays.
‡χοιλος, hollow; χρινον, a lily.

Palaechinida, McCoy.
Archaeocidaridæ, McCoy.
Palaechinus, McCoy, 1844 229
P. burlingtonensis, M. & W., 1860, pl. xvi, figs. 3 a–c., and fig. 23, p. 231 230–231

Mollusca.
Lamellibranchiata.
Aviculopecten, McCoy, 1851 231
A. burlingtonensis, M. & W., 1860, pl. xvi, figs. 1 a, b 231–232

INVERTEBRATE FOSSILS OF THE KEOKUK GROUP.
Protozoa.
Spongiæ.
Petrospongia.
Sphenopoterium, M. & W., 1860 233
S. obtusum, M. & W., 1860, pl. xvii, figs. 2 a-d (by error a on pl.), 2 e 233
S. compressum, M. & W., 1860, pl. xvii, figs. 1 a–c 234

Radiata.
Echinodermata.
Crinoidea.
Cyathocrinus, Miller, 1821 234
C. angulatus, M. & W., 1860, pl. xvii, fig. 4 234–236
C. saffordi, M. & W., 1860, pl. xvii, figs. 5 a and b, and fig. 24, p. 237 236–237
Poteriocrinus, Miller 237
Subgenus *Scaphiocrinus*, Hall, 1858 237
P. (Scaphiocrinus) decadactylus, M. & W., 1860, pl. xvii, fig. 6 and fig. 25, p. 240 .. 238–240
Zeacrinus, Troost 240
Z. planobrachiatus, M. & W., 1860, pl. xvii, fig. 5 240–241
Onychocrinus, Lyon & Casseday, 1859 242–244
O. monroensis, M. & W., 1861, pl. xvii, fig. 7 244–245
O. norwoodi, M. & W., 1860, pl. xvii, fig. 3, and fig. 26, p. 247 245–247
Echinoidea.
Perischœchinidæ.
Oligoporus, M. & W., 1860 247
Melonites multipora, fig. 27 248
Oligoporus danæ, M. & W., 1860, pl. xvii, fig. 8 and fig. 28, p. 248 248–251

Mollusca.
Brachiopoda.
Camarophoria, King, 1844.
C. subtrigonia, M. & W., 1860, pl. xviii, figs. 8 a–c 251–253
Chonetes, Fischer, 1837.
C. planumbona, M. & W., 1860, pl. xviii, figs. 1 a–d 253–254
Athyris, McCoy, 1844.
A. planosulcata, Phillips ? (sp.), 1836, pl. xxii, figs. 8 a–d 254–255
Lamellibranchiata.
Aviculopecten, McCoy, 1851 256
A. oweni, M. & W., 1860, pl. xviii, figs. 2 a–c 256
A. amplus, M. & W., 1860, pl. xviii, figs. 4 a–c 257–258
A. oblongus, M. & W., 1860, pl. xviii, figs. 3a and b 258–259
Gasteropoda
Pleurotomaria, Defrance, 1826 260
P. shumardi, M. & W., 1860, pl. xviii, figs. 6a and b 260–261

INVERTEBRATE FOSSILS OF THE ST. LOUIS GROUP.
Protozoa.
Spongiæ.
Petrospongia.
Sphenopoterium, M. & W.
S. cuneatum, M. & W., 1860, pl. xix, figs. 1 a–d 262–263
Radiata.
Echinodermata.
Crinoidea.
Dichocrinus, Munster.
D. constrictus, M. & W., 1860, pl. xix, figs. 2 a–c 263–264
Platycrinus, Miller, 1821 264
P. prattenanus, M. & W., 1860, pl. xx, fig. 2 264–265

---

* 'ολιγος, fow ; Πορος, a passage.

	Page.
P. penicillus, M. & W., 1860, pl. xix, figs. 6 *a–c*	266–267
P. plenus, M. & W., 1860, pl. xx, fig. 3	267–268
Taxocrinus, Phillips, 1843	268–270
Forbesiocrinus, de Koninck & Le Hon, 1854	270
Taxocrinus semioratus, M. & W., 1860, pl. xx, figs. 4 *a–b*	272–274
Blastoidea.	
Granatocrinus, Troost, 1850	274–275
G. cornutus, sp., M. & W., 1860, pl. xx, fig. 1	276
Asteroidea.	
Schœnaster, M. & W. [1860]	277–278
S. fimbriatus, M. & W., 1860, pl. xix, figs. 7 *a–d*	278–280
Mollusca.	
Brachiopoda.	
Productus, Sowerby, 1814	280
P. scitulus, M. & W., 1860, pl. xx, figs. 5 *a–d*	280–281
Lamellibranchiata.	
Myalina, de Koninck, 1844	281
M. concentrica, M. & W., 1860, pl. xix, figs. 3 *a–c*	281–282
Yoldia, Moller, 1842	282
Y.? levistriata, M. & W., 1860, pl. xx, figs. 7 *a* and *b*	282–283
Nuculana, Link, 1807	283
N.? curta, M. & W., 1861, pl. xx, figs. 6 *a* and *b*	283–284
Gasteropoda.	
Dentalium, Linnæus, 1758	284
D. venustum, M. & W., 1861, pl. xix, fig. 6	284
Straparollus, Montfort, 1810	285
S. similis, M. & W., 1861, pl. xix, figs. 4 *a–b*	285–286
S. similis var. *planus*, M. & W., 1861, pl. xix, fig. 5 *a–c*	286
Cephalopoda.	
Orthoceras, Auct	286
O. expansum, M. & W., 1860, pl. xx, figs. 8 *a–c*	286–287

INVERTEBRATE FOSSILS OF THE CHESTER GROUP.

Radiata.
Echinodermata.
Crinoidea.

Pterotocrinus, Lyon & Casseday, 1859	288–290
P. crassus, M. & W., 1860, pl. xxiii, figs. 2 *a*, *b*, and fig. xxix, p. 292	290–292
P. chesterensis, M. & W., 1860, pl. xxiii, figs 1 *a–c*, and fig. 30, p. 293	292–293

Echinoidea.
Perischœchinidœ.

Archœocidaris, McCoy, 1844	294–295
A. mucronata, M. & W., 1860, pl. xxiii, figs. 3 *a–c*	295–296

Mollusca.
Brachiopoda.

Productus, Sowerby, 1812	297
P. parvus, M. & W., 1860, pl. xxiii, figs. 4 *a–e*	297–298
Spirifer, Sowerby, 1815	298
Subgenus *Martinia*, McCoy, 1844	298
S. (Martinia) glaber var. *contractus*, M. & W., pl. xxiii, figs. 5 *a*, *b*	298–299

Lamellibranchiata.

Myalina de Koninck, 1844	300
M. angulata, M. & W., 1860, pl. xxiii, figs. 7 *a*, *b*	300–301
Schizodus, King, 1844	301
S. chesterensis, M. & W., 1865, pl. xxiii, figs. 6 *a*, *b*	301–302

Gasteropoda.

Straparollus, Montfort, 1810	302
S. planidorsatus, M. & W., 1860, pl. 24, figs. 2 *a–c*	302–303
Pleurotomaria, Defrance, 1826	303
P. chesterensis, M. & W., 1860, pl. xxiv, figs. 1 *a–c*	303–304

Cephalopoda.

Orthoceras, Auct	304
O. annulato-costatum, M. W., 1861, pl. xxiv, figs. 3 *a*, *b*	304–305

* σχοινος, a rope; αστηρ, a star.

	Page.
Nautilus, Linnæus, 1758	305
N. globatus, Sowerby? 1825, pl. xxiv, figs. 5 *a, b*	305–306
N. chesterensis, M. & W., 1860, pl. xxiv, figs. 4 *a, b*	306–307
Subgenus *Endolobus*,* M. & W., 1865	307–308
N. (Endolobus) spectabilis, M. & W., 1860, pl. xxv, figs. 1 *a, b*	308–309

INVERTEBRATE FOSSILS OF THE COAL MEASURES.

Radiata.
Crinoidea.

Cyathocrinus, Miller, 1821	310
C.? sangamonensis, M. & W., 1861, pl. xxvi, figs. 1 *a, b,* and fig. 31, p. 311	310–312
Zeacrinus, Troost	312
Z. discus, M. & W., 1860, pl. xxvi, figs. 3 *a, b*	312–313
Z.? crassus, M. & W., 1860, pl. xxvi, figs. 2 *a, b,* and fig. 32, p. 315	314–315
Erisocrinus, M. &. W., 1865	315–317
E. typus, M. & W., 1865, fig. 33, p. 317, and fig. 34 *a–c*, p. 318	317–318
E. conoideus, M. & W., 1865, figs. 35 *a, b*	318–319
E. tuberculatus, M. & W., 1865	319–320

Mollusca.
Brachiopoda.

Productus, Sowerby, 1812	320
P. nanus, M. & W., 1860, pl. xxvi, figs. 4 *a–d*	320–321
Syntrielasma, M. & W., 1865	321–323
S. hemiplicata, Hall, (sp.), 1862, figs. 36 *a–c*, p. 322, and fig. 37 *a, b,* p. 324	322–325

Lamellibranchiata.

Aviculopecten, McCoy, 1852	326
A. cozanus, M. & W., 1860, pl. xxvi, fig. 6 *a, b*	326–327
A. pellucidus, M. & W., 1860, pl. xxvi, fig. 5 *a, b*	327–328
A. koninckii, M. & W., 1860, pl. xxvi, fig. 8	328–329
A. interlineatus, M. & W., 1860, pl. xxvi, fig. 7 *a, b*	329–330
A. occidentalis, Shumard? 1855 pl. xxvii, figs. 4, 5, 5 *a*	331–332
Streblopteria, McCoy, 1851	332–334
S.? tenuilineata, M. & W., 1860, pl. xxvi, figs. 9 *a, b*	334–336
Eumicrotis, Meek, 1864	336–337
E. hawni var. *sinuata*, M. & W. [1865], pl. xxvii, figs. 12 *a,* 12 *b,* 13, 14	338–339
Pterinea, Goldfuss, 1833	339
Subgenus *Monopteria*,† M. & W., 1866	339–340
P. (Monopteria) gibbosa, M. & W., 1866, pl. xxvii, figs. 11, 11 *a, b*	340–341
Myalina, de Koninck, 1844	341
M. swallovi, McChesney, 1860, pl. xxvii, figs. 1 *a–d*	341–342
M. meliniformis, M. & W., 1806, pl. xxvii, fig. 3	343–344
M. recurvirostris, M. & W., 1860, pl. xxvi, figs. 9 *a–c*	344–345
Schizodus, King, 1844	345
Schizodus, ——, sp.?, pl. xxvii, fig. 7	345–346
Edmondia, de Koninck, 1844	346
E. unioniformis, sp., Phillips, 1836, pl. xxvii, figs. 6, 6 *a* and *b*	346–347
Pleurophorus, King, 1844	347
P. subcostatus, M. & W., 1865, pl. xxvii, figs. 2, 2 *a*	347–349
Solenomya, Lamarck, 1819	349
S. radiata, M. & W., 1860, pl. xxvi, fig. 10 *a, b*	349–350
Allorisma, King, 1844	350
Allorisma, sp. undt., pl. xxvii, figs. 9, 9 *a*	350

Gasteropoda.

Pleurotomaria, Defrance, 1836	351
P. subconstricta, M. & W., 1860, pl. xxviii, figs. 6 *a–c*	351–352
P. speciosa, M. & W., 1860, pl. xxviii, figs. 5 *a–c*	352–353
P. scitula, M. & W., 1860, pl. xxviii, figs. 9 *a–d*	353–354
P. brazoensis, Shumard? 1860, pl. xxviii, figs. 1 *a–d*	354–355
P. tenuicincta, M. & W., 1860, pl. xxviii, fig. 3 *a–d*	355–356
P. granulo-striata, M. & W., 1860, pl. xxviii, fig. 2 *a–d*	356–357
P. pratteni, M. & W., 1860, pl. xxviii, fig. 7 *a–d*	357–358
P. subsinuata, M. & W., 1860, pl. xxvii, figs. 4 *a–d*	358–359
P. turbiniformis, M. & W., 1860, pl. xxviii, figs. 8 *a–c*	359–360

* ενδον, within; λοβος, a lobe.   † μονος, solitary; πτερον, a wing.

	Page.
P. subscalaris, M. & W., 1860, pl. xxviii, figs. 10 a, b (by error on pl., 9 a, 9 b)	360–361
P.? tumida, M. & W., 1860, pl. xxxi, figs. 1 a, b	361–362
Straparollus, Montfort, 1810	362
S. umbilicatus, M. & W., 1860, pl. xxix, figs. 1 a–c	362–363
Naticopsis, McCoy, 1844	363–365
Naticopsis nana, M. & W., 1860, pl. xxxi, figs. 4 a, b	365–366
Subgenus *Trachydomia*,* M. & W, 1866	366
N. (Trachydomia) nodosa, M. & W., 1860, pl. xxxi, figs. 2 a, b	366–367
N. (Trachydomia) nodosa, var. *hollidayi*, M. & W., 1860, pl. xxxi, figs. 3 a, b	367
Macrocheilus, Phillips, 1841	367–369
M. medialis, M. & W., 1865, pl. xxxi, figs. 5 a, b	370
M. intercalaris, M. & W., 1860, pl. xxxi, figs. 6 a, b	371
Macrocheilus, sp. undt., pl. xxvii, fig. 10	372
Polyphemopsis, Portlock, 1843	372–373
P. inornata, H. & W., 1860, pl xxxi, figs. 8 a–c	374
P. nitidula, M. & W., 1860, pl. xxxi, figs. 9 a, b	374–375
P. peracuta, M. & W., 1860, pl. xxxi, figs. 7 a, b	375–376
Loxonema, Phillips, 1841	377
L. scitula, M. & W., 1860, pl. xxxi, figs. 10 a–c	377
L. rugosa, M. & W., 1860, pl. xxxi, figs. 11 a–c	378
L. multicostata, n. s., M. & W., 1866, pl. xxxi, figs. 12 a–c	378–379
L. cerithiformis, M. & W., 1860, pl. xxxi, figs. 13 a–c	379–380
{*Orthonema*, M. & W., 1861	380–381
O. salteri, M. & W., 1860, pl. xxxi, figs. 14 a–c	381
Turritella, Lamarck, 1799	382
T. ? ? stevensana, n. s., M. & W., 1860, pl. xxvii, figs. 8 and 8 a	382–383
§*Soleniscus*, M. & W., 1860	383–384
S. typicus, M. & W., 1860, pl. xxxi, figs. 15 a and b	384
Bellerophon, Montfort, 1808	385
B. crassus, M. & W., 1860, pl. xxxi, figs. 16 a and b	385
Cephalopoda.	
Nautilus, Linnæus, 1758	386
N. planorbiformis, M. & W., 1860, pl. xxix, figs. 4 a–c	386
N. sangamonensis, M. & W., 1860, pl. xxix, figs. 3, 3 a, b	386–388
Cyrtoceras, Goldfuss, 1832	388
C. (Aploceras) curtum, M. & W., 1860, pl. xxx, figs. 1 a–c	388–389
C. ? dilatatum, M. & W., 1860, pl. xxix, fig. 2	389
Goniatites, de Haan, 1825	390
G. globulosus, M. & W., 1860, pl. xxx, figs. 2 a–c, and fig. 38, p. 390, and fig. 39, p. 391	390–392
G. iowensis, M. & W., 1860, pl. xxx, fig. 3 a–c	392–393
Articulata.	
Crustacea.	
Entomostraca.	
Xiphosura.	
Bellinurus, Kœnig	393–395
B. danæ, M. & W., 1865, pl. xxxii, figs. 2, 2 a	395–398
Tetradecapoda.	
Isopoda.	
(Anisopoda.)	
Acanthotelson, M. & W., 1860	399–401
A. stimpsoni, M. & W., 1865, pl. xxxii, figs. 6, 6 a–f	401–402
A. inæqualis, M. & W., 1865, pl. xxxii, figs. 7, 7 a	403
?*Palæocaris*, M. & W., 1865	403–404
P. typus, M. & W., 1865, pl. xxxii, figs. 5, 5 a–d	405–406
Decapoda.	
Macrura.	
Anthrapalæmon, Salter, 1861	406
A. gracilis, M. & W., 1865, pl. xxxii, figs. 4, 4 a–c	407–408
Myriapoda.	
‖ ? *Anthracerpes*, M. & W. [1865]	409
A. typus, M. & W., 1865, pl. xxix, figs. 1, 1 a	409–410

*τραχος, rough; δωμα, a house.   ‡παλαιος, ancient; χαρις, a shrimp.
Ɩορθος, straight; νημα, a thread.   §σωληνισχος, a little channel or gutter.
‖ ανθραξ, coal; 'ερπω, to creep.

Insecta.
Lepidoptera.
Palaeocampa, 'M. & W. [1865]... 401
P. anthrax, M. & W., 1865, pl. xxxii, fig. 3.. 410–411

45.

MEEK, F. B. Check-lists of the Invertebrate fossils of North America. Cretaceous and Jurassic. < Smithsonian Miscellaneous Collections (No. 177), pp. 1–40. 1864. Washington, 1867.

Contains, besides list, "Notes and explanations" of generic and specific characters of much importance.

The following species are discussed in the notes and explanations:

CRETACEOUS.

	Page.
Planularia cuneata, Mort. = *Phonemus (Flabellina) cuneatus*, (Mort.) Meek............	31
Palmula sagittaria, Lea = *Phonemus (Flabellina) sagittarius*, (Lea) Meek...............	31
Orbitulites texanus, Rœmer = *Tinoporus (Orbitolina) texanus*, (Rœmer) Meek..........	31
Grammostomum phyllodes, Ehrenburg = *Textularia phyllodes*, (Ehrenburg) Meek......	31
Plagiostoma echinatum, Mort. = *Spondylus echinatus*, (Mort.) Meek........................	31
Ctenoides acutilineata, Con. = *Lima acutilineata*, (Con.) Meek.................................	31
Ctenoides denticulata, Gabb. = *Lima denticulata*, (Gabb.) Meek...............................	31
Plagiostoma pelagicum, Mort. = *Lima pelagica*, (Mort.) Meek...................................	31
Ctenoides squarrosa, Gabb. = *Lima squarrosa*, (Gabb.) Meek..................................	31
Syncyclonema, Meek..	31
Pecten rigida, H & M. = *Syncyclonema ? rigida*, (H. & M.) Meek.............................	31
Leda bisulcata, M. & H. = *Nuculana bisulcata*, (M. & H.) Meek................................	31
Leda longifrons, Con. = *Nuculana longifrons*, (Con.) Meek....................................	32
Leda pinnæformis, Gabb. = *Nuculana pinnæformis*, (Gabb.) Meek	32
Leda protexta, Gabb. = *Nuculana protexta*, (Gabb.) Meek	32
Leda slackiana, Gabb. = *Nuculana slackiana*, (Gabb.) Meek..................................	32
Leda subangulata, Gabb. = *Nuculana subangulata*, (Gabb.) Meek	32
Avicula abrupta, Con. = *Pteria abrupta*, (Con.) Meek...	32
Avicula convexo-plano, Rœmer = *Pteria convexo-plano*, (Roemer) Meek..............	32
Avicula cretacea, Con. = *Pteria cretacea*, (Con.) Meek...	32
Avicula haydeni, H. & M. = *Pteria haydeni* (H. & M.) Meek......................................	32
Avicula iridescens, Shumard = *Pteria iridescens*, (Shum.) Meek............................	32
Avicula laripes, Mort. = *Pteria laripes*, (Morton) Meek	32
Avicula linguiformis, E. & S. = *Pteria linguiformis*, (E. & S.) Meek...........................	32
Avicula nebrascana, E. & S. = *Pteria nebrascana*, (E. & S.) Meek.........................	32
Avicula pedernalis, Rœmer = *Pteria pedernalis*, (Rœmer) Meek............................	32
Avicula petrosa, Conrad = *Pteria petrosa*, (Con.) Meek	32
Avicula planisulca, Rœmer = *Pteria planisulca*, (Rœmer) Meek.............................	32
Avicula subgibbosa, M. & H. = *Pteria subgibbosa*, (M. & H.) Meek	32
Avicula triangularis, E. & S. = *Pteria triangularis*, (E. & S.) Meek	32
Actinoceramus, Meek	32
Inoceramus sulcatus, Parkinson = *Inoceramus (Actinoceramus) costellatus*, Con..	32
Volsella attenuata, M. & H. = *Modiola attenuata*, M. & H	32
Volsella concentrico-costellata, Rœmer = *Modiola concentrico-costellata*, Rœmer.	32
Volsella cretacea, Con. = *Modiola cretacea*, Conrad...	32
Volsella juliæ, Lea = *Modiola juliæ*, Lea...	32
Volsella meekii, Evans & Shumard = *Modiola meekii*, E. & S	32
Volsella pedernalis, Rœmer = *Modiola pedernalis*, Rœmer...................................	32
Volsella saffordi, Gabb. = *Modiola saffordi*, Gabb..	32
Modiola granulato-cancellata, Rœmer = *Crenella granulato-cancellata*, (Rœmer) Meek..	32
Liopistha, Meek	32
Cardium elegantulum, Rœmer = *Papyridea (Liopistha) elegantula*, (Rœm.) Con	32
Corbula sp. ind., Owen = *Papyridea (Liopistha) rostrata*, Meek	33
Cardium ? sancti-sabæ, Rœmer = *Papyridea ? sancti-sabæ*, (Rœm.), Meek...........	33
Cyprina arenaria, M. & H. = *Cyrena arenaria*, (M. & H.) Meek	33
Cyprina humilis, M. & H. = *Venilia humilis*, (M. & H.) Meek	33
Venilia quadrata, Gabb. = *Venilia gabbana* Meek..	33

* παλαιος, ancient; χαμπη, a caterpillar.

48   BULLETIN NO. 30, UNITED STATES NATIONAL MUSEUM.

	Page.
Cyprina subtumida, M. & H. = *Venilia subtumida*, (M. & H.) Meek	33
Cyprina laphami, Shumard = *Venilia laphami*, (Shum.) Meek	33
Venus ? circularis, M. & H. = *Cyclina ? circularis*, (M. & H.) Meek	33
Callista eufalensis, Con. = *Dione eufalensis*, (Con.) Meek	33
Cytherea deweyi, M. & H. = *Dione Deweyi*, (M. & H.) Meek	33
Cytherea leonensis, Con. = *Dione leonensis*, (Con.) Meek	33
Cytherea lamarensis, Shum. = *Dione lamarensis*, (Shumard) Meek	33
Cytherea missouriana, Mort. = *Dione missouriana*, (Morton) Meek	33
Cytherea nebrascensis, M. & H. = *Dione nebrascensis*, M. & H	33
Cytherea orbiculata, H. & M. = *Dione orbiculata*, (H. & M.) Meek	33
Cytherea owenana, M. & H. = *Dione owenana*, (Meek & Hayden) Meek	33
Cytherea pellucida, M. & H. = *Dione ? pellucida*, (M. & H.) Meek	33
Cytherea texana, Conrad = *Dione texana*, (Con.) Meek	33
Cytherea tippana, Conrad = *Dione tippana*, (Con.) Meek	33
Venus meckiana, Gabb. = *Dione [?] meekiana*, (Gabb) Meek	33
Venus ripleyana, Gabb. = *Dione [?] ripleyana*, (Gabb) Meek	33
Cytherea tenuis, H. & M. = *Dione [?] tenuis*, (H. & M.) Meek	34
Tellina formosa, M. & H. = *Abra ? formosa*, (M. & H.) Meek	34
Solen irradians, Rœmer = *Linearia irradians*, (Rœmer) Meek	34
Psammobia cancellato-sculpta, Rœmer = *Linearia cancellato-sculpta*, (Rœmer) Meek	34
Cymella, Meek	34
Pholadomya undata, M. & H. = *Pholadomya (Cymella) undata*, (M. & H.)	34
Leda fibrosa, Evans & Shumard = *Nevra fibrosa*, (E. & S.) Meek	34
Goniochasma, Meek	34
Xylophaga stimpsoni, M. & H. = *Goniochasma stimpsoni*, (M. & H.) Meek	34
Xylophagella, Meek	34
Xylophaga elegantula, M. & H. = *Xylophagella elegantula*, (M. & H.) Meek	34
Ringicula pulchella, Shumard = *Cinulia (Avellana) pulchella*, (Shum.) Meek	34
Actæonina naticoides, Gabb. = *Cinulia naticoides*, (Gabb) Meek	35
Ringicula subpellucida, Shumard = *Ringinella subpellucida*, (Shum.) Meek	35
Ringicula acutispira, Shumard = *Riginella acutispira*, (Shum.) Meek	35
Actæonina biplicata, Gabb. = *Solidula biplicata*, (Gabb) Meek	35
Scalpellum inequicostatum, Shumard = *Anisomyon ? inæquicostatus*, (Shum.)	35
Capulus occidentalis, H. & M. = *Tectura ? occidentalis*, (H. & M.) Meek	35
Phasianella haleana, d'Orbigny = *Eutropia haleana*, (d'Orb.) Meek	35
Phasianella perovata, Shumard = *Eutropia perovata*, Shumard	35
Phasianella punctata, Gabb. = *[Eutropia] punctata*, Gabb	35
Architectonica abotti, Gabb. = *Margaritella abotti*, (Gabb) Meek	35
Solarium abyssinus, Gabb. = *Margarita abyssinus*, (Gabb) Meek	35
Nerita (Nereis) densata, Conrad = *Neritella (Nereis) densata*, (Con.) Meek	35
Natica ambigua, M. & H. = *Vanikoro ambigua*, (M. & H.) Meek	35
Natica tuomeyana, M. & H. = *Neritopsis ? tuomeyana*, (M. & H.) Meek	35
Spironema, Meek	35
Turbo tenuilineata, M. & H. = *Spironema tenuilineata*, (M. & H.) Meek	35
Tuba ? bella, Conrad = *Spironema bella*, (Con.) Meek	35
Rostellaria bianguiata, M. & H. = *Anchura ? bianguiata*, (M. & H.) Meek	35
Aporrhais parca, M. & H. = *Anchura ? parca*, (M. & H.) Meek	35
Aporrhais sublævis, M. & H. = *Anchura ? sublævis*, (M. & H.) Meek	35
Drepanocheilus, Meek	35
Rostellaria americana, Evans & Shum. = *Anchura (Drepanoch[e]ilus) americana*, (Evans & Shumard) Meek	35–36
Aporrhais decemlirata, Conrad = *Anchura (Drepanoch[e]ilus) decemlirata*, (Con.) Meek	36
Rostellaria nebrascensis, Evans & Shumard = *Anchura (Drepanoch[e]ilus) nebrascensis*, (E. & S.) Meek	36
Rostellaria rostrata, Gabb. = *Anchura (Drepanoch[e]ilus) rostrata*, (Gabb) Meek	36
Isopleura, Meek	36
Rimella curvilirata, Conrad = *Isopleurus curviliratus*, (Con.) Meek	36
Chemnitzia meekiana, Gabb. = *Isopleurus meekianus*, (Gabb) Meek	36
Pterocerella, Meek	36
Harpago tippana = *Pterocerella tippana*, (Con.) Meek	36
Chemnitzia, Conrad, 1860 (not d'Orb) = *Chemnitzia corona* (Gou.), Meek	36
Scalaria texana, Rœmer = *Chemnitzia ? texana*, (Rœmer) Meek	36
Scalaria forshayii, Shumard = *Scala forshayii*, (Shumard) Meek	36
Natica acutispira, Shumard = *Lunatia ? acutispira*, (Shumard) Meek	36
Volutilithes bella, Gabb. = *Rostellites bellus*, (Gabb) Meek	36

	Page.
Volutilithes bipliteata, Gabb.= *Rostellites biplicatus*, (Gabb) Meek	36
Volutilithes conradi, Gabb.= *Rostellites conradi*, (Gabb) Meek	36
Volutilithes nasuta, Gabb.= *Rostellites nasutus*, (Gabb), Meek	36
Trachytriton, Meek	37
Fusus? vinculum, H. & M. = *Trachytriton vinculum*, (H. & M.) Meek	37
Pleurotomaria texana, Shumard = *Turris texanus*, (Shumard) Meek	37
Picstochilus, Meek, n. g. 1864	37
Fusus scarboroughi, M. & H.= *Clavellithes (Piestochilus) scarboroughi*, (M. & H.) Meek	37
Fusus vaughani, M. & H. = *Cantharus? vaughani*, (M. & H.) Meek	37
Fusus? flexicostatus, M. & H. = *Pyrifusus? flexicostatus*, (M. & H.) Meek	37
Fusus haleanus, d'Orbigny = *Pyrifusus? haleanus*, (d'Orb.) Meek	37
Neptunea impressa, Gabb.= *Pyrifusus? impressus*, (Gabb) Meek	37
Fusus intertextus, M. & H. = *Pyrifusus intertextus*, (M. & H.) Meek	37
Fusus newberryi, M. & H. = *Pyrifusus newberryi*, (M. & H.) Meek	37–38
Fusus subturritus, M. & H. = *Pyrifusus subturritus*, (M. & H.) Meek	38
Fusus? tenuilineatus, H. & M. = *Tritonifusus? tenuilineatus*, (H. & M.) Meek	38
Pleurotomaria mullicaensis, Gabb = *Fusus mullicaensis*, Gabb	38
Pyrula bairdi, M. & H.= *Tudicla (Pyropsis) bairdi*, (M. & H.) Meek	38
Fusus? dakotensis, M. & H. = *Tudicla? dakotensis*, (M. & H.) Meek	38
Hamites leai, Troost = *Ptychoceras leai*, (Troost) Meek	38
Hamites verneuilii, Troost = *Ptychoceras verneuilii*, (Troost) Meek	38
Turrilites sp. undt. = *Heteroceras oweni*, Meek	38
Helicoceras tortum, M. & H. = *Heteroceras tortum*, (M. & H.) Meek	38
Helicoceras? angulatum, M. & H. = *Heteroceras? angulatum*, (M. & H.) Meek	38
Turrilites cheyennensis, M. & H. = *Heteroceras? cheyennensis*, (M. & H.) Meek	38
Nautilus orbiculatus, Tuomey = *?Aturia orbiculata*, (Tuomey) Meek	38
Belemnites paxillosa, Lamarck = *Belemnitella paxillosa*, (Lamarck) Meek	38
Vermetus rotula, Morton = *Spirulæa rotula*, (Morton) Meek	38

JURASSIC.

Camptonectes, Agassiz, MSS. [1864]	39
Pecten bellistriatus, Meek = *Camptonectes bellistriatus*, Meek	39
Pecten extenuatus, M. & H. = *Camptonectes? extenuatus*, (M. & H.) Meek	39
Avicula? curta, Hall = *Eumicrotis curta*, (Hall) Meek	39
Oxytoma, Meek	39
Avicula munsteri, Brown = *Pteria (Oxytoma) munsteri*, (Brown?) Meek	39
Modiola pertenuis, M. & H. = *Volsella pertenuis*, (M. & H.) Meek	40
Modiola (Perna) formosa, M. & H. = *Volsella formosa*, (M. & H.) Meek	40
Venus unionides, Rœmer = *Myacites unionoides*, (Rœmer) Meek	40
Neritina nebrascensis, Meek = *Neritella nebrascensis*, M. & H.	40
Lioplacodes, Meek	40
Melania (Potadoma) veterna, M. & H. = *Lioplacodes veterna*, (M. & H.) Meek	40

46.

MEEK, F. B. Check lists of the Invertebrate fossils of North America. Miocene. <Smithsonian Miscellaneous Collections (No. 183), pp. 1-32, 1864. Washington, 1867.

Contains, besides the list, "Notes and explanations" of generic and specific characters.

MIOCENE.

	Page.
Columnaria? sexradiata, Lonsdale = *Septastrea (?) sexradiata*, (Lonsd.) Meek	25
Lithodendron lineata, Conrad = *Cladocora [?] lineata*, (Con.) Meek	25
Orbicula lugubris, Conrad = *Discina lugubris*, (Con.) Meek	25
Orbicula multilineata, Conrad = *Discina multilineata*, (Con.) Meek	25
Terebratula nitens, Conrad = *Rhynchonella nitens*, (Con.) Meek	25
Hinnites giganteus, Gray = *Hinnites crassis*, Conrad	26
Janira affinis, Tuomey & Holmes = *Pecten affinis*, (T. & H.) Meek	26
Janira bella, Conrad = *Pecten bella*, (Con.) Meek	26
Amussium propatulum = *Pecten propatulus*, Conrad	26
Nucula impressa, Conrad = *Yoldia impressa*, (Con.) Meek	27
Leda willamettensis, Shumard = *Nuculana willamettensis*, (Shum.) Meek	27
Nucula penita, Conrad = *Nuculana penita*, (Con.) Meek	27
Leda oregona, Shumard = *Nuculana oregona*, (Shum.) Meek	27

	Page.
Nucula divaricata, Conrad = *Nucula (Leila) Conradi*, Meek	27
Pectunculus nitens, Conrad = *Limopsis nitens*, (Con.) Meek	27-28
Arca canalis, Conrad = *Anadara ? canalis*, (Con.) Meek	28
Arca congesta, Conrad = *Anadara ? congesta*, (Con.) Meek	28
Arca incile, Say = *Anadara incile*, (Say) Meek	28
Arca microdonta, Conrad = *Anadara microdonta*, (Con.) Meek	28
Anomalocardia trigintinaria, Conrad = *Anadara trigintinaria*, (Con.) Meek	28
Arca protracta, H. D. & W. B. Rogers = *Anadara protracta*, (Rogers) Meek	28
Arca trilineata, Conrad = *Anadara trilineata*, (Con.) Meek	28
Avicula multangula, H. C. Lea = *Pteria [?] multangula*, (H. C. Lea) Meek	28
Perna montana, Conrad = *Melina montana*, (Con.) Meek	28
Perna t[e]rta, Say = *Melina torta*, (Say) Meek	28
Modiola contracta, Conrad = *Volsella contracta*, (Conrad) Meek	28
Modiola spiniger, H. C. Lea = *Volsella [?] spinigera*, (H. C. Lea) Meek	28
Modiola ducatelli, Conrad = *Volsella ducatelli*, (Conrad) Meek	28
Mytilus inflatus, Tuomey & Holmes = *Volsella inflata*, (T. & H.) Meek	28
Cardita abbreviata, Conrad = *Venericardia (Pteromeris) abbreviata*, (Con.) Meek	28
Cardita radians, Conrad = *Venericardia (Pteromeris) radians*, (Con.) Meek	29
Cardita carinata, Emmons = *Venericardia (Cardiocardites) carinata*, (Emm.) Meek	29
Cardita subtenta, Conrad = *Venericardia (Cardiocardites) sub[t]enta*, (Con.) Meek	29
Cardita monilicosta, Gabb. = *Venericardia (Cardiocardites) monilicosta*, (Gabb.) Meek	29
Cardita occidentalis, Conrad = *Venericardia (Cardiocardites) occidentalis*, (Con.) Meek	29
Lucina occidentalis, Conrad	29
Cyclas permacra, Conrad = *Lucina permaera*, (Con.) Meek	29
Venus bisecta, Conrad = *Thyatira ? bisecta*, (Conrad) Meek	29-30
Isocardia fraterna, Say = *Glossus fraterna*, (Say) Meek	30
Isocardia markoei, Conrad = *Glossus markoei*, (Conrad) Meek	30
Cardium modestum, Conrad = *(Cardium (Cerastoderma) modestum*, Conrad	30
Venus (Trigona) tantilla, Gould = *Psephis tantilla*, (Gould) Gabb	30
Venus athleta, Conrad = *Chione, (Lirophora) athleta*, (Con.) Meek	30
Venus alveata, Conrad = *Chione (Lirophora) alveatus*, (Con.) Meek	30
Venus latilirata, Conrad = *Chione (Lirophora) latilirata*, (Con.) Meek	30
Venus angustifrons, Conrad = *Dione angustifrons*, (Con.) Meek	30
Venus brevilineata, Conrad = *Dione ? brevilineata*, (Con.) Meek	30
Meretrix d·cisa, Conrad = *Dione decisa*, (Conrad) Meek	30
Cytherea oregonensis, Conrad = *Dione oregonensis*, (Con.) Meek	30
Meretrix tularana, Conrad = *Dione tularana*, (Con.) Meek	30
Meretrix uniomeris, Conrad = *Dione uniomeris*, (Con.) Meek	30
Cytherea vespertina, Conrad = *Dione vespertina*, (Con.) Meek	30
Solemya protexta = *Donax protexta*, Conrad	30
Glycimeris estrellanus, Conrad = *Panopea estrellana*, (Con.) Meek	30
Solen curtus, Conrad = *Ensis curtus*, (Conrad) Meek	31
Bulla petrosa, Conrad = *Cylichna petrosa*, (Conrad) Meek	31
Tornatella elliptica, Trask = *Actæon ellipticus*, (Trask) Meek	31
Helonyx thallus, (Conrad) Meek = *Dentalium thallus*, Conrad	31
Diodora crucibuliformis, Conrad = *Cemoria crucibuliformis*, Conrad	31
Narica diegoana, Conrad = *Vanikoro diegoana*, (Con.) Meek	31
Crepidula prærupta, Conrad = *Crypta prærupta*, (Con.) Meek	31
Turbo glabra, H. C. Lea = *Viviparus glaber*, (H. C. Lea) Meek	31
Natica inezana, Conrad = *Natica inezana*, Conrad	31
Sinum scopulosum, Meek ! = *Sigaretus scopulosus*, Conrad	31-32
Sycotypus ocoyanus, Conrad = *Ficus [??] ocoyanus*, (Conrad) Meek	32
Pyrula modesta, Conrad = *Ficus modestus*, (Conrad) Meek	32
Oliva ancillariæformis, H. C. Lea = *Olivella ancillariformis*, (H. C. Lea) Meek	32
Fusus Oregonensis, Conrad = *Busycon ? Oregonensis*, (Conrad) Meek	32
Colus arctatus, Conrad = *Fusus arctatus*, (Conrad) Meek	32
Fusus migrans, Conrad = *Tritonifusus migrans*, (Con.) Meek	32
Nautilus angustatus, Conrad = ! *Aturia angustatus*, (Con.) Meek	32

47.

MEEK, F. B. Note on Bellinurus Danæ, from the Illinois Coal-measures. <Am. Journ. Sci., vol. xliii, 2d ser., pp. 257, 258. 1867. New Haven, 1867.

In this note the author expresses the opinion that Bellinurus danæ, Meek and Worthen, properly belongs to the recently proposed genus Prestwichia, Woodward.

48.

MEEK, F. B. Note on a new genus of fossil Crustacea. <Am. Journ. Sci., vol. xliii, 2d ser., pp. 394, 395. 1867. New Haven, 1867.

Genus Eupröops. This genus was afterward fully described and illustrated in vol. iii of Worthen's Illinois Geological Reports.

49.

MEEK, F. B. On the punctate shell-structure of Syringothyris. <Am. Journ. Sci., vol. xliii, 2d ser., pp. 407, 408. 1867. New Haven, 1867.

50.

MEEK, F. B. Remarks on Professor Geinitz's views respecting the Upper Paleozoic rocks and Fossils of Southeastern Nebraska. <Am. Journ. Sci., vol. xliv, 2d ser., pp. 170-187; continued, pp. 327-339; notes to the same pp. 282, 283. 1867. New Haven, 1867.

This is an extended discussion and criticism of Dr. Geinitz's "Carbon formation and Dyas in Nebraska." The following fossils are discussed especially:

	Page
Serpula (Spirorbis) planorbites, Münster	177
Euomphalus rugosus, Hall	177
Spirorbis helix, King	178
Murchisonia subtæniata, Geinitz	178
Orthonema subtæniata, Geinitz, sp	178
Bellerophon interlineatus, Portlock	178
Bellerophon marcouanus, Geinitz	178
Macrocheilus pallianus, Geinitz	178
Allorisma elegans, King	178
Solenomya biarmica, de Verneuil	178-179
Astarte gibbosa, McCoy	179
Astarte nebrascensis, Geinitz	179
Astarte mortonensis, Geinitz	179
Astarte vallisnerianus, King	179
Schizodus obscurus, Sowerby	179
Schizodus rossicus, de Verneuil	179
Arca striata, Schlotheim	179-180
Macrodon tenuistriata, M. & W	179-180
Nucula kazanensis, de Verneuil	180
Leda bellistriata, Stevens	180
Nucula beyrichi, Schlotheim	180
Clidophorus pallasi, M. V. & King	180-181
Clidophorus (Pleurophorus) simplus, Keyserling	181
Pleurophorus subcuneatus, M. & H	181
Clidophorus solenoides, Geinitz	181
Avicula Hausmanni (Goldf.), Geinitz	182
Myalina swallovi, McChesney	182
Mytilus concavus? (Swallow), Geinitz	182
Myalina perattenuata (M. & H.), Geinitz	182
Myalina subquadrata (Shumard), Geinitz	182
Avicula speluncaria (Schlotheim), Geinitz	182
Monotis hawni, M. & H	182
Pseudomonotis hawni, M. & H	182
Pseudomonotis sinuata	182
Avicula pinnæformis, Geinitz	182
Gervillia parva (M. & H.), Geinitz	182-183
Gervillia longa, Geinitz	183
Gervillia (avicula) sulcata, Geinitz	183
Pecten neglectus, Geinitz	183
Pecten Missouriensis? (Shumard), Geinitz	183
Pecten hawni, Geinitz	183
Pecten broadheadii, Swallow	183
Aviculopecten hawni	183

	Page.
Rhynchonella angulata, (Linnæus), Geinitz	183
Camarophoria globulina, (Phillips), Geinitz	183
Rhynchonella uta, Marcou	183
Retzia mormonii (Marcou sp.), Geinitz	184
Athyris subtilita, Hall	184
Spirifer moosakpaliensis, Davidson	184
Spirifer cameratus (Morton), Geinitz	184
Spirifer laminosus (McCoy), Geinitz	184
Spiriferina kentuckensis, Shumard	184
Orthisina missouriana, Swallow	185
Plicatula striato-costata, Cox	185
Productus horrescens	185
Productus rogersii	185
Productus nebrascensis, Owen	185
Productus flemingii	185
Productus prattenianus, Norwood	185
Productus koninckianus (de Verneuil), Geinitz	185
Productus cancrini (de Verneuil), Geinitz	186
Productus orbignianus (de Koninck), Geinitz	186
Productus longispinus, Sowerby	186
Productus horridus (Sowerby), Geinitz	186
Chonetes mucronata (M. & W.), Geinitz	186
Chonetes glabra, Geinitz	186
Cyathocrinus ramosus (Schlotheim), Geinitz	186
Cyathocrinus inflexus, Geinitz	186
Poteriocrinus hemisphericus, Shumard	186
Eocidaris hallianus, Geinitz	187
Polypora marginata (McCoy), Geinitz	187
Polypora biarmica (Keyserling), Geinitz	187
Synocladia virgulacea (Phillips), Geinitz	187
Aviculopinna pinnæformis	282
Aviculopinna Americana, n. s., Meek, 1867	283
Cyathaxonia	328
Allorisma elegans, King	328
Schizodus rossicus, de Verneuil	328
Aucella hausmanni, Goldf	329
Spirifer laminosus, McCoy	329
Lima retifera, Shumard	329
Stenophora columnaris, Schlot., sp	329–330
Synocladia virgulacea, Geinitz, not Phillips	330
Synocladia biserialis, Swallow	330

51.

MEEK, F. B. Note on the genus Palæacis, Haime, 1860 (=Sphenopoterium, M. & W., 1866.) <Am. Journ. Sci., vol. xliv, 2d ser., pp. 419, 420. 1867. New Haven, 1867.

The author here takes the view that Sphenopoterium, originally published in the Illinois Geol. Reports, is identical with Palæacis.

	Page.
Palæacis obtusa, M. & W	419–420
Palæacis umbonata, Seebach	420
Palæacis cymbia, Seebach	420

52.

MEEK, F. B. Fossils from the west coast of Kennedy Channel. <Hayes's "Open Polar Ocean," London, 1867, p. 341.

Describes various fossils collected by Dr. Hayes on the west coast of Kennedy's Channel, from deposits of lower Helderberg age. *Zaphrentis hayesii* and *Lozonema ? kanei* are described as new species. A preliminary notice appeared in Am. Journ. Sci. and Arts, ser. 2, vol. xl, pp. 31–34. New Haven, 1865.

For list of species see entry number 33, p. 36.

53.

MEEK, F. B. Preliminary notice of a remarkable new genus of corals, probably typical of a new family; forwarded for study by Prof. J. D. Whitney, from the Silurian rocks of Nevada Territory. <Am. Journ. Sci., vol. xlv, 2d ser., pp. 62-64. 1868. New Haven.

	Page.
Ethmophyllum, n. g., Meek, 1868	62-64
Ethmophyllum whitneyi, n. s., Meek, 1868	64
Ethmophyllum gracile, n. s., ? Meek, 1868	64

54.

MEEK, F. B. Note on the shell-structure and family affinities of the genus Aviculopecten. <Am. Journ. Sci., vol. xlv, 2d ser., pp. 64, 65. 1868. New Haven, 1868.

The author shows that by the shell-structure the Avieulopectens are allied to Avicula rather than to Pecten.

55.

MEEK, F. B., and WORTHEN, A. H. Preliminary notice of a Scorpion, a Eurypterus? and other fossils, from the Coal-measures of Illinois. <Am. Jour. Sci., vol. xlvi, 2d ser., pp. 19-28. 1868. New Haven, 1868.

Afterward fully described and illustrated in one of the Illinois Geological Reports, vol. iii.

	Page.
Eurypterus (Anthraconectes) mazonensis, n. s., M. & W., 1868	19-21
Adelopthalmus mazonensis?	22
Ceratiocaris? sinuatus, n. s., M. & W., 1868	22
Buthus?? carbonarius, n. s., M. & W., 1868	22-24
Scorpio carbonarius, n. s., M. & W., 1868	22-24
Eoscorpius, n. g.? M. & W., 1868	24-25
Euphoberia, n. g., M. & W., 1868	25-26
Euphoberia armigera, n. s., M. & W., 1868	26
Euphoberia major, n. s., M. & W., 1868	26
Acanthotelson	27-28
Acanthotelson eveni, n. s., M. & W., 1868	28
Palæocaris	28
Gampsonyx	28
Anthracerpes	28

56.

MEEK, F. B. Note on Ethmophyllum and Archeocyathus. <Am. Journ. Sci., vol. xlvi, 2d ser., p. 144. 1868. New Haven, 1868.

The author abandons his formerly proposed genus Ethmophyllum, believing it to be identical with Archeocyathus of Billings.

57.

MEEK, F. B., and WORTHEN, A. H. Notes on some points in the Structure and Habits of the Paleozoic Crinoidea. <Proc. Acad. Nat. Sci. Phila., vol. xx, pp. 323-334. 1868. Philadelphia, 1868.

Afterward republished in the Illinois Geological Reports, vol. v, and in Amer. Jour. Sci. and Canad. Nat. See entries Nos. 61 and 64.

	Page.
Synbathocrinus, Phillips	323
Goniasteroidocrinus, Lyon & Casseday	323-324
Cyathocrinus, Miller	324-325
Convoluted support of the digestive sack, in the *Actinocrinidæ*	325-327
Ambulacral canals passing under the vault of the *Actinocrinidæ*	327-334

58.

MEEK, F. B., and WORTHEN, A. H. Remarks on some types of Carboniferous Crinoidea, with descriptions of new Genera and Species of the same, and of one Echinoid. <Proc. Acad. Nat. Sci. Phila., vol. xx, pp. 335-359. 1868. Philadelphia, 1868.

Genera Barycrinus, Nipterocrinus. Afterward republished in the Illinois Geological Reports, vol. v.

	Page.
Cyathocrinites, Miller	336-337
Cyathocrinites fragilis, n. s., M. & W., 1868	337
Cyathocrinites tenuidactylus, n. s., M. & W., 1868	337-338
Barycrinus, n. g., Wachsmuth MS	338-340
Barycrinus magnificus, n. s., M. & W., 1868	340-341
Barycrinus hoveyi var. *herculeus*, M. & W., 1868	341
Nipterocrinus, n. g., Wachsmuth, MS	341-342
Nipterocrinus wachsmuthi, n. s., M. & W., 1868	342-343
Catillocrinus, Troost	343
Catillocrinus bradleyi, n. s., M. & W., 1868	343
Dichocrinus, Munster	343
Dichocrinus expansus, n. s., M. & W., 1868	343-344
Dorycrinus, Roemer	344-345
Dorycrinus roemeri, n. s., M. & W., 1868	346
Dorycrinus quinquelobus, var. *intermedius*, M. & W., 1866	346-347
Amphoracrinus, Austin	347-348
Amphoracrinus divergens, Hall, sp., 1860	348-349
Batocrinus, Casseday	349-352
Subgenus *Eretmocrinus*, Lyon & Casseday	351
Subgenus *Alloprosallocrinus*, Casseday & Lyon	352
Batocrinus quasillus, n. s., M. & W., 1868	352-353
Batocrinus cassedayanus, n. s., M. & W., 1868	353-354
Batocrinus trochiscus, n. s., M. & W., 1868	354-355
Batocrinus (Eretmocrinus) neglectus, n. s., M. & W., 1868	355-356
Pentremites, Say	356
Pentremites (Troostocrinus?) woodmani, n. s., M. & W., 1868	356
Agelacrinites Vanuxem	357
Agelacrinites (Lepidodiscus) squamosus, n. s., M. & W., 1868	357-358
Echinoidea	358
Oligoporus, M. & W	358
Oligoporus nobilis, n. s., M. & W., 1868	358-359

59.

MEEK, F. B., and WORTHEN, A. H. Paleontology. <Geological Survey of Illinois, vol. iii, pp. 291-565, plates 1-20. 1868. Springfield, 1868.

Published by authority of the legislature of Illinois, 1868.
Silurian, Devonian, and Carboniferous.

LOWER SILURIAN SPECIES.

TRENTON GROUP.	Page.
Radiata	291
Echinodermata	291
Cystoidea	291
Comarocystites, Billings, 1854	291
C. shumardi, M. & W., 1865, fig. —, p. 292, and pl. i, figs. 1 a, b	292-294
C. shumardi, var. *obconicus*, M. & W., 1865, pl. i, figs. 2 a, b	294
Mollusca	294
Lamellibranchiata	294
Modiolopsis, Hall, 1847	294
M. modioliformis, n. s., M. & W., 1868, pl. i, figs. 7 b and 8	294-295
M. orthonota, n. s., M. & W., 1868, pl. i, fig. 7 a	295-296
Cypricardites, Conrad, 1841	297
Vanuxemia, Billings, 1858	297
V. ? dixonensis, M. & W., 1866, pl. i, figs. 5 a, b	297-298

*βαρύς, heavy; Κρίνον, a lily.   †νιπτήρ, a washing vessel; Κρίνον, a lily.

	Page.
Cephalopoda	298
Orthoceras, Auct	298
O. (Ormoceras) backii, Stokes? 1837, pl. i, fig. 4	298–299
Articulata:	
Crustacea	299
Lichas, Dalman, 1827	299
L. cucullus, M. & W, 1865	299–300

FOSSILS OF THE GALENA BEDS.

Protozoa	301
Receptaculites, Defrance, 1827	301
R. globularis, Hall, 1861, pl. ii, figs. 2 *a, b*	301
R. ——? pl. ii, figs. 1 *a, b*	301–302
R. oweni, Hall, 1861, pl. ii, fig. 3	302–303
Radiata	304
Zoophyta	304
? *Chaetetes*, Fischer, 1837	304
C. petropolitanus, Pander? sp., 1830, pl. ii, figs. 8 *a, b*	304–305
Mollusca	305
Brachiopoda	305
Lingula, Brugiere, 1792	305
L. quadrata, Eichwald, 1829, pl. ii, fig. 4 *a–c*	305–306
Lamellibranchiata	306
Ambonychia, Hall, 1847	306
A. intermedia, n. s., M. & W., 1868, pl. ii, figs. 5 *a, b*	306–307
Tellinomya, Hall, 1847 (not *Tellinya*, Brown, 1827=*Tellinomya*, Agassiz, 1846)	307
T. ventricosa, Hall, 1861, pl. ii, figs. 7 *a, b, c*	307–308
T. alta, Hall, 1861, pl. ii, figs. 6 *a, b*	309
Cypricardites, Conrad, 1841	309–310
C. ——? pl. iii, figs. 9 *a, b, c, d*	311
C. obliquus, n. s., M. & W., 1868, pl. ii, figs. 9 *a, b*	311–312
Gasteropoda	312
Bellerophon, Montfort, 1808	312
B. (Bucania?) platystoma, n. s., M. & W., 1868, pl. iii. figs. 8 *a, b*	312–313
Ophileta, Vanuxem, 1842	313
O. owenana, n. s., M. & W., 1868, pl. iii, figs. 6 *a, b*	313–314
Trochonema, Salter, 1857	314
T. umbilicata, Hall? sp., 1847, pl. iii, figs. 5 *a, b*	314–315
Raphistoma, Hall, 1847	316
R. lenticularis, Conrad, sp., 1842, pl. iii, figs. 7 *b*, (*a, c?*)	316–317
Murchisonia, d'Archiac & d'Verueuil, 1841	317
M. bicincta, Hall? 1847, pl. iii, fig. 4	317–318
Cephalopoda	318
Orthoceras, Auct	318
O. anellum, Conrad, 1843, pl. iii, fig. 3	318–320
Articulata	320
Crustacea	320
Illaenus, Dalman, 1826	320
I. taurus, Hall, 1861, pl. iii. fig. 2	320–322
I. crassicauda, Wahleub.?? 1821, pl. iii, figs. 1 *a, b*	322–323

FOSSILS OF THE CINCINNATI GROUP.

Radiata	324
Echinodermata	324
Crinoidea	324
Heterocrinus, Hall, 1847	324
H. crassus, M. & W., 1865, pl. iv, figs. 1 *a, b, c*	324–325
H. subcrassus, M. & W., 1865, pl. iv, figs. 5 *a–d*	325–326
Hybocrinus, Billings, 1866	327
Subgenus *Anomalocrinus*, M. & W	328
Hybocrinus? incurvus, M. & W., 1865, pl. iv, fig. 3 *a, b*, and p. 327, fig. —	327–329
Porocrinus, Billings, 1856	329–330
P. crassus, M. & W., 1865, p. 330, fig *a, b*, and pl. iv, figs. 2 *a, b*	330–332
P. pentagonius, M. & W., 1865, pl. i, fig. 3	332–333
Dendrocrinus, Hall, 1852	333
D. osweguensis, n. s., M. & W., 1868, p. 333, fig. —, and pl. iv, fig. 4	333–334

	Page.
Mollusca	335
Brachiopoda	335
Strophomena, Rafinesque, 1820 ?	335
S. unicostata, n. s., M. & W., 1868, pl. iv, figs. 11 *a, b*	335–337
Lamellibranchiata	337
Ambonychia, Hall, 1847	337
Subgenus *Megaptera*, M. & W., 1866	337
Megaptera casei, M. & W., 1866, pl. iv, figs. 9 *a, b*	337–338
Dolabra, McCoy, 1844 ?	339
D. sterlingensis, M. & W., 1866, pl. iv, figs. 10 *a, b, c*	339–340
Gasteropoda	340
Cyrtolites, Conrad, 1838	340
C. imbricatus, n. s., M. & W., 1868, pl. iv, fig. 12	340–341
Pteropoda	341
Tentaculites, Schlotheim, 1820	341
T. tenuistriatus, M. & W., 1865, pl. iv, figs. 7 *a, b*	341–342
T. oswegoensis, M. & W., 1865, pl. iv, fig. 6 *a*	342–343
T. sterlingensis, M. & W., 1865, pl. iv, fig. 8	343

UPPER SILURIAN SPECIES.

FOSSILS OF THE NIAGARA GROUP.

Protozoa	344
Spongiæ	344
Astylospongia, Rœmer	344
A. ?? christiani, n. s., M. & W., 1868, pl. v, figs. 3 *a, b, c*	344–345
Pasceolus, Billings, 1859	345
P. ? dactylioides, Owen, sp., 1844, pl. v, figs. 2 *a, b, c*	345–346
Radiata	347
Echinodermata	347
Saccocrinus, Hall, 1852	347
S. christyi, Hall ? sp., 1863, pl. v, fig. 1	347–349
Mollusca	349
Brachiopoda	349
Hemipronites, Pander, 1830	349
H. subplanus, Conrad ? sp., 1842, pl. vi, figs. 6 *a, b*	349–351
Obolus, Eichwald, 1829	351
O. [Trimerella ?] conradi, Hall, 1868, pl. v, figs. 7 *a, b*	351–352
Centronella, Billings, 1859	352
C. billingsiana, n. s., M. & W., 1868, p. 352, figs. a–c, and pl. vi, figs. 5 a–c	352–353
Meristella, Hall, 1860	354
Meristella ? sp., pl. vi, figs. 4 *a, b*	354
Lamellibranchiata	354
Pterinea, Goldfuss, 1832	354
P. thebesensis, n. s., M. & W., 1868, page 354, fig. —, and pl. vi, fig. 3	354–355
Ambonychia, Hall, 1847	356
A. acutirostris, Hall ? 1865, pl. v, figs. 8 *a, b*, and 9 *c*	356–357
Amphicœlia, Hall, 1864	357–358
A. neglecta, McChesney, 1861, pl. v, figs. 9 *a, b* (not *c*)	358–359
Gasteropoda	359
Pleurotomaria, Defrance, 1824	359
P. casii, n. s., M. & W., 1868, pl. v, fig. 5	359–360
P. cyclonemoides, n. s., M. & W., 1868, pl. v, fig. 4	360–361
Subulites (Conrad), Emmons, 1842	361–362
S. (Polyphemopsis) brevis, Winchell & Marcy, 1865, pl. v, fig. 6	362–363
Articulata	363
Crustacea	363
Dalmanites, Auct	363
D. danæ, M. & W., 1865, pl. vi, fig. 1 *a–f*	363–367

FOSSILS OF THE LOWER HELDERBERG GROUP.

Radiata	368
Zoophyta	368
Striatopora, Hall, 1852	368–369
S. missouriensis, n. s., M. & W., 1868, pl. vii, fig. 4	369–370
Echinodermata	370

	Page.
Edriocrinus, Hall, 1859	370
E. pocilliformis, Hall, 1859, pl. vii, figs. 5 *a, b*	370-371
Mollusca	371
Brachiopoda	371
Orthis, Dalman, 1828	371
O. hybrida, Sowerby ? 1839, pl. vii, figs. 7 *a-d*	371-372
O. subcarinata, Hall, 1857, pl. vii, figs. 6 *a-d*	373-374
Strophomena, Rafinesque, 1820	374
S. (*Strophodonta*) *cavumbona*, Hall ? 1857, pl. vii, fig. 10 *a, b*	374-376
Merista, Suess, 1851	376
M. lævis, Vanuxem ? sp., 1843, pl. vii, figs. 8 *a, c*	376-377
Zygospira, Hall, 1862	377-380
Z. subconcava, n. s., M. & W., 1868, pl. vii, fig. 1 *a-d*	380-381
Trematospira, Hall, 1859	381
T. ? imbricata, Hall, 1857, pl. vii, fig. 2 *a-e*	381-382
Cyrtina, Davidson, 1858	383
C. dalmani, Hall, sp., 1857, pl. vii, figs. 3 *a, b*	383
Spirifer, Sowerby, 1815	384
Subgenus *Trigonotreta*, Kœnig, 1825	384
S. perlamellosus, Hall, 1857, pl. vii, figs. 9 *a, b*	384
Gasteropoda	384
Platyceras, Conrad, 1840	384-387
P. subundatum, n. s., M. & W., 1868, pl. vii, figs. 13 *a, b*, and 14 *a, b*	387-388
P. spirale, Hall, 1859, pl. vii, fig 12 *a-c*	389
P. (*Orthonychia*) *pyramidatum*, Hall ? 1859, pl. vii, fig. 11	389-390
Articulata	390
Crustacea	390
Acidaspis, Murchison, 1839	390
Acidaspis hamata, Conrad, sp., 1841, pl. vii, fig. 15	390-391
Dalmanites, Anct.	391
D. tridentiferus, Shumard, 1855, pl. vii, fig. 16	391-392

Devonian Species.

Fossils of the Oriskany Group.

Mollusca	393
Brachiopoda	393
Leptæna, Dalman, 1827	393
L. ? nucleata, Hall, 1859, pl. viii, figs. 8 *a-d*	393-394
Rhynchonella, Fischer, 1809	394
R. speciosa, Hall, 1857, pl. viii, fig. 9	394-395
Eatonia, Hall, 1857	395
E. peculiaris, Conrad, sp., 1841, pl. viii, figs. 2 *a-d*	395-396
Leptocœlia, Hall, 1857	397
L. flabellites, Conrad, sp., 1841, pl. viii, figs. 3 *a-c*	397-398
Spirifer, Sowerby, 1815	398
Subgenus *Trigonotreta*, Kœnig, 1825	398
S. engelmanni, n. s., M. & W., 1868, pl. viii, figs. 5 *a-d*	398-399
S. hemicyclus, n. s., M. & W., 1868, pl. viii, figs. 6 *a-d*, and 7 *a, b ?*	399-401
Rensselæria, Hall, 1859	401
R. condoni, McChesney, 1861, pl. viii, figs. 4 *a, b*	401-402
Stricklandinia, Billings, 1863	402
S. elongata var. *curta*, M. & W., 1868, pl. viii, figs. 1 *a-c*, and pl. ix, fig. 5 ?	402-404
Gasteropoda	404
Strophostylus, Hall, 1859	404
S. cancellatus, n. s., M. & W., 1868, p. 404, figs. and pl. viii, fig. 12 (11 *a, b?*)	404-405
Platyceras, Conrad, 1840	406
P. spirale, Hall ? 1859, pl. viii, fig. 10	406

Fossils of the Corniferous Group.

Radiata	407
Zoophyta	407
Pleurodictyum, Goldfuss, 1829	407
P. problematicum, Goldf. ? 1859, pl. ix, figs. 1 *a-c*	407-409
Baryphyllum, E. & H., 1850	409

	Page.
B. ?? arenarium, n. s., M. & W., 1868, pl. ix, figs. 2 a, b	409–410
Zaphrentis, R. & C., 1820	410
Zaphrentis, sp. undt., pl. ix, fig. 3 a, b	410
Mollusca	410
Brachiopoda	410
Orthis, Dalman, 1828	410
Orthis, undt., pl. ix, fig. 4	410–411
Strophomena, Raf., 1820	411
S. (Strophodonta), sp. ? pl. ix, figs. 9 (and 7 a?)	411–412
S. (Strophodonta), sp., pl. vi, figs. 6 a, b	412
Productus, Sowerby, 1814	412
P. exanthematus, Hall ?? 1857, pl. x, figs. 3 a–c	412–413
Spirifer, Sowerby, 1815	414
Subgenus *Trigonotreta*, Kœnig, 1825	414
S. perextensus, n. s., M. & W., 1868, pl. x, figs. 1 a–d	414–415
S. paradoxus, Schlot. ?? sp., 1813, pl. x, fig. 2	415–416
Articulata	416
Crustacea	416
Dalmanites, Auct	416
Subgenus *Odontocephalus*, Conrad, 1840	416
Odontocephalus ——— ? pl. ix, fig. 10	416–417
Dalmanites (Odontocephalus) ægeria, Hall ? sp., 1861, pl. x, figs. 4 a–c	417–418

FOSSILS OF THE HAMILTON GROUP.

Protozoa	419
Spongiæ	419
Astræospongia, Rœmer, 1854	419
A. hamiltonensis, M. & W., 1866, pl. x, fig. 6	419
Radiata	420
Zoophyta	420
Microcyclus. n. g., M. & W., 1866	420
M. discus, n. s., M. & W., 1868, pl. xi, figs. 7 a, b	420–421
Echinodermata	421
Taxocrinus, Phillips, 1843	421
T. gracilis, M. & W., 1865, page 421, fig. —, and pl. xiii, fig. 3	421–423
Mollusca	423
Brachiopoda	423
Orthis, Dalman, 1827	423
O. mcfarlani, Meek, 1868, pl. xiii, figs. 10 a–d	423–424
O. iowensis var. *furnarius*, Hall, pl. xiii, figs. 9 a, b	424–425
Strophomena, Raf., 1820	426
S. rhomboidalis, Wahlemb., sp., 1821, pl. x, fig. 7 a, b	426–427
Tropidoleptus, Hall, 1857	427
T. carinatus, Conrad, sp., 1839, pl. xiii, figs. 2 a–c	427–428
Pentamerus, Sowerby, 1813	428
P. comis, Owen ? sp., 1855, pl. xiii, figs. 6 a–c	428–429
P. subglobosus, n. s., M. & W., 1868, pl. xiii, figs. 5 a–c	429–430
Atrypa, Dalman, 1827	430
A. aspera, Schlotheim, sp., 1813, pl. xiii, figs. 7 a–d	430–431
A. reticularis, Linnæus, sp., 1767, pl. xiii, fig. 11	432–433
Spirifer, Sowerby, 1815	433
Subgenus *Trigonotreta*, Kœnig, 1825	433
S. furnacula, Hall, 1857, pl. xiii, figs. 8 a–c	433–434
S. subundiferus, n. s., M. & W., 1868, pl. x, fig. 5 a–e	434–435
Cyrtina, Davidson, 1858	436
C. triquetra, Hall, sp., 1858, pl. xiii, fig. 4 a–d	436
Lingula, Bruguière, 1792	437
L. subspatulata, n. s., M. & W., 1868, pl. xiii, fig. 1	437
Lamellibranchiata	437
Pterinea, Goldf., 1832	437
P.? subpapyracea, M. & W., 1866, pl. xi, fig. 5	437–438
Modiolopsis, Hall, 1847	438
M.? perovata, M. & W., 1865, pl. xi, fig. 2	438–439

* μικρος, small ; χυχλος, a circle.

	Page.
Grammysia, de Verneuil, 1847	439
G. ? rhomboidalis, M. & W., 1865, pl. xi, figs. 5 *a, b*	439–441
Gasteropoda	441
Platyceras, Conrad, 1840	441
P. ventricosum, Conrad, 1840, pl. xi, figs. 4 *a, b*	441–442
Isonema,* M. & W., 1865	442–443
I. depressa, M. & W., 1865, p. 443, figs. A, B, and pl. xi, figs. 6 *a, b*	443
Cephalopoda	444
Gomphoceros, Sowerby, 1839	444
G. turbiniforme, M. & W., 1866, pl. xii, figs. 2 *a, b*	444
Cyrtoceras, Goldf. 1832	445
C. sacculum, M. & W., 1866, pl. xii, figs. 3 *a–c*	445–446
Gyroceras, de Koninck, 1844	446
G. constrictum, n. s., M. & W., 1866, pl. xii, figs. 1 *a, b*	446–447
Articulata	447
Crustacea	447
Phacops, Emmerich, 1839	447
P. rana, Green, sp., 1832. pl. xi, figs. 1 *a–e*	447–449

CARBONIFEROUS SPECIES.

FOSSILS OF THE KINDERHOOK GROUP.

Mollusca	450
Brachiopoda	450
Rhynchonella, Fisher, 1809	450
R. missouriensis, Shumard, 1855, pl. xiv, figs. 7 *a–d*	450–452
Lamellibranchiata	453
Pernopecten, Winchell, 1865	453
P. shumardianus, Winchell, 1865, pl. xiv, figs. 6 *a, b*	453–455
Pterinea, Goldf., 1832	456
P. undulata, n. s., M. & W., 1868, pl. xiv, fig. 5	456
Gasteropoda	457
Platyceras, Conrad, 1840	457
P. (Orthonychia?) subplicatum, M. & W., 1866, pl. 14, figs. 4 *a–c*	457
P. haliotoides, M. & W., 1866, pl. xiv, figs. 3 *a, b*	458
Porcellia, Leveille, 1835	458
P. nodosa, Hall, 1860, pl. xiv, figs. 1 *a, b*	458–459
Gyroceras, de Koninck, 1844	459
G. rockfordensis, M. & W., 1866, pl. xiv, fig. 2 *a*	459–460
Articulata	460
Crustacea	460
Proetus, Steininger, 1830 ?	460
P. ellipticus, M. & W., 1865, pl. xiv, fig. 8	460–462

FOSSILS OF THE BURLINGTON GROUP.

Radiata	463
Echinodermata	463
Belemnocrinus, White, 1862	463
B. whitii, M. & W., 1866, p. 463, fig. —, and pl. xviii, fig. 4 *a–c*	463–464
Catillocrinus, Troost, 1850	465
C. wachsmuthi, M. & W., 1866, pl. xviii, fig. 5	465–466
Platycrinus, Miller, 1821	466
P. scobina, M. & W., 1861, p. 466, fig. —, and pl. xvi, fig. 9	466–467
P. planus, Owen & Shumard ? 1850, pl. xvi, fig. 6	467–468
P. (Pleurocrinus) asper, M. & W., 1861, p. 468, fig. —, and pl. xviii, fig. 9	468–469
Actinocrinus, Miller, 1821	470
A. (Saccocrinus?) amplus, M. & W., 1861, p. 470, fig. —, and pl. xvi, fig. 2	470–472
A. (Batocrinus) pistillus, M. & W., 1865, pl. xvi, figs. 4 *a, b*	472–474
Steganocrinus, M. & W., 1866	474
S. pentagonus, Hall, sp., 1858, pl. xvi, fig. 8	474–475
Rhodocrinus, Miller, 1821	476
R. nanus, M. & W., 1866, p. 476, figs. —, and pl. xviii, figs. 2 *a, b*	476–478
Bursacrinus,† M. & W., 1865	478
B. wachsmuthi, M. & W., 1861, pl. xvii, fig. 6, and p. 479, fig. —	479–480

---

* ισος, equal; νημα, a thread.   † βυρσα, a purse; χρινον, a lily.

	Page.
Cyathocrinus, Miller, 1821	481
C. enormis, M. & W., 1865, pl. xvi, figs. 3 *a*, *b*	481-482
C. wachsmuthi, M. & W., 1861, p. 482, fig. —, and pl. xvi, fig. 5	482-484
Poteriocrinus, Miller, 1821	484
P. tenuibrachiatus, M. & W., 1861, p. 484, fig. —, and pl. xvi, fig. 1	484-485
P. subimpressus, M. & W., 1861, p. 485, fig. —, and pl. xviii, figs. 1 *a*, *b*	485-486
P. carinatus, M. & W., 1861, p. 486, fig. —, and pl. xvii, fig. 1	486-488
Subgenus *Scaphiocrinus*, Hall, 1858	488
S. wachsmuthi, M. & W., 1861, p. 488, fig. —, and pl. xvi, figs. 7 *a*, *b*	488-489
Poteriocrinus (Scaphiocrinus) tenuidactylus, M. & W., 1865, p. 490, fig. —, and pl. xviii, fig. 10	490-491
Onychocrinus, Lyon & Casseday, 1859	492
O. diversus, M. & W., 1866, page 492, fig. —, and pl. xvii, figs. 5 *a*, *b*	492-495
Taxocrinus, Phillips, 1843	495
Forbesiocrinus, de Kon. & Le Hon, 1854	495
F. agassizi, var. *giganteus*, M. & W., 1861, pl. xviii, fig. 3	495
Granatocrinus (Troost), Hall, 1852	496
G. projectus, M. & W., sp., 1861, page 496, fig. —, and pl. xviii, fig. 7	496
G. norwoodi, O. & S. ? sp., 1860, pl. xviii, fig. 8	496-497
G. shumardi, M. & W., 1866, page 498, fig. —, and pl. xviii, figs. 6 *a*, ? *b*	498-499
Asteroidea	499
Schœnaster, M. & W., 1860	499
S. wachsmuthi, M. & W., 1866, pl. xvii, fig. 4	499-500
Mollusca	501
Polyzoa	501
Evactinopora, M. & W., 1865	501
E. radiata, M. & W., 1865, page 502, fig. —, and pl. xvii, figs. 2 *a*, *b*	502
E. sexradiata, n. s., M. & W., 1868, pl. xvii, fig. 3	502
E. grandis, n. s., M. & W., 1868, p. 503, fig. —, and pl. xv, figs. 2 *a*, *b*	503
Fenestella, Lonsdale, 1839	504
Subgenus *Lyropora*, Hall, 1856	504
Fenestella (Lyropora) retrorsa, n. s., M. & W., 1868, pl. xv, fig. 1	504
Brachiopoda	505
Chonetes, Fischer, 1837	505
C. illinoisensis, Worthen, 1860, pl. xv, figs. 8 *a*, *b*	505-506
Gasteropoda	506
Metoptoma, Phillips, 1836	506
M. ? umbella, M. & W., 1866, pl. xv, figs. 6 *a*, *b*, *c*; and 7	506-507
Platyceras, Conrad, 1840	508
P. [?] reversum, Hall, 1860, p. 508, fig. —, and pl. xv, figs. 4 *a*, *b*	508-509
P. biserialis, Hall, 1860, pl. xv, figs. 3 *a*, *b*	509
P. (Orthonychia) quincyense, McChesney, 1861, pl. xv, figs. 5 *a*, *b*	510

FOSSILS OF THE KEOKUK GROUP.

Radiata	511
Echinodermata	511
Platycrinus, Miller, 1821	511
P. hemisphæricus, M. & W., 1865, p. 511, fig. —, and pl. xx, figs. 2 *a*, *b*	511-513
P. niotensis, M. & W., 1865, p. 513, fig. —, and pl. xx, fig. 3	513-514
Poteriocrinus, Miller, 1821	515
P. indianensis, M. & W., 1865, pl. xx, fig. 4 ; and p. 515, fig. —	515-516
Cyathocrinus, Miller, 1821	517
C. farleyi, M. & W., 1866, p. 517, fig. —, and pl. xx, figs. 1 *a*, *b*, and 6 *c*	517-518
C. ? sp. undt., pl. xx, figs. 5 *a*-*c*	518-519
C. quinquelobus, M. & W., 1865, p. 519, fig. —; and pl. xx, figs. 6 *a*, *b* (not *c*)	519-520
C. arboreus, M. & W., 1865, p. 520, fig. —	520-522
Echinoidea	522
Perischoechinidæ	522
Lepidesthes,* n. g., M. & W., 1868	522-524
L. coreyi, n. s., M. & W., 1868, p. 524, fig. A	524-525
Melonites multipora, p. 524, fig. B	524
Oligoporus danæ, p. 524, fig. C	524

* λεπις, a scale; εσθης, a garment.

	Page.
Asteroidea	526
Onychaster,* n. g., M. & W., 1868	526
O. flexilis, n. s., M. & W., 1868, p. 526, figs. A, B, C, D	526–528
Mollusca	528
Brachiopoda	528
Productus, Sowerby, 1814	528
P. magnus, M. & W., 1861, pl. xx, figs. 7 a–c	528–530
Spirifer, Sowerby, 1815	530
S. propinquus, Hall, 1858, pl. xix, figs. 8 a–c	530–532
Lamellibranchiata	532
Aviculopecten, McCoy, 1851	532
A. indianensis, M. & W., 1866, pl. xix, figs. 6a, b	532–534
Anthracoptera, Salter, 1862	534
A. ? fragilis, M. & W., 1866, pl. xix, fig. 4	534–535
Pleurophorus, King, 1844	535
P. costatiformis' M. & W., 1865, p. 535, fig. —, and pl. xix, fig. 8 ?	535–536
Lithophaga, Lamarck, 1812	536
L. lingualis, Phillips ?, sp., 1836, pl. xix, figs. 1, 2	536–537
Sedgwickia, McCoy, 1844	537
S. (Sanguinolites) subarcuata, M. & W., 1865, pl. xix, fig. 3 b (not 3 a)	537–538
Allorisma, King, 1844	538
A. (Chænomya ?) hybrida, M. & W., 1865, pl. xix, fig. 3 a (not 3 b)	538–539

ARTICULATE FOSSILS OF THE COAL MEASURES.

Crustacea	540
Entomostraca	540
Gnathostomata	540
Phyllopoda	540
Ceratiocaris, McCoy, 1849	540
C.? sinuatus, M. & W., 1868, p. 540, fig. A	540–541
Leaia, Jones, 1862	541
L. tricarinata, n. s. M. & W., 1868, woodcut, figs. B 1, 2, 3 (and C ?), p. 540	541–543
Merostomata	544
Eurypteridæ	544
Eurypterus, De Kay, 1825	544
E. (Anthraconectes) mazonensis, M. & W., 1868, figs. —, p. 544	544
Ziphosura	547
Euproops, Meek, 1867	547
E. danæ, M. & W., 1865, p. 547, figs. A, B	547–549
Tetradecapoda	549
Isopoda	549
? Acanthotelson, M. & W., 1860	549
A. stimpsoni, M. & W., 1860, p. 549, figs. A, B	549–550
A. eveni, M. & W., 1868, p. 551, figs. A, B, C, D	551
Decapoda	552
Macrura	552
Palæocaris, M. & W., 1865	552
P. typus, M. & W., 1865, p. 552, figs. A, B	552–553
Gampsonix fimbriatus, p. 552, figs. C, D	552
Anthrapalæmon, Salter, 1861	554
A. gracilis, M. & W., 1865, p. 554, figs. A, B	554–555
Myriapoda	556
Euphoberia, M. & W., 1868	556
E. armigera, M. & W., 1865, p. 556, figs. A, B, C, D	556–558
E. ?? major, M. & W., 1868, p. 558, fig. —	558–559
Arachnida	560
Pulmonaria	560
Eoscorpius, M. & W., 1868	560
E. carbonarius, M. & W., 1868, p. 560, figs. a, c, d, m, p	560–562
Mazonia,† n. g. M. & W., 1868	563
M. woodiana, M. & W., 1868, p. 563, figs. A, B, C, D	563–565
Note on the genus Palæocampa	565

---

* ονυξ, a claw ; αστηρ, a star.   † Mazon, name of stream.

60.

MEEK, F. B. Remarks on the geology of the valley of Mackenzie River, with figures and descriptions of fossils from that region, in the Museum of the Smithsonian Institution, chiefly collected by the late Robert Kennicott, Esq. <Trans. Chicago Acad. of Sci., vol. i, pp. 61-114, plates xi, xv. 1868. Chicago, 1867-1869.

Devonian.

CORALS.

Cyathophyllidæ. Page.
   Cyathophyllum, Goldfuss, 1826 79
   C. articum, n. s., Meek, 1868, pl. xi, fig. 8 79-80
   Cysteophyllum, Lonsdale, 1839 80
   C. americanum var. articum, Meek, pl. xi, fig. 6 80-81
   Aulophyllum, Edwards & Haime, 1850 81
   A.? richardsoni, n. s., Meek, 1868, pl. xi, fig. 3 81-82
   Zaphrentis, Rafinesque, 1820 82
   Z. recta, n. s., Meek, 1868, pl. xi, fig. 1 82
   Z. mcfarlanei, n. s., Meek, 1868. pl. xi, fig. 2 83
   Smithia, Edwards & Haime, 1851 83
   S. verrilli, n. s., Meek, 1868, pl. xi, fig. 7 83-84
   Combophyllum, Edwards & Haime, 1858 84
   C. multiradiatum, n. s., Meek, 1868, pl. xi, fig. 4 84-85
Fungidæ.
   Palæocyclus, Edwards & Haime, 1849 85
   P. kirbyi, n. s., Meek, 1868, pl. xi, fig. 5 85
Favositidæ.
   Favosites, Lamarck, 1816 86
   F. polymorpha, Goldfuss, sp., pl. xi, fig. 10 86
   Alveolites, Lamarck, 1801 86
   A. vallorum, n. s., Meek, 1868, pl. xi, fig. 9 86-87
Brachiopoda.
Lingulidæ.
   Lingula, Bruguière 87
   L. minuta, n. s., Meek, 1868, pl. xiii, fig. 1 87
Strophomenidæ.
   Strophomena, Rafinesque 87
   S. (Strophodonta) demissa, Conrad, 1842, pl. xiii, fig. 6 87-88
   S. (Strophodonta) subdemissa, Hall, 1856, pl. xiii, fig. 7 88
   Orthis, Dalman, 1828 88
   O. mcfarlanei, n. s., Meek, 1868, pl. xii, fig. 1 88-90
   O. iowensis, Hall? 1858, pl. xii, fig. 2 90-91
Productidæ.
   Productus, Sowerby 91
   P. dissimilis, Hall? 1858, pl. xiii, fig. 3 91
   Productus ——? Meek, 1869, pl. xiii, fig. 4 91-92
   Productus ——? Meek, 1869, pl. xiii, fig. 5 92
   Chonetes, Fischer 93
   C. pusilla, Hall? 1857, pl. xiii, fig. 2 93
Rhynchonellidæ.
   Rhynchonella, Fischer, 1809 93
   R. castanea, n. s., Meek, 1868, pl. xiii, fig. 9 93-95
   Rhynchonella ——? Meek, 1869, pl. xv, fig. 4 95
   Pentamerus, Sowerby, 1812 95
   P. borealis, n. s., Meek, 1868, pl. xiii, fig. 11 95-96
   Atrypa, Dalman, 1827 96
   A. aspera, Schlotheim, sp., 1820, pl. xiii, fig. 12 96-97
   A. reticularis, Linn, sp., 1767, pl. xiii, fig. 13 97
Spiriferidæ.
   Cyrtina, Davidson, 1858 97
   C. billingsi, n. s., Meek, 1868, pl. xiv, fig. 6 97-99
   C. hamiltonensis, Hall, 1857, pl. xiv, fig. 10, and figs. 5 and 7? 99-100
   C. panda, n. s., Meek, 1868, pl. xiv, fig. 8 100-101
   Spirifer, Sowerby, 1815 101
   S. kennicotti, n. s., Meek, 1868, pl. xiv, fig. 9 101-102
   S. compactus, n. s., Meek, 1868, pl. xiv, fig. 11 102-103

	Page.
Subgenus *Martinia*, McCoy, 1844	103
S. (Martinia) sublineatus, n. s., Meek, 1868, pl. xiv, fig. 1	103–104
S. (Martinia) richardsoni, n. s., Meek, 1868, pl. xiv, fig. 2	104–105
S. (Martinia) meristoides, n. s., Meek, 1868, pl. xiv, fig. 3	106–107
S. (Martinia) franklinii, n. s., Meek, 1868, pl. xiv, fig. 12	107–108
Rensselæria, Hall, 1859	108
R. lævis, n. s., M. & W., 1868, pl. xiii, fig. 8, and pl. xiv, fig. 4	108–109
Gasteropoda.	
Pleurotomariidæ.	
Pleurotomaria, Defrance, 1826	110
Pleurotomaria ——— ? Meek, 1869, pl. xv, fig. 3	110
Cephalopoda.	
Nautilidæ.	
Gyroceras, Koninck, 1844	110
G. logani, n. s., Meek, 1868	110–111

61.

MEEK, F. B., *and* WORTHEN, A. II. Notes on some points in the Structure and Habits of the Paleozoic Crinoidea. <Am. Journ. Sci., vol. xlviii, 2d ser., pp. 23–40. 1869. New Haven, 1869.

A reprint from the Proc. Acad. Nat. Sci. Phila., vol. xx, pp. 323–334. (See No. 57.) Afterward republished in the Illinois Geological Reports, vol. v, and in the Canad. Nat., new series, vol. iv, pp. 434–452. (See No. 64.) For list of species see No. 57, p. 53.

62.

MEEK, F. B., *and* WORTHEN, A. II. Descriptions of new Crinoidea and Echinoidea from the Carboniferous rocks of the Western States, with a note on the Genus Onychaster. <Proc. Acad. Nat. Sci. Phila., vol. xxi, pp. 67–83. 1869. Philadelphia, 1869.

Afterward republished in the Illinois Geological Reports, vol. v.

	Page.
Synbathocrinus, Phillips, 1836	67
S. wachsmuthi, M. & W., 1866	67–68
S. brevis, n. s., M. & W., 1869	68–69
Dichocrinus, Münster, 1839	69
D. lineatus, n. s., M. & W., 1869	69
D. pisum, n. s., M. & W., 1869	69–70
Erisocrinus, M. & W., 1865	70
E. antiquus, n. s., M. & W., 1869	71–72
E. whitei, n. s., M. & W., 1869	72
Calceocrinus, Hall, 1852	72–73
C. ? bradleyi, n. s., M. & W., 1869	73–74
C. ? wachsmuthi, n. s., M. & W., 1869	74–75
Gilbertsocrinus, Phillips	75
Subgenus *Goniasteroidocrinus*, Lyon & Casseday, 1859	75
G. (Goniasteroidocrinus) tenuiradiatus, n. s., M. & W., 1869	75–76
G. (Goniasteroidocrinus) obovatus, n. s., M. & W., 1869	76–77
Lepidocentrus, Müller (?), 1856	77–78
L. irregularis, n. s., M. & W., 1869	78–79
Eocidaris ? squamosa, n. s., M. & W., 1869	79–81
Palæchinus gracilis, n. s., M. & W., 1869	82
Onychaster, M. & W.	82–83

63.

MEEK, F. B., *and* A. II. WORTHEN. Remarks on the Blastoidea, with Descriptions of New Species. <Proc. Acad. Nat. Sci. Phila., vol. xxi, pp. 83–91. 1869. Philadelphia, 1869.

Afterward republished in the Illinois Geological Reports, vol. v.

	Page.
Granatocrinus, Troost	88
G. melonoides, n. s., M. & W., 1869	88–89
G. pisum, n. s., M. & W., 1869	89–90
G. neglectus, n. s., M. & W., 1869	90–91
G. glaber, n. s., M. & W., 1869	91

64.

MEEK, F. B., and WORTHEN, A. H. Notes on some points in the Structure and Habits of the Palæozoic Crinoidea. <Canad. Nat., new ser., vol. iv, pp. 434–452. 1869.

Reprinted from the Proc. Acad. Nat. Sci. Phila., 1868. (See entry numbers 61 and 57.)

65.

MEEK, F. B., and WORTHEN, A. H. Note on the Relations of Synocladia, King, 1849, to the Proposed Genus Septopora, Prout, 1858. <Proc. Acad. Nat. Sci. Phila., vol. xxxi, pp. 15–18. 1870. Philadelphia, 1870.

The author regards these forms as congeneric.

66.

MEEK, F. B., and WORTHEN, A. H. Descriptions of new Species and Genera of Fossils from the Paleozoic rocks of the Western States. <Proc. Acad. Nat. Sci. Phila., vol. xxii, pp. 22–56. 1870. Philadelphia, 1870.

Silurian and Carboniferous: Genera Codonites, Carbonarca, Clinopistha, Solenocheilus, Temnocheilus. Afterward republished and illustrated in the Illinois Geological Reports, vol. vi.

	Page.
Foraminifera.	
Receptaculites formosus, n. s., M. & W., 1870	22–23
Echinodermata.	
Barycrinus spectabilis, n. s., M. & W., 1870	23–24
Cyathocrinites? poterium, n. s., M. & W., 1870	24–26
Poteriocrinites (Zeacrinus?) concinnus, n. s., M. & W., 1870	26–27
Scaphiocrinus depressus, n. s., M. & W., 1870	27
Zeacrinus? armiger, n. s., M. & W., 1870	27–28
Zeacrinus (Hydreionocrinus?) acanthoporus, n. s., M. & W., 1870	28–29
Eupachycrinus boydii, n. s., M. & W., 1870	30
Homocrinus angustatus, n. s., M. & W., 1870	30–31
Codonites, n. g., M. & W., 1870	31–32
C. gracilis, n. s., M. & W., 1870	32–33
Pentremites burlingtonensis, n. s., M. & W., 1870	33–34
Oligoporus coreyi, n. s., M. & W., 1870	34
Brachiopoda.	
Chonetes?? millepunctata, n. s., M. & W., 1870	35–36
Spirifer fastigatus, n. s., M. & W., 1870	36–37
Stricklandinia deformis, n. s., M. & W., 1870	37–38
Lamellibranchiata.	
Monotis? gregaria, n. s., M. & W., 1870	38
Aviculopecten spinuliferus, n. s., M. & W., 1870	39
Carbonarca, n. g., M. & W., 1870	39
C. gibbosa, n. s., M. & W., 1870	40
Macrodon delicatus, n. s., M. & W., 1870	40
Modiolopsis subnasuta, n. s., M. & W., 1870	41
Schizodus amplus, n. s., M. & W., 1870	41–42
Seacrinus (Prisconaia) perelegans, n. s., M. & W., 1870	42–43
Clinopistha, n. g., M. & W., 1870	43–44
C. radiata, var. *levis*, n. s., M. & W., 1870	44–45

	Page.
Gasteropoda.	
Dentalium annulostriatum, n. s., M. & W., 1870	45
Straparollus (Euomphalus) pernodosus, n. s., M. & W., 1870	45-46
S. (Euomphalus) subquadratus, n. s., M. & W., 1870	46-47
Subulites inflatus, n. s., M. & W., 1870	47
Cephalopoda.	
Nautilus, Auct	48
Subgenus *Solenochilus*,* M. & W., 1870	48
N. (Solenochilus) collectus, n. s., M. & W., 1870	48-49
Subgenus *Temnochilus*, McCoy	49
N. (Temnochilus) latus, n. s., M. & W., 1870	49
N. (Temnochilus) winslowi, n. s., M. & W., 1870	50
N. (Temnochilus) cozanus, n. s., M. & W., 1870	50-51
Lituites graftonensis, n. s., M. & W., 1870	51-52
Crustacea.	
Phillipsia tuberculata, n. s., M. & W., 1870	52
Phillipsia (Griffithides) bufo, n. s., M. & W., 1870	52-53
Asaphus (Isotelus) vigilans, n. s., M. & W., 1870	53-54
Illaenus (Bumastus) graftonensis, n. s., M. & W., 1870	54-55
Dithyrocaris carbonarius, n. s., M. & W., 1870	55-56

67.

MEEK, F. B. Descriptions of Fossils collected by the U. S. Geological Survey under the charge of Clarence King, Esq. <Proc. Acad. Nat. Sci. Phila., vol. xxii, pp. 56-64. 1870. Philadelphia, 1870.

Silurian, Devonian, and Tertiary. Afterward republished and illustrated in vol. iv of Mr. King's series of final reports, 1877.

TERTIARY SPECIES.

	Page.
Sphaerium rugosum, n. s., Meek, 1870	56-75
S. idahoense, n. s., Meek, 1870	57
Ancylus undulatus, n. s., Meek, 1870	57-58
Melania (Goniobasis?) sculptilis, n. s., Meek, 1870	58
Melania (Goniobasis) subsculptilis, n. s., Meek, 1870	58-59
Carinifex binneyi, n. s., Meek, 1870	59
Carinifex (Vorticifex) tryoni, n. s., Meek, 1870	59-60
Carinifex tryoni var. *concava*, n. s., Meek, 1870	60

DEVONIAN SPECIES.

Spirifer (Trigonotreta) pinonensis, n. s., Meek, 1870	60-61

LOWER SILURIAN SPECIES.

Euomphhlus (Raphistoma?) rotuliformis, n. s., Meek, 1870	61
E. (Raphistoma) trochiscus, n. s., Meek, 1870	61-62
Paradoxides? nevadensis, n. s., Meek, 1870	62-63
Conocoryphe (Oonocephalites) kingii, n. s., Meek, 1870	63-64

68.

MEEK, F. B. Geology of the Line of the Great Pacific Railroad. [In a letter to Dr. J. J. Bigsby.] <Geological Magazine, Decade I, vol. vii, pp. 163-164. 1870. London, 1870.

Notes the fossils obtained by Mr. Clarence King along the line of the Pacific Railway.

---

* σωλην, a channel; χειλος, lip.

69.

MEEK, F. B. A Preliminary List of Fossils collected by Dr. Hayden in Colorado, New
Mexico, and California, with Brief Descriptions of a few of the New Species.
<Proc. Am. Philos. Soc., vol. xi, pp. 425-431. 1870. Philadelphia, 1871.

Silurian, Carboniferous, Jurassic, Cretaceous, and Tertiary.

SILURIAN SPECIES.

	Page.
Orthis coloradoensis, n. s., Meek, 1871	425-426
Bucanella nana, n. s., Meek, 1871	426

CRETACEOUS SPECIES.

Ammonites serrato-carinatus, n. s., Meek, 1871	429-430

TERTIARY SPECIES.

Ostrea soleniscus, n. s., Meek, 1871	430
Unio belliplicatus, n. s., ? Meek, 1871	430
Cyrena (Corbicula) durkeei, n. s., Meek, 1871	431

70.

MEEK, F. B. Preliminary notice of a new species of Trimerella from Ohio. <Am.
Journ. Sci., vol. i, 3d ser., pp. 305-306. 1871. New Haven, 1871.

Trimerella ohioensis. Republished and illustrated in Paleontology of Ohio (Newberry).

71.

MEEK, F. B. On some new Silurian Crinoids and Shells. <Am. Journ. Sci., vol. ii,
3d ser., pp. 295-302. 1871. New Haven, 1871.

This article consists of descriptions of species, together with some extended remarks on the
genus Lichenocrinus of Hall.

	Page.
Dendrocrinus casei, n. s., Meek, 1871	295-296
Lepocrinites moorei, n. s., Meek, 1871	296-297
Anodontopsis ? milleri, n. s., Meek, 1871	297-299
Anodontopsis ? unionoides, n. s., Meek, 1871	299
Remarks on the genus *Lichenocrinus*	299-302

72.

MEEK, F. B. Descriptions of new species of invertebrate fossils from the Carboniferous and Devonian rocks of Ohio. <Proc. Acad. Nat. Sci. Phila., vol. xxiii,
pp. 57-93. 1871. Philadelphia, 1871.

Echinodermata.

	Page.
Dolatocrinus ornatus, n. s., Meek, 1871	57-59

Lamellibranchiata.

Aviculopecten crenistriatus, n. s., Meek, 1871	60-61
Aviculopecten (streblopteria ?) hertzeri, n. s., Meek, 1871	61-62
Lucina (Paracyclas) ohioensis, n. s., Meek, 1871	62-63
[*Ptilodictya (Stictopora) gilberti*,* n. s., Meek, 1871	63-64]
Conocardium ohioense, n. s., Meek, 1871	65-66
Solenomya (jancia) retusta, n. s., Meek, 1871	66-67
Clinopistha antiqua, n. s., Meek, 1871	67-68
Sanguinolites ! sanduskyensis, n. s., Meek, 1871	68-69
Sanguinolites? obliquus, n. s., Meek, 1871	69-70
Allorisma (Sedgwickia?) pleuropistha, n. s., Meek, 1871	70-71

---

* This description was accidentally inserted by Mr. Meek in this place

	Page.
Grammysia? rhomboides, n. s., Meek, 1871	72–73
Grammysia ventricosa, n. s., Meek, 1871	73
Gasteropoda.	
Platyceras multispinosum, n. s., Meek, 1871	73–75
Platyceras attenuatum, n. s., Meek, 1871	75–76
Naticopsis levis, n. s., Meek, 1871	76
Naticopsis (Platyostoma?) æquistriata, n. s., Meek, 1871	76–77
Bellerophon newberryi, n. s., Meek, 1871	77–78
Bellerophon propinquus, n. s., Meek, 1871	78
Cyclonema crenulata, n. s., Meek, 1871	79
Ixonema humilis, n. s., Meek, 1871	79–80
Orthonema newberryi, n. s., Meek, 1871	81
Trochita? antiqua, n. s., Meek, 1871	82
Trochonema tricarinata, n. s., Meek, 1871	82–84
Pteropoda.	
Conularia micronema, n. s., Meek, 1871	84
Conularia elegantula, n. s., Meek, 1871	85–86
Cephalopoda.	
Cyrtoceras ohioense, n. s., Meek, 1871	86–87
Gyroceras (Trochoceras?) ohioense, n. s., Meek, 1871	87–88
Gyroceras (Nautilus?) inelegans, n. s., Meek, 1871	89
Crustacea.	
Proetus planimargitus, n. s., Meek, 1871	89–91
Dalmanites ohioensis, n. s., Meek, 1871	91–93

73.

MEEK, F. B. Descriptions of new species of fossils from Ohio and other Western States and Territories. <Proc. Acad. Nat. Sci. Phila., vol. xxiii, pp. 159–184. 1871. Philadelphia, 1871.

This paper contains descriptions of fossils, mostly Carboniferous, from Ohio, Illinois, and Texas, with a Melantho and Viviparus from Wyoming.

OHIO COLLECTIONS.

	Page.
Fenestella delicata, n. s., Meek, 1871	159–160
Ptilodictya (Stictopora) carbonaria, n. s., Meek, 1871	160–161
Aviculopecten sanduskyensis, n. s., Meek, 1871	161–162
Pterinea (Pteronites?) newarkensis, n. s., Meek, 1871	162–163
Cypricardina? carbonaria, n. s., Meek, 1871	163–165
Schizodus medinœnsis, n. s., Meek, 1871	165–166
Schizodus subtrigonalis, n. s., Meek, 1871	166
Allorisma winchelli, n. s., Meek, 1871	167–168
Allorisma ventricosa, n. s., Meek, 1871	168–169
Platyostoma? trigonostoma, n. s., Meek, 1871	169–170
Platyceras (Orthonychia?) lodiense, n. s., Meek, 1871	170–171
Platyceras tortum, n. s., Meek, 1871	171–172
Holopea (Cyclora) nana, n. s., Meek, 1871	172
Orthoceras? isogramma, n. s., Meek, 1871	172–173

ILLINOIS COLLECTIONS.

Streptacis whitfieldi, n. s., Meek, 1871	173–174
Loxonema attenuata, var. *semicostata*, Meek, 1871	174–175
Murchisonia obsoleta, n. s., Meek, 1871	175
Pleurotomaria textiligera, n. s., Meek, 1871	176–177
Pleurotomaria gurleyi, n. s., Meek, 1871	177–178

COLLECTIONS FROM MISSOURI, WYOMING, TEXAS, ETC.

Aviculopecten? williamsi, n. s., Meek, 1871	178–179
Spirifer (Trigonotreta?) texanus, n. s., Meek, 1871	179–181

	Page.
Campeloma (*Melantho*) *macrospira*, n. s., Meek, 1871	181-182
Viviparus? wyomingensis, n. s., Meek, 1871	182-183
Isocardia? hodgei, n. s., Meek, 1871	183-184

74.

MEEK, F. B. Notice of a new Brachiopod, from the lead-bearing rocks at Mine La Motte, Missouri. <Proc. Acad., Nat. Sci. Phila., vol. xxiii, pp. 185-187. 4 woodcuts. 1871. Philadelphia, 1871.

	Page.
Lingulella lamborni, n. s., Meek, 1871, p. 185, fig. 1	185-187
Lingulella davisii (Salter), Davidson, figs. 2 and 3	185-187
Lingulepis pinniformis, Hall, fig. 4	185-187

75.

MEEK, F. B. Descriptions of new Western Palæozoic fossils, mainly from the Cincinnati Group of the Lower Silurian series of Ohio. <Proc. Acad. Nat. Sci. Phila., vol. xxiii, pp. 308-336. 1872. Philadelphia, 1871.

Afterward redescribed and illustrated in the Paleontology of Ohio (Newberry).

Radiata.
Echinodermata.

	Page.
Heterocrinus exiguus, n. s., Meek, 1872	308-310
Heterocrinus subcrassus, M. & W., 1865	310
Poteriocrinites (*Dendrocrinus*) *dyeri*, n. s., Meek, 1872	310-312
Poteriocrinites (*Dendrocrinus*) *cincinnatiensis*, n. s., Meek, 1872	312-314
Poteriocrinus (*Dendrocrinus*) *polydactylus*, Shumard, sp., 1867	314
Glyptocrinus dyeri, n. s., Meek, 1872	314-316
Glyptocrinus dyeri, var. *subglobosus*, Meek, 1872	316-317

Mollusca.
Polyzoa.

Ptilodictya (*Stictopora*) *shafferi*, n. s., Meek, 1872	317-318

Brachiopoda.

Retzia (*Trematospira*) *granulifera*, n. s., Meek, 1872	318-319

Lamellibranchiata.

Ambonychia (*Megaptera*) *alata*, n. s., Meek, 1872	319-321
Megambonia jamesi, n. s., Meek, 1872	321-322
Sedgwickia? fragilis, n. s., Meek, 1872	323
Sedgwickia? compressa, n. s., Meek, 1872	324-325
Sedgwickia (*Grammysia?*) *neglecta*, n. s., Meek, 1872	325-326
Dolabra? carinata, n. s., Meek, 1872	326-327
Cardiomorpha?? obliquata, n. s., Meek, 1872	327-328

Gasteropoda.

Macrocheilus klipparti, n. s., Meek, 1872	328-330

Cephalopoda.

Orthoceras ortoni, n. s., Meek, 1872	330-331

Articulata.
Crustacea.

Cythere cincinnatiensis, n. s., Meek, 1872	331-332
Ceratiocaris (*Colpocaris*) *bradleyi*, n. s., Meek, 1872	332-333
Ceratiocaris (*Colpocaris*) *elytroides*, n. s., Meek, 1872	334
Ceratiocaris (*Solenocaris*) *strigota*, n. s., Meek, 1872	335
Archæocaris vermiformis, n. s., Meek, 1872	335-336

76.

MEEK, F. B. Descriptions of some new types of Palæozoic shells. <Am. Journ. Conch., vol. vii, pp. 4-10, 1 plate. 1871-1872. Philadelphia, 1872.

Carboniferous and Cretaceous? Genera *Promacrus*, *Prothyris*.

	Page.
Sanguinolites, McCoy	4
Subgenus *Promacrus*, Meek, 1871	4-5

	Page.
S. (Promacrus) nasutus, n. s., Meek, 1871, pl. i, fig. 1	
S. (Promacrus) missouriensis, Swallow, 1860, pl. i, fig. 2	6-7
Prothyris, Meek, 1869	8
P. elegans, n. s., Meek, 1871, pl. 1, fig. 3	8-9
Martesia? ræssleri, n. s., Meek, 1871, pl. i, figs. 4, 4 *a*	9-10

77.

MEEK, F. B. List of Carboniferous fossils from West Virginia, with descriptions of new species. <Appendix B, Report of Regents of West Virginia University for 1870, pp. 67-73 or 1-7. 1871. Wheeling, 1871.

	Page.
Macrodon obsoletus, n. s., Meek, 1871	5
Nucula? anodontoides, n. s., Meek, 1871	5-6
Yoldia stevensoni, n. s., Meek, 1871	6
Yoldia (Palæoncio?) carbonaria, n. s., Meek	6-7
Phillipsia stevensoni, n. s., Meek, 1871	7

78.

MEEK, F. B. Remarks on the Genus Lichenocrinus. <Ann. and Mag. Nat. Hist., ser. 4, vol. viii, pp. 341-345. 1871. London, 1871.

A reprint from Amer. Journ. Sci. & Arts, 1871. See entry number 71.

79.

MEEK, F. B. Supplementary note on the Genus Lichenocrinus. <Ann. and Mag. Nat. Hist., ser. 4, vol. ix, pp. 247-248. 1872. London, 1872.

An additional description of the characters of Lichenocrinus, founded on a number of fresh specimens. The author concludes that it is an aberrant type of Cystoidea, representing a distinct family. See entry number 80.

80.

MEEK, F. B. Supplementary Note on the Genus Lichenocrinus. <Am. Journ. Sci., vol. iii, 3d ser., pp. 15-17. 1872. New Haven, 1872.

This is supplementary to the article at page 299 of vol. ii. See entry numbers 71 and 79.

81.

MEEK, F. B. Descriptions of two new starfishes, and a Crinoid, from the Cincinnati group of Ohio and Indiana. <Am. Journ. Sci., vol. iii, 3d ser., pp. 257-262. 1872. New Haven, 1872.

These descriptions, with illustrations, are republished in the Paleontology of Ohio (Newberry).

	Page.
Palæaster? dyeri, n. s., Meek, 1872	257-258
Stenaster grandis, n. s., Meek, 1872	258-259
Glyptocrinus baeri, n. s., Meek, 1872	260-261
Note on the Genus *Lichenocrinus*	261-262

82.

MEEK, F. B. Descriptions of New Species of Fossils from the Cincinnati Group of Ohio. <Am. Journ. Sci., vol. iii, 3d ser., pp. 423-428. 1872. New Haven, 1872.

These have since been redescribed and figured in the Paleontology of Ohio (Newberry).

	Page.
Anomalocystites (Ateleocystites?) balanoides, n. s., Meek, 1872	423-424
Dalmanites curleyi, n. s., Meek, 1872	424-426
Proetus spurlocki, n. s., Meek, 1872	426-428

83.

MEEK, F. B. Descriptions of a few new species and one new genus of Silurian fossils from Ohio. <Am. Journ. Sci., vol. iv, 3d ser., pp. 274-281. 1872. New Haven, 1872.

Genus *Dicraniscus*, afterward fully described and illustrated in the Paleontology of Ohio (Newberry).

	Page.
Protaster? granuliferus, n. s., Meek, 1872	274-275
Palæaster incomptus, n. s., Meek, 1872	275-277
Rhynchonella neglecta var. *scobina*, Meek, 1872	277-278
Pleurotomaria (Scalites?) tropidophora, n. s., Meek, 1872	278-279
*Dicraniscus, n. g., Meek, 1872	279-280
D. ortoni, n. s., Meek, 1872	280-281

84.

MEEK, F. B. Preliminary Paleontological report consisting of lists of fossils, with descriptions of some new types, &c. <Prelim. Rep. U. S. Geol. Surv. of Wyoming and Portions of Contiguous Territories, pp. 287-318. 1870. Washington, 1871.

Silurian, Carboniferous, Jurassic, Cretaceous, and Tertiary, Genera Arcopagella, Crassatellina, Leptesthes, Pyrgulifera.

	Page.
General remarks	287-295
Pyrgulifera, n. g., Meek, 1871	294
Lists of fossils collected	295-299
Descriptions of new species and genera	299-318

CARBONIFEROUS SPECIES.

	Page.
Edmondia aspenwallensis, n. s., Meek, 1871	299-300

CRETACEOUS FORMS.

Crassatellina, n. g., Meek, 1871	300, 301
C. oblonga, n. s., Meek, 1871, figs. A & B, p. 301	301
Pachymya? truncata, n. s., Meek, 1871	301-302
Inoceramus altus, n. s., Meek, 1871	302-303
Unio (Baphia?) nebrascensis, n. s., Meek, 1871	303
Arca? parallela, n. s., Meek, 1871	303-304
Yoldia microdonta, n. s., Meek, 1871	304
Corbicula nuculis, n. s., Meek, 1871	304-305
Corbicula? subtrigonalis, n. s., Meek, 1871	305-306
Cardium paupereulum, n. s., Meek, 1871	306
Cardium (Protocardia) salinense, n. s., Meek, 1871	306-307
Cardium kansasense, n. s., Meek, 1871	307-308
Mactra? cañonensis, n. s., Meek, 1871	308
Arcopagella, n. g., Meek, 1871	308
A. mactroides, n. s., Meek, 1871, figs. A and B, p. 309	309-310
Tellina subscitula, n. s., Meek, 1871	310
Tapes wyomingensis, n. s., Meek, 1871	310-311
Leptosolen conradi, n. s., Meek, 1871	311-312
Anisomyon centrale, n. s., Meek, 1871	312
Turritella kansasensis, n. s., Meek, 1871	312-313
Turbo mudgeanus, n. s., Meek, 1871	313

TERTIARY SPECIES.

Unio leanus, n. s., Meek, 1871	313-314
U. washakiensis, n. s., Meek, 1871	314
Corbicula? fracta, n. s., Meek, 1871	314-315
Corbicula crassatelliformis, n. s., Meek, 1871	315-316
Goniobasis chrysallis, n. s., Meek, 1871	316
Goniobasis nodulifera, n. s., Meek, 1871	316-317
Bythinella gregaria, n. s., Meek, 1871	317-318

---

* Diminutive of δικρανος a two-pronged fork.

85.

MEEK, F. B. Preliminary list of the fossils collected by Dr. Hayden's exploring expedition of 1871 in Utah and Wyoming Territories, with descriptions of a few new species. <Prelim. Rep. of U. S. Geol. Surv. of Montana and portions of adjacent Territories. [Report for 1871,] pp. 373–377. 1872. Washington, 1872.

	Page.
Silurian fossils	373
Carboniferous fossils	373–374
Platycrinites (Eucladocrinus) montanaensis, n. s., Meek, 1872	373–374
Jurassic species	374–375
Aviculopecten (Pseudomonotis?) idahoensis, n. s., Meek, 1872	374–375
Cretaceous species	375–376
Ostrea idriaensis, Gabb?!	375
Anomia? gryphorhynchus, n. s., Meek, 1872	375–376
Tertiary species	376

MEEK, F. B. Report on the Paleontology of Eastern Nebraska, with some remarks on the Carboniferous rocks of that district. <Final Rep. of the U. S. Geol. Surv. of Nebraska and portions of the adjacent Territories, pp. 83–239. 11 plates. Washington, 1872.

Carboniferous fossils only. Genus Rhombopora.

	Page.
Introductory remarks	83
Statement of a boring made in the Missouri Valley at Omaha City, by the Union Pacific Railroad Company, starting 22 feet above low-water mark of the Missouri	87–88
Section of beds exposed at Bellevue, with an enumeration of the fossils found in each	89
Section of the beds exposed on the north side of Platte River, between three and four miles from its mouth, with an enumeration of the fossils found in each bed	90–91
Section of the beds exposed at Plattsmouth, with the names of the fossils found in each	93
Section of the beds exposed at Rock Bluff, on the Missouri, with a statement of the fossils found in each	95–96
Section of the rocks seen at Cedar Bluff	98
Section at Wyoming, with an enumeration of the imbedded fossils	99
Section at Bennett's mill	100
Section of beds exposed at the Nebraska City landing, with an enumeration of the fossils found in each	101–102
Section 1¾ and 2¾ miles due west of Nebraska City	103
Mr. Croxton's boring at Nebraska City	105–107
Section of the beds exposed at Otoe City	107–108
Sections of the various beds exposed at Brownville	110
Sections one and a half miles below Brownville	111–112
Sections of beds exposed at Aspinwall	112–113
Section two miles above Rulo, on the Missouri	114
Shaft and boring one and one-fourth miles south of Rulo	115
Section of a boring two miles south of Saint Joseph, 60 feet above high water of the Missouri	117–118
Section of the rocks exposed at Riverside, Kansas, and along the river bluff between there and the Atchison Landing	119–120
Boring at Atchison, Kansas, commencing 22½ feet above high-water mark of the Missouri; made by the Atchison Coal Company, 1865–'66	121–122
Tabular list, illustrating the geological and geographical range of the fossils of Eastern Nebraska	124–127
Remarks on the probability of finding valuable beds of coal within profitable working distance of the surface in Eastern Nebraska	134–139

DESCRIPTIONS OF FOSSILS.

Protozoa.
  Foraminifera.
    Fusulina, Fischer.
      F. cylindrica, Fischer, 1837, pl. i, fig. 2; pl. ii, fig. 1; pl. v, figs. 3 *a, b*; pl. vii, figs. 8 *a, b* .. 140–141

Radiata.
  Polypi.
    Rhombopora, Meek, n. g., 1872 141
    R. lepidodendroides, n. s., Meek, 1872, pl. vii, figs. 2 *a–f* 141–143
    Fistulipora, McCoy 143

	Page.
F. nodulifera, n. s., Meek, 1872, pl. v, figs. 5 a–d	143–144
Syringopora, Goldfuss	144
S. multattenuata, McChesney, 1860, pl. i, figs. 5 a–d	144
Lophophyllum, Edwards & Haime	144
L. proliferum, McChesney, sp., 1860, pl. v, figs. 4 a, b	144–145
Campophyllum, Edwards & Haime	145
C. torquium, Owen, sp., 1852, pl. i, figs. 1 a–d	145–146

Echinodermata.

Eriocrinus, M. & W	146
E. typus, M. & W., 1865, pl. i, figs. 3 a, b, and fig. 1, p. 146	146–147
Scaphiocrinus, Hall	147
S. ? hemisphœricus, Shumard, sp., 1858, pl. v, figs. 1 a, b; pl. vii, figs. 1 a–c, and fig. 2, p. 148	147–149
Zeacrinus, Troost	149
Z. ? mucrospinus, McChesney, 1860, pl. v, figs. 2 a–c	149–150
Eupachycrinus, M. & W	150
E. verrucosus, White & St. John, 1869, figs. 3 and 4 a–d, p. 150	150–151
Archœocidaris, McCoy	151
A. ? triserrata, n. s., Meek, 1872, pl. i, figs. 6 a–c	151–152
Eocidaris, Desor	152
E. hallianus, Geinitz, 1866, pl. vii, figs. 9 a–d	152

Mollusca.

Polyzoa.

Fenestella, Lonsdale	152
Fenestella, sp., pl. i, figs. 4 a, b	152–153
F. shumardi, Prout ?, 1858, pl. vii, figs. 3 a–c	153–154
Polypora, McCoy	154
P. submarginata, n. s., Meek, 1872, pl. vii, figs. 7 a, b	154–155
Polypora, sp. undt., Meek, pl. vii, figs. 6	155
Synocladia, King	156
S. biserialis, Swallow, 1858, pl. vii, figs. 5 a–e	156–157
Glauconome, Goldfuss	157
G. trilineata, Meek, n. s., 1872, pl. vii, figs. 4 a–d	157–158

Brachiopoda.

Lingula, Bruguière	158
L. scotica, var. *nebrascensis*, Meek, 1872, pl. viii, figs. 3 a, b	158
Orbiculoidea, d'Orbigny	158
Orbiculoidea, sp., pl. iv, fig. 3	158–159
Productus, Sowerby	159
P. costatus, Sowerby? sp., 1827, pl. vi, figs. 6 a, b	159–160
P. semireticulatus, Martin, sp., 1809, pl. v, figs. 7 a, b	160–161
P. longispinus, Sowerby, ? 1814, pl. vi, fig. 7, and pl. viii, figs. 6 a–c	161–163
P. prattenianus, Norwood, 1854, pl. ii, figs. 5 a–c, pl. v, fig. 13, and pl. viii, figs. 10 a, b	163–164
P. pertenuis, n. s., Meek, 1872, pl. i, figs. 14 a–c, and pl. viii, figs. 9 a–d	164–165
P. nebrascensis, Owen, 1852, pl. ii, fig. 2, pl. iv, fig. 6, and pl. v, figs. 11 a–c	165–167
P. symmetricus, McChesney, 1860, pl. v, figs. 6 a, b, and pl. viii, fig. 13	167–168
P. punctatus, Martin, sp., 1809, pl. ii, fig. 6, and pl. iv, fig. 5	169
Chonetes, Fischer	170
C. verneuiliana, Norwood & Pratten, 1854, pl. i, figs. 10 a, b	170
C. granulifera, Owen, 1855, pl. iv, fig. 9, pl. vi, fig. 10, pl. viii, fig. 7	170–171
C. glabra, Geinitz, 1866, pl. iv, fig. 10, pl. viii, figs. 8 a, b	171–172
Orthis, Dalman	173
O. carbonaria, Swallow, 1858, pl. i, figs. 8 a–c	173
Hemipronites	173
H. crassus, M. & H., 1858, pl. v, figs. 10 a–c, and pl. viii, fig. 1	174–175
Meekella, White & St. John	175
M. striato-costata, Cox, sp., 1857, pl. v, figs. 12 a–c, and figs. 5 a, b, and 6, p. 175	175–177
Syntrielasma, M. & W	177
S. hemiplicata, Hall, sp., 1852, pl. vi, figs. 1 a, b, and pl. viii, figs. 12 a, b, and figs. 7 a–c, figs. 8 a, b, p. 177	177–178
Rhynchonella, Fischer	179
R. osagensis, Swallow, 1858, pl. i, figs. 9 a, b, and pl. vi, figs. 2 a, b	179–180
Athyris, McCoy	180
A. subtilita, Hall, sp., 1852, pl. i, fig. 12, pl. v, fig. 8, and pl. viii, fig. 4	180–181
Retzia, King	181

	Page.
R. punctulifera, Shumard, 1858, pl. i, fig. 13, and pl. v, fig. 7	181-183
Spirifer, Sowerby	183
S. cameratus, Morton, 1836, pl. vi, fig. 12, and pl. viii, fig. 15	183-184
S. (Martinia) planoconvexus, Shumard, 1855, pl. iv, figs. 4 *a*, *b*, and pl. viii, figs. 2 *a*, *b*	184-185
Spiriferina, d'Orbigny	185
S. kentuckensis, Shumard, 1855, pl. vi. figs. 3 *a–d*, and pl. viii, figs. 11 *a*, *b*	185-186
Terebratula, Lhwyd	187
T. bovidens, Morton, 1836, pl. i, figs. 7 *a–d*, and pl. ii, fig. 4	187-188
Lamellibranchiata.	
Lima, Bruguière	188
L. retifera, Shumard, 1859, pl. ix, fig. 5	188-189
Entolium, Meek	189
E. aviculatum, Swallow, sp., 1858, pl. ix, figs. 11 *a–f*	189-191
Aviculopecten, McCoy	191
A. occidentalis, Shumard, sp., 1855, pl. ix, fig. 10	191-193
A. neglectus, Geinitz, sp., 1866, pl. ix, figs. 1 *a*, *b*	193
A. carboniferus, Stevens, sp., 1858, pl. iv, fig. 8 and pl. ix, figs. 4 *a*, *b*	193-195
A. whitei, n. s., Meek, 1872, pl. iv, figs. 11 *a–c*	195
A. coxanus, M. & W., 1860, pl. ix, figs. 2 *a*, *b*	196
Aviculopinna, Meek	197
A. americana, Meek, 1867, pl. ix, figs. 12 *a–d*	197-198
Pinna, Linnæus	198
P. peracuta, Shumard, 1858, pl. vi, fig. 11 *a*, *b*	198
Avicula (Klein), Brug	199
A. longa, Geinitz, sp., 1866, pl. ix, fig. 8	199
A. ? sulcata, Geinitz, 1866, pl. ix, fig. 9	200
Pseudomonotis, Beyrich	200
Pseudomonotis, sp., pl. ii, fig. 11	200-201
P. radialis, Phillips ?? sp.,1834, pl. ix. fig. 3	201
Myalina, de Koninck	201
M. [?] swallovi, McChesney, 1860, pl. ix, figs. 7 *a*, *b*	201-202
M. subquadrata, Shumard, 1855, pl. iv, fig. 12 and pl. ix, fig. 6	202-203
Nucula, Lamarck	203
N. beyrichi, v. Schauroth? 1854, pl. x, fig. 18	203-204
N. ventricosa, Hall, 1858, pl. x, figs. 17 *a–c*	204-205
Yoldia, Möller	205
Y. subscitula, M. & H., ? 1858, pl. x, fig. 10	205-206
Nuculana, Link	206
N. bellistriata var. *attenuata*, pl. x, figs. 11 *a*, *b*	206-207
Macrodon, Lycett	207
M. tenuistriata, M. & W., 1867, pl. x, figs. 20, *a*, *b*	207-208
Schizodus, King	208
S. curtus, M. & W., ? 1866, pl. x, figs. 13 *a–c* (*d ?*), *e*	208-209
S. wheeleri, Swallow, sp., 1862, pl. x, figs. 1 *a–d* (and *e*, *f* ?)	209-210
Schizodus, undt. pl. x, fig. 2	210-211
Modiola, Lamarck	211
M. ? subelliptica, Meek, 1867, pl. x, fig. 5	211-212
Pleurophorus, King	212
P. oblongus, n. s., Meek, 1872, pl. x, fig. 4 *a–c*	212
P. occidentalis, M. & H., ? 1858, pl. x, fig. 12	212-213
Edmondia, de Koninck	213
E. reflexa, n. s., Meek, 1872, pl. x, figs. 6 *a*, *b*, and pl. iv, fig. 7 ?	213-214
E. ? glabra n. s., Meek, 1872, pl. x, figs. 7 *a*, *b*	214
E. ? nebrascensis, Geinitz, sp., 1866, pl. x, figs. 8 *a*, *b*	214-215
E. subtruncata n. s., Meek, 1872, pl. ii, fig. 7	215-216
E. aspinwallensis, Meek, 1871, pl. iv, figs. 2 *a–c*	216
Chænomya, M. & H	216
C. leavenworthensis, M. & H., 1858, pl. ii, fig. 9	216-217
C. minehaha, Swallow, sp., 1858, pl. — figs. 13 *a*, *b*	217
Allorisma, King	217
A. (Sedgwickia) reflexa, n. s., Meek, 1872, pl. x, fig. 15	217-218
A. (Sedgwickia) geinitzii, Meek, 1867, pl. x, figs. 16 *a*, *b*	219
A. (Sedgwickia ?) subelegans n. s., Meek, 1872, pl. x, fig. 14	220
A. (Sedgwickia) granosa, Shumard, sp., 1858, pl. ii, fig. 8	220-221
A. subcuneata, M. & H., 1858, pl. ii, figs. 10 *a*, *b*	221-222

	Page.
Prothyris, Meek	223
P. elegans, Meek, 1871, pl. x, figs. 9 *a, b*	223
Solenopsis, McCoy	223
S. solenoides, Geinitz, sp., 1866, pl. x, fig. 3	223-224
Gasteropoda, Cuvier.	
Dentalium, Linnæus	224
D. meekianum, Geinitz, 1866, pl. xi, figs. 16 *a, b*	224
Bellerophon, Montfort	224
B. carbonarius, Cox, 1857, pl. iv, fig. 10, pl. xi, figs. 11 *a–c*	224-225
B. montfortianus, Norwood & Pratten, 1855, pl. xi, fig. 15, and 12?	225-226
B. marcouianus, Geinitz, 1866, pl. iv, fig. 17, and pl. xi, figs. 13 *a, b*	226-227
B. percarinatus, Conrad, 1842, pl. xi, fig. 14	227
Platyceras, Conrad	227
P. nebrascense, n. s., Meek, 1872, pl. iv, figs. 15 *a, b*	227-228
Macrocheilus, Phillips	228
M. intercalaris var. *pulchellus*, M. & W., 1860, pl. vi, fig. 8	228
Orthonema, M. & W	228
O. subtaeniata, Geinitz, sp., 1866, pl. xi, fig. 10	228-229
Aclis, Lovén	229
A. swalloviana, Geinitz, sp., 1866, pl. xi, figs. 7 *a, b*	229-230
Straparollus, Montfort	230
S. (Euomphalus) rugosus, Hall, 1858, pl. vi, figs. 5 *a, b*, and pl. xi, figs. 4 *a, b*	230-231
Pleurotomaria, Defrance	231
P. haydeniana, Geinitz, 1866, pl. xi, fig. 5	231
P. perhumerosa, n. s., Meek, 1872, pl. iv, figs. 13 *a, b*	232
P. inornata, n. s., Meek, 1872, pl. iv, fig. 14	232-233
P. grayvillensis, Norwood & Pratten, 1855, pl. xi, fig. 9	233
P. marcouiana, Geinitz, 1866, pl. xi, fig. 8	233
P. subdecussata, Geinitz, 1866, pl. xi, fig. 10	233
Murchisonia, de Verneuil	234
M. nebrascensis, Geinitz, 1866, pl. xi, fig. 6	234
Cephalopoda.	
Orthoceras, Auct	234
O. cribrosum, Geinitz, 1866, pl. xi, figs. 18 *a, b*	234
Nautilus, Linn	234
N. occidentalis, Swallow, 1858, pl. xi, fig. 17	234-236
N. ponderosus, White, M. S., 1872, pl. iii, figs. 7 *a, b*	236

Articulata.
Crustacea.

Cythere, Müller	237
C. nebrascensis, Geinitz, 1866, pl. xi, figs. 2; and 3 *a, b*?	237
Cythere, sp., pl. xi, figs. 1 *a–d*	237
Phillipsia, Portlock	237
Phillipsia, sp., Geinitz, 1866, pl. iii, figs. 1 *a, b*	237-238
P. scitula, M. & W., 1865, pl. vi, fig. 9	238
P. major, Shumard, 1858, pl. iii, figs. 2 *a–c*	238-239

MEEK, F. B. [Geological reports on Miller, Morgan, and Saline Counties, Missouri.] <Reports on the Geological Survey of the State of Missouri, 1855–'71, by G. C. Broadhead, F. B. Meek, and B. F. Shumard, chapters vii–ix, pp. 111-188. Jefferson City, 1873.

Geological maps of Miller and Morgan Counties accompany these reports.

CHAPTER VII.

	Page.
Miller County	112
Streams	112
Springs	114
Timber	115
Geological structure of Miller County	115
Quaternary system.—Alluvium	117
Carboniferous system	118
Lower Silurian rocks.—Saccharoidal sandstone	118

	Page.
Second magnesian limestone	118
Second sandstone	121
Third magnesian limestone	123
Third sandstone	127
Fourth magnesian limestone	127
Economical geology	128
Soil, clays for the manufacture of bricks	128
Building stones	128
Road material, limestone for quicklime, sand	129
Iron ore	130
Lead	131
Coal	132
Sulphate of baryta	133

CHAPTER VIII.

Morgan County	135
Springs	136
Streams	137
Timber	138
Geological structure	138
Quaternary system.—Alluvium	139
Carboniferous rocks.—Coal measures	140
Encrinital limestone	140
Chouteau limestone	141
Silurian system	141
First magnesian limestone	141
Saccharoidal sandstone	142
Second magnesian limestone	143
Second sandstone	144
Third magnesian limestone	145
Third sandstone	147
Fourth magnesian limestone	148
Economical geology	149
Soils	149
Coal	149
Lead	152
Iron ore	155
Heavy spar	156
Building materiel	156

CHAPTER IX.

Saline County	157
Streams	158
Timber	159
Geological structure	160
Quaternary deposits	161
Alluvium of the Missouri Flats	161
Bluff or Loess Deposit	161
Drift	162
Carboniferous rocks.—Coal measures	163
Ferruginous sandstone	172
Archimedes limestone	172
Encrinital limestone	173
Chouteau limestone	174
Cooper marble	176
Semi-crystalline limestone	176
Silurian system	178
Trenton limestone ?	178
Saccharoidal sandstone	178
Economical geology	179
Soil	179
Coal	180
Building stone	180
Limestones for making lime, sand, clays for bricks, &c.	181
Springs	181

88.

MEEK, F. B. Spergen Hill fossils identified among specimens from Idaho. <Am. Journ. Sci., vol. v, 3d ser., pp. 383, 384. 1873. New Haven, 1873.

The author identifies, among some collections made by Prof. F. H. Bradley, some of the minute species of Mollusca, for which the locality in Washington County, Indiana, known as Spergen Hill, is noted. A list of the species identified is given, but no discussion of them is made.

89.

MEEK, F. B. Preliminary Paleontological Report, consisting of lists and descriptions of fossils, with remarks on the ages of the rocks in which they were found, &c., &c. <Sixth Ann. Report of the U. S. Geol. Survey of the Territories, pp. 431-518. 1873. Washington, 1873.

Genera Admetopsis, Velatella.

	Page.
General remarks	431
Silurian age	431–432
Carboniferous age	432–434
Jurassic age	434–435
Cretaceous age	435–438
Section from about 1½ miles northeast of Coalville, in a northwesterly direction, to Echo Cañon, fig. 52	439–440
Section of the rocks exposed on Sulphur Creek, near Bear River, Wyoming, fig. 53	451–452
Tertiary age	462
Lists of fossils collected	463

Descriptions of new species of fossils.

SILURIAN FORMS.

Iphidea (??) sculptilis, n. s., Meek, 1873	479
Asaphus (Megalaspis?) goniocercus, n. s., Meek, 1873	480
Bathyurus serratus, n. s., Meek, 1873	480–482
B. ? haydeni, n. s., Meek, 1873	482–484
Bathyurellus (Aspiscus) bradleyi, n. s., Meek, 1873	484–485
Conocoryphe (Ptychoparia) gallatinensis, n. s., Meek, 1873	485–487

CRETACEOUS FORMS.

Ostrea soleniscus, Meek, 1870	487–488
Ostrea anomioides, n. s., Meek, 1873	488
Avicula (Pseudoptera) propleura, n. s., Meek, 1873	489–490
Avicula (Pseudoptera) rhytophora, n. s., Meek, 1873	490–491
Avicula (Oxytoma) gastrodes, n. s., Meek, 1873	491–492
Modiola (Brachydontes) multilinigera, n. s., Meek, 1873	492–493
Trapezium micronema, n. s., Meek, 1873	493
Corbicula (Veloritina) inflexa, n. s., Meek, 1873	493–494
Corbicula (Cyrena ?) securis, n. s., Meek, 1873	494–495
Corbicula æquilateralis, n. s., Meek, 1873	495
Cyrena carletoni, n. s., Meek, 1873	495–496
Pharella ? pealei, n. s., Meek, 1873	496
Corbula nematophora, n. s., Meek, 1873	496–497
Neritina (Dostia ?) bellatula, n. s., Meek, 1873	497–498
Neritina (Dostia ?) patelliformis, n. s., Meek, 1873	498–499
Neritina (Dostia ?) carditoides, n. s., Meek, 1873	499
Neritina (Neritella) bannisteri, n. s., Meek, 1873	499–500
Neritina (Neritella) pisum, n. s., Meek, 1873	500
Neritina pisiformis, n. s., Meek, 1873	500–501
Admete ? rhomboides, n. s., Meek, 1873	501
Admete ? gregaria, n. s., Meek, 1873	501–502
Admete ? subfusiformis, n. s., Meek, 1873	502
Turritella coalvillensis, n. s., Meek, 1873	502–503
Turritella spironema, n. s., Meek, 1873	503–504
Turritella (Aclis ?) micronema, n. s., Meek, 1873	504
Fusus (Neptunea ?) gabbi, n. s., Meek, 1873	504–505
Fusus (Neptunea ?) utahensis, n. s., Meek, 1873	505

	Page.
Turbonilla (Chemnitzia ?) coalvillensis, n. s., Meek, 1873	505–506
Eulima funicula, n. s., Meek, 1873	506
Eulima chrysalis, n. s., Meek, 1873	506
Eulima ? inconspicua, n. s., Meek, 1873	507
Melampus antiquus, n. s., Meek, 1873	507
Valvata nana, n. s., Meek, 1873	507
Physa carletoni, n. s., Meek, 1873	508

Species from the Bitter Creek series.

Ostrea wyomingensis, Meek, 1872	508–509
Anomia (Placunopsis ?) gryphorhyncus, Meek, 1871	509–511
Corbicula (Veloritina) cytheriformis, M. & H., 1860	511
Corbicula ? fracta, var. *crassiuscula*, Meek, 1873	512–513
Corbicula (Veloritina) bannisteri, n. s., Meek, 1873	513
Corbula undifera, n. s., Meek, 1873	513–514
Corbula tropidophora, n. s., Meek, 1873	514–515
Goniobasis ? insculpta, n. s., Meek, 1873	515–516
Melania (Goniobasis?) wyomingensis, n. s., Meek, 1873	516

Tertiary forms.

Physa bridgerensis, n. s., Meek, 1873	516–517
Limnæa (Limnophysa?) compactilis, n. s., Meek, 1873	517
Pupa ? leidyi, n. s., Meek, 1873	517–518

90.

MEEK, F. B. Descriptions of Invertebrate fossils of the Silurian and Devonian systems. <Geol. Surv. of Ohio, vol. i, part ii. Palæontology, pp. 1–243, plates 1–23, and 3 plates of diagrams of Crinoids. 1873. Columbus, 1873.

FOSSILS OF THE CINCINNATI GROUP.

Radiata.
Echinodermata.
Crinoidea.

	Page.
Heterocrinus, Hall, 1847	1–2
H. constrictus, Hall, 1866, pl. i, figs. 10 *a, b* (and 11 ?)	3–5
H. exilis, Hall ? 1868, pl. i, fig. 12	5–7
H. simplex, Hall, 1847, pl. i, figs. 4 *a, b*; 5 *a, b* (with 6 *a, b* and 7 *a–c*?)	7–10
H. juvenis, Hall, 1866, pl. i, figs. 3 *a–c*	10–12
H. heterodactylus, Hall ? 1847, pl. i, figs. 1 *a, b* (and 2 *a, b*?)	12–14
H. laxus, Hall, 1872, pl. i, figs. 8 *a, b*	14, 15
H. (Iocrinus) subcrassus, M. and W., 1865, pl. i, figs. 9 *a, b*	15–17
Anomalocrinus, M. and W., 1868	17
A. incurvus, M. and W., 1865, pl. ii, figs. 6 *a–f*	17–20
Poteriocrinites, Miller, 1821	20
Subgenus (*Dendrocrinus*) Hall, 1852	20
P. (Dendrocrinus) cincinnatiensis, Meek, 1872, pl. iii bis, figs. 5 *a, b*	20–22
P. (Dendrocrinus) polydactylus, Shumard, sp. 1867, pl. iii bis, fig. 9	23
P. (Dendrocrinus) posticus, Hall, 1872, pl. iii bis, figs. 4 *a–c*	22–24
P. (Dendrocrinus) dyeri, Meek, 1872, pl. iii bis, figs. 3 *a, b*	24–25
P. (Dendrocrinus) caduceus, Hall, 1866, pl. 3 bis, figs. i, *a–d*	26–27
P. (Dendrocrinus) casei, Meek, 1871, pl. iii bis, figs. 2 *a–c*	28–30
Glyptocrinus, Hall, 1847	30
G. decadactylus, Hall, 1847, pl. ii, figs. 5 *a, b*	30–32
G. dyeri, Meek, 1872, pl. ii, figs. *a, b*	32–34
G. dyeri var. *sub-globosus*, Meek, 1872, pl. ii, fig. 2, *c*	34
G. nealli, Hall, 1866, pl. ii, figs. 3 *a–c*	34–36
G. parvus, Hall, 1872, pl. ii, figs. 4 *a, b*	36–37
G. bacri, Meek, 1872, pl. ii, figs. 1 *a, b*	37–39
Cystoidea.	
Lepacrinites, Conrad, 1840	39
L. moorei, Meek, 1871, pl. iii, figs. 4 *a–c*	39–41
Anomalocystites, Hall, 1859	41
A. (Ateleocystites?) balanoides, Meek, 1872, pl. iii bis, figs. 6 *a–c*	41–44
Lichenocrinus, Hall, 1866	44–51
L. dyeri, Hall, 1866, pl. iii, figs. 2 (and 3 *a, b*?)	51

	Page
L. crateriformis, Hall, 1866, pl. iii, figs. 1 a-t	51–52
Hemicystites, Hall	52
H. stellatus, Hall, 1866, pl. iii, figs. 8 a, b	52–54
H. (Cystaster) granulatus, Hall, 1872, pl. iii, figs. 9 a, b	54
Agelacrinites, Vanuxem, 1842	55
A. (Lepidodiscus) cincinnatiensis, Roemer, 1851, pl. iii, figs. 6 a, b	55–56
A. pileus, Hall, 1866, pl. iii, fig. 5	56 57
A. vorticellata, Hall, 1866, pl. iii, figs. 7 a, b	57–58
Asteroidea.	
Palaeaster, Hall	58
P.? dyeri, Meek, 1872, pl. iv, figs. 2 a-f	58–60
P. granulosus, Hall? 1866, pl. iv, figs. 3 a-c	60–61
P.? jamesii, Dana, 1863, pl. iv, fig. 4	62–64
P. incomptus, Meek, 1872, pl. iv, figs. 5 a, b	64–65
P. shaefferi, Hall, 1868, pl. iv, fig. 1	66
Stenaster Billings, 1858	66
S. grandis, Meek, 1872, pl. iii bis, fig. 7 a-c	66–67
Ophiuroidea.	
Protaster?, Forbes, 1849	68
P. granuliferus, Meek, 1872, pl. iii bis, figs. 8 a, b	68–69
Mollusca.	
Polyzoa.	
Ptilodictya, Lonsdale, 1839	69
P. (Stictopora) shafferi, Meek, 1872, pl. v, figs. 1 a-c	69–70
Brachiopoda.	
Leptaena, Dalman, 1828	70
L. sericea, Sowerby ? 1839, pl. v, figs. 3 a-h	70–72
Strophomena, Rafinesque, 1827	73–75
• *S. rhomboidalis*, Wilckens sp., 1769, pl. v, figs. 6 a-e	75–77
Subgenus *Hemipronites*, Pander	77
(Resupinate species.)	
S. (Hemipronites) nutans, James, 1871, pl. vi, figs. 1 a-f	77–79
S. (Hemipronites) planumbona, Hall, 1847, pl. vi, figs. 3 a-h	79–81
S. (Hemipronites) plicata, James, 1871, pl. vi, figs. 4 a-h	81–82
S. (Hemipronites) plano-convexa, Hall, 1847, pl. ii, figs. 2 a-h	82–83
S. (Hemipronites) filitexta, Hall, 1847, pl. vi, figs. 5 a-e	83–85
S. (Hemipronites) sulcata, de Verneuil, 1848, pl. v, figs. 4 a-e	85–86
S. (Hemipronites) sinuata, James, 1871, pl. v, figs. 5 a-f	87–88
(Non-resupinate species.)	
S. (Hemipronites) alternata (Conr.), Emmons, 1838, pl. vii, figs. 1 a-g	88–91
Orthis, Dalman, 1828	92
(Resupinate species.)	
O. retrorsa, Salter ? 1858, pl. xi, figs. 7 a-e	92–94
O. subquadrata, Hall, pl. ix, figs. 2 b-g	94–96
O. occidentalis, Hall, 1847, pl. ix, figs. 3 a-h	96–99
O. insculpta, Hall, 1847, pl. ix, figs. 1 a-h	99–101
O. borealis, Billings, 1859, pl. viii, figs. 4 a-f	101–103
O. bellula, James, 1871, pl. viii, figs. 5 a-f	103–104
O. (?) ella, Hall, 1861, pl. viii, figs. 9 a-d	105–106
(Non-resupinate species.)	
O. fissicosta, Hall, 1847, pl. viii, figs. 6 a-h	106–107
O. plicatella, Hall, 1847, pl. viii, figs. 7 a-h	108–109
O. emacerata, Hall, 1860, pl. viii, figs. 1 a-d, and figs. 2 a-g	109–111
O. emacerata var. *multisecta*, James, 1871, pl. viii, figs. 3 a-d	112
O. (Platystrophia) biforata, Schlotheim, sp., 1820, pl. x	112–114
Var. 1. *O. (Platystrophia) lynx*, Von Buch., pl. x, figs. 1 a-e	114–116
Var. 2. *O. (Platystrophia) laticosta*, James, 1871, pl. x, figs. 4 a-f	116–117
Var. 3. *O. (Platystrophia) dentata*, Pander ? ? James, 1871, pl. x, figs. 3 a-d	117–119
Var. 4. *O. (Platystrophia) acutilirata*, Con., sp., 1842, pl. x, figs. 5 a-g	119–121
Rhynchonella, Fischer de Waldh., 1809	121
R. dentata, Hall, 1847, pl. xi, figs. 3 a-d	121–122
R. capax, Conrad, sp., 1842, pl. xi, figs. 6 a-f	123–124
Zygospira, Hall, 1862	125
Z. modesta, Say, sp., pl. xi, figs. 4 a-d	125–126
Z. cincinnatiensis, James, 1871, pl. xi, figs. 5 a-c	126

	Page.
Z. headi, Billings, sp., 1862, pl. xi, figs. 1 a–d	127–128
Retzia, King, 1850	128
R. (Trematospira) granulifera, Meek, 1872, pl. xi, figs. 6 a–c	128–129
Pholidops, Hall, 1860	130
P. cincinnatiensis, Hall, 1872, pl. v, figs. 2 a, b	130
Lamellibranchiata.	
Ambonychia, Hall, 1847	130
A. costata, James, 1871, pl. xii, figs. 5 a–c	130–131
A. (Megaptera) alata, Meek, 1872, pl. xi, fig. 9, and pl. xii, fig. 10	131–132
A. (Megaptera) casei, M. & W. ?, 1866, pl. xi, fig. 8	133
Cypricardites, Conrad, 1841	133
C. sterlingensis, M. & W. ? sp., 1866, pl. xi, figs. 12 a, b	133–135
C. ? carinata, Meek, 1872, pl. xii, figs. 6 a, b	135–136
Megambonia, Hall, 1859	136
M. jamesi, Meek, 1871, pl. xii, figs. 9 a, b	136–138
Clidophorus, Hall, 1847	138
C. (Nuculites?) fabula, Hall, sp., 1845, pl. xi, figs. 10 a, b	138–139
Tellinomya, Hall, 1847	139
T. ? obliqua, Hall, sp., 1845, pl. xi, figs. 11 a–c	139
Anodontopsis, McCoy, 1851	140
A. ? milleri, Meek, 1871, pl. xii, figs. 1 a–d	140–141
A. (Modiolopsis?) unionoides, Meek, 1871, pl. xii, figs. 2 a, b	141–142
Sedgwickia, McCoy, 1844	142
S. (? Grammysia) neglecta, Meek, 1872, pl. xii, fig. 8	142–143
S. ? fragilis, Meek, 1872, pl. xii, figs. 3 a, b	143–144
S. ? compressa, Meek, 1872, pl. xii, figs. 7 a, b	144–145
Cardiomorpha, de Koninck, 1844	146
C. ?? obliquata, Meek, 1872, pl. xii, figs. 4 a, b	146–147
Gasteropoda.	
Cyrtolites, Conrad, 1838	147
C. (Microceras) inornatus, Hall, sp., 1845, pl. xiii, figs. 4 a, b	147–148
C. ornatus, Conrad, 1838, pl. xiii, figs. 3 a, b	148–149
C. dyeri, Hall, 1871, pl. xiii, figs. 2 a–e	149–150
C. ? costatus, James, 1872, pl. xiii, figs. 1 a–c	150
Cyclonema, Hall, 1852	151
C. bilix, Conrad sp., 1842, pl. xiii, figs. 5 a, c, d, g, and 5 e, f ?	151–152
Cyclora, Hall, 1845	152
C. minuta, Hall, 1845, pl. xiii, figs. 7 a–c	152–153
C. ? parvula, Hall, sp., 1845	154
Pleurotomaria, Defrance, 1826	154
P. (Scalites?) tropidophora, Meek, 1872, pl. xiii, figs. 6 a–c	154–155
Cephalopoda.	
Orthoceras, Auct.	
O. ortoni, Meek, 1872, pl. xiii, fig. 8	155–156
Trochoceras, Barrande, 1847	157
T. ? baeri, M. & W., 1865, pl. xiii, fig. 9	157–158
Articulata.	
Crustacea.	
Entomostraca.	
Cythere, Muller, 1785	158
C. cincinnatiensis, Meek, 1872, pl. xiv, figs. 1 a–d	158–159
Asaphus, Brongniart, 1822	159
A. (Isotelus) megistos, Locke ? 1842, pl. xiv, fig. 13	159–160
Proetus, Steininger, 1831.	
P. spurlocki, Meek, 1872, pl. xiv, fig. 12	161–162
Ceraurus, Green, 1833	622
C. icarus, Billings, 1859, pl. xiv, figs. 11 a–c	162–165
Acidaspis, Murchison, 1839	165
A. crosotus, Locke, ? 1842, pl. xiv, figs. 10 a, b	165–167
A. cincinnatiensis, Meek, n. sp., 1873, pl. xiv, fig. 3	167–169
A. ceralepta, Anthony, sp.?, 1838, pl. xiv, figs. 8, 9	169–170
Dalmanites, Barrande, 1852	170
D. carleyi, Meek, 1872, pl. xiv, figs. 2 a–d	170–173
Calymene, Brongniart, 1822	173
C. senaria, Conrad, 1841, pl. xiv, figs. 14 a–f	173–175

FOSSILS OF THE NIAGARA AND CLINTON GROUPS.

Mollusca.
  Brachiopoda.
     Triplesia, Hall, 1859 176–177
     T. ortoni, Meek, 1872, pl. xv, figs. 1 a–k.. 178–179
     Rhynchonella, Fischer, 1809... 179
     R. neglecta, Hall, 1852, pl. xv, figs. 3 a–d..................................... 179–180
     Meristella, Hall, 1860.. 180
     M. ? (Meristina) cylindrica, Hall, sp., 1852, pl. xv, figs. 2 a–d............... 180–182
     Trimerella, Billings, 1862... 182
     T. grandis, Billings, 1862, pl. xvi, figs. 2 a, b................................ 182
     T. ohioensis, Meek, 1871, pl. xvi, figs. 1 a–c, and figs. a, b, p. 183.......... 183–185
  Gasteropoda.
     Platyostoma, Conrad, 1842.. 185
     P. niagarensis var. trigonostoma, pl. xvi, figs. 3 a–c, and fig. p. 186......... 185–186
  Cephalopoda.
     Lituites, Montfort, 1808... 186
     L. ? ortoni, Meek, n. s., 1873, pl. xv, fig. 4................................... 186–187
Articulata.
  Crustacea.
     Leperditia, Rouault, 1851.. 187
     L. alta, Conrad, sp., 1843, pl. xvii, figs. 2 a, b............................... 187–188
     Illaenus, Dalman, 1826... 189
     I. (Bumastus) insignis, Hall, ? 1864, pl. xv, figs. 5 a–c, and figs. a and b, p. 189..... 189–193

FOSSILS OF THE CORNIFEROUS GROUP.

Mollusca.
  Polyzoa.
     Ptilodictya, Lonsdale, 1839.. 104
     P. (Stictopora) gilberti, Meek, 1871, pl. xviii, figs. 1 a–c.................... 194–195
  Brachiopoda.
     Rhynchonella, Fischer, 1809.. 196
     R. carolina, Hall, 1867, pl. xviii, figs. 8 a–e.................................. 196–197
  Lamellibranchiata.
     Aviculopecten, McCoy, 1851... 197
     A. parilis, Conrad, ? 1842, pl. xviii, figs. 6 a, b.............................. 197–199
     Lucina, Bruguiere, 1792.. 199
     L. (Paracyclas) ohioensis, Meek, 1871, pl. xviii, fig. 7 a, b................... 199–200
     L. lirata, Conrad, sp., fig. --, p. 200.. 200
     Conocardium, Brown, 1835. ?.. 201
     C. trigonale, Hall, 1843, figs. A–C, p. 201, and figs. A, B, p. 204............. 201–203
     C. ohioense, Meek, 1871, pl. xviii, fig. 9, and figs. A and B, p. 204........... 203–206
     Solemya, Lamarck, 1818... 206
     S. (Janeia) vetusta, Meek, 1871, pl. xviii, fig. 4............................... 206–207
     Clinopistha, M. and W., 1870... 208
     C. antiqua, Meek, 1871, pl. xviii, figs. 5 a, b.................................. 208
     Sanguinolites, McCoy, 1844... 209
     S. ? sanduskyensis, Meek, 1871, pl. xviii, fig. 3................................ 209
  Gasteropoda.
     Platyceras, Conrad, 1842... 210
     P. multispinosum, Meek, 1871, pl. xx, figs. 7 a, b............................... 210–211
     P. dumosum var. attenuatum, Meek, 1871, pl. xx, figs. 2 a, b.................... 212–213
     Cyclonema, Hall, 1852.. 213
     C. crenulata, Meek, 1871, pl. xix, figs. 2 a–d................................... 213
     Naticopsis, McCoy, 1844.. 214
     N. ? (Isonema) humilis, Meek, 1871, pl. xix, figs. 1 a–c, and fig. p. 215....... 214–215
     N. levis, Meek, 1871, pl. xix, figs. 4 a, b...................................... 215–216
     N. acquistriata, Meek, n. s., 1873, figs. a, b, p. 216........................... 216–217
     Orthonema, M. and W., 1861... 217
     O. newberryi, Meek, 1871, pl. xx, figs. 3 a, b................................... 217–218
     Trochonema, Salter, 1859... 218
     T. tricarinata, Meek, 1871, pl. xix, figs. 5 a, b................................ 218–219
     Euomphalus, Sowerby, 1814.. 220
     E. decewi, Billings, 1861, pl. xix, figs. 3 a, b, and pl. xx, fig. 1............ 220–221
     Xenophora, Fischer, 1806... 221
     X. ? (Pseudophorus) antiqua, Meek, 1871, pl. xvii, figs. 1 a–e.................. 221–222

	Page.
Bellerophon, Montfort, 1808	222
B. newberryi, Meek, 1871, pl. xx, fig. 5, and fig. —, p. 223	222-225
B. patulus, Hall, fig. —, p. 223	223
B. propinquus, Meek, 1871, pl. xx, figs. 4 *a–b*	226
Pleurotomaria, Defrance, 1826	226
P. lucina, Hall?, 1862, pl. xx, fig. 6	226-227
Conularia, Miller, 1818, MS	228
C. elegantula, Meek, 1871, pl. xxiii, fig. 4	228-229
Cephalopoda.	
Cyrtoceratites, Goldfuss, 1830?	229
C. ohioensis, Meek, 1871, pl. xxiii, figs. 2 and 2 *b*	229-230
Gyroceratites, Meyer, 1831	230
G. (? *Trochoceras*) *ohioensis*, Meek, 1871, pl. xxii	230-231
G. (? *Nautilus*) *inelegans*, Meek, pl. xxi	232
Articulata.	
Crustacea.	
Proetus, Steininger, 1831	233
P. planimarginatus, Meek, 1871, pl. xxiii, figs. 3 *a*, *b*	233-234
Dalmanites, Barrande, 1852	234
D. ohioensis, Meek, 1871, pl. xxiii, fig. 1	234-236

91.

MEEK, F. B., *and* WORTHEN, A. H. Paleontology. Descriptions of Invertebrates from the Carboniferous System. <Geological Survey of Illinois, vol. v, pp. 323-619, pls. i-xxxii. 1873. Springfield, 1873.

Published by authority of the legislature of Illinois, 1873.
Genera Physetocrinus, Nipterocrinus, Codonites.

LOWER CARBONIFEROUS SPECIES.

FOSSILS OF THE BURLINGTON GROUP.

	Page.
Echinodermata	323
Notes on the structure and habits of the Palaeozoic Crinoidea	323
Synbathocrinus, Phillips	324
Goniasteroidocrinus, Lyon & Casseday	324-325
Cyathocrinus, Miller	325-327
Convoluted support of the digestive sack in the *Actinocrinidœ*	327-329
Ambulacral canals passing under the vault in the *Actinocrinidœ*	329-339
Actinocrinites, Miller	339-341
Actinocrinites. Section (*a*)	342
A. penicillus, M. & W., 1869, pl. viii, fig. 2	342-343
A. sculptus, Hall, 1858, pl. iv, fig. 2	343
A. delicatus, M. & W., 1869, pl. viii, fig. 2	343-345
Actinocrinites. Section (*b*)	345
A. longus, M. & W., 1869, pl. viii, fig. 1	345-347
Strotocrinus, M. & W., 1866	347-349
S.? asperrimus, M. & W., 1869, pl. viii, fig. 3	349-351
Subgenus *Physetocrinus*, M. & W	351
S. (*Physetocrinus?*) *asper*, M. & W., 1869, pl. vii, fig. 1. Section (*b*), p. 253	351-353
S. ectypus, M. & W., 1869, pl. vii, fig. 5	353-355
S. liratus, Hall, sp., 1860, pl. vii, fig. 2	355-357
S. perumbrosus, Hall, sp., 1860, pl. viii, fig. 4	357-360
S. umbrosus, Hall, sp., 1858, pl. viii, fig. 5	360-362
S. (*Physetocrinus*) *dilatatus*, M. & W., 1869, pl. x, fig. 6	363-364
Batocrinus, Casseday	364-368
B. quasillus, M. & W., 1869, pl. v, fig. 2	369-370
B. (*Eretmocrinus*) *remibrachiatus*, Hall's sp., 1861, pl. x, fig. 5	370
B. cassedayanus, M. & W., 1868, pl. v, fig. 1	370-372
B. trochiscus, M. & W., 1868, pl. v, fig. 6	372-374
B. pyriformis, Shumard sp., 1855, pl. v. fig. 5	375-377
B. (*Eretmocrinus?*) *neglectus*, M. & W., 1868, pl. v, fig. 3	377-379
B. christyi, Shumard's sp., pl. v, fig. 4	379

	Page.
B. verneuilianus, Shum. sp., 1855, pl. iv, figs. 3 and 4	370
Dorycrinus, Ræmer	379–380
D. canaliculatus, M. & W., 1869, pl. vi, fig. 4	381–383
D. unicornis, O. & S., sp., 1850, pl. vi, fig. 2	383
D. ræmeri, M. & W., 1868, pl. 10, fig. 3	383–385
D. quinquelobus var. *intermedius*, M. & W., 1868, pl. x, fig. 4	385–386
Amphoracrinus, Austin	386–388
A. divergens, Hall, sp., 1860, pl. vi, fig. 6	388–389
A. ? spinobrachiatus, Hall, sp., 1860, pl. vi, fig. 5	389
Gilbertsocrinus, Phillips	389
Subgenus *Goniasteroidocrinus*, Lyon & Casseday, 1859	389
G. (Goniasteroidocrinus) tenuiradiatus, M. & W., 1869, pl. xi, fig. 1	389–390
G. (Goniasteroidocrinus) obovatus, M. & W., 1868, pl. iv, fig. 6	391–392
Megistocrinus, O. & S., 1850	393–396
M. parvirostris, M. & W., 1869, pl. vi, fig. 7	396–397
M. (Saccocrinus) whitei, Hall, 1861, pl. vi, fig. 1	397
Agaricocrinus, Troost	397
A. nodosus, M. & W., 1869, pl. x, fig. 7	397–399
Taxocrinus, Phillips, 1843	399
T. thiemei, Hall, sp., 1861, pl. iv, fig. 1	399
Cyathocrinites, Miller, 1821	400–401
C. sculptilis, Hall, pl. iv, fig. 5	401
C. fragilis, M. & W., 1868, pl. ii, fig. 14	401–403
C. tenuidactylus, M. & W., 1868, pl. ii, fig. 15	403–405
Poteriocrinites, Miller	405
P. ? perplexus, M. & W., 1869, pl. ii, fig. 12	405–406
Scaphiocrinus, Hall	407
S. delicatus, M. & W., 1869, pl. i, fig. 5	407–408
S. clio, M. & W., 1869, pl. i, fig. 10	408–409
S. notabilis, M. & W., 1869, pl. i, fig. 9	410–412
S. rudis, M. & W., 1869, pl. i, fig. 1	412–413
S. penicillus, M. & W., 1869, pl. i, fig. 7	414–415
S. macrodactylus, M. & W., 1869, pl. ii, fig. 9	415–416
S. juvenis, M. & W., 1869, pl. ii, fig. 8	417–418
S. striatus, M. & W., 1869, pl. ii, fig. 11	418–419
S. tethys, M. & W., 1869, pl. ii, fig. 13	419–421
S. scalaris, M. & W., 1869, pl. ii, fig. 10	421–423
S. nanus, M. & W., 1869, pl. i, fig. 8	423–424
S. fiscellus, M. & W., 1869, pl. i, fig. 3	424–426
Subgenus *Zeacrinus*	426
S. (Zeacrinus) scobina, M. & W., 1869, pl. i, fig. 2	426–428
S. (Zeacrinus) serratus, M. & W., 1869, pl. i, fig. 6	428–430
S. (Zeacrinus) asper, M. & W., 1869, pl. i, fig. 7	430–432
S. (Zeacrinus) lyra, M. & W., 1869, pl. i, fig. 11	432–433
Nipterocrinus, Wachsmuth	434–435
N. wachsmuthi, M. & W., 1868, pl. ii, fig. 4	435–436
N. arboreus, Worthen MSS., pl. iv, fig. 8	436–437
Synbathocrinus, Phillips, 1836	437
S. wachsmuthi, M. & W., 1869, pl. ii, fig. 5	437–439
S. brevis, M. & W., 1869, pl. ii, fig. 6	439
Dichocrinus, Munster, 1839	440
D. lineatus, M. & W., 1869, pl. iii, fig. 1	440–441
D. pisum, M. & W., 1869, pl. iii, fig. 2	441–442
Calceocrinus, Hall, 1852	442–443
C. ? wachsmuthi, M. & W., 1869, pl. ii, fig. 1	444–445
Erisocrinus, M. & W., 1865	445–447
E. antiquus, M. & W., 1869, pl. ii, fig. 3	447–448
E. whitei, M. & W., 1869, pl. ii, fig. 2	448–449
Platycrinites, Miller	450
P. tenuibrachiatus, M. & W., 1869, pl. iii, fig. 4	450–452
P. planus, O. & S., 1852, pl. iii, fig. 5	452
P. subspinosus, Hall, 1858, pl. ii, fig. 2	452
P. burlingtonensis, O. & S., 1850, pl. iii, fig. 6	452–454
P. halli, Shumard ? 1866, pl. iii, fig. 3	454–456
P. æqualis, Hall, 1861, pl. iii, fig. 8	456–458

	Page
P. incomptus, White, 1863, pl. iii, fig. 7	459–461
Pentremites, Say	461
P. burlingtonensis, M. & W., 1869, pl. viii, fig. 7	461–462
Codonites, M. & W., 1869	463–464
C. stelliformis, O. & S., sp., 1850, pl. ix, fig. 5, 5a, b	464–466
O. gracilis, n. s., M. & W., 1873, pl. viii, fig. 6	466–468
Granatocrinus, Troost	468
G. melonoides, M. & W., 1869, pl. ix, fig. 1	468–470
G. pisum, M. & W., 1869, pl. ix, fig. 4	470–471
G. neglectus, M. & W., 1869, pl. ix, fig. 3	471–473
G. norwoodi, O. & S., sp., 1850, pl. ix, fig. 2	473
Palæchinus, McCoy	473
P. gracilis, M. & W., 1869, pl. x, fig. 2	473–474
Onychaster, M. & W	474–475
O. barrisi, Hall, sp., 1861, pl. x, fig. 1	476
Oligoporus, Meek & Worthen	476
O. nobilis, M. & W., 1868, pl. xi, fig. 3	476–478
Eocidaris	478
E. ? squamosus, M. & W., 1869, pl. ix, fig. 15	478–482

FOSSILS OF THE KEOKUK GROUP.

Barycrinus, Wachs	483
B. magnificus, M. & W., 1868, pl. xii, fig. 2	483–485
B. geometricus, M. & W., pl. xii, fig. 3	485
B. hoveyi, var. *hurculeus*, M. & W., 1868, pl. xiii, fig. 2	485–486
B. hoveyi, Hall, sp., 1861, pl. xiii, fig. 1	486
B. mammatus, Worthen MSS., pl. xv, fig. 4	486
B. pentagonus, Worthen MSS., pl. xv, fig. 3	487
B. subtumidus, M. & W., pl. xiii, fig. 3	487–488
Cyathocrinites? poterium, M. & W., 1870, pl. xii, fig. 4	489–490
Poteriocrinites	490
P. (Zeacrinus?) concinnus, M. & W., 1870, pl. xiv, fig. 3	490–492
Subgenus *Scaphiocrinus*	492
P. (Scaphiocrinus) depressus, M. & W., 1870, pl. xiv, fig. 8	492–493
P. (Scaphiocrinus) unicus, Hall, 1861, pl. xv, fig. 5	493
P. (Scaphiocrinus) æqualis, Hall, 1861, pl. xv, fig. 6	494
P. (Scaphiocrinus) coreyi, M. & W., 1869, pl. xv, fig. 1	494–495
P. (Scaphiocrinus) incadamsi, Worthen MSS., pl. xv, fig. 2	495–496
Forbesiocrinus wortheni, Hall, 1858, pl. xiv, fig. 2, and pl. xv, fig. 7	496–498
Onychocrinus exculptus, L. & C., 1859, pl. xiv, fig. 4	498
Agaricocrinus, Troost	499
A. whitfieldi, Hall, 1868, pl. xii, figs. 1 a, b, and pl. xv, fig. 8	499–500
Dichocrinus, Munster	500
D. expansus, M. & W., 1868, pl. xiv, fig. 1	500–502
D. ficus, C. & L., 1860, pl. xiv, fig. 5	502
Calceocrinus, Hall	502
C. bradleyi, M. & W., 1869, pl. xiv, fig. 9	502–504
Catillocrinus, Troost	504
C. bradleyi; M. & W., 1868, pl. xiv, fig. 10	504–505
Platycrinites	506
P. hemisphericus, M. & W., 1865, pl. xvi, figs. 6 a–c	506
Pentremites	506
P. wortheni, Hall? 1858, pl. xiv, fig. 11	506
P. (Troostocrinus?) woodmani, M. & W., 1868, pl. xvi, fig. 4	506–508
Granatocrinus	508
G. granulosus, M. & W., 1865, pl. xv, fig. 10	508–509
Protaster, Forbes	509
P. ? gregarius, M. & W., 1869, pl. xvi, fig. 5	509–510
Onychaster, M. & W	510
O. flexilis, M. & W., pl. xvi, fig. 3	510
Pholidocidaris, M. & W., 1869	510–511
P. irregularis, M. & W., 1869, pl. xv, fig. 9	512–513
Agelacrinites, Vanuxem	513
A. (Lepidodiscus) squamosus, M. & W., 1868, pl. xvi, fig. 1	513–515

Mollusca.
Gasteropoda.

Page.

Platyceras, Conrad 516
P. uncum, M. & W., 1866, pl. xvii, fig. 1 516–517
P. infundibulum, M. & W., 1866, pl. xvii, fig. 3 517–518
P. equilatera, Hall, 1860, pl. xvii, fig. 2 518–519
P. fissurella, Hall, 1859, pl. xvii, fig. 4 519–520

Pteropoda.
Conularia.
  Conularia subcarbonaria, M. & W., 1865, pl. xix, fig. 4 520–522

Cephalopoda.
  Nautilus, Linn 522
  N. (Discites) disciformis, M. & W., 1865, pl. xviii, fig. 1 522–523
  N. (Temnocheilus) niotensis, M. & W., 1865, pl. xix, fig. 3 523–524
  N. (Solenocheilus) leidyi, M. & W., 1865, pl. xviii, fig. 2 524–525

Articulata.
  Phillipsia, Portlock 525
  P. (Griffithides) portlockii, M. & W., 1865, pl. xix, fig. 6 525–528
  P. (Griffithides) bufo, M. & W., 1870, pl. xv, fig. 10 528–529

FOSSILS OF THE SAINT LOUIS GROUP.

Echinodermata.
  Barycrinus 530
  B. spectabilis, M. & W., 1869, pl. xx, fig. 8 530–533
  Poteriocrinites 533
  P. hardinensis, Worthen MSS., pl. xx, fig. 10 533
  Subgenus *Scaphiocrinus* 534
  P. (Scaphiocrinus) huntsvillæ, Worthen MSS., pl. xx, fig. 1 534
  Subgenus *Zeacrinus* 534
  P. (Zeacrinus) arboreus, Worthen MSS., pl. xx, fig. 5 534–535
  P. (Zeacrinus) cariniferous, Worthen MSS., pl. xx, fig. 4 535–536
  P. (Zeacrinus) compactilis, Worthen MSS., pl. xxi, fig. 5 536–537
  Dichocrinus, Munster 537
  D. cornigerus, Shumard ? 1860, pl. xx, fig. 6 537
  Granatocrinus 537
  G. glaber, M. & W., 1869, pl. xx, fig. 11 537–539

Mollusca.
Lamellibranchiata.
  Lithophaga ? pertenuis, M. & W., 1865, pl. xxii, fig. 1 539–540
  Myalina 540
  M. St. ludovici, Worthen MSS., pl. xxii, fig. 3 540
  Chænomya, M. & W 540
  C. ? rhomboidea, M. & W., 1865, pl. xxii, fig. 4 540–541

Pteropoda.
  Conularia 541
  C. missouriensis, Swallow ? 1860, pl. xxii, fig. 5 541–542

Cephalopoda.
  Nautilus 543
  N. (Temnocheilus) coxanus, M. & W., 1869, pl. xxiii, fig. 1 543–544
  N. (Solenocheilus) collectus, M. &. W., 1870, pl. xxiii, figs. 3 and 4 544–545

FOSSILS OF THE CHESTER GROUP.

Echinodermata.
  Poteriocrinites 546
  P. bisselli, Worthen MSS., pl. xxi, fig. 4 546–547
  Subgenus *Zeacrinus* 547
  P. (Zeacrinus ?) armiger, M. & W., 1870, pl. xxi, fig. 3 547–548
  P. (Zeacrinus) subtumidus, Worthen MSS., pl. xxi, fig. 1 548–549
  P. (Zeacrinus) formosus, Worthen MSS., pl. xxi, fig. 2 549
  Subgenus *Scaphiocrinus* 550
  P. (Scaphiocrinus) bauensis, M. & W., 1865, pl. xx, fig. 2 550–551
  P. (Scaphiocrinus) randolphensis, Worthen MSS., pl. xxi, fig. 14 .. 551–552
  Onychocrinus 552
  O. whitfieldi, Hall sp., 1858, pl. xx, fig. 3 552–554
  Eupachycrinus 554

	Page
E. boydii, M. & W., 1870, pl. xxi, fig. 6	554-555
Platycrinites	555
P. parvulus, M. & W., 1865, pl. xx, fig. 7	555-556
Agassizocrinus	556
A. pentagonus, Worthen MSS., pl. xxi, fig. 10	556-557
A. conicus, O. & S., 1851, pl. xxi, fig. 8	557
A. gibbosus, Worthen MSS., pl. xxi, fig. 12	557-558
A. gibbosus, Hall, 1858, pl. xxi, fig. 11	558
A. chesterensis, Worthen MSS., pl. xxi, fig. 9	558
Pterotocrinus, L. & C	559
P. depressus, L. & C.? 1860, pl. xxi, fig. 13	559
Graphiocrinus	559
G. dactylus, Hall, 1860, pl. xx, fig. 9	559

FOSSILS OF THE COAL-MEASURES.

Foraminifera.
Fusulina, Fischer	560
F. gracilis, Meek? 1864, pl. xxiv, fig. 7	560
F. ventricosa, M. & H., 1864, pl. xxiv, fig. 8	560

Radiata.
Lophophyllum, Edwards & Haime	560
L. proliferum, McC. sp., 1860, pl. xxiv, fig. 1	560
Erisocrinus	561
E. typus, M. & W., 1865, pl. xxiv, fig. 6	561
Poteriocrinites	561
P. macoupinensis, Worthen MSS., pl. xxiv, fig. 3	561
Subgenus *Scaphiocrinus*	561
P. (Scaphiocrinus?) hemisphericus, Shum. sp., 1858, pl. xxiv, fig. 5	561
P. (Scaphiocrinus) carbonarius, M. & W., 1861, pl. xxiv, fig. 2	562
Subgenus *Zeacrinus*	563
P. (Zeacrinus?) mucrospinus, McC., 1859, pl. xxiv, fig. 12	563
P. (Zeacrinus [Hydreionocrinus?]) acanthophorus, M. & W., 1870, pl. xxiv, fig. 11	563-565
Eupachycrinus	565
E. fayettensis, Worthen MSS., pl. xxiv, fig. 10	565-566
E. tuberculatus, M. & W., 1866, pl. xxiv, fig. 9	566
Agassizocrinus carbonarius, Worthen MSS., pl. xxiv, fig. 4	566

Mollusca.
Brachiopoda.
Chonetes, Fischer	566
C. millepunctata, M. & W., 1870, pl. xxv, fig. 3	566-569
Productus, Sowerby	569
P. nebrascensis, Owen, 1852, pl. xxv, fig. 8	569
P. longispinus, Sowerby? 1814, pl. xxv, fig. 10	569
P. punctatus, Martin, 1809, pl. xxv, fig. 13	569
P. lasallensis, Worthen MSS., pl. xxv, fig. 9	569-570
Chonetes smithii, N. & P., 1855, pl. xxv, fig. 11	570
Hemipronites, Pander	570
H. crassus, M. & W., 1858, pl. xxv, fig. 12	570
Athyris, McCoy	570
A. subtilita, Hall's sp., pl. xxv, fig. 14	571
Syntrielasma, M. & W	571
S. hemiplicata, Hall's sp., 1852, pl. xxvi, fig. 20	571
Meekella, White & St. John	571
M. striato-costata, Cox sp., 1857, pl. xxvi, fig. 21	571
Rhynchonella, Fischer	571
R. osagensis, Swallow, 1858, pl. xxvi, fig. 22	571
Orthis, Dalman	571
O. carbonaria, Swallow	571
Terebratula, Llhwyd	572
T. bovidens, Morton, 1836, pl. xxv, fig. 15	572
Discina, Lamarck, 1819	572
D. nitida, Phillips sp.? pl. xxv, fig. 1	572
Lingula, Bruguière, 1789	572
L. mytiloides, Sowerby? 1813, pl. xxv, fig. 2	572
Spirifer fultonensis, Worthen, MSS. pl. xxv, fig. 5	572-573
Spirifer cameratus, Morton, pl. xxv, fig. 7	573

Lamellibranchiata.

	Page.
Monotis	573
M. ? gregaria, M. & W., 1870, pl. xxvi, fig. 5	573-574
Macrodon, Lycett	575
M. ? delicatus, M. & W., 1870, pl. xxvi, fig. 3	575
M. ? tenuistriatus, M. & W., 1867, pl. xxvi, fig. 4	576
Avicula	576
A. morganensis, M. & W., 1866, pl. xxvi, fig. 14	576-578
A. longa, Geinitz sp., 1866, pl. xxvi, fig. 1	578
Placunopsis	578
P. carbonaria, M. & W., 1866, pl. xxvii, fig. 2	578-579
Schizodus, King	579
S. amplus, M. & W., 1870, pl. xxvii, fig. 6	579-580
S. (Prisconia) perelegans, M. & W., 1870, pl. xxvi, fig. 19	581
S. curtus, M. & W., 1866, pl. xxvi, fig. 16	582
Myalina, de Koninck	582
M. perattenuata, M. & W., 1858, pl. xxvi, fig. 11	582
Edmondia	583
E. ? peroblonga, M. & W., 1866, pl. xxvii, fig. 4	583-584
Clinopistha, M. & W.	584
C. radiata, var. *levis*, M. & W., 1870, pl. xxvii, fig. 7	584-585
Allorisma, King	585
A. costata, M. & W., 1869, pl. xxvi, fig. 15	585-586
A. geinitzii, Meek, 1867, pl. xxvi, fig. 23	586
Chænocardia, M. & W	586
C. ovata, M. & W., 1869, pl. xxvii, fig. 5	586-587
Chænomya, M. & W	588
C. minnehaha, Swallow sp., 1858, pl. xxvii, fig. 3	588
Cardiomorpha	588
C. missouriensis, Shum., 1860, pl. xxvii, fig. 8	588
Entolium, Meek	588
E. aviculatum, Swallow sp., 1858, pl. xxvi, fig. 12	588
Lima, Bruguière	588
L. retifera, Shum., 1858, pl. xxvi, fig. 2	588
Aviculopecten neglectus, Geinitz sp., 1866, pl. xxvi, fig. 7	589
Pleurophorus, King	589
P. oblongus, Meek ? 1872, pl. xxvi, fig. 6	589
Nucula, Lamarck	589
N. parva, McC., 1860, pl. xxvi, fig. 8	589
N. beyrichi, v. Schauroth, 1854, pl. xxvi, fig. 9	589

Gasteropoda.

Dentalium, Linnæus	589
D. ? annulostriatum, M. & W., 1870, pl. xxix, fig. 7	589-590
D. meekianum, Geinitz ? 1866, pl. xxix, fig. 8	590
Orthonema, M. & W., 1861	590
O. conica, M. & W., 1866, pl. xxix, fig. 5	590-591
Naticopsis, McCoy	592
N. ventrica, N. & P., 1855, sp., pl. xxviii, fig. 13	592-593
Macrocheilus, Phillips	593
M. altonensis, Worthen MSS., pl. xxviii, fig. 8	593-594
M. newberryi, Stevens sp., pl. xxviii, fig. 14	594
Acteonina, d'Orbigny	594
A. minuta, Stevens sp., pl. xxix, fig. 2	594
Platyceras, Conrad	594
P. spinigerum, Worthen MSS., pl. xxviii, fig. 4	594-595
Naticopsis subcatus, Worthen MSS., pl. xxviii, fig. 9	595
Naticopsis wherleri, Swallow sp., 1860, pl. xxviii, fig. 3	595
Naticopsis altonensis, McC., 1865, pl. xxviii, fig. 11	595
Streptacis	596
S. whitfieldi, Meek, 1871, pl. xxix, fig. 1	596
Loxonema	596
L. semicostata, Meek. 1871, pl. xxix, fig. 2	596
Aclis, Loven	596
A. robusta, Stevens, 1858, pl. xxix, fig. 6	596
Polyphemopsis, Portlock	596
P. chrysallis, M. & W., 1865, pl. xxviii, fig. 7	596-597

	Page.
Anomphalus, M. & W	597
A. rotulus, M. & W., 1866, pl. xxix, fig. 10	597
Microdoma, M. & W	598
M. conica, M. & W., 1866, pl. xxviii, fig. 2	598
Murchisonia inornata, M. & W., 1866, pl. xxviii, fig. 6	599-600
Pleurotomaria coxana, M. & W., 1866, pl. xxviii, fig. 15	600
Pleurotomaria spironema, M. & W., 1866, pl. xxviii, fig. 5	601-602
Pleurotomaria valvatiformis, M. & W., 1866, pl. xxix, fig. 9	602-603
Pleurotomaria cunoides, M. & W., 1866, pl. xxviii, fig. 1	603-604
Straparollus, Montfort	604
S. (Euomphalus) pernodosus, M. & W., 1870, pl. xxix, fig. 14	604-605
S. (Euomphalus) subquadratus, M. & W., 1870, pl. xxix, figs. 12 and 13	605-607
S. (Euomphalus) subrugosus, M. & W., pl. xxix, fig. 11	607
Chiton, Linnæus	608
C. carbonarius, Stevens, 1859, pl. xxix, fig. 15	608
Cephalopoda.	
Nautilus (Temnocheilus) latus, M. & W., 1870, pl. xxx, fig. 2	608-609
N. (Temnocheilus) winslowi, M. & W., 1870, pl. xxxii, fig. 2	609-610
N. lasallensis, M. & W., 1866, pl. xxxi, fig. 1	610-611
Goniatites compactus, M. & W., 1865, pl. xxxi, fig. 2	611-612
Orthoceras	612
O. rushensis, McC.? 1860, pl. xxx, fig. 4	612
Articulata.	
Phillipsia, Portlock	612
P. (Griffithides) scitula, M. & W., 1865, pl. xxxii, fig. 3	612-615
P. (Griffithides ?) sangamonensis, M. & W., 1865, pl. xxxii, fig. 4	615-618
Dithyrocaris, Scouler	618
D. carbonarius, M. & W., 1869, pl. xxxii, fig. 1	618-619

92.

MEEK, F. B. Notes on some of the Fossils figured in the recently issued Fifth volume of the Illinois State Geological Report. <Am. Jour. Sci., vol. vii, 3d ser., pp. 189-193; continued on pp. 369-376 and 484-490 and 580-584. 1874. New Haven, 1874.

In this series of articles Mr. Meek revises and extends the descriptions of a large number of the species embraced in the fifth volume of Illinois Geological Report, and also presents some very important philosophical discussions of the relations of the species and of the higher groups.

	Page.
Actinocrinites sculptilis, Hall	190-191
Taxocrinus thiemei, Hall	191-192
Batocrinus pyriformis, Shumard	192
Actinocrinites delicatus, M. & W	192-193
Actinocrinites Cyathocrinites, Codonites, etc	369-374
Codonites stelliformis	374-375
Pholidocidaris irregularis, M. & W	375
Pentremites (Troostocrinus ?) woodmani, M. & W	375-376
Agassizocrinus, Troost	484
Fusulina gracilis	484
Fusulina ventricosa	484
Zeacrinus (Hydreionocrinus ?) acanthophorus, M. & W	485-486
Archæocidaris ?, sp. undt	486
Septopora cestriensis, Prout	486-488
Synocladia virgulacea var. *biserialis*, Swallow	486-488
Aviculopecten neglectus, Geinitz	488-489
Aviculopecten carboniferus, Stevens sp	489
Nuculana, sp. undt., Meek, 1874	489-490
Edmondia ovata, ...	580
Schizodus rossicus, de Verneuil	580-580
Solenomya sp. undt., Meek, 1874	582-583
Placunopsis carbonaria, M. & W	583
Anomphalus rotulus, M. & W	583
Euomphalus rugosus, Hall	583-584
Pleurotomaria gurleyi	584

93.

MEEK, F. B. The new genus Euchondria. <Am. Journ. Sci., vol. vii, 3d ser., p. 445. 1874. New Haven, 1874.

Mr. Meek, in a brief note, proposes the generic name Euchondria, of which the *Pecten neglectus* of Geinitz is the type.

94.

MEEK, F. B. [Descriptions of] Pleurotomaria taggarti. <Hayden's Ann. Rep. U. S. Geol. and Geog. Surv. of the Terr. for 1873, p. 231, foot-note. Washington, 1874. Carboniferous.

Page.
Pleurotomaria taggarti, n. s., Meek, 1874... 231

95.

MEEK, F. B. Notes on some fossils from near the eastern base of the Rocky Mountains, west of Greeley and Evans, Colorado, and others from about two hundred miles farther eastward, with descriptions of a few new species. <Bulletin U. S. Geol. and Geog. Surv. of the Terr., 2d ser., No. 1, pp. 39–47. 1875. Washington, 1875.

These fossils are from the Fox Hills and Laramie (Lignitic) Group.

Page.
Anomia micronema, n. s., Meek, 1875 43
Corbicula ? (Leptesthes) planumbona, n. s., Meek, 1875 43–45
Cyrena ? holmesi, n. s., Meek, 1875 45–46
Sphæriola ? obliqua, n. s., Meek, 1875 46
Rhynchonella endlichi, n. s., Meek, 1875 46–47

96.

MEEK, F. B. Description of Unios supposed to be of Triassic age. <Ann. Rep. Geogr. Expls. and Survs. West of the 100th Merid., by G. M. Wheeler, Appendix L L of the Ann. Rep. Chief of Engineers for 1875, pp. 83, 84. Washington, 1875.

Page.
Unio cristonensis, n. s., Meek, 1875 83–84
Unio gallinensis, n. s., Meek, 1875 84
Unio terra-rubræ, n. s., Meek, 1875 84

97.

MEEK, F. B. Description of Olenellus gilberti and O. howelli. <Rep. Geogr. and Geol Expls. and Survs. West of the 100th Merid., 4to, vol. iii, Geology, pp. 182, 183. 1875. Washington, 1875.

These two species are fully described and illustrated in White's Report on Invertebrate Paleontology, part i, Vol. iv, Wheeler's Expl. and Surv. West of the 100th Meridan.

Page.
Oleanus (Olenellus) gilberti, n. s., Meek, 1875 182–183
Oleanus (Olenellus) howelli, n. s., Meek, 1875 183

98.

MEEK, F. B. A report on some of the Invertebrate fossils of the Waverly group and Coal-Measures of Ohio. <Rep. Geol. Surv. of Ohio, vol. ii, part ii, Paleontology, pp. 269–347, plates x–xx. 1875. Columbus, 1875.

WAVERLY GROUP SPECIES.
Mollusca.
Polyzoa.

Page.
Fenestella, Lonsdale, 1837.. 273
F. delicata, Meek, 1871, pl. x, figs. 2 *a, d* 273–274
F. multiporata ? var. *indicensis*, Meek, pl. x, figs. 1 *a, c*........................ 274–275

Brachiopoda.

	Page.
Lingula, Bruguière, 1792	275
L. (*Lingulella?*) *membranacea*, Winchell, 1863, pl. xiv, fig. 4	275
L. melie, Hall? 1864, pl. xiv, fig. 3	276
L. scotica, Davidson? 1868, pl. xiv, fig. 9	276–277
Discina, Lamarck, 1819	277
Subgenus *Orbiculoidea*, d'Orbigny, 1847	277
D. (*Orbiculoidea*) *newberryi*, Hall, 1864, pl. xiv, figs. 1 *a–d*	277–278
D. (*Orbiculoidea?*) *pleurites*, Meek, n. s., 1875, pl. xiv, figs. 2 *a, b*	278–279
Strophomena, Rafinesque, 1827	279
Subgenus *Hemipronites*, Pander, 1830	279
S. (*Hemipronites*) *crenistria*, Phillips? sp., 1836, pl. x, figs. 5 *a–d*	279–283
Productus, Sowerby, 1814	
Productus sp., Meek, 1875, pl. x, figs. 4 *a–d*	282–283
Productus sp., Meek, 1875, pl. x, fig. 3	283
Athyris, McCoy, 1844	283
A. lamellosa, Leveille? sp., 1835, pl. xiv, figs. 6 *a, b*	283–285
Spirifer, Sowerby, 1815	285
S. carteri, Hall, 1857, pl. xiv, figs. 7 *a–c*, (*d?*)	285–288
Subgenus *Trigonotreta*, King, 1825	289
S. (*Trigonotreta*) *striatiformis*, n. s., Meek, 1875, pl. xiv, figs. 8 *a–e*	289–290
S. (*Trigonotreta*) *biplicatus*, Hall?! 1858, pl. xiv, fig. 5	290–292

Lamellibranchiata.

Entolium, Meek	292
E. shumardianum Winchell! sp., 1865, pl. xv, figs. 4 *a, b*	292–294
Aviculopecten, McCoy, 1851	295
A. crenistriatus, Meek, 1871, pl. xv, figs. 7 *a, b*	295–296
A. winchelli n. s., Meek, 1875, pl. xv, figs. 5 *a* and 5 *b?*	296–298
Palæoneilo, Hall, 1870?	298
P. bedfordensis n. s., Meek, 1875, pl. xv, figs. 3 *a–c*	298
Schizodus, King, 1844	299
S. medinensis, Meek, 1871, pl. xv, figs. 1 *a–c*	299–300
Grammysia, De Verneuil, 1847	300
G.? hannibalensis, Shumard sp., 1855, pl. xvi, figs. 5 *a–c*	300–301
G.? rhomboides, Meek, 1871, pl. xvi, figs. 7 *a, b*	302–303
G. ventricosa, Meek, 1871, pl. xvi, figs. 6 *a, b* (and pl. xiii, figs. 5 *a, b, var.*)	303
Edmondia, De Koninck, 1844	304
E. tapesiformis n. s., Meek, 1875, pl. xiii, fig. 6	304
Cardiomorpha, De Koninck, 1844	304
C. subglobosa n. s., Meek, 1875, pl. xv, figs. 6 *a, b*	304–305
Prothyris, Meek, 1869	305
P. meeki, Winchell, MS., 1872, pl. xv, fig. 2	305–306
Sanguinolites, McCoy, 1844	306
S.? obliquus, Meek, 1871, pl. xvi, figs. 2 *a, b*	306–307
S. aeolus, Hall, Whitfield, 1870?, pl. xvi, figs. 1 *a–c*	307–308
Promacrus, Meek, 1871	308
P. andrewsi, Meek, 1871, pl. xvii, figs. 1 *a, b*	308–309
Allorisma, King, 1844	309
A. (*Cercomyopsis*) *pleuropistha*, Meek, 1871, pl. xiii, figs. 4 *a–c*	309–311
A. winchelli, Meek, 1871, pl. xvi, figs. 3 *a–c*	311–312
A. ventricosa, Meek, 1871, pl. xvi, figs. 4 *a, b*	312–313

Gasteropoda.

Platyceras, Conrad, 1840	313
P. (*Orthonychia?*) *lodiense*, Meek, 1871, pl. xiii, figs. 1 *a, b*	313–314
Pleurotomaria, Defrance, 1826	314
P. textiligera, Meek, 1871, pl. xiii, figs. 7 *a, b*	314–315

Pteropoda.

Conularia, Miller, 1818	316
C. micronema, Meek, 1871, pl. xviii, figs. 1 *a–d*	316
C. newberryi, Hall, pl. xviii, figs. 2 *a, b*	316–317

Crustacea.

Entomostraca.

Ceratiocaris, McCoy	317
? Subgenus *Colpocaris*, Meek, 1872	317
C. (*Colpocaris*) *bradleyi*, Meek, 1872, pl. xviii, figs. 6 *a–e*	318–319

	Page.
C. (Colpocaris) clytroides, Meek, 1872, pl. xviii, figs. 5 a–e	319–320
Subgenus *Solenocaris*, Meek, 1872	320–321
C. (Solenocaris) strigata, Meek 1872, pl. xviii, figs. 4 a–e	321

Tetradecopoda.

? *Archæocaris*, Meek, 1872	321
A. vermiformis, Meek, 1872, pl. xviii, fig. 7	321–322

Trilobita.

Phillipsia, Portlock, 1843	323
P. (Griffithides ?) lodiensis, n. s., Meek, 1875, pl. xviii, fig. 3	323–325

COAL MEASURE SPECIES.

Mollusca.

Polyzoa.

Synocladia, King, 1849	326
S. biserialis, Swallow, 1858, pl. xx, figs. 5 a, b	326–327
Ptilodictya, Lonsdale, 1839	327
P. (Stictopora), sereata, n. s., Meek, 1875, pl. xx, fig. 4	327–328
P. (Stictopora) carbonaria, Meek, 1871, pl. xx, figs. 3 a, b	328

Brachiopoda.

Spirifer, Sowerby, 1815	329
S. (Trigonotreta) opinus, Hall / 1858, pl. xix, figs. 14 a–d (e ?)	329–330

Mollusca (Proper).

Lamellibranchiata.

Aviculopecten, McCoy, 1851	330
A. (Streblopteria ?) hertzeri, n. s., Meek, 1875, pl. xix, figs. 13 a–c	330–331
Placunopsis, Morris & Lycett, 1853	331
P. recticardinalis, n. s., Meek, 1875, pl. xix, fig. 12	331–333
Posidonomya, Brown, 1837	333
P. fracta, n. s., Meek, 1875, pl. xix, figs. 7 a, b	333–334
Macrodon, Lycett, 1845	334
M. obsoletus, Meek, 1871, pl. xix, fig. 9	334–335
Yoldia, Möller, 1842	335
Y. stevensoni, Meek, 1871, pl. xix, figs. 4 a, b	335
Y. (Palæoneilo ?) carbonaria, Meek, 1871, pl. xix, fig. 5	336
Schizodus, King, 1844	336
S. cuneatus, n. s., Meek, 1875, pl. xx, fig. 7	336–337
Aviculopinna americana, Meek, 1867, pl. xx, fig. 2	337–338
Pleurophorus, King, 1844	338
P. tropidophorus, n. s., Meek, 1875, pl. xix, figs. 10 a, b	338–339
Solenomya, Lamarck, 1818	339
S. ?! anodontoides, n. s., Meek, 1875, pl. xix, fig. 11	339–340
Astartella, Hall, 1858	340
A. newberryi, n. s., Meek, 1875, pl. xix, fig. 3	340–341
A. varica, McChesney, 1860, pl. xix, fig. 2	341
Astartella, sp., Meek, pl. xix, fig. 1 a, b	341–342
Cypricardina, Hall, 1860	342
C. ? carbonaria, Meek, 1871, pl. xix, fig. 8 a, b	342–343
Allorisma, King, 1844	343
A. costata, M. & W., 1860, pl. xix, figs. 6 a, b	344–345

Gasteropoda.

Platyceras, Conrad, 1840	345
P. tortum, Meek, 1871, pl. xx, figs. 1 a–e	345
Macrocheilus, Phillips, 1841	346
M. klipparti, Meek, 1872, pl. xx, figs. 6 a–c	346–347

99.

MEEK, F. B., *and* A. H. WORTHEN. Paleontology of Illinois. Descriptions of Invertebrates. <Geological Survey of Illinois, vol. vi, section ii, pp. 491–532, plates 23–32. 1875. Springfield, 1875.

Published by authority of the legislature, 1875.

Genus Carbonarca. A portion of these descriptions are by Mr. Worthen alone, and are not mentioned in the following list:

LOWER SILURIAN SPECIES.

Echinodermata.   Page.
  Homocrinus, Hall 492
  H. angustatus, M. & W., 1870, pl. xxiii, fig. 8 492–493
  Heterocrinus crassus, M. & W., 1865, pl. xxiii, fig. 1 493
Mollusca.
  Lamellibranchiata.
    Modiolopsis subnasuta, M. & W., 1870, pl. xxiii, figs. 9 *a*, *b* 494–495
  Gasteropoda.
    Subulites, Conrad 495
    S. inflatus, M. & W., 1869, pl. xxiii, fig. 5 495–496
  Articulata.
    Asaphus, Brongniart 497
    A. (Isotelus) vigilans, M. & W., 1870, pl. xxiii, fig. 6 497–498

UPPER SILURIAN SPECIES.

Spongiæ.
  Astylospongia præmorsa, Goldf. ? sp., 1826, pl. xxv, figs. 2, 2 *a* 499
Foraminifera ?
  Receptaculites, Defrance 500
  R. formosus, M. & W., 1870, pl. xxiv, fig. 1 500
Brachiopoda.
  Stricklandinia, Billings 502
  S. deformis, M. & W., 1870, pl. xxiv, figs. 5 *a*, *b* 502–503
Cephalopoda.
  Orthoceras, Auct 503
  O. crebristriatum, M. & W., 1865, pl. xxvi, fig. 2 503–504
  O. medullare, Hall ? 1860, pl. xxvi, fig. 1 504
  O. angulatum, Wahl, 1827, pl. xxiv, fig. 8 504
  - O. jolietensis, M. & W., 1865, pl. xxvi, fig. 5 505
  Cyrtoceras, Goldfuss 506
  C. dardanus, Hall ? 1861, pl. xxv, fig. 6 506
  Lituites, Breyn 507
  L. graftonensis, M. & W., 1869, pl. xxv, tig. 1 507–508
Crustacea.
  Lichas, Dalman 508
  L. boltoni, Bigsby, sp., 1825, pl. xxv, fig. 5 508
  Illaenus, Dalman 508
  I. (Bumastus) graftonensis, M. & W., 1869, pl. xxv, fig. 4 508–510
  Sphærexochus, Beyrich 510
  S. romingeri, Hall, 1862, pl. xxiv, fig. 4 510
Mollusca.
  Cephalopoda.
    Phragmoceras 511
    P. walshii, M. & W., 1866, pl. xxviii, figs. 2 *a*, *b* 511–512
    Orthoceras 512
    O. winchelli, M. & W., 1866, pl. xxviii, fig. 1 512–513

LOWER CARBONIFEROUS SPECIES.

Echinodermata.
  Synbathocrinus, Phillips 514
  S. robustus, Shumard, 1866, pl. xxix, fig. 4 514
  Dichocrinus, Munster 515
  D. ficus, Casseday & Lyon, 1860, pl. xxix, fig. 7 515
  Poteriocrinus 516
  Subgenus *Scaphiocrinus*, Hall 519
  P. (Scaphiocrinus) unicus, Hall, 1861, pl. xxix, fig. 1 519
  Pentremites, Say 521
  P. (Tricoelocrinus) obliquatus Roemer, sp., 1852, pl. xxxi, fig. 4 521
  Spirifer, Sowerby 521
  S. fastigatus, M. & W., 1870, pl. xxx, fig. 3 521–523
  S. neglectus, Hall, 1858, pl. xxx, figs. 2 *a*, and 1 *c* 523
  S. suborbicularis, Hall, 1858, pl. xxx, fig. 1 523–524

COAL MEASURE SPECIES.

	Page.
Axophyllum, Edwards & Haime	525
A. rudis, White & St. John, pl. xxxii, figs. 6 *a–c*	525
Conocardium, Brown	529
C. obliquum, M. & W., 1865, pl. xxxiii, fig. 4	529
Pleurophorus? King	529
P. ? angulatus, M. & W., 1865, pl. xxxiii, fig. 5	529–530
Carbonarca, Meek & Worthen, 1870	530
C. gibbosa, M. & W., 1876, pl. xxxiii, fig. 6	531
Nautilus, Linnaeus	531
N. (Cryptoceras) capax, M. & W., 1865, pl. xxxiii, fig. 1	532

100.

MEEK, F. B. Notice of a very large Goniatite from Eastern Kansas (Carboniferous). <Bulletin U. S. Geol. and Geog. Surv. of the Terr. No. 6, 2d ser., vol. i, p. 445. Washington, 1875.

> The author regards it as at most only a variety of *G. globulosus*, Meek & Worthen, although attaining so great size.

	Page.
Goniatites globulosus var. *excelsus*, Meek, 1875	445

101.

MEEK, F. B. Descriptions and illustrations of fossils from Vancouver's and Sucia Islands, and other Northwestern localities. <Bulletin U. S. Geol. and Geog. Surv. of the Terr., vol. ii, No. 4, pp. 351–374, 6 plates. 1876. Washington, 1876.

> Carboniferous, Cretaceous, and Tertiary; mostly Cretaceous. A large part of the species embraced in this paper were originally described by the author in 1856 in vol. iv of Transactions Albany Institute, and are here redescribed with others and illustrated.

CARBONIFEROUS SPECIES.

Brachiopoda.

	Page.
Productus, Sowerby	354
P. latissimus, Sowerby, 1822, pl. i, fig. 1	354–355
Spirifer, Sowerby	355
S. keokuk, Hall ? 1858, pl. i, figs. 3 and 3 *a*	355
Athyris, McCoy, 1844	355
A. subtilita, Hall, sp., 1852, pl. i, figs. 2 and 2 *a*	355–356

CRETACEOUS SPECIES.

Lamellibranchiata.

	Page.
Nucula, Lamarck	356
N. traskana, Meek, 1857	356
Grammatodon, Meek	356
G. ? vancouverensis, Meek, 1857, pl. iii, figs. 5 and 5 *a*	356–357
Arca, Sim	357
A. ? equilateralis, Meek, 1857, pl. ii, figs. 6, 6 *a*	357
Inoceramus, Sowerby	358
I. cripsii ? var. *subundatus*, Meek, 1861, pl. iii, figs. 1, 1 *a* and 3, 3 *a*	358–359
I. barabini, Morton, 1834, fig. —	358
Inoceramus ——?, pl. i, fig. 6	359
Trigonia, Bruguière	359
T. evansi, Meek, 1857, pl. ii, figs. 7 *a, b*	359–360
Protocardia, Beyrich	360
P. scitula, Meek, 1857, pl. iii, figs. 4 and 4 *a*	360–361
Cyprimeria, Conrad	361
C. ? tenuis, Meek, 1861, pl. ii, figs. 5 *a, b*	361
Pholadomya, Sowerby	362
P. subelongata, Meek, 1857, pl. ii, figs. 1 *a*	362
Goniomya, Agassiz	362
G. borealis, Meek, 1857, pl. ii, fig. 2	362–363
Thracia, Leach	363
T. ? occidentalis, Meek, 1857, pl. ii, figs. 3 *a*	363
T. ? subtruncata, Meek, 1857, pl. ii, figs. 4 and 4 *a*	363–364

Gasteropoda.

	Page.
Dentalium, Linnæus	364
D. komooksense, Meek, 1857, pl. iii, fig. 6	364

Cephalopoda.

Baculites, Lamarck	364
B. chicoensis, Trask ? 1856, pl. iv, figs. 2 *a–c*	364–365
B. occidentalis, Meek, 1861, pl. iv, figs. 1 *a–b*	366–367
Heteroceras, d'Orbigny	367
H. cooperi, Gabb sp., 1864, pl. iii, figs. 7 *a*	367
Ammonites, Bruguière	367
A. newberryanus, Meek, 1857, pl. iv, figs. 3 *a, b*	367–368
A. complexus var. *suciensis*, Meek, 1861, pl. v, figs. 2 *a–c*	369–370
Placenticeras, Meek	370
P. ? vancouverense, Meek, 1861, pl. vi, figs. 1 *a–c*	370–371
Phylloceras, Suess	371
P. ? ramosus, Meek, 1857, pl. v, figs. 1 *a–b*	371–373
Nautilus, Linnæus	373
N. campbelli, Meek, 1861, pl. vi, figs. 2, 2 *a*	373

TERTIARY SPECIES. ?

Mactra, Linnæus	374
M. gibbsana, Meek, 1861, pl. ii, figs. 8 *a–b*	374

102.

MEEK, F. B. Note on the new genus Uintacrinus, Grinnell. <Bulletin U. S. Geol. and Geog. Surv. of the Terr., vol. ii, No. 4, pp. 375–378, 2 wood cuts. 1876. Washington, 1876.

This paper consists largely of a redescription and rectification of the genus.

	Page.
Uintacrinus, n. g., Grinnell, 1876	375–378
U. socialis, n. s., Grinnell, 1876 ?, figs. A, B, p. 375	375

103.

MEEK, F. B. Descriptions of the Cretaceous Fossils collected on the San Juan exploring expedition under Capt. J. N. Macomb, U. S. Engineers. <Report of the Exploring Expedition from Santa Fé, New Mexico, to the junction of the Grand and Green rivers of the Great Colorado of the West, in 1859, pp. 121–133, pls. i and ii. Washington, 1876.

The exploration was made in 1859, but the report was not published until 1876, when Mr. Meek revised the work in accordance with his views at the time of publication.

Lamellibranchiata.

	Page.
Ostrea, Linnæus	123
O. lugubris, Conrad, 1857, pl. i, figs. 1 *a–d*	123–124
O. (Gryphæa ?) uniformis, n. s., Meek, 1876, pl. i, figs. 2 *a–c*	124
Exogyra, Say	124
E. columbella, n. s., Meek, 1876, pl. i, figs. 3 *a–d*	124–125
Anomia, Linnæus	125
A. nitida, n. s., Meek, 1876, pl. i, figs. 4 *a–b*	125
Caprotina, d'Orbigny	126
C. (Requienia ?) bicornis, Meek, 1876, pl. i, figs. 7 *a–b*	126
Plicatula, Lamarck	126
P. arenaria, n. s., Meek, 1876, pl. i, figs. 5 *a–c*	126–127
Inoceramus, Sowerby	127
I. fragilis, H. & M., 1856 ?, pl. i, fig. 6	127
Crassatella, Lamarck	127
C. shumardi, n. s., Meek, 1876, pl. ii, figs. 7 *a–c*	127–128
Cyprimeria, Conrad	128
C.? crassa, n. s., Meek, 1876, pl. i, figs. 8 *a–d*	128
Cardium, Linnæus	128
C. bellulum, n. s., Meek, 1876, pl. ii, figs. 6 *a, b*	128–129

Gasteropoda.

	Page.
Actæon, Montfort	129
A. intercalaris, n. s., Meek, 1876, pl. ii, figs. 4 a–o	129
Anchura, Conrad	129
A.? newberryi, n. s., Meek, 1876, pl. ii, fig. 5	129–130

Cephalopoda.

Baculites, Lamarck	130
B. anceps var. obtusus, Meek, 1876, pl. ii, figs. 1 a–h	130–132
Prionocyclus, Meek	132
P.? macombi, n. s., Meek, 1876, pl. ii, figs. 3 a–d	132–133

104.

MEEK, F. B. Report on the Paleontological collections of the expedition, <Report Expl. Great Basin of the Terr. of Utah, in 1859. By J. H. Simpson. Appendix J, pp. 337–373, pls. i–v. Washington, 1876.

Devonian, Carboniferous, Jurassic, Cretaceous, and Tertiary. The explanations were made and the fossils collected nearly eighteen years before the publication of this report, but the paleontology was corrected in accordance with the views of the author at the time of publication.

Descriptions of new species.

DEVONIAN FOSSILS.

Mollusca.
Brachiopoda.

	Page.
Productus, Sowerby	345
P. subaculeatus, Murchison? 1840, pl. i, figs. 3 a–c	345
Spirifer, Sowerby	345
S. utahensis, Meek, 1860, pl. i, figs. 4 a–c	345–346
S. engelmanni, Meek, 1860, pl. i, figs. 1 a–c	346–347
S. strigosus, Meek, 1860, pl. i, figs. 5 a–d	347
Atrypa, Dalman	347
A. reticularis (Lin.) Dalm., 1767, pl. i, figs. 6 a–b	347–348
A. aspera, Schloth, 1813, pl. i, figs. 2 a–b	348

CARBONIFEROUS FOSSILS.

Mollusca.
Polyzoa.

Archimedipora, D'Orbigny	348
Archimedipora, ——? Meek, pl. i, fig. 11	348

Brachiopoda.

Chonetes, Fischer	348
C. verneuiliana var. utahensis, Meek, 1876, pl. ii, figs. 2 a–c	348–349
Productus, Sowerby	349
P. semistriatus, Meek, 1860, pl. i, figs. 7 a–b	349
P. multistriatus, Meek, 1860, pl. i, figs. 8 a–b	350
Athyris, McCoy	350
A. subtilita, Hall, sp., 1852, pl. ii, figs. 4 a–b	350–351
Spirifer, Sowerby	351
S. (Spiriferina?) scobina, Meek, 1860, pl. ii, figs. 5 a–c	351–352
S. (Spiriferina) pulcher, Meek, 1860, pl. ii, figs. 1 a–h	352
S. cameratus, Morton, 1836, pl. ii, figs. 3 a, b	353

Lamellibranchiata.

Aviculopecten, McCoy	354
A. utahensis, Meek, 1860, pl. i, figs. 9 a–c	354

Cephalopoda.

Orthoceras, Auct	354
O. baculum, Meek, 1860, pl. i, figs. 10 a–b	354–355

JURASSIC SPECIES.

Radiata.
Echinodermata.

Pentacrinites, Miller	355
Pentacrinites, undt. sp., Meek, pl. iii, figs. 5 a–c	355

Mollusca.
Lamellibranchiata.

Ostrea, Linnæus	355

	Page.
O. engelmanni, Meek, 1860, pl. iii, fig. 6	355–356
Gryphœa calceola, Quenstedt? 1856, pl. iii, fig. 2	356
Camptonectes, Agassiz	356
O. bellistriata, Meek, 1860, pl. iii, figs. 3 a–d	356–357

Gasteropoda.
| *Dentalium*, Linn | 357 |
| *D.? subquadratum*, Meek, 1860, pl. iii, figs. 1 a–c | 357 |

Cephalopoda.
| *Belemnites*, Lamarck | 358 |
| *B. densus*, M. & H., 1858, pl. iii, figs. 4 a, b | 358 |

CRETACEOUS FOSSILS.

Lamellibranchiata.
Inoceramus, Sowerby	358
I. problematicus, Schloth., 1820, pl. iv, figs. 1 a and 1 b, c (?)	358–359
Anomia, Linn	359
A. concentrica, Meek, 1860, pl. iv, fig. 3	359
Inoceramus simpsoni, Meek, 1860, pl. iv, fig. 4	360

BEAR RIVER FRESH-WATER OR ESTUARY BEDS.
Mollusca.
Lamellibranchiata.
Unio, Retzius	361
U. retusus, Meek, 1860, pl. v, figs. 12 a, b	361
Corbula, Bruguière	361
C. (Anisorhynchus) pyriformis, Meek, 1860, pl. v, figs. 9 and 10	361–362
C. engelmanni, Meek, 1860, pl. v, figs. 13 a, b	362

Gasteropoda.
Pyrgulifera, 1871, Meek	363
P. humerosa, Meek, 1860, pl. v, figs. 6 a–c	363
Limnœa nitidula, Meek, 1860, pl. v, fig. 14	363–364
Rhytophorus	364
R. priscus, Meek, 1860, pl. v, figs. 4 a, b	364

TERTIARY FOSSILS.
Mollusca.
Lamellibranchiata.
Unio, Retzius	364
U. haydeni, Meek, 1860, pl. v, figs. 11 a, b	364–365
Goniobasis, Lea	365
G. simpsoni, Meek, 1860, pl. v, figs. 1 a–e	365–366
G. arcta, Meek, 1860, pl. v, fig. 5	366
Planorbis, Müller	366
P. spectabilis, Meek, 1860, pl. v, figs. 7 a–d	366–367
P. spectabilis var. *utahensis*, Meek, 1860, pl. v, figs. 8 a–c	367
Limnœa retusta, Meek, 1860, pl. v, figs. 3 a, b	367
L. similis, Meek, 1860, pl. v, figs. 2 a, b	367
Catalogue of the organic remains contained in the collection	368–373
Devonian species	368
Carboniferous species	368–371
Permian forms	371–372
Jurassic species	372
Cretaceous species	372–373
Fossils of the Bear River Fresh or Brackish water beds	373
Tertiary species	373

105.

MEEK, F. B. A report on the Invertebrate Cretaceous and Tertiary Fossils of the Upper Missouri Country. <Rep. U. S. Geol. & Geogr. Surv. of the Terr. 4°. vol. ix, pp. i–xiv, 1–629, pls. i–xlv. Washington, 1876.

This great work contains descriptions and illustrations of nearly 300 species; more than 200 genera and subgenera are fully diagnosed, besides which full diagnoses of the families which embrace them are given; also philosophical discussion of many important questions. The greater part of the species embraced in this volume were previously, from time to time, described and published, mainly in the publications of the Acad. Nat. Sc. Phila.

CONTENTS.

	Page.
List of woodcuts	xiii
List of errata	xv
Letter of Dr. F. V. Hayden	xvii
Introductory remarks	xix
Cretaceous formation	xxi
Section of Cretaceous formation on the Missouri	xxiii
General section of the Cretaceous rocks of Nebraska	xxiv
Subdivisions of the Upper Missouri undoubted Cretaceous series, and their geographical extension west of the Mississippi, individually considered	xxvi
Dakota group	xxvi
Fort Benton group	xxviii
Niobrara group	xxx
Fort Pierre group	xxxii
Fox Hills group	xxxv
New Mexican Cretaceous section	xxxvii
Relations of the Upper Missouri Cretaceous beds to those east of the Mississippi	xxxviii
Section of Cretaceous rocks of Mississippi	xxxviii
Alabama section	xxxix
New Jersey section	xli
Parallelism of the subdivisions of the Upper Missouri Cretaceous series with those of the same in Europe	xliii
Fresh and brackish water lignite deposits of the Upper Missouri	xlviii
Judith River group	xlvii
Section of the Judith River group	xlviii
Fort Union group	lv
Section of the Fort Union	lix
Tertiary rocks of the Wind River and White River groups	lxi
Wind River group	lxi
White River group	lxi
Section of the White River group	lxii
Invertebrate palæontology	1
Cretaceous species	1
Species of the fresh and brackish water lignite beds	509
Fossils of the Wind River Tertiary	593
Fossils of the White River Tertiary	598
Appendix	607

CRETACEOUS.

Radiata.
  Polypi.
    Actinaria.
      Fungiidæ.

Micrabacia, Edwards & Haime, 1849	1
M. americana, M. & H., 1860, pl. xxviii, figs. 1 a–d	1–2

  Alcyonaria.
    Gorgoniidæ.

Websteria, Edwards & Haime, 1854	2–3
W. cretacea, Meek, 1864, pl. xxviii, figs. 3 a–c	3–4
* *Microstizia*, n. g., Meek, 1876	4
M. millepunctata, n. s., Meek, 1876, pl. xxviii, figs. 2 a–c	4

Echinodermata.
  Echinoidea.
    Spatangidæ.

Hemiaster, Desor., 1847	5
H. humphreysanus, M. & H., 1857, pl. x, figs. 1 a–g	5–6

Mollusca.
  Brachiopoda.
    Lyopomata.
      Lingulidæ.

Lingula, Bruguière, 1792	7–9
L. nitida, M. & H., 1861, pl. xxviii, figs. 18 a, b	9–10

* μικρός, small : στίξις, puncture.

Lamellibranchiata.
Monomyaria.
Ostreidæ.

	Page.
Ostrea, Linnæus, 1758	10–12
Alectryonia, Fischer	11
Gryphæostrea, Conrad	11
Ostrea, sp. undt., pl. ii, figs. viii a and b	12–13
O. congesta, Conrad, 1843, pl. ix, figs. 1 a–f	13–14
O. inornata, M. & H., 1860, pl. x, fig. 4	14
O. pellucida, M. & H. 1860, pl. xxviii, figs. 4 a and b	15
O. (Gryphæostrea?) subalata, Meek, pl. xxviii, fig. 5	15–16
O. (Gryphæa?) patina, M. & H., 1856, pl. x, figs. 2 a, b; a, b (bis), and 3 e, f; also pl. xi varieties	16–18
Gryphæa, Lamarck, 1801	10
G. vesicularis, Lam. ? 1806, pl. xi, figs. 2 a–c, and pl. xvi, figs. 8 a, b	20–21

Anomiidæ.

Anomia, Linnæus, 1767	21–22
A.? obliqua, M. & H., 1860, pl. ix, fig. 2	22
A.? subtrigonalis, M. & H., 1860, pl. xvi, figs. 4 a, b	22–23

Pectinidæ.

Chlamys, Bolten, 1798	23–25
C. nebracensis, M. & H., 1856, pl. xvi, figs. 6 a–c	25–26
Syncyclonema, Meek, 1864	26–27
S. rigida, H. & M., 1854, pl. xvi, figs. 5 a, b	27–28

Heteromyaria.
Pteriidæ.

Pteria, Scopoli, 1777	28–32
Electroma, Stoliczka	29
Pseudoptera, Meek	29
Oxytoma, Meek	29
P. linguiformis, Evans & Shumard, sp., 1854, pl. xvi, figs. 1 a–d	32–33
P. linguiformis var. *subgibbosa*, pl. xxviii, fig. 12	33
P. haydeni, H. & M., 1854, pl. xvi, figs. 2 a, b	33–34
P. (Oxytoma) nebrascana, E. & S., 1857, pl. xvi, figs. 3 a, b, and pl. xxviii, fig. 11	34–36
P. (Pseudoptera) fibrosa, M. & H., 1856, pl. xvii, figs. 17 a–d	36–37
Inoceramus, Sowerby, 1819	38–41
Subgenus *Inoceramus*	38–39
Mytiloides, Brongniart	39
Catillus (Brongniart), Chenu	39
Actinoceramus, Meek	39
Volviceramus, Stoliczka	40
I. (Inoceramus) fragilis, H. & M., 1854, pl. v, fig. 5, and figs. 1 and 2, p. 42	42–43
I. (Inoceramus) altus, Meek, 1871, pl. xiv, figs. 1 a, b	43–44
I. (Volviceramus) umbonatus, M. & H., 1858, pl. iii, figs. 1 a–c, and pl. iv, figs. 1 a–b and 2 a, b	44–46
I. (Volviceramus) exogyroides, M. & H., 1862, pl. v, figs. 3 a–c	46–47
Subgenus *Catillus*, Brong.	
I. (Catillus) pertenuis, M. & H., 1856, pl. xxxvii, figs. 3 a, b, and pl. xxxviii, figs. 3 a, b	47–48
I. (Catillus) cripsii? var. *subcompressus*, M. & H., pl. xxxviii, fig. 2 bis	48–49
I. (Catillus) cripsii? var. *barabini*, Morton, 1834, pl. xiii, figs. 1 a–c, and pl. xii, fig. 3, figs. 1–4	49–50
I. (Catillus) convexus, H. & M., 1854, pl. xii, figs. 5 a and b	51–52
I. (Catillus) sagensis var. *nebrascensis*, Owen, 1852, pl. xiii, figs. 2 a, b	52–53
I. (Catillus) proximus, Tuomey? 1854, pl. xii, figs. 7 a, b	53–55
I. (Catillus) proximus? var. *subcircularis*, Meek, pl. xii, figs. 2 a, b	55–56
I. (Catillus) balchii, M. & H., 1860, pl. xv, figs. 1 a, b	56
I. (Catillus) tenuilineatus, H. & M., 1854, pl. xii, fig. 6	57
I. (Catillus) vanuxemi, M. & H., 1860, pl. xiv, figs. 2 a, b	57–58
I. (Catillus) sublævis, H. & M., 1854, pl. xii, figs. 1 a and b	58–59
I. (Catillus) tenuirostris, M. & H., 1862, fig. 5	59
I. (Catillus) undabundus, M. & H., 1862, pl. iii, figs. 2 a, b	60–61
I. (Catillus) incurvatus, M. & H., 1856, pl. xii, figs. 4 a and b	61
Subgenus *Mytiloides*, Brong.	
I. (Mytiloides) problematicus, Schlot., pl. ix, figs. 3 a, b	62

	Page
I. (Mytiloides) problematicus var. *aviculoides*, M. & H., 1860, pl. ix, fig. 4	63-64
Gervillia, Defrance.	
G. subtortuosa, M. & H., 1856, pl. xvi, figs. 7 a–c	65-66
G. recta, M. & H., 1861, pl. xxix, figs. 1 a, b	66-67
Mytilidæ.	
Mytilus, Linnæus, 1758	67-68
Aulacomya, Mörch (= *Hormomya*, Mörch)	68
Stavella, Gray	68
M. subarcuatus, M. & H., 1856, pl. xxxviii, figs. 2 a, b	69
Volsella, Scopoli, 1777	69-71
Brachydontes, Swainson	70
V. meekii, E. & S., sp., 1857, pl. xv, figs. 3 a–c	72
V. galpiniana, E. & S., sp., 1854, pl. xxviii, figs. 7 a, b	73
V. attenuata, M. & H., 1856, pl. xxviii, figs. 8 a, b	74
Crenella, Brown, 1827	74-75
Modiolaria, Beck	75
C. elegantula, M. & H., 1861, pl. xxviii, figs. 6 a–c	75-76
Dimyaria.	
Arcidæ.	
Barbatia, Gray, 1840	76-80
Polynema, Conrad	78
Acar, Gray	78
Calloarca, Gray	78
Striarca, Conrad	78
Plagiarca, Conrad	78
? Granoarca, Conrad (= *? Cucullæarca*, Conrad)	78
B. (Polynema ?) parallela, Meek, 1872, pl. ii, fig. 10	80-81
Nemodon, Conrad, 1870	81-82
N. sulcatinus, Evans & Shum., sp., 1857, pl. xv, figs. 6 a, b	82-83
Cucullæa, Lamarck, 1801	83-85
Idonearca, Conrad	84
Latiarca, Conrad	84
C. (Idonearca) shumardi, M. & H., 1856, pl. xxviii, figs. 15, a–g, and pl. xxix, fig. 4	86-87
C. (Idonearca) nebrascensis, Owen, 1852, pl. xxix, figs. 5 a, b	88-89
C. (Idonearca ?) cordata, M. & H., 1856, pl. xxix, figs. 6 a, b	89-90
Trigonarca, Conrad, 1862	90-91
Breviarca, Conrad	91
T. (Breviarca ?) siouxensis, H. & M., 1854, pl. i, fig. 6	92
T. (Breviarca ?) saliniensis, Meek, pl. ii, figs. 1 a–c	92-93
T. (Breviarca) exigua, M. & H., 1856, pl. xv, figs. 2 a–f	93-94
Axinæa, Poli, 1791	94-95
A. subimbricata, M. & H., 1860, pl. xxviii, figs. 14 a–c	95-96
Limopsis, Sassi, 1827	96
L. parvula, M. & H., 1856, pl. xxviii, figs. 17 a–c	97-98
Nuculidæ.	
Nucula, Lamarck, 1799	98-99
Acila, H. & A., Ad	98
N. subplana, M. & H., 1856, pl. xvii, figs. 7 a, b	99-100
N. obsoletistriata, M. & H., 1856, pl. xv, figs. 10 a, b	100-101
N. planimarginata, M. & H., 1856, pl. xv, figs. 8 a, b, and pl. xxviii, fig. 16	101-102
N. cancellata, M. & H., 1856, pl. xxviii, figs. 13 a–e	102-103
Ledidæ.	
Nuculana, Link, 1807	103-104
N. bisulcata, M. & H., 1864, pl. xv, figs. 4 a, b	104-105
N. subnasuta, H. & M., 1854, pl. xv, fig. 9	105-106
N. ? equilateralis, M. & H., 1856, pl. xv, figs. 7 a, b	106
Yoldia, Müller, 1842	107-108
Portlandia, Mörch	
Y. microdonta, Meek, 1872, pl. ii, fig. 2	109
Y. scitula, M. & H., 1856, pl. xxviii, fig. 9	110
Y. evansi, M. & H., 1856, pl. xxviii, figs. 10 a–c	111
Y. ventricosa, H. & M., 1854, pl. xv, figs. 5 a, b	112
Unionidæ.	
Margaritana, Schumacher, 1817	112-114
Abumadonta, Say	113

WRITINGS OF F. B. MEEK. 99

	Page.
Complanaria, Swainson	113
Unionopsis, Swainson (= *Calceola*, Swainson, 1840; not Lam., 1799)	113
M. nebrascensis, Meek, 1871, pl. i, figs. 5 *a–c*	114–115
Crassatellidæ.	
Crassatella, Lamarck, 1801	115–117
Subgenus *Pachythœrus*, Conrad, 1870	116
C. (Pachythœrus) evansi, H. & M., 1854, pl. xvii, figs. 6 *a–d*	117–118
Crassatellina, Meek, 1871	118–120
C. oblonga, Meek, 1871, pl. ii, figs. 3 *a–e*	120–121
Eriphyla, Gabb, 1864	121–124
E. gregaria, M. & H., 1856, pl. xvii, figs. 9 *a*, *b*, figs. 6 and 7, p. 124	124–126
Solemyidæ.	
Solemya, Lamarck, 1818	126–129
S. subplicata, M. & H., 1856, pl. xxviii, fig. 10	129
Lucinidæ.	
Lucina, Bruguière, 1792	130–133
Myrtea, Turton, 1822	130
Cyclas, H. & A. Adams, 1857 (not Brug.)	131
Milthea, H. & A. Adams, 1857	131
L. subundata, Hall & Meek, 1854, pl. xvii, figs. 2 *a–e*	133–134
L. occidentalis, Morton, 1842, pl. xvii, figs. 4 *a–d*	134–135
L. occidentalis var. *ventricosa*, M. & H., 1860, pl. xvii, figs. 3 *a–c*	135–136
Sphœriola, Stoliczka, 1871	136–137
S. ? cordata, M. & H., 1857, pl. xxix, figs. 3 *a–c*	137–138
S. ? warrenana, Meek	138
S. ? endotrachys, Meek, pl. xxix, fig. 2	139
Tancredidæ.	
Tancredia, Lycett, 1850	140–142
T. americana, M. & H., 1856, pl. xxxviii, figs. 1 *a–h*	142–144
Glossidæ.	
Cyprina, Lamarck, 1812	144–146
C. ovata, M. & H., 1857, pl. xxix, figs. 7 *a–c*, and fig. 8, p. 146	146–147
C. ovata var. *compressa*, M. & H., pl. xxx, fig. 11	147
Veniella, Stoliczka, 1870 (= *Venilia*, Morton; not of Duponchel, or Alder & Hancock)	147–152
Venilicardia, Stoliczka, 1870	149
V. conradi, Morton, figs. 9–11	148
V. goniophora, n. s., Meek, 1876, pl. iv, fig. 4 and fig. 12, p. 152	152–153
V. mortoni, M. & H., 1862, pl. iv, figs. 3 *a*, *b*	154
V. subtumida, M. & H., 1857, pl. xvii, figs. 5 *a*, *b*	154–150
V. (Venilicardia ?) humilis, M. & H., 1860, pl. xxx, figs. 5 *a–c*	155–156
Cyrenidæ.	
Cyrena, Lamarck, 1818	157–158
Egeta, H. & A. Adams, 1858	157
Cyrena dakotensis, M. & H., pl. i, figs. 1 *a–f*	159–160
Corbicula, Mühlfeldt, 1811	160–163
Veloritina, Meek	161
Leptesthes, Meek	161
C. ? nucalis, Meek, 1872, pl. ii, figs. 5 *a*, *c*	163–164
C. ? subtrigonalis, Meek, 1872, pl. ii, fig. 6	164–165
Cardiidæ.	
Cardium, Linnæus, 1758 (= *Acanthocardium*, Gray)	165–168
Pectunculus (Adanson). Stoliczka; but not H. & A. Adams and others	166
Trachycardium, Mörch	166
Criocardium, Conrad	166
Tropidocardium, Roemer (*Cardium* proper of most authors)	166
Cerastoderma (Poli), Mörch	166
Nemocardium, Meek	167
C. (Criocardium) speciosum, M. & H., 1856, pl. xxxvii, figs. 4 *a–c*	169–170
C. kansasense, Meek, 1871, pl. ii, figs. 14 *a–d*	170–171
Protocardia, Beyrich, 1845	171
Pachycardium, Conrad	172
Leptocardia, Meek	172
Subgenus, *Protocardia*	172
P. (Protocardia) salinensis, Meek, 1871, pl. ii, figs. 13 *a–c*	174

	Page.
Subgenus *Leptocardia*, Meek, Section (a)	175
P. (*Leptocardia*) *subquadrata*, E. & S., sp., 1857, pl. xxix, figs. 8 a–e	175
P. (*Leptocardia*) *cara*, E. & S. sp., 1857, pl. xvii, figs. 1 a–c	176
P. (*Leptocardia ?*) *pertenuis*, M. & H., 1861, figs. 13 and 14, p. 176	176–177
Veneridæ	
Callista Poli, 1791	177
Callista, Poli, typical (= *Chione* Gray; not of Mühlfeldt)	178
Dione, Gray	178
Macrocallista Meek	179
Pitar, Roem. (= *Caryatis*, Roemer)	179
Aphrodina, Conrad	179
? Dosiniopsis, Conrad	179
C. (*Dosiniopsis*) *deweyi*, M. & H., 1856, pl. xvii, figs. 15 a–e	182–183
C. (*Dosiniopsis*) *owenana*, M. & H., 1856, pl. xxxvii, fig. 1	183–184
C. (*Dosiniopsis*) *nebrascensis*, M. & H., 1856, figs. 15–17, p. 184	184–186
C. (*Dosiniopsis*) *orbiculata*, H. & M., 1854, pl. v, figs. 2 a–c	186–187
C. ? pellucida, M. & H., 1856, pl. xvii, figs. 10 a–e and figs. 12 a–c	187–188
C. (*Aphrodina ?*) *tenuis*, H. & M., pl. v, figs. 1 a–d	188–189
Thetis, Sowerby, 1826	189–190
T. ? circularis, M. & H., 1856, pl. xvii, figs. 8 a–e and figs. 18, 19, p. 190	190–191
Tellinidæ	
Tellina, Linnæus, 1758	192–193
Tellinella, Gray	193
Peronæoderma (Poli), Mörch	193
Mœra, H. & A. Adams (= *Donacilla*, Gray; not Lam)	193
Palæomœra, Stoliczka	193
Phylloda, Schumacher	193
Angulus, Schumacher (= *Tellinula*, Chem., *Fabulina*, Gray)	193
Tellinides, Lam	193
Homalina, Stoliczka	194
Peronæa, Poli (= *Omala*, Schum., corrected *Homala*, by Agassiz; also *Homala*, H. & A. Adams)	194
Metis, H. & A. Adams	194–195
T. (*Eme ?*) *subcitula*, Meek, 1871, pl. ii, figs. 11 a and b	195–196
T. (*Peronæa ?*) *equilateral s*, M. & H., 1856, pl. xxxix, figs. 5 a–c	196–197
T. (*Peronæa ?*) *scitula*, M. & H., 1856, pl. xxx, figs. 1 a, b	197–198
Linearia, Conrad, 1860	198–199
L. ? formosa, M. & H., 1860, pl. xxx, fig. 2	199–200
Areopagella, Meek, 1871	200–202
A. mactroides, Meek, 1871, pl. ii, figs. 4 a–d	202
A. ? macrodonta. n. s., Meek, 1876, pl. i, fig. 2	202–203
Mactridæ	
Mactra, Linnæus, 1767	203–204
Mactra, Linn. (typical *Trigonella*, da Costa)	204
Cymbophora, Gabb	204
Schizodesma, Gray	204–206
Mactra (*Cymbophora) ? sionxensis*, M. & H., 1860, pl. i, figs. 7 a–c	206
M. (*Cymbophora ?*) *formosa*, M. & H., 1856, pl. xxxix, fig. 7	207
M. (*Cymbophora ?*) *warrenana*, M. & H., 1856, pl. xxx, figs. 7 a–d	208
M. (*Cymbophora ?*) *gracilis*, M. & H., 1860, pl. xvii, figs. 18 a, b	209
M. (*Cymbophora*) *alta*, M. & H., 1856, pl. xxxvii, figs. 2 a, b	210
M. (*Cymbophora ?*) *nitidula*, M. & H., 1861, pl. xxx, figs. 6 a–c	211–213
Pholadomyidæ	
Pholadomya, Sowerby, 1823	213–214
Procardia, Meek	215–216
P. papyracea, M. & H., 1862, pl. v, figs. 4 a, b	217
P. subcentricosa, M. & H., 1857, pl. xxxix, figs. 8 a, b	217–218
Subgenus *Procardia*	219
P. (*Procardia*) *hodgii*, Meek, 1871, pl. xlii, figs. 3 a, b	219
Goniomya, Agassiz, 1838	220–221
G. americana, M. & H., 1856, pl xxx, figs. 12 a, b	221–222
Anatinidæ	
Thracia Leach 1819	222–223
T. ? subtortuosa, M. & H., 1856, pl. xxxvii, fig. 5	223–224
T. gracilis, M. & H., 1856, pl. xxxix, figs. 6 a, b	224–225

	Page
T. ? prouti, M. & H., 1860, pl. xxxvii, figs. 6 a, b	225-226
Liopistha, Meek, 1864	227-236
Cymella, Meek	229
Psilomya, Meek	229
L. protexta, Conrad, figs. 20-24	227
Cymella bella, Conrad, figs. 25-30	228
Subgenus *Cymella*, Meek	236
Liopistha (Cymella) undata, M. & H., 1856, pl. xxxix, figs. 1 a, b	236-237
Neaera, Gray, 1834	237-238
N. ventricosa, M. & H., 1856, pl. xxx, figs. 3 a-e	238-239
N. moreauensis, M. & H., 1856, pl. xvii, figs. 11 a-c	239-240
Corbulidæ.	
Corbula, Bruguière, 1792	240-244
Anisorhynchus, Conrad	241
Pachyodon (Pachydon, Gabb, and Anisothyris, Conrad)	241
Corbula crassimarginata, M. & H., 1860, pl. xvii, figs. 14 a-e	244-245
C. inornata, M. & H., 1856, pl. xxx, figs. 4 a-d	245-246
Corbulamella, M. & H., 1857	246
C. gregaria, M. & H., 1857, pl. xvii, figs. 13 a-d	247
Saxicavidæ.	
Glycimeris, Lamarck, 1799	248-249
G. occidentalis, M. & H., 1856, pl. xxxix, figs. 9 a, b	250
Solenidæ.	
Pharella, Gray, 1854	250-251
P. ? dakotensis, M. & H., 1860, pl. i, fig. 3	251-252
Leptosolen, Conrad, 1867	252-253
L. conradi, Meek, 1872, pl. ii. figs. 12 a, b	253-254
Pholadidæ.	
Turnus, Gabb, 1864	254-256
Goniochasma, Meek	255
Xylophagella, Meek	255
T. (Goniochasma) stimpsoni, M. & H., 1857, pl. xxx, figs. 9 a, b	256-257
T. (Xylophagella) elegantulus, M. & H., 1857, pl. xxx, figs. 10 a-e	257-258
Martesia, Leach, 1824	258-259
M. cuneata, M. & IL, 1856, pl. xxx, figs. 8 a, b	259-260
Teredidæ.	
Teredo, Linnaeus, 1758	260-262
Calobates, Gould	261
T. selliformis, M. & H., 1860, pl. xvii, figs. 19 a-d	262-263
T. globosa, M. & H., 1858, pl. xxx, fig. 13 (burrows), figs. 31, 32, p. 264	264-265
Gasteropoda.	
Solenoconcha.	
Dentaliidæ.	
Dentalium, Linn., 1758	266
D. gracile, H. & M, 1854, pl. xviii, figs. 13 a-d	266-267
Entalis, Sowerby, 1839	267-268
E. pauperculа, M. & H., 1860, pl. xviii, fig. 14	269
Tectibranchiata.	
Bullidæ.	
Haminea, Leach, 1847	270-271
H. occidentalis, M. & H., 1856, pl. xviii, figs. 11 a, b, and 12 a, b	271-272
H. subcylindrica, M. & H., 1856, pl. xviii, figs. 10 a. b	272-273
H. minor, M. & H., 1856, pl. xxxi, figs. 1 a, b	273
Cylichnidæ.	
Cylichna, Lovén, 1846	274-275
Mnestia, H. & A. Adams	274
C. ? volvaria, M. & H., 1856, pl. xxxi, figs. 2 a, b	275-276
C. scitula, M. & H., 1860, pl. xxxi, figs. 3 a, b	276-277
Actæonidæ.	
Actæon, Montfort, 1810	277-280
A. subellipticus, M. & H., 1856, pl. xix, fig. 16	280-281
A. attenuatus, M. & H., 1858, pl. xix, figs. 17 a, b	281-282
Ringiculidæ.	
Cinulia, Gray, 1840	282-283
Oligoptycha, Meek	283

	Page
Arellana, d'Orbigny, 1843	283
Cinulia (*Oligoptycha*) *concinna*, H. & M., sp., 1854, pl. xxxi, 6 bis. *a–c*	284
Pulmonata.	
Siphonariidæ	
Anisomyon, M. & H., 1860	285–288
A. borealis, M. & H., 1860, pl. xviii, figs. 9 *a–e*	288–289
A. shumardi, M & H., 1860, pl. xviii, figs. 7 *a–c*	289–290
A. patelliformis, M. & H., 1860, pl. xviii, figs. 5 *a–f* (not *d* and *e?*)	290–291
A. suboratus, M. & H., 1856, pl. xviii, figs. 5 *d* and 6	291–292
A. alcedus, M. & H., 1860, pl. xviii, figs. 4 *a, b*	292
A. sexsulcatus, M. & H., 1860, pl. xviii, figs. 8 *a, b*	293
Docoglossa.	
Acmæidæ.	
Acmæa, Esch.	
A. occidentalis, M. & H., 1860, pl. xviii, figs. 3 *a, b*	295–296
A. ? parva, M. & H., 1860, pl. xviii, figs. 1 *a – c*, and fig. 2	296
A. ? papillata, M. & H., 1860, pl. xxxi, figs. 4 *a, b*	296–297
Rhipidoglossa.	
Trochidæ.	
Margarita, Leach, 1819	298
M. nebrascensis, M. & H., 1860, pl. xix, figs. 8 *a, b*, 9 *a, b*	298, 299
M. mudgeana, Meek, 1871, pl. ii, figs. 9 *a, b*	300
Margaritella, M. & H., 1860	300–302
M. flexistriata, E. & S., 1854, pl. xix, figs. 11 *a–d*	302
Pectinibranchiata.	
Tritonidæ.	
Trachytriton, Meek, 1864	303–304
T. vinculum, H. & M., sp., 1856, pl. xix, figs. 7 *a–d*	304–306
* *Closteriscus*, n. g., Meek, 1876	306–307
C. tenuilineatus, H. & M., sp., 1856, pl. xix, figs. 10 *a. b*, and 9 *c*	308–309
Naticidæ.	
Gyrodes, Conrad, 1860	309–310
G. conradi, Meek, figs. 33–36, p. 310	310–311
Lunatia, Gray, 1847	311–314
L. concinna, H. & M., sp., 1854, pl. xxxii, figs. 11 *a–c*	314–315
L. occidentalis, M. & H., 1856, pl. xxxii, figs. 12 *a–c*	315–316
L. subcrassa, M. & H., 1856, pl. xxxix, figs. 3 *a–c*	316–317
Amauropsis, Mörch, 1857	317–318
A. paludiniformis, H. & M., sp., 1854, pl. xix, figs. 15 *a–c*	318–319
Aporrhaidæ.	
Aphorrhais, Dillwyn, 1823	320–322
Alipes, Conrad, 1865 (= *Gonioekeila*, Gabb)	320
Arrhoges, Gabb, 1868	321
Aporrhais biangulata, M. & H, 1856, pl. xix, figs. 6 *a–c*	322–323
Anchura, Conrad, 1860	324
Drepanochilus, Meek, 1864 (= *Perissoptera*, Tate, in part)	324
A. (*Drepanochilus*) *americana*, E. & S., sp., 1857, pl. xxxii, figs. 8 *a, b*	325–326
A. (*Drepanochilus*) *nebrascensis*, E. & S., sp., 1854, pl. xix, figs. 5 *a–c*	326–327
A. ? sublevis, M. & H., 1860, pl. xix, figs. 3 *a, b*	327–328
A. ? parva, M. & H., 1860, pl. xix, figs. 4 *a, b*	328
Vanikoridæ.	
Vanikoro, Quoy & Gaim. 1832	329
V. ambigua, M. & H., 1856, pl. xix, figs. 12 *a–d*	330, 331
Vanikoropsis,† n. g., Meek, 1876	331
V. tuomeyana, M. & H., sp., 1856, pl. xxxix, figs. 2 *a, b*	332
Turritellidæ.	
Mesalia, Gray, 1842	332–333
M. kansasensis, Meek, 1871, pl. ii, figs. 7 *a, b*	333–334
Cerithiopsidæ	
Cerithiopsis, Forbes & Hanley, 1849	334–335
Alaba, H. & A. Adams, 1853	335
Seila, A. Adams, 1861	335
Cerithiopsis moreauensis, M. & H., 1856, pl. xxxi, fig. 4 (not 4 *a, b*), fig. 38, p. 336	336–337

---

* κλωστήρ, a spindle (diminutive of).   † *Vanikoro*; ὄψις, form.

Pyramidellidæ.
  Ohemnitzia, d'Orbigny, 1850.. 337-339
  O. cerithiformis, M. & H., sp., 1856, pl. xxxii, figs. 10 a, b................... 339-341
Littorinidæ.
  Spironema, Meek, 1864... 341-342
  S. tenuilineata, M. & H., sp., 1856, pl. xxxii, figs. 9 a-c...................... 342-343
Muricidæ.
  Pyrifusus, Conrad, 1858... 343-345
    Neptunella, Meek, 1864 (not Gray)... 344
  Pyrifusus (Neptunella) newberryi, M. & H., 1856, pl. xxxi, fig. 6 a-f, fig. 39, p. 346. 346-347
  P. (Neptunella) subturritus, M. & H., sp., 1857, pl. xxxii, figs. 3 a, b, and fig. 40,
    p. 347... 347-348
  P. (Neptunella) intertextus, M. & H., sp., 1857, pl. xix, figs. 14 a, b.......... 348-349
Buccinidæ.
  ? Pseudobuccinum, M. & H. 1856.. 349-350
  P. nebrascense, M. & H., 1856, pl. xxxi, figs. 5 a-d............................. 350-351
  ? Odontobasis,* n. g., Meek, 1876... 351-352
  O. constricta, H. & M., sp., 1856, figs. 41, 42, p. 353.......................... 352-354
Fasciolariidæ.
  Fasciolaria, Lamarck, 1799.. 355-358
  • Terebrispira, Conrad, 1862.. 356
    Piestochilus, Meek, 1864.. 356
    Mesorhytis, Meek.. 356
    Cryptorhytis, Meek.. 356
  Fasciolaria buccinoides, M. & H., 1856, pl. xxxi, figs. 8 a-d.................... 358-359
  F. (Piestochilus) scarboroughi, M. & H., 1857, pl. xxxii, figs. 4 a-d............ 359-360
  F. (Piestochilus) culbertsoni, M. & H., 1856, pl. xxxii, figs. 1 a-f, fig. 44, p. 360.... 360-362
  F. (Piestochilus) galpiniana, M. & H., sp., 1856, pl xxxii, figs. 2 a, b......... 362-363
  F. (Piestochilus) cretacea, M. & H., 1856, pl. xxxi, figs. 11 a, b............... 363-364
  F. ? (Mesorhytis) gracilenta, Meek, fig. 45, p. 364............................... 364-365
  F. ? (Cryptorhytis) cheyennensis, M. & H., sp., 1860, pl. xix, figs. 13 a, b..... 365-366
  F. ? (Cryptorhytis) flexicostata, M. & H., sp., pl. xix, fig. 2 and fig. 46, p. 367..... 367-368
  Pyropsis, Conrad, 1860.. 368-36
  P. bairdi, M. & H., 1856, pl. xxxi, figs. 10 a, b................................. 369-379
  P. bairdi var. rotula, Meek, pl. xxxi, fig. 10 and fig. 47, p. 371............... 371-372
  Fusus, Bruguière, 1789.. 372-374
    Serrifusus, Meek.. 373
    Sinistralia, H. & A. Adams, 1853.. 373
  Fusus ? (Serrifusus) dakotensis, M. & H., 1856, pl. xxxi, fig. 11 and pl. 32, figs.
    6 a, b... 374, 375
  F. ? (Serrifusus) dakotensis, var., pl. xxxii, fig. 7 a and 7 b ?................. 375-377
  Cantharus, Bolten, 1798... 377-379
    Tritonidea, Swainson, 1840.. 378
    Cantharulus, Meek... 378
  C. (Cantharulus) vaughani, M. & H., sp., pl. xxxii, figs. 5 a, b, and fig. 48, p. 379. 379-380
Pleurotomidæ.
  Turris, Bolten, 1798.. 380-384
    Surcula, H. & A. Adams, 1853 (= Turricula, Schum.; not of others)............... 381
    Surculites, Conrad, 1865.. 382
    Genota, H. & A. Adams, 1853 (not Adanson)....................................... 382
  T. minor, E. & S., sp. 1857, pl. xxxi, figs. 9 a-c................................ 384-385
  T. (Surcula)? contortus, M. & H., 1856, pl. xxxi, figs. 7 a-c, fig. 49, p. 385... 385-386
  T. (Surcula)? hitzi, Meek, fig. 50, p. 387.. 386-388
Cephalopoda.
  Tetrabranchiata.
    Baculitidæ.. 388-391
    Baculites, Lamarck, 1799.. 391-394
    Cyrtochilu, Meek... 392
    B. ovatus, Say, 1821, pl. xx, figs. 2 a, b, d, and 1 a, b, and fig. 52, p. 397. 394-397
    B. grandis, H. & M., 1854, pl. xxxiii, figs. 1 a-c and fig. 53, p. 399 and fig. 54, p. 400. 398-400
    B. compressus, Say, 1821, pl. xx, figs. 3 a-c and figs. 55, 56, p. 403......... 400-404
    B. asper, Morton ? 1834, pl. xxxix, figs. 10 a, d (not b, c)................... 404-405
    B. anceps var. obtusus, figs. 57-60, p. 406.................................... 406-408

* ὀδούς, a tooth; βάσις, a base.

Ancyloceratidæ.

	Page.
Ancyloceras, d'Orbigny, 1841	408–409
A. ? uncum, M. & H., 1858, pl. xxi, figs. 1 *a, b*	409–410

Ptychoceratidæ.

Ptychoceras, d'Orbigny, 1841	410–412
P. mortoni, M. & H., 1857, pl. xx, figs. 4, *a-c*	412–413

Scaphitidæ.

Scaphites, Parkinson, 1811	413–418
Macroscaphites, Meek	414
Scaphites, Parkinson	414
Discoscaphites, Meek, 1872	15
S. larvæformis, M. & H., 1856, pl. vi, figs. 6 *a-c*	418–419
S. warreni, M. & H., 1860, pl. vi, fig. 5 and figs. 61, 62, p. 421	420–423
S. vermiformis, M. & H., 1862, pl. vi, figs. 4 *a, b*	423–425
S. ventricosus, M. & H., 1862, pl. vi, figs. 7 *a, b,* and figs. 8 *a, b*	425–426
S. nodosus var. *brevis,* pl. xxv, figs. 1 *a-c*	426–428
S. nodosus var. *quadrangularis,* pl. xxv. figs. 3 *a-c,* 2 *a-c* and fig. 4	428–429
S. nodosus var. *plenus,* pl. xxvi, figs. 1 *a-c*	429–430
S. (Discoscaphites) conradi, Morton, sp., 1834, pl. xxxvi, figs. 2 *a-c*	430–432
S. (Discoscaphites) conradi, var. *gulosus,* Morton,1834, pl. xxxvi, fig. 1	432–433
S. (Discoscaphites) conradi, var. *intermedius,* pl. xxxiv, figs. 3 *a-c*	433–435
S. (Discoscaphites) nicoletii, Morton, sp., 1841, pl. xxxiv, figs. 4 *a-c* and 2 *a-b*	435–436
S. (Discoscaphites) cheyennensis, Owen, sp., 1852, pl. xxxv, figs. 3 *a-i*	437–441
S. (Discoscaphites) abyssinus, Morton, sp., 1841, pl. xxxv, figs. 2 *a, b* and 4	441–443
S. (Discoscaphites) mandanensis, Morton, sp., 1841, pl. xxxv, figs. 1 *a-c*	443–444

Ammonitidæ.

Ammonites, Brug., 1789	445–447
A. complexus, H. & M., 1854, pl. xxiv, figs. 1 *a-c*	447–448
Mortoniceras, n. g., Meek, 1876	448–449
M. shoshonense, Meek, pl. vi, figs. 3 *a, c* and 6 *b*	449–450
M. ? vermilioneuse, M & H., 1860, pl. vii, fig. 2 *a, b*	450–452
Prionocyclus, Meek, 1872	452–455
Prionotropis, Meek	453
P. (Prionotropis) woolgari, Mantell, sp., 1822, pl. vii, figs. 1 *a-h,* and pl. vi, fig. 2	455–457
Phylloceras, Suess, 1865	458
P. ? halli, M. & H., 1856, pl. xxiv, figs. 3 *a-c* and fig. 64, p. 458	458–462
Placenticeras, Meek, 1870	462–464
Sphenodiscus, Meek, 1872	463
P. placenta, DeKay, sp., 1828, pl. xxiv, figs. 2 *a, b* and fig. 65, p. 466	465–468
P. placenta var. *intercalare,* pl. xxiii, figs. 1 *a-c*	468–472
P. (Sphenodiscus) lenticulare, Owen, sp., 1852, pl. xxxiv, figs. 1 *a-c,* fig. 66, p. 473	473–476

Turrilitidæ.

Heteroceras, d'Orbigny, 1849	477–478
H. ? cochleatum, H. & M., pl. xxii, figs. 2 *a, b*	478–479
H. ? nebrascense, M. & H., 1856, pl. xxii, figs. 1 *a-c*	480–481
H. tortum, M. & H., 1858, pl. xxii, figs. 4 *a-c*	481–482
H. ? umbilicatum, M. & H., 1858, pl. xxii, fig. 5	482–483
H. ? cheyennense, M. & H., 1856, pl. xxi, figs. 2 *a, b*	483–484
H. ? angulatum, M. & H., 1860, pl. xxi, figs. 3 *a-c*	484–485
Helicoceras, d'Orbigny, 1840	485–487
Patoceras, Meek	485
Helicoceras mortoni var. *tenuicostatum,* pl. xxii, figs. 3 *a-c*	487–489

Nautilidæ.

Nautilus, Linnæus, 1758	489–495
Temnochilus, McCoy, 1844	490
Trematodiscus, M. & W., 1861	491
Discites (De Hann), McCoy, 1825	491
Solenochilus, M. & W., 1870 (= *Cryptoceras,* d'Orbigny)	491
Hercoglossa, Conrad, 1866 (= *Aganides* Montfort ?)	491
Pseudonautilus, Meek	491
N. dekayi, Morton, 1834, pl. xxvii, figs. 1 *a-c* and fig. 67, p. 496	496–498
N. dekayi var. *moanonensis,* pl. xxvii, figs. 2 *a-e*	498
N. elegans, Sowerby, 1816, pl. viii, figs. 2 *a-c*	499–501

---

* *ναρος,* a path; κερας, a horn.

WRITINGS OF F. B MEEK. 105

Dibranchiata.
Belemnitidæ.   Page.
  Belemnitella, d'Orbigny, 1840 501–503
  B. bulbosa, M. & H., 1856, pl. xxxiii, figs. 2 *a-e* 504
Teuthidæ.
  Phylloteuthis, M. & H., 1860 505
  P. subovata, M. & H., 1860, pl. xxxiii, fig. 3 505–506
Articulata.
Annulata.
Tubicola.
Serpulidæ.
  Serpula, Linnæus, 1758 506–507
  S. ? tenuicarinata, M. & H., 1857, pl. vi, fig. 1 507–508

SPECIES OF THE FRESH AND BRACKISH WATER LIGNITE BEDS.
Mollusca.
Lamellibranchiata.
Monomyaria.
Ostreidæ.
  Ostrea, Linnæus 509
  O. subtrigonalis, E. & S. ? 1857, pl. xl, figs. 1 *a-d* 510
Dimyaria.
Unionidæ.
  Unio, Retzius, 1788 511–513
    Boriosta, Raf., 1820 (= *Potamida,* Swainson) 513
    Naidea, Swainson, 1840 514
    Obovaria, Raf., 1819 (= *Rhipidodonta,* Mörch) 514
    Nicæa, Swainson, 1837 514
    Hyridella, Swainson 514
    Lampsilis, Raf., 1820 (= *Truheilla, Pleurobema, Syntoxia, Sealenaria,* and
      Plagiola, Raf.; *Crenodonta,* Schlüt.; *Eglia,* Swainson) 514
    Canthyria, Swainson, 1840 514
    Iridea, Swainson, 1840 (= *Tritigonia* and *Orthonymus,* Agassiz) 514
    Rotundaria, Raf., 1820 (— *Cyprogenia,* Agassiz) 514
    Quadrula, Raf., 1820 (= *Theliderma,* Swainson) 514
    Diplodon, Spix, 1827 (= *Cucumaria,* Conrad, and *Naia,* Swainson) 514
    Dysnomya, Agassiz, 1852 514
    Metaptera, Raf., 1820 (= *Proptera,* Raf, and *Lymnadia* and *Megadomus,* Swain-
      son) 515
  U. priscus, M. & H., 1856, pl. xliii, figs. 8 *a-d* 516–517
  U. danæ, M. & H., 1857, pl. xli, figs. 3 *a-c* 517–518
  U. subspatulatus, M. & H., 1857, pl. xli, figs. 1 *a, b* 518–519
  U. deweyanus, M. & H., 1857, pl. xli, figs. 2 *a-c* 519
Cyrenidæ.
  Corbicula, Mühlfeldt 520
  C. cytheriformis, M. & H., 1860, pl. xl, figs. 5 *a-c* 520–521
  C. occidentalis, M. & H., 1856, pl. xl, figs. 6 *a-c* 521–522
  C. nebrascensis, M. & H., 1860, pl. lxiii, figs. 2 *a, b* (not 2 *c*) 522
  Subgenus *Leptesthes* 523
  C. (*Leptesthes*) *subelliptica,* M. & H., 1856, pl. xliii, figs. 9 *a-c* 523–524
  C. (*Leptesthes*) *subelliptica* var. *moreauensis,* pl. xliii, figs. 1 *a, b* and 2 *c* ... 524
  Sphærium, Scopoli, 1777 525
  S. planum, M. & H., 1860, pl. xliii, figs. 6 *a, b* 526
  S. formosum, M. & H., 1860, pl. xliii, figs. 4 *a-c* 526–527
  S subellipticum, M. & H., 1860, pl. xliii, figs. 5 *a, b* 527
  S. recticardinale, M. & H, 1860, pl. xliii, figs. 3 *a, b* 527–528
Corbulidæ.
  Corbula, Bruguière 528
  Pachyodon, Gabb 528
  C. (*Pachyodon*) *mactriformis,* M. & H., 1856, pl. xliii, figs. 7 *a-f* 528–529
  C. (*Pachyodon*) *subtrigonalis,* M. & H., 1856, pl. xl, figs. 3 *a, b* 529–530
  C. (*Pachyodon*) *perundata.* M. & H., 1856, pl. xl, figs. 4 *a-d* 530–531
Gasteropoda.
Pulmonata.
Limnæidæ.
  Limnæa, Lamarck, 1799 531–533
    Radix, Montfort, 1810 (= *Gulnaria,* Leach) 532

	Page.
Polyrhytis, Meek	532
Bulimnea, Haldeman, 1842	532
Limnophysa, Fitzinger, 1833 (=*Stagnicola*, Leach & *Galba*, Schranck)	533
Omphiscola, Raf. (*Leptolimnea*. Swainson)	533
Acella, Haldeman, 1842	533
Pleurolimnea, Meek, 1866	533
L. (*Pleurolimnæa*) *tenuicostata*. M. & H., 1856, pl. xliv, figs. 13 *a–c*	534
Planorbis, Müller, 1776	534–536
Helisoma, Swainson, 1840	535
Planorbella, Haldeman. 1842	536
Taphius, H. & A. Adams, 1856	536
Menetus, H. & A. Adams. 1856 (—*Anisus*, Beck, not Fitz.)	536
Anisus, Fitzinger, 1833 (— *Tropidiscus*, Stein.)	536
Bathyomphalus, Agassiz, 1837 (=*Spirorbis*, Swainson, not Lamarck)	536
Gyraulus, Agassiz (= *Nautilina*, Stein.)	536
P. convolutus, M. & H., 1856, pl. xlii, figs. 12 *a*, *b*	536–537
P. convolutus var., pl. xlii, figs. 11 *a–e*	538
P. (*Bathyomphalus*) *planoconvexus*. M. & H., 1857, pl. xliv, figs. 9 *a–c*	538–539
P. (*Bathyomphalus*) *amplexus*, M. & H., 1857, pl lxii, figs. 16 *a–e*	539
Physidæ.	
Bulinus, O. F. Müller, 1781	540
B. subelongatus, M. & H., 1856, pl. lxii, figs. 13 *a*, *b*	540–541
B. longiusculus, M. & H., 1856, pl. xliii, figs. 16 *a*, *b*	541–542
B. rhomboideus, M. & H., 1856, pl. xliii, fig. 17	542
Ancylidæ.	
Acroloxus, Beck, 1837	543
A. minutus, M. & H., 1856, pl. xliv, fig. 10	543–544
Vitrinidæ.	
Vitrina, Draparnaud, 1801	544
V.? obliqua, M. & H., 1857, pl. xlii, figs. 16 *a*, *b*	545
Hyalina, Férussac, 1819	545–547
H.? occidentalis, M. & H., 1857, pl. xlii, figs. 6 *a–d*	547–548
H.? evansi, M. & H., 1860, figs. 68, 69, 70, p. 548	548–549
Helicidæ.	
Helix, Linnæus, 1758	549–551
Galaxias, Beck, 1837	549
Camœna, Albers, 1850	550
Helix retuxta, M. & H., 1860, pl. xlii, figs. 7 *a*, *b*	552
Thaumastus, Albers, 1860	553
T. limnæiformis, M. & H., 1856, pl. xliv, figs. 8, *a–d*	553–554
Columna, Perry, 1811	554
Rhodea, H. & A. Adams, 1855	555
Columna teres, M. & H., 1856, pl. xliv, figs. 11 *a*, *b*	555–556
O. vermicula, M. & H., 1856, pl. xliv, figs. 12 *a*, *b*	556–557
O. vermicula var. *contraria*, Meek, 1866	557
Pectinibranchiata.	
Cerithiidæ.	
Cerithidea, Swainson, 1840	558
Pirenella, Gray, 1847	558
Cerithidea (*Pirenella?*) *nebrascensis*, M. & H., 1856, pl. xliii, figs. 9 *a–c* (*bis*)	559
Ceriphasiidæ.	
Goniobasis, Lea, 1862	560–561
G. convexa, M. & H., 1856, pl. xlii, figs. 2 *a*, *b*, and figs. 71, p. 562 and 72, p. 563	562–563
G. convexa var. *impressa*, M. & H., 1857, pl. xlii, figs. 2 *e*, *d*	563–564
G. invenusta, M. & H., 1857, pl. xlii, figs. 1 *a–e*	564–565
G. nebrascensis, M. & H., 1856, pl. xliii, figs. 12 *a–h*, and fig. 73, p. 565	565–566
G. tenuicarinata, M. & H., 1857, pl. xliii, figs. 14 *a–c*	566–567
G. subtorvis, M. & H., 1857, pl. xlii, figs. 5 *a*, *b*	567
G.? omitta. M. & H., 1857, pl. xlii, figs. 4 *a–c*	568
G. gracilenta, Meek, pl. xlii, fig. 3 and fig. 74, p. 569	568–569
G.? subtortuosa, M. & H. 1857, pl. xlii, figs. 17, *a–b*, and figs. 75, 76, p. 569	569–570
Rissoidæ.	
Hydrobia, Hartmann, 1821	571
H. anthonyi, M. & H., 1856, pl. xliii, figs. 10 *a–d*	571–572
H. warrenana, M. & H., 1857, pl. xliii, figs. 1 *a–c*	572–573

	Page.
H. subconica, Meek, fig. 77, p. 573	573
H. ? eulimoides, Meek. fig. 78, p. 573	573–574
Micropyrgus, Meek, 1866	574–575
M. minutulus, M. & H., 1856, pl. xliii, figs. 18 *a*, *b*	575
Viriparidæ.	
Viviparus, Montfort, 1810	576–577
V. leai, M. & H., 1856, pl. xliv, figs. 6 *a–d*	577–578
V. retusus, M. & H., 1856, pl. xliv, figs. 5 *a–f*	578–579
V. conradi, M. & H., 1856, pl. xlii, figs. 15 *a–d*	579–580
V. peculiaris, M. & H., 1856 fig. 79.	580
V. trochiformis, M. & H., 1856, pl. xliv, fig. 2 *a–e*	580–582
V. leidyi, M. & H., 1856, pl. xliv, fig. 4.	582–583
V. leidyi var. *formosus*, pl. xliv, figs. 3 *a*, *b*	583
V. raynoldsanus, M. & H., 1861, pl. xliv, figs. 7 *a*, *b*	584–585
Campeloma, Ratinesque	
C. multilineata, M. & H., 1856, pl. xliv, figs. 1 *a*, *b*	586–587
C. vetula, M. & H., 1856 pl. xliv, figs. 14 *a*, *b*	587–588
C. multistriata, M. & H., 1856, pl. xliii, figs. 15 *a–e*, and fig. 80, p. 588	588–589
Valvatidæ.	
Valvata, Müller, 1774	589–590
Tropidina, H. & A. Adams, 1856	590
V. subumbilicata, M. & H., 1856, pl. xliii, figs. 13 *a–c*	590–591
V. parvula, M. & H., 1856	591
V. ? montanaensis, Meek, figs. 81, 82, 83, p. 591	591–592

FOSSILS OF THE WIND RIVER TERTIARY.

Mollusca.
Gasteropoda.
Pulmonata.
Vitrinidæ.

Macrocyclis, Beck, 1837	593–594
Ampelita, Beck, 1837	594
M. spatiosa, M. & H., 1861, pl. xlii, figs. 9 *a–e*	594–595
Helicidæ.	
Helix, Linn	596
H. ? reterna, M. & H., 1861, pl. xlii, figs. 8 *a*, *b*, and figs. 84, 85, p. 596	596–597

FOSSILS OF THE WHITE RIVER TERTIARY.

Mollusca.
Gasteropoda.
Pulmonata.
Limnæidæ.

Limnæa, Lamarck	598
L. meekiana, E. & S., pl. xlv, figs. 5 *a–c*	598–599
L. shumardi, Meek, pl. xlv, figs. 6 *a*, *b*	599
Planorbis, Müller	599
P. leidyi, M. & H., 1860, pl. xlv, figs. 3 *a–d*	599–600
Subgenus *Menetus*, H. & A. Adams ?	600
P. (Menetus) nebrascensis, E. & S., 1854, pl. xlv, figs. 2 *a*, *b*	600–601
P. (Menetus) retusus, M. & H., 1860, pl. xlv, figs. 1 *a–c*	601–602
Physidæ.	
Physa, Draparnaud, 1801	603–604
Physella, Haldeman, 1842	603
Physodon, Haldeman, 1842	603
Isidora, Ehrenb., 1831 (= *Diastropha*, Guilding)	604
Costella, Dall, 1870	604
P. secalina, E. & S., 1854, pl. xlv, figs. 4 *a*, *b*	604
Helicidæ.	
Helix, Linn	604
H. leidyi, H. & M., 1854, pl. xlv, figs. 7 *a*, *b*	604–605

APPENDIX.

Tellina (Arcopagia) ? cheyennensis, M. & H., 1856, pl. xvii, fig. 16	607
Ammonites ? ? mullananus, M. & H., 1862, pl. viii, figs. 1 *a–c*	607–609

106.

MEEK, F. B. Palæontology. < Rep. Geol. Expl. 40th Parallel, vol. iv, part i, pp. 1-197, pls. i-xvii. Washington, 1877.

Silurian, Devonian, Carboniferous, Triassic, Jurassic, Cretaceous, Tertiary. Genera *Eutomoceras* (Hyatt), *Eudiscoceras* (Hyatt), *Polyrhytis*, *Rhytophorus*, *Pyrgulifera*.

DESCRIPTIONS OF FOSSILS.

SILURIAN SPECIES.

	Page.
Mollusca.	
Gasteropoda.	
Solariidæ.	
? *Ophileta*, Vanuxem	17
O. complanata var. *nana*, Meek, 1870, pl. i, figs. 1, 1 *a*, 1 *b*	17–18
Raphistoma, Hall	18
R. ? *rotuliformis*, Meek, 1870, pl. i, figs. 2, 2 *a*, *b*	18–19
R. ? *trochiscus*, Meek, 1870, pl. i, figs. 3, 3 *a*, *b*	19
Articulata.	
Crustacea.	
Paradoxidæ.	
Ornocoryphe, Corda	20
C. (*Ptychoparia*) *kingii*, Meek, 1870, pl. i, fig. 4	20–23
Paradoxides, Brongniart	23
P. ? *nevadensis*, Meek, 1870, pl. i, fig. 5	23–25

DEVONIAN SPECIES.

Radiata.	
Polypi.	
Favositidæ.	
Alveolites, Lamarck	25
A. multilamella, n. s., Meek, 1877, pl. ii, figs. 7, 7 *a*, *b*	25–26
Alveolites, undt. sp., Meek, 1877	26–27
Favosites, Lamarck	27
Favosites, undt. sp., Meek, 1877, pl. i, fig. 6	27
F. polymorpha, Goldf.? var. Meek, 1877, pl. ii, fig. 3	27–28
Syringopora, Goldfuss	28
Syringopora, undt. sp., Meek, 1877	28
Cyathophyllidæ.	
Ptychophyllum, E. & H.	28
P. ? *infundibulum*, n. s., Meek, 1877, pl. ii, figs. 1, 1 *a*, *b*	28–29
Diphyphyllum, Lonsdale	29
D. fasciculum, n. s., Meek, 1877, pl. ii, figs. 4, 4 *a*, *b*	29–31
Acervularia, Schweigger	31
A. pentagona, Goldfuss, sp., 1826, pl. ii, figs. 5, 5 *a*	31–32
Smithia, E. & H	32
S. hennahii, Lonsdale, sp., 1840, pl. 2, figs. 6, 6 *a*	32–33
Cyathophyllum, Goldfuss	33
C. palmeri, n. s., Meek, 1877, pl. ii, fig. 2	33–34
Mollusca.	
Brachiopoda.	
Strophomenidæ.	
Hemipronites, Pander	35
H. chemungensis var. *arctostriata*, Conrad, sp., pl. iii, fig. 2	35–36
Productidæ.	
Productus, Sowerby	36
P. subaculeatus, Murchison ? 1840, pl. iii, figs. 7, 7 *a*, *b*	36–37
Rhynchonellidæ.	
? *Atrypa*, Dalman	38
A. reticularis, Linnæus, sp., 1767, pl. i, figs. 7 and 7 *a*, and pl. 3, figs. 6 ? 6 *a*	38, 39
Spiriferidæ.	
Spirifer, Sowerby	39
S. utahensis, Meek, 1860, pl. iii, figs. 1, 1 *a–e*	39–41
S. engelmanni, Meek, 1860, pl. iii, figs. 3 *a–e* (and 3 *d–f*?)	41–42
S. (*Trigonotreta*) *argentarius*, n. s., Meek, 1877, pl. iii, figs. 4 and 4 *a*, *b*	42–43
S. (*Trigonotreta*) *strigosus*, Meek, 1860, pl. iii, figs. 5, 5 *a*, *b*	43–45
S. (*Trigonotreta*) *piñonensis*, Meek, 1870, pl. i, figs. 9, 9 *a*, *b*	45–46
Lamellibranchiata.	

Anatinidæ. | Page.
Edmondia, de Koninck 46
E. ? piñonensis, n. s., Meek, 1877, pl. i, figs. 8, 8 *a* 46–47
Cephalopoda.
  Orthoceratitidæ.
    Orthoceras, Auct..
    O. kingii, n. s., Meek, 1877, pl. ii, fig. 8 47–48
    Orthoceras, undt. sp., Meek, 1877, pl. ii, fig. 9 48
Articulata.
  Crustacea.
    Phacopsidæ.
      Dalmanites, Auct.. 48
      Dalmanites, sp. undt., Meek, 1877, pl. i, figs. 11, 11 *a* 48–49
    Proetidæ.
      Proetus, Steininger... 49
      P. (Phæton) denticulatus, n. s., Meek, 1877, pl. i, figs. 10, 10 *a* and *b* 49–50

CARBONIFEROUS SPECIES.

Radiata.
  Polypi.
    Favositidæ.
      Syringopora, Goldfuss.. 50
      Syringopora, undt. sp., Meek, 1877, pl. vi, figs. 2, 2 *a* 50–51
    Cyathophyllidæ.
      Zaphrentis, Rafinesque & Clifford 52
      Z. excentrica, n. s., Meek, 1877, pl. iv, figs. 1, 1 *a–d* 52–53
      Z. ? multilamella, Hall ? 1852, pl. vi, figs. 4, 4 *a* and *b* 53–54
      Z. ? stansburii, Hall ? 1852, pl. vi, figs. 3, 3 *a–c* 54–56
      Campophyllum, E. & H 57
      Campophyllum, undt. sp., Meek, 1877, pl. v, figs. 2, 2 *a* and *b* 57–58
      Lithostrotion, Fleming.. 58
      L. whitneyi, Meek, n. s., 1875, pl. vi, figs. 1, 1 *a–c* 58–59
      Cyathophyllum, Goldfuss... 60
      C. (Campophyllum ?) nevadense, n. s., Meek, 1877, pl. v, figs. 3, 3 *a* and *b* ... 60
      C. subcæspitosum, n. s., Meek, 1877, pl. v, figs. 4, 4 *a* and *b* 60–61
Mollusca.
  Brachiopoda.
    Strophomenidæ.
      Hemipronites, Pander.. 62
      H. crenistria, Phillips, sp., 1836, pl. vii, fig. 2 62
      Orthis, Dalman 63
      O. michelini, L'Éveillé, var. Meek, 1877, pl. vii, figs. 1, 1 *a–c* 63–64
    Productidæ.
      Productus, Sowerby... 64
      P. nevadensis, n. s., Meek, 1877, pl. viii, figs. 2, 2 *a–e* 64–67
      Productus, undt. sp., Meek, 1877, pl. vii, figs. 6, 6 *a* and *b* 67–68
      P. semireticulatus, Martin, 1709, pl. vii, fig. 5 69
      P. costatus, Sowerby ? var. Meek, 1877, pl. vii, figs. 4, 4 *a* and *b* 69–72
      P. prattenianus, Norwood, 1854, pl. vii, fig. 7 72–73
      P. semistriatus, Meek, 1860, pl. vii, figs. 8, 8 *a* 74–75
      P. subhorridus, n. s., Meek, 1877, pl. vii, figs. 3, 3 *a* and *b* 75–76
      P. multistriatus, Meek, 1860, pl. viii, figs. 3, 3 *a–e* 76–78
      P. longispinus, Sowerby, 1814, pl. viii, figs. 4, 4 *a* 78–79
    Rhynchonellidæ.
      Leiorhynchus, Hall... 79
      L. quadricostatus, Vanuxem? sp., 1842, pl. iii, figs. 9, 9 *a* and *b* 79–80
    Spiriferidæ.
      Athyris, McCoy.. 81
      A. ? persinuata, n. s., Meek, 1877, pl. ix, figs. 4, 4 *a* and *b* 81–82
      A. roissyi, l'Éveillé, sp., 1835, pl. ix, figs. 3, 3 *a* and *b* 82–83
      A. subtilita, Hall, 1852, pl. viii, figs. 6, 6 *a* 83–84
      Spiriferina d'Orbigny... 84
      Spiriferina, undt. sp, Meek, 1877, pl. viii, figs. 5, 5 *a, b* 84–85
      S. pulchra, Meek, 1860, pl. viii, figs. 1, 1 *a–e*, and pl. xii, figs. 12. 12 *a–d ?* .. 85–87
      Spirifer, Sowerby... 87
      S. cuspidatus, Martin ? sp., 1796, pl. iii, figs. 11, 11 *a* 87–88
      Spirifer (Trigonotreta) opimus, Hall ! 1858, pl. ix, fig. 6 88–90
      S. (Trigonotreta) scobina, Meek, 1860, pl. ix, figs. 1, 1 *a–d* 90–91
      S. (Trigonotreta) cameratus, Morton, 1836, pl. ix, figs. 2, 2 *a* 91–92

Lamellibranchiata.
Pteriidæ.
? *Posidonomya*, Bronn 92
P. ? fragosa, n. s., Meek, 1877, pl. iii, figs. 8, 8 a 92-93
Aviculopecten, McCoy 93
A. cactutus, n. s., Meek, 1877, pl. iii, figs. 10, 10 a, and 10 b 1............ 93-05
A. utahensis, Meek, 1860, pl. ix, figs. 7, 7 a and b (and 7 c, and d ?)........ 95-96
A. occidaneus, n. s., Meek, 1877, pl. xii, figs. 13, 13 a and b................ 96-97
Cephalopoda.
Goniatitidæ.
Goniatites, De Haan 98
G. goniolobus. n. s., Meek, 1877, pl. ix, figs. 5, 5 a, b 98-99

UPPER TRIASSIC SPECIES.
Mollusca.
Brachiopoda.
Discinidæ.
Discina, Lamarck 99
Discina, sp. undt., Meek, 1877, pl. x, figs. 3, 3 a 99-100
Lamellibranchiata.
Pteriidæ.
Halobia, Bronn 100
H. (Daonella) lommeli, Wissmann, 1841, pl. x, fig. 5 100-102
Lucinidæ.
Sphæra, Sowerby 102
S. whitneyi, n. s., Meek, 1877, pl. x, figs. 4, 4 a-c 102
Mytilidæ.
? *Modiomorpha*, H. & W ....
M. ? ovata, n. s., Meek, 1877, pl. x, figs. 1, 1 a 103
M. ? lata, n. s., Meek, 1877, pl. x, fig. 2 103-104
Cephalopoda.
Orthoceratitidæ.
Orthoceras, Auct 104
O. blakei, Gabb ? 1864, pl. x, fig. 11 104-105

AMMONITOID FORMS OF THE UPPER TRIAS OF NEVADA.

Clydonitidæ, n. f., Hyatt, 1877 107
Coroceras, n. g., Hyatt, 1877 107-108
Clydonites, Hauer 109
C. lævidorsatus, Hauer, sp., 1860, pl. x, fig. 7 109-110
Trachyceratidæ.
? *Gymnotoceras*, n. g., Hyatt, 1877 110-111
G. rotelliforme, n. s., Meek, 1877, pl. x, figs. 9, and 9 a 111-113
G. blakei, Gabb, sp., 1864, pl. x, figs. 10, 10 a-c, and pl. xi, figs. 6, 6 a. 113-116
Trachyceras, Laube 116
T. whitneyi, Gabb, sp., 1864, pl. xi, figs. 3, 3 a 116-118
T. judicaricum, Mojsisovics, 1869, pl. xi, figs. 1, 1 a 118
T. judicaricum, var. *subasperum*, Meek, 1877, pl. xi, figs. 2, 2 a and b .. 118-119
Arcestidæ.
Arcestes, Suess 119-120
A. ? perplanus, n. s., Meek, 1877, pl. xi, figs. 7 and 7 a 120-121
A. gabbi, n. s., Meek, 1877, pl. x. figs. 6, 6 a and b 121-123
Physanoidæ.
? *Acrochordiceras*, n. g., Hyatt, 1877 124
A. hyatti, n. s., Meek, 1877, pl. xi, figs. 5 and 5 a 124-126
Entomoceras, n. g., Hyatt, 1877 126
E. laubei, n. s., Meek, 1877, pl. 10, figs. 8, 8 a 126-128
Eudiscoceras, n. g., Hyatt, 1877 128
E. gabbi, n. s., Meek, 1877, pl. xi, figs. 3 and 3 a 128-129

JURASSIC SPECIES.
Mollusca.
Lamellibranchiata.
Limidæ.
Lima, Bruguière 130
L. (Limatula) erecta, n. s., Meek, 1877, pl. xii, fig. 2 130

---
* κόρυς, a helmet ; κερας, a horn.   ? ἀκροχορδών, **a wart**; **κερας, a horn.**
† γυμνός, naked ; νῶτος, back ; κερας, a horn.

	Page.
Pinnidæ.	
Pinna, Linnæus	131
P. kingii, n. s., Meek, 1877, pl. xii, figs. 9, 9 a	131–132
Mytilidæ.	
Volsella, Scopoli	132
V. scalprum, var. *isonema,* Meek, 1877, pl. xii, figs. 4, 4 a	132–133
Trigoniidæ.	
Myophoria, Bronn	133
M. lineata, Münster? 1834, pl. xii, figs. 3, 3 a	133–134
Arcidæ.	
Oucullæa, Lamarck	134
O. haguei, n. s., Meek, 1877, pl. xii, figs. 1, 1 a, b	134–135
Anatinidæ.	
Myacites, Auct	136
M. (Pleuromya) subcompressa, Meek, 1873, pl. xii, figs. 6, 6 a	136–137
M. inconspicuus, n. s., Meek, 1877, pl. xii, fig. 10	137
M. (Pleuromya) weberensis, n. s., Meek, 1877, pl. xii, figs. 11, 11 a	137–138
Cephalopoda.	
Belemnitidæ.	
Belemnites, Auct	138
B. nevadensis, n. s., Meek, 1877, pl. xii, figs. 7, 7 a, b (and 8 a, b ?)	138–139
CRETACEOUS FOSSILS.	
Mollusca.	
Ostreidæ.	
Ostrea, Linn	140
Ostrea, undt. sp., Meek, 1877, pl. xv, figs. 10, 10 a–c	140–141
Anomiidæ.	
Anomia rætiformis, n. s., Meek, 1877, fig. i, p. 141	141
Pteriidæ.	
Inoceramus, Sowerby	142
I. simpsoni, Meek, 1860, pl. xiii, fig. 3	142–143
I. problematicus, Schlot? 1820, pl. xiii, figs. 2 and 2 a	143–144
Inoceramus, sp. undt., Meek, 1877, pl. xiii, figs. 4, 4 a	144
I. erectus, n. s., Meek, 1877, pl. xiii, figs. 1, 1 a, and pl. xiv, fig. 3	145
I. deformis, Meek, 1872, pl. xiv, figs. 4, 4 a	146–148
Arcidæ.	
Oucullæa, Lamarck	148
O. (Trigonarea ?) obliqua, n. s., Meek, 1877, pl. xiv, figs. 1, 1 a–b	148–149
Axinæa, Poli	149
A. wyomingensis, n. s., Meek, 1877, figs. 2 and 3, p. 150	149–150
Corbulidæ.	
Corbula, Bruguière	150
Corbula, undt. sp., Meek, 1877, pl. xiv, fig. 2	150–151
Cardiidæ.	
Cardium, Linnæus	151
C. curtum, M. & H.? 1861, pl. xv, fig. 3 (not 3 a)	151–152
C. subcurtum, Meek, 1873, pl. xv, fig. 3 a (not 3)	152–153
Mactridæ.	
Mactra, Linnæus	153
M.? emmonsi, n. s., Meek, 1877, pl. xv, fig. 8	153–154
M. (Trigonella) ? arenaria, n. s., Meek, 1877, pl. xiv, fig. 5	154–155
M. (Cymbophora) ? utahensis, n. s., Meek, 1877, pl. xv, figs. 9, 9 a, b	155–156
Tellinidæ.	
Tellina, Linnæus	156
T.? isonema, n. s., Meek, 1877, pl. xv, fig. 6	156–157
T. modesta, n. s., Meek, 1877, pl. xv, figs. 4–5	157–158
Veneridæ.	
Cyprimeria, Conrad	158
C.? subalata, Meek, 1873, pl. xv, fig. 7	158–159
Gasteropoda.	
Naticidæ.	
Gyrodes, Conrad	159
G. depressa, n. s., Meek, 1877, pl. xv, figs. 1, 1 a	159–160
Aporrhaidæ.	
Anchura, Conrad	160
A.? fusiformis, n. s., Meek, 1877, pl. xv, figs. 2, 2 a	160–161

Siphonariidæ.
? *Anisomyon*, M. & H 162
A. sexsulcatus, M. & H.? 1856, figs. 4 and 5, p. 162 162

FOSSILS OF THE BEAR RIVER FRESH OR BRACKISH WATER BEDS.

Mollusca.
Lamellibranchiata.
Unionidæ.
Unio, Retzius 164
U. retusus, Meek, 1860, pl. xvi, figs. 5, 5 *a–c* 164–165
U. belliplicatus, Meek, 1870, pl. xvi, figs. 4, 4 *a* 165–167
Cyrenidæ.
Corbicula, Benson 167
C. (Veloritina) durkeei, Meek, 1870, pl. xvi, figs. 6 *a–g* 167–170
Corbulidæ.
Corbula, Brug 170
C. (Anisorhynchus) pyriformis, Meek, 1860, pl. xvii, figs. 2, 2 *a–d* 170–174
C. (Anisorhynchus?) engelmanni, Meek, 1860, pl. xvii, figs. 1, 1 *a* 174–175
Gasteropoda.
Auriculidæ.
Rhytiphorus, Meek, 1873 175
R. priscus, Meek, 1860, pl. xvii, figs. 6 and 6 *a* 175–176
Ceriphasiidæ.
Pyrgulifera humerosa, Meek, 1860, pl. xvii, figs. 19, 19 *a*, and fig. 6, p. 177 .. 176–178
Viviparidæ.
Viviparus, Montfort 178
V. conradi, M. & H., 1856, pl. xvii, figs. 18, 18 *a* 178–179
Campeloma, Rafinesque 179
C. macrospira, Meek ? 1872, pl. xvii, figs. 17 *a, b* 179–181
Campeloma (undt. sp.), Meek, 1877, pl. xvii, figs. 15 *a, b*, and fig. 16 *a–c* 181
Limnæidæ.
Limnæa, Lamarck 181
L. (Limnophysa) nitidula, Meek, 1860, pl. xvii, figs. 5, 5 *a* 181–182

SPECIES OF UNDOUBTED TERTIARY AGE.

Mollusca.
Lamellibranchiata.
Cyrenidæ.
Sphærium, Scopoli.
S. rugosum, Meek, 1870, pl. xvi, figs. 2, 2 *a, b* 182–183
S.? idahoense, Meek, 1870, pl. xvi, fig. 1, 1 *a* 183–184
Unionidæ.
Unio, Retzius 184
U. haydeni, Meek, 1860, pl. xvi, figs. 3, 3 *a, b* 184–185
Gasteropoda.
Limnæidæ.
Ancylus, Geoffroy 186
A. undulatus, Meek, 1870, pl. xvii, figs. 12 *a, b* 186
Carinifex, Binney 187
Subgenus *Vorticifex*, Meek, 1870 187
C. (Vorticifex) binneyi, Meek, 1870, pl. xvii, figs. 11, 11 *a* 187–188
C. (Vorticifex) tryoni, Meek, 1870, pl. xvii, figs. 10, 10 *a–c* 188–189
Planorbis, Guettard 189
P. spectabilis, Meek, 1860, pl. xvii, figs. 13, 13 *a–f* 189
P. spectabilis var. *utahensis*, Meek, 1860, pl. xvii, figs. 14, 14 *a–c* 190–191
Limnæa, Lamarck 191
L. (Limnophysa) retusa, Meek, 1860, pl. xvii, figs. 4, 4 *a, b* 191
L. similis, Meek, 1860, pl. xvii, figs. 3, 3 *a* 191–192
L. (Polyrhytis) kingii, n. s., Meek, 1877, figs. 6 and 7, p. 192 192–193
Ceriphasiidæ.
Goniobasis, Lea 193
G. simpsoni, Meek, 1860, pl. xvii, figs. 7, 7 *a–d* 193–195
Melaniidæ.
Melania, Auct 195
M. ? sculptilis, Meek, 1870, pl. xvii, fig. 8 195–196
M. ? subsculptilis, Meek, 1870, pl. xvii, fig. 9 196–197

PART II.

THE PUBLISHED WRITINGS OF CHARLES ABIATHAR WHITE, 1860–1884.

II.—THE PUBLISHED WRITINGS OF CHARLES ABIATHAR WHITE, A. M., M. D.

This catalogue is intended to embrace the titles and place of publication, not only of all the scientific writings of Dr. White, but his reviews of the writings of other authors, and his more popular articles also. In only a few instances however, are newspaper articles, of which he has written many, noticed on the following pages; but the intention has been to make entry of all his short published notes which contain any expression of his views upon scientific subjects, as well as of his more elaborate works. The annotations which accompany this catalogue are made up mainly from data furnished by the author of the works, and all expressions of opinion upon geologic and paleontologic subjects contained in them should be regarded as his own.

Charles A. White was born in North Dighton, Bristol County, Massachusetts, on January 26, 1826. He has held the following official positions in education and science, to which subjects most of his writings pertain: State Geologist of Iowa, by legislative appointment, from 1866 to 1869, inclusive; Professor of Natural History in the Iowa State University, from 1867 to 1873; Professor of Natural History in Bowdoin College, from 1873 to 1875; Paleontologist to the U. S. Geographical and Geological Surveys West of the 100th Meridian, in charge of Lieut. George M. Wheeler, in 1874; Geologist and Paleontologist to the U. S. Geological Survey of the Territories, in charge of Maj. J. W. Powell, in 1875; Geologist and Paleontologist to the U. S. Geological Survey of the Territories, in charge of Dr. F. V. Hayden, from 1876 to 1879; in charge of paleontological collections in the U. S. National Museum from 1879 to 1882; detailed in 1881 to act as chief of the Artesian Wells Commission upon the Great Plains, under the auspices of the U. S. Agricultural Department; Geologist to the U. S. Geological Survey in 1882; Paleontologist to the U. S. Geological Survey in 1883, which position he now holds. together with honorary curatorship in the U. S. National Museum; President of the Biological Society of Washington for the years 1883 and 1884.

A partial bibliography of Dr. White appeared in the "American Field" for March and April, 1885; this is by Charles Aldrich.

1.

WHITE, C. A. Observations upon the Geology and Paleontology of Burlington, Iowa, and its Vicinity. <Bost. Jour. Nat. Hist., (Bost. Soc. Nat. Hist.) vol. vii, pp. 209–235. Boston, 1861.

Same. Boston, 1860, 8vo., p. 209–235. Fifty separates printed without title-page, covers, or repaging.

The 50 separates appeared at the date mentioned, but Part II of vol. vii, according to a note inserted in it, did not appear until December, 1862, although it bears the imprint 1861.

This paper is divided into two parts; in the first, details of local geology are given, and attention is called to the fact that in this locality the change from Devonian to Carboniferous took place so gradually as to render it impossible to point out the exact line where one ends and the other begins. A section of the rocks at Burlington is given with a table showing the vertical range of shells, and a list of the genera discovered in the rocks at Burlington, showing the different beds in which they have been recognized.

The second part contains descriptions of seven new species of *Brachiopoda* from the Chemung rocks at Burlington, Iowa.

	Page.
Rhynchonella pustulosa, n. s., White, 1860	226
Nucleospira barrisii, n. s., White, 1860	227–228
Terebratula burlingtonensis, n. s., White, 1860	228–229
Athyris crassicardinalis, n. s., White, 1860	229–230
Productus lævicostus, n. s White, 1860	230–231
Orthis thiemei, n. s., White, 1860	231–232
Spirifer solidirostris, n. s., White, 1860	232–233
This part also contains a list of described fossils recognized in the Burlington beds	233–235

2.

WHITE, C. A., and WHITFIELD, R. P. Observations upon the Rocks of the Mississippi Valley which have been referred to the Chemung Group of New York, together with Descriptions of New Species of Fossils from the same horizon at Burlington, Iowa. <Proc. Bost. Soc. Nat. Hist., vol. viii, pp. 289–306. Boston, 1862.

Same. Boston, 1862, 8vo., pp. 289–306. Fifty separates printed without title-page, covers, or repaging.

Thirty-one species of Lower Carboniferous fossils are described in this paper.

Description of new species.

	Page.
Brachiopoda.	
Orthis, Dalman	292
O. subelliptica, n. s., White & Whitfield, 1862	292–293
Streptorhynchus, King	293
S. inflatus, n. s., White & Whitfield, 1862	293
Spirifer, Sowerby	293
S. hirtus, n. s., White & Whitfield, 1862	293–294
Retzia, King	294
R. sexplicata, n. s., White & Whitfield, 1862	294
Rhynchonella, Fischer	294
R. opposita, n. s., White & Whitfield, 1862	294–295
Pentamerus, Sowerby	295
P. lenticularis, n. s., White & Whitfield, 1862	295
Conchifera.	
Avicula-pecten, McCoy	295
A. limaformis, n. s., White & Whitfield, 1862	295–296
A. nodocostatus, n. s., White & Whitfield, 1862	296
Mytilus, Linn	296
M. febristriatus, n. s., White & Whitfield, 1862	296–297
M. occidentalis, n. s., White & Whitfield, 1862	297
Subgenus *Orthonota*, Conrad	297
M. (Orthonota) ventricosa, n. s., White & Whitfield, 1862	297–298
Nucula, Lamarck	298
N. iowensis, n. s., White & Whitfield, 1862	298
Leda, Schum	298
L. barrisii, n. s., White & Whitfield, 1862	298

	Page.
Macrodon, Lycett	298
M. parvus, n. s., White & Whitfield, 1862	299
Conocardium, Bronn	299
C. pulcellum, n. s., White & Whitfield, 1862	299
Cypricardia? rigida, u. s., White & Whitfield, 1862	300
Cypricardella, Hall	300
C. quadrata, n. s., White & Whitfield, 1862	300–301
Edmondia, Koninck	301
E. burlingtonensis, n. s., White & Whitfield, 1862	301
Gasteropoda.	
Euomphalus, Sowerby	301
E. ammon, n. s., White & Whitfield, 1862	301
Platyceras, Conrad	302
P. parahum, u. s., White & Whitfield, 1862	302
P. bivalve, n. s., White & Whitfield, 1862	302
Pleurotomaria, Defrance	302
P. mississippiensis, n. s., White & Whitfield, 1862	302
Murchisonia, D'Archiac	303
M.? prolixa, n. s., White & Whitfield, 1862	303
Porcellia, Léveillé	303
P. crassinoda, n. s., White & Whitfield, 1862	303
Bellerophon, Montfort	304
B. vinculatus, n. s., White & Whitfield, 1862	304
B. perelegans, n. s., White & Whitfield, 1862	304
B. bilabiatus, n. s., White & Whitfield, 1862	304–305
Cephalopoda.	
Goniatites, De Haan	305
G. opimus, n. s., White & Whitfield, 1862	305
Radiata.	
Lophophyllum, Edwards & Haime	305
L. calceola, n. s., White & Whitfield, 1862	305–306
Zaphrentis, Rafinesque	306
Z. acutus, n. s., White & Whitfield, 1862	306
Favosites, Lam	306
F.—[Whitfieldi]	306

3.

WHITE, C. A. Observations on the Summit Structure of Pentremites, the Structure and Arrangement of certain Parts of Crinoids, and Descriptions of New Species from the Carboniferous Rocks of Burlington, Iowa. < Boston Jour. Nat. Hist. (Boston Soc. Nat. Hist.), vol. vii, pp. 481–506. Boston, 1863.

Same. Boston, 1863. 8vo, pp. 581–506. Fifty separates printed without title-page, covers, or repaging.

The summit structure of *Pentremites norwoodi*, *P. stelliformis*, *P. lineatus*, and *P. clongatus* is noticed. Instances of the recuperative power of Crinoids are also given. The genus *Coeliocrinus* is proposed, and five species of Crinoids are described.

Some observations on certain modifications of the structure of the proboscis of *Actinocrinus* are given, p. 489-491.

	Page.
Coeliocrinus*, n. g., White, 1863	499–501
C. subspinosus, n. s., White, 1863	501–502
Platycrinus, Miller	502
P. verrucosus, n. s., White, 1863	502–503
P. incomptus, n. s., White, 1863	503–504
Cyathocrinus, Miller	504
C. lamellosus, n. s., White, 1863	504–505
Scaphiocrinus, Hall	505
S. rusticellus, n. s., White, 1863	505–506

* Κοιλία, venter; κρίνον, lilium.

4.

WHITE, C. A. Descriptions of new species of Fossils from the Devonian and Carboniferous Rocks of the Mississippi Valley. < Proc. Boston Soc. Nat. Hist., vol. ix, pp. 8–33. Boston, 1865.

Same. Boston, 1862. 8vo, pp. 8–33. Fifty separates printed without title-page, covers, or repaging.

    Forty-five species and two varieties of fossils are described, and the genera *Belemnocrinus* and *Acambona* are proposed. The author now regards the latter as identical with *Eunicrotis*, Hall.

	Page.
Echinodermata.	
Crinoidea.	
Cyathocrinus, Miller	8
C. rigidus, n. s., White, 1862	8
C. kelloggi, n. s., White, 1862	8–9
Poteriocrinus, Miller	9
P. ob-uncus, n. s., White, 1862	9
P. salignoideus, n. s., White, 1862	10
P. bursæformis, n. s., White, 1862	10–11
Bursacrinus, Meek & Worthen	11
B. confirmatus, n. s., White, 1862	11
Zeacrinus, Troost	11
Z. perangulatus, n. s., White, 1862	11–12
Z. sacculus, n. s., White, 1862	12–13
Z. sacculus var. *concinnus*, White, 1862	13
*Belemnocrinus, n. g., White, 1862	14
B. typus, n. s., White, 1862, figs. 1 and 2, p. 13	14–15
Actinocrinus, Miller	15
A. quadrispinus, n. s., White, 1862	15
A. wachsmuthi, n. s., White, 1862	15–16
A. nashvillæ, Troost, var. *subtractus*, White, 1862	16
Megistocrinus, Owen & Shumard	16
M. plenus, n. s., White, 1862	16–17
M. crassus, n. s., White, 1862	17
Platycrinus, Miller	17
P. pleurovimenus, n. s., White, 1862	17–18
P. quinquenodus, n. s., White, 1862	18–19
Dichocrinus, Munster	19
D. angustus, n. s., White, 1862	19
D. crassitestus, n. s., White, 1862	19–20
Pentremites, Say	20
P. sirius, n. s., White, 1862, fig. 3, p. 20	20–21
Mollusca.	
Gasteropoda.	
Porcellia, Léveillé	21
P. obliquinoda, n. s., White, 1862	21
Bellerophon, Montfort	21
B. panneus, n. s., White, 1862	21
B. scriptiferus, n. s., White, 1862	21–22
Euomphalus, Sowerby	22
E. roberti, n. s., White, 1862	22
Pteropoda.	
Conularia, Miller	22
C. byblis, n. s., White, 1862	22
C. victa, n. s., White, 1862	22–23
Brachiopoda.	
Rhynchonella, Fischer	23
R. caput-testudinis, n. s., White, 1862	23
R. ottumwa, n. s., White, 1862	23–24
Spirifer, Sowerby	24
S. glans-cerasus, n. s., White, 1862	24
Observations on the genus *Spiriferina*, d'Orbigny	24–25

---

    * βελεμνον, a dart; κρινον, a lily.

	Page.
S. ? subtexta, n. s., White, 1862	25
Oyrtia, Dalman	25
O. curvilineata, n. s., White, 1862	25-26
Amboecelia, Hall	26
A. (Spirifer?) minuta, n. s., White, 1862	26
*Acambona, n. g., White, 1872	27
A. prima, n. s., White, 1862, figs. 1 & 2, p. 27	27-28
Retzia, King, *Acambona*, White	28
R. (Acambona?) altirostris, n. s., White, 1862	28
Streptorhynchus, King	28
S. lens, n. s., White, 1862	28-29
Productus, Sowerby	29
P. viminalis, n. s., White, 1862	29
Chonetes, Fischer	29
O. geniculata, n. s., White, 1862	29
Crania, Retzius	29
O. sheldoni, n. s., White, 1862	29-30
O. reposita, n. s., White, 1862	30
Discina, Lamarck	30
D. capax, n. s., White, 1862	30
Lingula, Bruguière	30
L. halli, n. s., White, 1862	30
Conchifera.	
Aviculopecten, McCoy	31
A. gradocostus, n. s., White, 1862	31
Cardiomorpha, de Koninck, *Cardiopsis*, Meek & Worthen	31
C. (Cardiopsis?) parvirostris, n. s., White, 1862	31
Gervillia, Defrance	31
G. strigosa, n. s., White, 1862	31
Zoophyta.	
Zaphrentis, Rafinesqne et Clifford	31
Z. elliptica, n. s., White, 1862	31-32
Z. glans, n. s., White, 1862	32
Syringopora, Goldfuss	32
S. harveyi, n. s., White, 1862	32
Striatopora, Hall	32
S. carbonaria, n. s., White, 1862	32
Nullipora? obtexta, n. s., White, 1862	33

5.

WHITE, C. A. Fœtal hydrocephalus. <Chicago Medical Journal, vol. xxii, pp. 55–57. Chicago, 1865.

A report of an obstetrical case in which the child's cranium was too much enlarged to pass through the pelvic arch, and the serum was drawn off by an operation *in utero*.

6.

WHITE, C. A. Cerebro-spinal meningitis. <Chicago Medical Journal, vol. xxii, pp. 529–532. Chicago, 1865.

A report of a successful case, in which the application of dry heat to the extremities was a leading feature of the treatment.

7.

WHITE, C. A. The Soils of Iowa and their origin. <Report of the Secretary of the Iowa State Agricultural Society for the year 1865. pp. 245–267. Des Moines, 1866.

A popular lecture, delivered before the Society September 29, 1865. Some of the views expressed in the lecture have since been much modified by the author.

---

*ἀχη, a point; ἀμβων, umbo.

8.

WHITE, C. A. Observations on the genus Belemnocrinus. <Proc. Boston Soc. Nat. Hist., vol. x, p. 180. Boston, 1866.

> This note is merely a rectification of the generic formula of *Belemnocrinus* as it was originally published in volume ix.

9.

WHITE, C. A. First Annual Report of Progress, of Charles A. White, State Geologist. pp. 1–4. Des Moines, 1867.

> Probably not over fifty copies of this report were printed in its original form. It was reprinted on pages 5 to 8 of the first and second annual reports, Des Moines, 1868. See entry No. 20.

10.

WHITE, C. A. and ST. JOHN, O. H. Preliminary notice of new genera and species of fossils. By C. A. White, M. D., State Geologist, and O. H. St. John, Assistant. [pp. 1–3.] Iowa City, May 8, 1867.

> This small publication of only fifty copies was made by the Iowa State Geological Survey. Four species are described. The genus *Meekella* proposed, *Cryptacanthia* suggested; and the whole, together with other matter, was republished in Vol. I, Transactions of the Chicago Academy of Sciences. See entry No. 15.
> *Protozoa.*
> *Amphistegina*, White & St. John, 1867.
> *Mollusca.*
> *Aulosteges spondyliformis*, n. s., White & St. John, 1867.
> *Waldheimia compacta*, n. s., White & St. John, 1867.
> *Beyrichia lithofactor*, n. s., White & St. John, 1867.
> *Beyrichia lithofactor* var. *velata*, White & St. John, 1867.
> *Meekella*, n. g., White & St. John, 1867.

11.

WHITE, C. A. Observations upon the Drift phenomena of Southwestern Iowa. <Amer. Jour. Sci., 2d ser., vol. xliii, pp. 301–305. New Haven, 1867.
Same. New Haven, 1867. 8vo, pp. 301–305. Twenty-five separates printed without covers, title-page, or repaging.

> Mentions the occurrence of glacial striæ upon rocks *in situ.*

12.

WHITE, C. A. A Sketch of the Geology of Southwestern Iowa. <Amer. Jour. Sci., 2d ser., vol. xliv, pp. 23–31. New Haven, 1867.
Same. New Haven, 1867. 8vo, pp. 23–31. Twenty-five separates printed without title-page, covers, or repaging.

> It is shown that the limestones of the region discussed belong to the upper and not to the lower Carboniferous series, as had been supposed by some previous authors.

13.

WHITE, C. A. Drift phenomena in Southwestern Iowa. <Amer. Jour. Sci., 2d ser., vol. xliv, p. 119. New Haven, 1867.

> This is an additional note to the article of the preceding entry, which was inadvertently omitted by the printer.

14.

WHITE, C. A. Exogenous leaves in the Cretaceous rocks of Iowa. <Amer. Jour. Sci., 2d ser., vol. xliv, p. 119. New Haven, 1867.

> A note announcing the discovery of exogenous leaves, and showing that the "Nishnabotany sandstone" is identical with the Dakota group.

15.

WHITE, C. A., and ST. JOHN, O. H. Descriptions of new Subcarboniferous and Coal-Measure Fossils, collected upon the Geological Survey of Iowa, together with notice of new generic characters observed in two species of Brachiopods. <Trans. Chicago Acad. Sci., vol. i, pp. 115-127, figs. 1-12. Chicago, 1867.

Same. Chicago, 1867. pp. 115-127. Fifty separates printed without title-page, covers, or repaging.

> Fourteen species are described, four of which were previously described in the paper, entry No. 10. The genus *Tomoceras* is proposed, and the previously suggested genera *Meekella* and *Cryptacanthia* are proposed and illustrated.

	Page.
Protozoa.	
Polypi.	
Cyathophyllidæ.	
Axophyllum, Edwards & Haime	115
A. rudis, n. s., White & St. John, 1867	115–116
Amplexus, Sowerby	116
A. fragilis, n. s., White & St. John, 1867	116
Crinoidea.	
Hydreinocrinus, De Koninck	117
H. verrucosus, n. s., White & St. John, 1867, fig. 1, p. 117	117–118
Brachiopoda.	
Craniidæ.	
Crania, Retzius	118
C. modesta, n. s., White & St. John, 1867	118
Productidæ.	
Aulosteges, Helmersen	118
A. spondyliformis, n. s., White & St. John, 1867, fig. 2, p. 118	118
Terebratulidæ.	
Waldheimia, King	119
W. ? compacta, White & St. John, 1867, fig. 3, p. 119	119
Meekella, White & St. John, 1867	120
M. striatocostata, White & St. John, figs. 4 and 5, p. 120, fig. 6, p. 121	120–122
Pinnidæ.	
Pinna, Linnæus	122
P. hinrichsiana, n. s., White & St. John, 1867, fig. 7, a b, p. 122	122–123
Gasteropoda.	
Atlantidæ.	
Cyrtolites, Conrad	123
C. ? gillianus, n. s., White & St. John, 1867, fig. 8, p. 123	123
Cephalopoda.	
Nautilidæ.	
Nautilus divisus, n. s., White & St. John, 1867, fig. 9, p. 124	124
Nautilus (Cryptoceras) springeri, n. s., White & St. John, 1867, fig. 10, p. 124	124, 125
Crustacea.	
Cypridinidæ.	
Beyrichia, McCoy	125
B. petrifactor, n. s., White & St. John, 1867	125
B. petrifactor var. *velata*, White & St. John, 1867	126
B. fœtoidea, n. s., White & St. John, 1867, fig. 11, a b, p. 126	126, 127
Cypridæ.	
Cythere, Mueller	127
C. simplex, n. s., White & St. John, 1867	127

16.

WHITE, C. A. Character of the Unconformability of the Iowa Coal-measures upon the Older Rocks. <Am. Jour. Sci., 2d ser., vol. xlv, pp. 331-334. New Haven, 1868.

Same. New Haven, 1868, 8vo., pp. 331-334. Twenty-five separates printed without title-page, covers, or repaging.

17.

WHITE, C. A. On Coal in Nebraska, with reference to a paragraph in the Geological Report of Dr. Hayden. <Am. Jour. Sci., 2d ser., vol. xlv., pp. 399-400. New Haven, 1868.

> The paragraph referred to is on page 125 of the report of the Commissioner of the General Land Office for the year 1867.

18.

WHITE, C. A. Note on the shell-structure of certain Naiades. <Am. Jour. Sci., 2d ser., vol. xlv, pp. 400-401. New Haven, 1868.

> The outer prismatic layer of the shell is noticed as a family character, not as a new discovery, as was supposed by Meek.

19.

WHITE, C. A. Note on "Cone-in-cone." <Am. Jour. Sci., 2d ser., vol. xlv, pp. 401-402. New Haven, 1868.

20.

WHITE, C. A. First and Second Annual Report of Progress by the State Geologist, and the Assistant and Chemist; on the Geological Survey of the State of Iowa; together with the substance of Popular Letters contributed to the Newspapers of the State during the years 1866 and 1867, in accordance with law; also extracts originally contributed to Scientific Journals as a part of the work of the Survey. pp. 1-284. Des Moines; 1868.

> The brief first annual report (Entry No. 9) is reprinted in this volume, upon pages 5-8.

21.

WHITE, C. A. Lakes of Iowa, Past and Present. <American Naturalist, vol. ii, pp. 143-155. Salem, 1868.

Same. Salem, 1868, pp. 143-155. Thirty separates printed without title-page, covers, or repaging.

> The Drift lakes, including the so-called walled lakes, are described, and the origin of the "walls" explained. Also the Bluff deposit of the Missouri River valley is shown to be the deposit of an ancient lake.

22.

WHITE, C. A. The Iowa Drift. <American Naturalist, vol. ii, pp. 615-616. Salem, 1869.

> The derivation of the drift material from the underlying rocks, by their disintegration and comminution, is shown.

23.

WHITE, C. A. A trip to the Great Red Pipestone Quarry. <American Naturalist, vol. ii, pp. 644-653. Salem, 1869.

Same. Salem, 1869, 8vo, pp. 644-653. Thirty separates printed without title-page, covers, or repaging.

> The quarry, the region, and the formation which contains the pipestone, are described.

24.

WHITE, C. A. Are Unios sensitive to light? <Am. Jour. Sci., 2d ser., vol. xlvii, pp. 280-281. New Haven, 1869.

> An experiment is described, showing that the sensitiveness which Unios manifest when the sun's rays are suddenly intercepted by an opaque body, is due to an interruption of light rays, and *not* heat rays.

25.

WHITE, C. A. Announcement of the Existence of Cretaceous Rocks in Guthrie County, Iowa <Proc. Amer. Ass. Adv. Sci., 17th Meeting. Chicago, 1868, vol. xvii, pp. 326-327. Cambridge, 1869.

26.

WHITE, C. A. Observation on the Red Quartzite Bowlders of Western Iowa; and their original ledges of Red Quartzite in Iowa, Dakota, and Minnesota. <Proc. Amer. Ass. Adv. Sci., 17th Meeting. Chicago, 1868, vol. xvii, pp. 340-342. Cambridge, 1869.

Same. Cambridge, 1869, pp. 340-342. Thirty separates printed without title page, covers, or repaging.

27.

WHITE, C. A. Report on the Geological Survey of the State of Iowa, to the Thirteenth General Assembly, January, 1870, Containing results of Examinations and Observations made within the years 1866, 1867, 1868, and 1869. By Charles A. White, M. D., Geological Corps; Charles A. White, State Geologist; Orestes H. St. John, Assistant; Rush Emery, Chemist. Vols. I and II. Des Moines, 1870,

CONTENTS OF VOLUME I.

	Page.
Introduction, including popular explanation of geological subjects	7-27
Part First: Physical Geography and Surface Geology; Four plates and two maps	28-166
Chapter I. Surface features	28-81
Chapter II. Surface deposits	82-121
Chapter III. Soils, &c	122-138
Chapter IV. Climate (by T. S. Parvin)	139-164
Part Second: General Geology; Three plates and three sections	165-294
Chapter I. Azoic, Lower Silurian, Upper Silurian, and Devonian systems	167-188
(*Smithia* [*Pachyphyllum*] *woodmani* [n. s., White, 1870] is described on p. 188.)	
Chapter II. Carboniferous system	189-230
Chapter III. Carboniferous system continued	231-263
Chapter IV. Carboniferous system concluded	264-284
Chapter V. Cretaceous system	285-294
Part Third: County and Regional Geology	295-381
Chapter I. Geology of Southwestern Iowa; and one section	296-381
Glossary	383-386
Index	387-391

CONTENTS OF VOLUME II.

Part First: County and Regional Geology	1-274
Chapter I. Geology of the Middle Region of Western Iowa and other counties (by O. H. St. John); with one plate and one section	1-200
Chapter II. Northwestern Iowa; One plate	201-232
Chapter III. Middle Region of Northern Iowa	233-253
Chapter IV. Geology of the Coal counties; One plate	254-274
Part Second: Mineralogy, Lithology, and Chemistry	275-402
Chapter I. Peat and Petroleum; One plate	275-292
Chapter II. Gypsum and other sulphates of the Alkaline Earths	293-306
Chapter III. Building materials, metals, and miscellaneous substances	307-342
Chapter IV. Chemist's Report (by Rush Emery).	
Section I. Rocks and minerals	345-354
Section II. Waters	354-357
Section III. Coals	357-397
Section IV. Peats	397-402
Appendices:	
Appendix A. Elevation in feet of points along the lines of Iowa railroads, both completed and projected (by the chief engineers of the respective roads	405-418
Appendix B. Catalogue of the Birds of Iowa (by J. A. Allen)	419-427
Appendix C. Government Surveys of the Public Lands (by C. W. Irish)	428-435
One Geological map of the State, colored.	

28.

WHITE, C. A. Natural Science in our Common Schools. <Iowa School Journal, vol. xii, pp. 1-4. Des Moines, 1870.

> Methods of interesting school children in natural history are suggested.

29.

WHITE, C. A. Kjœkkenmœddings in Iowa. <American Naturalist, vol. iii, pp. 54-55. Salem, 1870.

> This is the first public announcement of the fact that the shell-heaps of the western rivers are true kjœkkenmœddings.

30.

WHITE, C. A. Lilies of the Rocks. <American Naturalist, vol. iii, pp. 553-554. Salem, 1870.

> A review of part of an article by G. Hinrichs, in the August No., vol. iii, of the Naturalist, entitled "Lilies of the Fields, of the Rocks, and the Clouds." This note is signed "Zoologicus."

31.

WHITE, C. A. Prairie Fires. <American Naturalist, vol. v, pp. 68-70. Salem, 1871.

> An incident of personal experience related.

32.

WHITE, C. A. Albino Flowers. <American Naturalist, vol. v, pp. 161-162. Salem, 1871.

> It is observed that a certain cluster of the common field clover produced white flowers one year, and those of the ordinary red color the next. Also, that a specimen of white *Liatris* was observed, the latter being deemed important because the rose-red color is common to the whole genus.

33.

WHITE, C. A. [Remarks on the "Geological History of the Gulf of Mexico" by Prof. E. W. Hilgard.] <American Naturalist, vol. v, pp. 519-520. Salem, 1871.

> These remarks were made before the American Association for the Advancement of Science, at the 20th (Indianapolis) meeting, 1871, with reference to the paper above cited.

34.

WHITE, C. A. [Remarks on the homologies of the Carpal and Tarsal bones in Birds.] <American Naturalist, vol. v, p. 525. Salem, 1871.

> These remarks were made before the American Association for the Advancement of Science, 20th meeting (Indianapolis), with reference to a paper read by E. S. Morse, "On the Carpal and Tarsal bones of Birds."

35.

WHITE, C. A. Mammoth Cave. <University Reporter, [a college paper of the Iowa State University.] 4to. Vol. iv, pp. 81-83. Iowa City, 1872.

> An account of a visit to Mammoth Cave, Kentucky, in company with the American Association for the Advancement of Science.

36.

WHITE, C. A. Manual of Physical Geography and Institutions of the State of Iowa. 4to. pp. 1-85, pls. i-iii and figs. Davenport, 1873.

> This book was prepared for use in the schools of the State. It embraces an account of the history, constitution, and school laws of the State ; its educational, charitable, and penal institutions, land surveys, elections, taxes, &c. In 1883 an edition was issued purporting to be the 15th and also to be revised, but it has never been revised by the author since the first edition.

37.

WHITE, C. A. Woodpeckers Tapping Sugar Trees <American Naturalist, vol. vii, p. 496. Salem, 1873.

Woodpeckers were observed to peck holes in the bark of young and sound sugar maples, evidently to get the sap.

38.

WHITE, C. A. Kjœkkenmœddings de l'Amérique du nord. <Congrès International d'Anthropologie et d'Archéologie. Préhistoriques; Compte rendu de la cinquième session à Bologne, 1871. pp. 379-389. Bologna (Italy), 1873.
Same. Bologna (Italy), 1873. 8vo, pp. 15. Thirty separates printed with title-page and repaging.

This paper embraces a general review of the subject of shell heaps in North America as known up to that date.

39.

WHITE, C. A. On Spontaneous fission? in Zaphrentis. <Amer. Jour. Sc., 3d ser., vol. v., p. 72. New Haven, 1873.

A specimen of *Zaphrentis spinulifera*, Hall is described, which seemed to be a case of spontaneous fission. The author now thinks it probable that it was the result of a twin polyp or that the fission took place at a very early stage in the formation of the corallite.

40.

WHITE, C. A. On the Eastern Limit of Cretaceous Deposits in Iowa. <Proc. Amer. Ass. Adv. Sci. ? 21st meeting (Dubuque), 1872, vol. xxi, pp. 187-192. Cambridge, 1873.
Same. Cambridge, 1873. 8vo. pp. 187-192. Fifty separates printed without title-page, covers, or repaging.

The discovery is announced of Cretaceous fossils in the drift or glacier-disturbed Cretaceous deposits in Howard, Black Hawk, and Johnson Counties, Iowa; showing that the Cretaceous deposits once extended as far eastward as Eastern Iowa and Southeastern Minnesota.

41.

WHITE, C. A. The proposed genus Anomalodonta of Miller identical with the earlier Megaptera of Meek. <Amer. Jour. Sci., 3d ser., vol. viii, pp. 218-219. New Haven, 1874.

Meek's claim to priority is defended against that of Miller. The note bears only the initials of the author, "C. A. W."

42.

WHITE, C. A. Artificial Shell-heaps of Fresh-water Mollusks. <Proc. Amer. Ass. Adv. Sci., 22d meeting, Portland, 1873, pp. 133-137. Salem, 1874.
Same. Salem, 1874. 8vo. pp. 133-137. Fifty separates printed without title-page, covers, or repaging.

It is herein shown that the fresh-water mollusks were extensively used as food by the aboriginal inhabitants.

43.

WHITE, C. A. Preliminary Report upon Invertebrate Fossils collected by the Expeditions of 1871, 1872, and 1873, with Descriptions of New Species. <Engineer Department, U. S. Army. Geographical and Geological Explorations and Surveys west of the 100th meridian. First Lieut. George M. Wheeler, Corps of Engineers, in charge. pp. 1-27. Washington, 1874.

Thirty-nine species are described as new, and five others are noticed. They are all redescribed and figured in part i, vol. iv, Report upon Geographical and Geological Explorations and Surveys West of the 100th Meridian. See entry No. 48 *Anchura nuptialis* is a Cretaceous species, and herein wrongly referred to the Jurassic.

LOWER SILURIAN

PRIMORDIAL PERIOD.

Plantæ.

	Page.
Cruziana, d'Orbigny	5
C. linnarsoni, n. s., White, 1874	5
C. rustica, n. s., White, 1874	5–6

Animalia.

Brachiopoda.
Acrotreta, Kutorga	6
A. ? subsidua, n. s., White, 1874	6
Trematis, Sharpe	6
T. pannulus, n. s., White, 1874	6

Pteropoda.
Hyolithes, Eichwald	6
H. primordialis, Hall?	6

Crustacea.
Agnostus, Brongniart	7
A. interstricta, n. s., White, 1874	7
Olenellus, Hall	7
O. gilberti, Meek MSS	7–8
O. powelli, Meek MSS	8

CANADIAN PERIOD.

Hydrozoa.
Phyllograptus, Hall	9
P. loringi, n. s., White, 1874	9

Brachiopoda.
Acrotreta, Kutorga	9
A. pyscidicula, n. s., White, 1874	9
Lingula, Bruguière	9
L. ? manticula, n. s., White, 1874	9–10
Strophomena, Rafinesque	10
S. fontinalis, n. s., White, 1874	10

Gasteropoda.
Bellerophon, Montfort	10
B. allegoricus, n. s., White, 1874	10

Cephalopoda.
Orthoceras, Breynius	10
O. colon, n. s., White, 1874	10–11

Crustacea.
Leperditia, Rouault	11
L. bivia, n. s., White, 1874	11
Megalaspis, Angelin	11
M. belemnurus, n. s., White, 1874	11–12
Dicellocephalus, Owen	12
D. flagricaudus, n. s., White, 1874	12

TRENTON PERIOD.

Hydrozoa.
Graptolithus, Linnæus	12
G. (Diplograptus) hypniformis, n. s., White, 1874	12–13
G. qua(r)drinucronatus, Hall ?	13
G. (Climacograptus ?) ramulus, n. s., White, 1874	13
Rhynchonella, Fischer	14
R. argenturbica, n. s., White, 1874	14

CARBONIFEROUS.

SUBCARBONIFEROUS PERIOD.

Polypi.
Favosites, Lamarck	15
F. whitfieldi, White & Whitfield MSS	15

Blastoidea.
Granatocrinus, Troost	15
G. lotoblastus, n. s., White, 1874	15

	Page.
Crinoidea.	
Platycrinus, Miller	15
Platycrinus, ——? White, 1874	15–16
Actinocrinus, Miller	16
A. viaticus, n. s., White, 1874	16
Brachiopoda.	
Spirigera, d'Orbigny	16
S. monticola, n. s., White, 1874	16

CARBONIFEROUS PERIOD.

Echinodermata.	
Archæocidaris, McCoy	17
A. trudifer, n. s., White, 1874	17–18
Polyzoa.	
Glauconome, Goldfuss	18
G. nereidis, n. s., White, 1874	18–19
Polypora, McCoy	19
P. stragula, n. s., White, 1874	19
Brachiopoda.	
Chonetes, Fischer	19
C. platynota, n. s., White, 1874	19
Rhynchonella, Fischer	19
R. wasatchensis, n. s., White, 1874	19–20
R. metallica, n. s., White, 1874	20
Spirifer, Sowerby	20
S. (Martinia) glaber var. contracta, Meek & Worthen	20–21
Spiriferina, d'Orbigny	21
S. spinosa, Norwood & Pratten, var. campestris, White, 1874	21
Dielasma, King	21
D. ? bovidens, Morton, sp	21
Conchifera.	
Aviculopecten, McCoy	21
A. coreyana, n. s., White, 1874	21–22
Monopteria, Meek & Worthen	22
M. marian, n. s., White, 1874	22
Gasteropoda.	
Macrocheilus, Phillips	22
M. angulifera, n. s., White, 1874	22–23
Dentalium, Linnæus	23
D. canna, n. s., White, 1874	23

MESOZOIC.

JURASSIC PERIOD.

Conchifera.	
Camptonectes, Agassiz	23
C. stygius, n. s., White, 1874	23
Gasteropoda.	
Neritina, Lamarck	24
N. phaseolaris, n. s., White, 1874	24
Anchura, Conrad	24
A. nuptialis, n. s., White, 1874	24

CRETACEOUS PERIOD.

Conchifera.	
Pinna, Linnæus	24
P. petrina, n. s., White, 1874	24–25
Camptonectes, Agassiz	26
C. platessa, n. s., White, 1874	25
Inoceramus, Sowerby	25
I. dimidius, n. s., White, 1874	25–26
Leiopistha, Meek	26
Subgenus Psilomya, Meek	26
L. (Psilomya) meekii, n. s., White, 1874	26
Gasteropoda.	
Cassiope, Coquand	27
C. whitfieldi, n. s., White, 1874	27

44.

WHITE, C. A. On the Equivalency of the Coal-Measures of the United States and Europe. < Proc. Amer. Ass. Adv. Sci., 23d Meeting (Hartford), 1874, pp. B, 35–38. Salem, 1875.

Same. Salem, 1875. 8vo, pp. 35–38. Thirty separates printed without title-page, covers, or repaging.

> An affirmative opinion is expressed.

45.

WHITE, C. A. The Great Western Exploring Parties and their Progress. < Boston Daily Advertiser (newspaper), vol. 126, No. 91. Boston, October 15, 1875.

> This is the first of a series of three articles written from the field. This one was written from Southern Wyoming, describing the country, and mentioning Powell's plan for having the arid region surveyed into irregular-shaped ranches, and not by the rectangular method in use in the humid regions.

46.

WHITE, C. A. The Far West. < Boston Daily Advertiser (newspaper), vol. 126, No. 108. Boston, November 4, 1875.

> The second of a series of articles written from the field. This one is from Northern Utah, describing the vegetable products of that region.

47.

WHITE, C. A. The Far West. < Boston Daily Advertiser (newspaper), vol. 126, No. 115. Boston, November 12, 1875.

> The third and last of a series of three articles written from the field. This was written from Northern Utah, describing the animals of that region.

48.

WHITE, C. A. Report upon the Invertebrate Fossils collected in portions of Nevada, Utah, Colorado, New Mexico, and Arizona, by Parties of the Expeditions of 1871, 1872, 1873, and 1874. < Report upon Geographical and Geological Explorations and Surveys West of the 100th Meridian, in charge of First Lieut. Geo. M. Wheeler, Corps of Engineers, U. S. Army. 4to. vol. iv. part i, Paleontology. pp. 1–219, pls. i–xxi. Washington, 1875.

Same. Washington: Government Printing Office, 1875. 4to, pp. 219, and 21 plates of illustrations. Two hundred and fifty separates printed and bound in paper covers, and 30 copies printed and bound separately in boards for the author, with a different imprint upon the back.

> This report comprises descriptions and illustrations of fossils and some general observations upon the periods which they represent. One hundred and seventy-five species are described and illustrated, from the Primordial, Silurian, Carboniferous, Jurassic, Cretaceous, and Tertiary rocks, the majority of which were formerly known. A few, however, are herein described for the first time, but the new forms of these collections were mostly described in the preliminary report (Entry No. 43). The genus *Lispodesthes* is here proposed.

LOWER SILURIAN AGE.

PRIMORDIAL PERIOD.

Plantæ.

Animalia.

Page

Brachiopoda.
- *Acrotreta*, Kutorga, 1848 34
- *A. ? subsidua*, White, 1874, pl. i, figs. 3 *a–d* 34–36
- *Trematis*, Sharpe, 1847 36
- *T. pannulus*, White, 1874, pl. i, figs. 4 *a* and *b* 36–37

Gasteropoda.
- *Hyolithes*, Eichwald, 1840 37
- *H. primordialis?* Hall? 1861, pl. i, figs. 5 *a–e* 37–38

Crustacea.
- *Agnostus*, Brongniart, 1821 38
- *A. interstrictus*, White, 1874, pl. ii, figs. 5 *a* and *b* 38–40
- *Conocoryphe*, Corda, 1847 40
- Subgenus *Ptychoparia*, Corda, 1847 40
- *C. (Ptychoparia) kingii*, Meek, 1870, pl. ii, figs. 2 *a–e* 40–43
- *Asaphiscus*, Meek, 1872 43
- *A. wheeleri*, Meek, 1872, pl. ii, figs. 1 *a–f* 43–44
- *Olenellus*, Hall, 1861 44
- *O. gilberti*, Meek, MSS., 1874, pl. ii, figs. 3 *a–e* 44–46
- *O. howelli*, Meek, MSS., 1874, pl. ii, figs. 4 *a* and *b* 47–48
- *Vestigia*, White, 1875, pl. i, figs. 6, *a* and *b* 49

CANADIAN PERIOD.

Rhizopoda.
- *Receptaculites*, Defrance, 1827 50
- *Receptaculites* —— (?) 50

Hydrozoa.
- *Phyllograptus*, Hall, 1858 51
- *P. loringi*, White, 1874, pl. iii, figs. 1 *a* and *b* 51–52

Brachiopoda.
- *Lingula*, Bruguière, 1789 52
- *L. ? manticula*, White, 1874, pl. iii, figs. 2 *a* and *b* 52–53
- *Acrotreta*, Kutorga, 1848 53
- *A. pyxidicula*, White, 1874, pl. iii, figs. 3 *a* and *b* 53–54
- *Strophomena*, Rafinesque, 1827 54
- *S. fontinalis*, White, 1874, pl. iii, figs. 4 *a–c* 54–55
- *Orthis*, Dalman, 1828 55
- *O. electra*, Billings (?) 55

Gasteropoda.
- *Bellerophon*, Montfort, 1808 55
- *B. allegoricus*, White, 1874, pl. iii, figs. 6 *a–c* 55–56

Cephalopoda.
- *Orthoceras*, Breynius, 1732 56
- Subgenus *Camaroceras*, Conrad, 1842 56
- *O. (Camaroceras) colon*, White, 1874, pl. iii, figs. 5 *a–d* 56–57
- *Cyrtoceras*, Goldfuss, 1833 57
- *C.* —— (?), White, 1875 57–58

Crustacea.
- *Leperditia*, Rouault, 1851 58
- *L. biria*, White, 1874, pl. iii, figs. 7 *a–d* 58–59
- *Megalaspis*, Angelin, 1854 59
- *M. belemnurus*, White, 1874, pl. iii, fig. 9 59–60
- *Dicellocephalus*, Owen, 1852 60
- *D. ? flagricaudus*, White, 1874, pl. iii, figs. 8 *a* and *b* 60–61

TRENTON PERIOD.

Hydrozoa.
- *Graptolithus*, Linnæus, 1736 62
- Subgenus *Climacograptus*, Hall, 1865 62
- *G. (Climacograptus) ramulus*, White, 1874, pl. iv, figs. 3 *a–c* 62–63
- Subgenus *Diplograptus*, M'Coy, 1850 63
- *G. (Diplograptus) hymniformis*, White, 1874, pl. iv, figs. 4 *a* and *b* 63–64
- *G. (Diplograptus) pristis*, Hall (?), 1847, pl. iv, figs. 2 *a* and *b* 65
- *G. quadrimucronatus*, Hall (?), 1865, pl. iv, figs. 1 *a* and *b* 65–66

Actinozoa.
- *Monticulipora*, d'Orbigny, 1850 66
- *M. dalii*, Edwards & Haime, 1851, pl. iv, fig. 5 66–67

	Page
Favosites, Lamarck, 1816	67
F. —— (?) White, 1875	67
Favistella, Hall, 1847	67
F. stellata, Hall, 1847, pl. iv, figs. 6 *a–e*	67–68
Zaphrentis, Rafinesque et Clifford, 1820	68
Z. —— (?), White, 1875	68

Brachiopoda.

Strophomena, Rafinesque, 1827	69
S. filitexta, Hall, 1847, pl. iv, fig. 8	69–70
Leptœna, Dalman, 1828	70
L. sericea, Sowerby (?), pl. iv, fig. 7	70
Orthis, Dalman, 1828	70
O. occidentalis, Hall, 1847, pl. iv, figs. 2 *a* and *b*	70–72
O. testudinaria, Dalman (?)	72
O. plicatella, Hall (?), 1847, pl. iv, figs. 10 *a–d*	72–74
O. biforata, Schlotheim, var. *lynx.*, White, 1875, pl. iv, figs. 9 *a* and *b*	74–75
Rhynchonella, Fischer, 1809	75
R. argenturbica, White, 1874, pl. iv, figs. 12 *a–e*	75–76

Conchifera.

Modiolopsis, Hall, 1847	76
M. —— (?), White, 1875	76

Gasteropoda.

Maclurea, Le Sueur, 1818	77
M. —— (?), White, 1875	77
Raphistoma, Hall, 1847	77
R. trochiscus, Meek, 1870, pl. iv, figs. 13 *a–e*	77–78

CARBONIFEROUS AGE.

SUBCARBONIFEROUS PERIOD.

Actinozoa.

Favosites, Lamarck, 1816	79
F. divergens, White & Whitfield, 1862, pl. v, fig. 4	79–80
Syringopora, Goldfuss, 1826	80
S. harveyi, White (?)	80

Echinodermata.

Granatocrinus, Troost, 1850	80
G. lotoblastus, White, 1874, pl. v, figs. 3 *a* and *b*	80–81
Platycrinus, Miller, 1821	81
P. —— (?), White, 1875, pl. v, fig. 2	81–82
Actinocrinus, Miller, 1821	82
A. viaticus, White, 1874, pl. v, fig. 1	82–83

Brachiopoda.

Productus, Sowerby, 1812	83
P. parvus, M. & W., 1866, pl. v, figs. 6 *a* and *b*	83–84
Strophomena, Rafinesque, 1827	85
S. rhomboidalis, Wilckins, sp., 1767, pl. v, fig. 5	85–86
Spirifer, Sowerby, 1815	86
S. centronatus, Winchell, 1865, pl. v, figs. 8 *a–c*	86–87
S. striatus, Martin, sp., pl. v, fig. 10 *a*	88
S. extenuatus, Hall, 1858, pl. v, figs. 9 *a–d*	88–90
Subgenus *Martinia*, McCoy, 1844	90
S. (Martinia) peculiaris, Shumard, 1855, pl. v, figs. 7 *a* and *b*	90–91
Spirigera, d'Orbigny, 1847	91
S. monticola, White, 1874, pl. v, figs. 11 *a–d*	91–92
S. obnaxima, McChesney, 1860, pl. v, fig. 12	92–93
Terebratula, Llwhyd, 1698	93
Subgenus *Dielasma*, King, 1859	93
T. (Dielasma) burlingtonensis, White, 1860	93

Conchifera.

Conocardium, Brown, 1835	94
C. —— (?), White, 1875	94

Gasteropoda.

Euomphalus, Sowerby, 1815	94
E. luxus, n. s., White, 1875, pl. v, figs. 13 *a* and *b*	94–95

WRITINGS OF C. A. WHITE. 131

Protozoa.
    Rhizopoda.   CARBONIFEROUS PERIOD.   Page.

        Fusulina, Fischer, 1837 96
        F. cylindrica, Fischer, 1837, pl. vi, figs. 6 a and b 96-98
    Actinozoa.
        Chætetes, Fischer, 1837 98
        C. milleporaceus, Troost, sp., 1849 ?, pl. vi, fig. 2 a 98-99
        Rhombipora, Meek, 1872 99
        R. lepidodendroides, Meek, 1872, pl. vi, figs. 5 a-d 99-100
        Syringopora, Goldfuss, 1826 100
        S. multattenuata, McChesney (?) 100-101
        Zaphrentis, Rafinesque & Clifford 101
        Z. excentrica, Meek, 1872, pl. vi, fig. 3 a 101
        Lophophyllum, Edwards & Haime, 1850 101
        L. proliferum, McChesney, sp., var. sauridens, White, 1875, pl. vi, figs. 4 a-d ... 101-103
        Lithostrotion, Fleming, 1828 103
        L. whitneyi, Meek, 1875, pl. vi, figs. 1 a-c 103
    Echinodermata.
        Archæocidaris, McCoy, 1844 104
        A. ornatus, Newberry, 1861, pl. vi, fig. 7 104
        A. trudifer, White, 1874, pl. vi, figs. 8 a and b 104-105
    Polyzoa.
        Glauconome, Goldfuss, 1826 105
        G. nereidis, White, 1874, pl. vii, figs. 5 a-e 105-107
        Synocladia, King, 1849 107
        S. biserialis, Swallow, 1858, pl. vii, figs. 3 a-c 107-108
        Polypora, McCoy, 1844 108
        P. stragula, White, 1874, pl. vii, figs. 4 a and b 108-109
    Brachiopoda.
        Productus, Sowerby, 1812 109
        P. costatus, Sowerby (?), 1827, pl. viii, figs. 2 a-d 109-111
        P. semireticulatus, Martin, sp., 1809, pl. viii, figs. 1 a-c 111-113
        P. prattenianus, Norwood, 1854, pl. vii, figs. 1 a-c 113-114
        P. punctatus, Martin, sp., 1809, pl. vii, figs. 2 a-c 114-116
        P. nebrascensis, Owen, 1852, pl. viii, figs. 3 a-d 116-117
        P. longispinus, Sowerby, 1814, pl. viii, figs. 5 a-d 118-119
        P. muricatus, Norwood & Pratten, 1854, pl. viii, figs. 4 a-c 120
        P. mexicanus, Shumard (?), 1858, pl. viii, figs. 6 a-c 120-121
        Chonetes, Fischer, 1837 121
        C. platynota, White, 1874, pl. ix, figs. 6 a-e 121-122
        C. granulifera, Owen, 1855, pl. ix, figs. 8 a-c 122-123
        C. mesoloba, Norwood & Pratten, 1854, pl. ix, fig. 7 a 123
        Hemipronites, Pander, 1830 124
        H. crinistria, Phillips, sp., pl. x, fig. 9 a 124-125
        Orthis, Dalman, 1828 125
        O. pecosii, Marcou, 1858, pl. ix, figs. 5 a-e 125-126
        Meekella, White & St. John, 1867 126
        M. striatocostata, Cox, sp., 1857, pl. ix, figs. 4 a-e 126-128
    Rhynchonellidæ.
        Rhynchonella, Fischer, 1809 128
        R. uta, Marcou, sp., 1858, pl. ix, figs. 2 a-c 128-129
        R. metallica, White, 1874, pl. x, figs. 10 a-d 129-130
        R. wasatchensis, White, 1874, pl. ix, figs. 3 a-d 130-131
        R. rockymontana, Marcou, 1858, pl. ix, figs. 1 a-d 131-132
        Spirifer, Sowerby, 1815 132
        S. cameratus, Morton, 1836, pl. x, figs. 1 a-d 132-134
        S. striatus, Martin, sp 134
        S. rockymontanus, Marcou, 1858, pl. xi, figs. 9 a-d 134-135
        Subgenus Martinia, McCoy, 1844 135
        S. (Martinia) planoconvexus, Shumard, 1855, pl. x, figs. 3 a-c 135-136
        S. (Martinia) glaber var. contracta, Meek & Worthen, 1866, pl. x, figs. 2 a-c ... 136-138
        Spiriferina, d'Orbigny, 1847 138
        S. kentuckensis, Shumard, 1855, pl. x, figs. 4 a-c 138-139
        S. octoplicata, Sowerby, 1827, pl. x, figs. 8 a-c 139-140
        Retzia, King, 1850 141
        R. mormonii, Marcou, 1858, pl. x, figs. 7 a-c 141

	Page.
Spirigera, d'Orbigny, 1847	141
S. subtilita, Hall, 1852, pl. x, figs. 6 *a–c*	141–143
S. planosulcata, Phillips, sp. 1836, pl. x, figs. 5 *a–d*	143–144
Terebratulidæ.	
Terebratula Llhwyd, 1698	144
Subgenus *Dielasma*, King, 1859	144
T. (Dielasma) bovidens, Morton, 1836, pl. xi, figs. 10 *a–c*	144–146

Mollusca vera.
Conchifera.
Monomyaria.
Pectinidæ.

Aviculopecten, McCoy, 1852	146
A. occidentalis, Shumard, 1855, pl. xii, figs. 8 *a* and *b*	146–147
A. coreyanus, White, 1874, pl. xi, figs. 1 *a* and *b*	147–148
A. mccoyi, M. & H., 1864, pl. xi, fig. 2 *a*	149
A. interlineatus, M. & W., 1860, pl. xi, fig. 3 *a*	149–150
Pinnidæ.	
Pinna, Linnæus, 1758	151
P. peracuta, Shumard (?), pl. xi, fig. 5 *a*	151
Pteriidæ.	
Monopteria, M. & W., 1866	151
M. marian, White, 1874, pl. xi, figs. 4 *a–c*	151–152
Myalina, de Koninck, 1844	152
Myalina (?), White, 1875	152
M. ? swallovi, McChesney, pl. xi, fig. 8 *a*	152
Bakevellia, King, 1848	153
B. parra, M. & H., 1858, pl. xi, figs. 7 *a* and *b*	153

Dymyaria.
Trigonidæ.

Schizodus, King, 1844	154
S. wheeleri, Swallow, 1862, pl. xi, figs. 6 *a* and *b*	154
Allorisma, King, 1850	155
A. subcuneata, var. M. & H., pl. xii, figs. 7 *a* and *b*	155

Gasteropoda.
Prosopocephala.
Solenoconcha.
Dentaliidæ.

Dentalium, Linnæus, 1740	156
D. canna, White, 1874, pl. xii, figs. 6 *a* and *b*	156

Diœca.
Rhiphidoglossa.
Dicranobranchia.
Bellerophontidæ.

Bellerophon, Montfort, 1808	157
B. crassus, M. & W., 1860, pl. xii, fig. 1 *a*	157

Podophthalma.
Euomphalidæ.

Euomphalus, Sowerby, 1815	158
E. pernodosus, M. & W., 1870, pl. xii, figs. 2 *a* and *b*	158

Pectinibranchiata.
Tænioglossa.
Naticidæ.

Naticopsis, McCoy, 1844	159
N. nana, M. & W., 1866, pl. xii, figs. 4 *a* and *b*	159

Copulidæ.

Platyceras, Conrad, 1840	159
P. nebrascense, Meek, 1872, pl. xii, figs. 5 *a–d*	159–160

Macrocheilidæ.

Macrocheilus, Phillips, 1841	160
M. anguliferus, White, 1874, pl. xii, figs. 3 *a–f*	160–161

Cephalopoda.
Tetrabranchiata.
Goniatitidæ.

Goniatites, de Haan, 1825	161

Nautilidæ.

Nautilus, Breynius, 1732	161

MESOZOIC AGE.

JURASSIC PERIOD.

Radiata.
Echinodermata.
Crinoidea.
Pentacrinidæ.
    Pentacrinus, Miller, 1821 162
    P. asteriscus, M. & H., 1864, pl. xiii, figs. 6 a, b 162–163
Mollusca.
Conchifera.
Monomyaria.
Ostreidæ.
    Ostrea, Linnæus, 1758 163
    O. strigilecula, n. s., White, 1875, pl. xiii, figs. 3 a–d 163–164
Pectinidæ.
    Camptonectes, Meek (Agassiz), 1864 164
    C. stygius, White, 1874, pl. xiii, figs. 2 a–c 164–165
    C. bellistriatus, M. & H 165
Ptereidæ.
    Inoceramus, Sowerby, 1814 166
    I. crassalatus, n. s., White, 1875, pl. xiii, figs. 4 a–c 166
Dimyaria.
Trigoniidæ.
    Myophoria, Brown, 1830.
    M. ambilineata, n. s., White, 1875, pl. xiii, figs. 5 a and b 166–167
    Trigonia, Bruguière, 1789 167
    T. ——— (?), White, 1875 167
Gasteropoda.
Diœca.
Rhiphidoglossa.
Podophthalma.
Neritidæ.
    Neritina, Lamarck, 1809 167
    N.? phascolaris, White, 1874, pl. xiii, figs. 1 a–e 167–168

CRETACEOUS PERIOD.

Mollusca.
Molluscoidea.
Brachiopoda.
Lyopomata.
Lingulidæ.
    Lingula, Bruguière, 1792 169
    L. subspatulata, H. & M., 1856, pl. xv, fig. 4 a 169
Mollusca vera.
Conchifera.
Monomyaria.
Ostreidæ.
    Ostrea, Linnæus, 1[7]58 170
    O. cortex, Conrad, 1857, pl. xv, figs. 2 a–c 170
    O. prudentia, n. s., White, 1875, pl. xiv, figs. 2 a–d 171
    Gryphea, Lamarck, 1801 171
    G. pitcheri, var. Morton, pl. xvii, figs. 1 a–f 171–172
    Exogyra, Say, 1819 172
    E. ponderosa, Rœmer, 1852, pl. xiv, figs. 1 a–c 172–173
    E. læviuscula, Rœmer, 1852, pl. xvii, figs. 2 a–d 173–174
    E. costata, Say, var. *fluminis*, White, 1875, pl. xvii, figs. 3 a–d 174–175
Pectinidæ.
    Camptonectes, Meek (Agassiz), 1864 176
    C. platessa, White, 1874, pl. xvii, fig. 5 a 176
Limidæ.
    Lima, Bruguière, 1791 176
    L. wacoensis, Rœmer, 1852, pl. xvii, figs. 4 a–c 176–177
Pteriidæ.
    Inoceramus, Sowerby, 1814 177
    I. problematicus, Schlotheim sp., 1820, pl. xvi, fig. 3 a 177–178
    I. fragilis, Hall & Meek, 1856, pl. xv, fig. 3 a 178

	Page.
I. flaccidus, n. s., White, 1875, pl. xvi, figs. 1 *a* and *b*	178-179
I. deformis, Meek, 1872, pl. xv, figs. 1 *a* and *b*	179-180
I. barabini, Morton, 1834, pl. xvi, fig. 4 *a*	180-181
I. dimidius, White, 1874, pl. xvi, figs. 2 *a–d*	181-182

Pinnidæ.

Pinna, Linnæus, 1758	182
P. petrina, White, 1874, pl. xiii, figs. 7 *a* and *b*	182-183

Dymyaria.
Arcidæ.

Idonearca, Conrad, 1862	183
I. depressa, n. s., White, 1875, pl. xviii. figs. 13 *a* and *b*	183-184

Lucinidæ.

Lucina, Bruguière, 1792	184
L. subundata, Hall & Meek, 1856, pl. xviii, fig. 12 *a*	184

Glossidæ.

Veniella, Stoliczka, 1870	185
V. goniophora, Meek, 1875	185

Mactridæ.

Mactra, Linnæus	185
M. ? incompta, n. s., White, 1875, pl. xvii, fig. 6 *a* and *b*	185

Anatinidæ.

Leiopistha, Meek, 1864	186
Subgenus *Psilomya*, Meek, MSS., 1874	186
L. (*Psilomya*) *meeki*, White, 1874, pl. xviii, figs. 14 *a–d*	186-187
Subgenus *Cymella*, Meek, 1864	187
L. (*Cymella*) *undata*, M. & H., sp., 1856, pl. xviii, fig. 15 *a*	187

Corbulidæ.

Corbula, Bruguière, 1792	188
C. nematophora, Meek, 1872, pl. xvii, figs. 7 *a–c*	188-189

Gasteropoda.
Diœca.
Rhiphidoglossa.
Neritidæ.

Neritina, Lamarck, 1809	189
Subgenus (*Velatella*) Meek, 1872	189
N. (*Velatella*) *carditoides*, Meek, 1872, pl. xviii, figs. 7 *a–c*	189-190

Pectinibranchiata.
Tænioglossa.
Aporrhaidæ.

Anchura, Conrad, 1860	190
A. ? fusiformis, Meek, MSS., 1874, pl. xviii, fig. 4 *a*	190-191
* *Lispodesthes*, n. g., White, 1875	191
L. nuptialis, White, 1874, pl. xviii, figs. 3 *a* and *b*	192
L. lingulifera, n. s., White, 1875, pl. xviii, figs. 2 *a* and *b*	192-193

Tecturnidæ.

Anisomyon, M. & H., 1860	193
A. borealis, Morton, sp., 1842, pl. xviii, figs. 9 *a* and *b*	193-194
A. centrale, Meek, 1870, pl. xviii, figs. 8 *a* and *b*	194-195

Turritellidæ.

Turritella, Lamarck, 1801	195
T. uvasana, Conrad, 1856, pl. xviii, figs. 11 *a* and *b*	195
Cassiope, Coquand, 1865	196
C. whitfieldi, White, 1874, pl. xviii, fig. 1 *a*	196
Eulimella, Forbes, 1846	197
E. funicula, Meek, 1872, pl. xviii, fig. 6 *a*	197

Pyramidellidæ.

Turbonilla, Leach, 1825	197
Subgenus *Chemnitzia*, Conrad, 1860	197
T. (*Chemnitzia*) *melanopsis*, Conrad ? pl. xviii, fig. 10 *a*	197-198

Toxoglossa.
Admetidæ.

Admete, Müller, 1842	198
Subgenus *Admetopsis*, Meek, 1872	198
A. (*Admetopsis*) *gregaria*, Meek, 1872, pl. xviii, figs. 5 *a* and *b*	198-199

---

* Λισπος, smooth, and ἐσθης, a garment.

Cephalopoda.
  Tetrabranchiata.
    Baculitidæ.
      Baculites, Lamarck, 1801 199
      B. ovatus, Say. pl. xix, figs. 4 a-c, and 5 a-c 199-200
    Scaphitidæ.
      Scaphites, Parkinson, 1811 200
      S. warreni, M. & H., 1860, pl. xix, fig. 3 a 200-201
    Ammonitidæ.
      Ammonites, Bruguière, 1789 201
      A. lævianus, n. s., White, 1875, pl. xix, figs. 1 a and b 201-202
      A. placenta, DeKay, var. intercalaris M. & H. 202
      Buchiceras, Hyatt, 1875 202
      B. swallovi, Shumard, 1860, pl. xx, figs. 1 a-c 202-203
    Turrilitidæ.
      Helicoceras, d'Orbigny, 1842 203
      H. pariense, n. s., White, 1875, pl. xix, figs. 2 a-d 203-204
Articulata.
  Vermes.
    Tubicola.
      Serpulidæ.
        Serpula, Linnæus, 1758 205
        S. intrica, n. s., White, 1875, pl. xv, fig. 5 a 205

CENOZOIC AGE.

TERTIARY PERIOD.

Mollusca.
  Conchifera.
    Dimyaria.
      Unionidæ.
        Unio, Retzius, 1788 206
        U. vetustus, Meek, 1860, pl. xxi, figs. 12 a-d 206-207
      Cyrenidæ.
        Cyrena, Lamarck, 1818 207
        Subgenus Veloritina, Meek, 1873 207
        C. (Veloritina) durkeei, Meek, 1870, pl. xxi, figs. 13 a and b 207-208
        Sphærium, Scapoli, 1777 208
        Sphærium —— (?), White, 1875 208
Gasteropoda.
  Pulmonifera.
    Basommatophora.
      Limnæidæ.
        Planorbis, Guettard, 1756 209
        P. utahensis, Meek, 1860, pl. xxi. fig. 8 a 209
        Planorbis —— (?), White, 1875 210
      Physidæ.
        Physa, Draparnaud, 1801 210
        P. bridgerensis, Meek ? 1872, pl. xxi, fig. 2 a 210
        P. pleromatis, n. s., White, 1875, pl. xxi, figs. 1 a and b 211
    Geophila.
      Helicidæ.
        Helix, Linnæus, 1758 211
        H. leidyi, H. & M., 1856, pl. xxi, figs. 3 a-c 211
  Diœca.
    Pectinibranchiata.
      Tænioglossa.
        Melaniidæ.
          Goniobasis, Lea, 1862 212
          G. tenuicarinata, M. & H., 1857, pl. xxi, figs. 10 a and b 212
          G. tenera, Hall, sp., 1845, pl. xxi, figs. 11 a-c 212-213
          G. nebrascensis, M. & H., 1856, pl. xxi, figs. 9 a-c 213-214
        Viviparidæ.
          Viviparus, Montfort, 1810 214
          V. trochiformis, M. & H., 1856, pl. xxi, figs. 4 a-c 214
          V. trochiformis, var. White, 1875, pl. xxi, figs. 5 a and b 214-215

	Page.
V. ionicus, n. s., White, 1875, pl. xxi, figs. 6 *a* and *b*	215
Viriparus —— (?), White, 1875, pl. xxi, figs. 7 *a* and *b*	215

Articulata.
  Crustacea.
    Ostracoda.
      Cypridinidæ.

Cypris, Müller, 1785	216
Cypris —— (?), White, 1875	216

49.

WHITE, C. A. Invertebrate Paleontology of the Plateau Province, together with notice of a few species from localities beyond its limits in Colorado. <United States Geological and Geographical Survey of the Territories: Report on the Geology of the Uintah Mountains. By J. W. Powell. 4to, pp. 74–135. Washington, 1876.

Same. Washington, 1876. 4to, pp. 74–135. Fifty separates printed without title-page, covers, or repaging.

This memoir embraces some general geological discussions, catalogues of fossils, and descriptions of 48 new species.

In subsequent works the author has referred to the Laramie group some of the fossils herein assigned, respectively, to the Cretaceous and Tertiary. Most of the species have since been illustrated by the author in various works.

General observations, pp. 75–87.

Catalogue of the fossils collected by the various parties in the field during the years 1868 to 1875, inclusive, pp. 88–107.

Descriptions of new species of Invertebrate fossils from strata of the Carboniferous, Jurassic, Cretaceous, and Tertiary Periods, pp. 107–135.

CARBONIFEROUS PERIOD.

Radiata.
  Actinozoa.   Page.

Amplexus, Sowerby	107
A. zaphrentiformis, n. s., White, 1876	107–108

Echinodermata.

Eupachycrinus, M. & W	108
E. platybasis, n. s., White, 1876	108
Archæocidaris, McCoy	109
A. cratis, n. s., White, 1876	109

Mollusca.
  Gasteropoda.

Naticopsis, McCoy	109
N. remex, n. s., White, 1876	109

JURASSIC PERIOD.

Mollusca.
  Conchifera.

Unio, Retzius	110
U. stewardi, n. s., White, 1876	110

Gasteropoda.

Neritina, Lamarck	110
N. (?) powelli, n. s., White, 1876	110–111

CRETACEOUS PERIOD.

Mollusca.
  Conchifera.

Ostrea, Linnæus	112
Subgenus *Alectryonia*, Fischer	112
O. (Alectryonia) sannionis, n. s., White, 1876	112
O. insecura, n. s., White, 1876	112–113
Plicatula, Lamarck	113
P. hydrotheca, n. s., White, 1876	113
Inoceramus, Sowerby	113
I. gilberti, n. s., White, 1876	113–114
I. howelli, n. s., White, 1876	114–115
Aricula, Klein	115
A. parkensis, n. s., White, 1876	115

	Page.
Arca, Linnæus	115
A. ? coalvillensis, n. s., White, 1876	115–116
Unio, Retzius	116
U. gonionotus, n. s., White, 1876	116–117
Cyrena, Lamarck	117
Subgenus *Veloritina*, Meek	117
C. (Veloritina) erecta, n. s., White, 1876	117
Turnus, Gabb	117
T. sphenoideus, n. s., White, 1876	117–118

Gasteropoda.

Rhytophorus, Meek	118
R. meekii, n. s., White, 1876	118–119
Planorbis, Guettard	119
Subgenus *Bathyomphalus*, Agassiz	119
P. (Bathyomphalus) kanabensis, n. s., White, 1876	119
Physa, Draparnaud	119
P. kanabensis, n. s., White, 1876	119
Helix, Linnæus	120
H. kanabensis, n. s., White, 1876	120
Anchura, Conrad	120
A. ruida, n. s., White, 1876	120–121
A. prolabiata, n. s., White, 1876	121–122
Lunatia, Gray	122
L. utahensis, n. s., White, 1876	122
Goniobasis, Lea	122
G. cleburni, n. s., White, 1876	122–123
G. chrysaloidea, n. s., White, 1876	123
Viviparus, Montfort	123
V. panguitchensis, n. s., White, 1876	123–124
Odontobasis, Meek	124
O. buccinoidea, n. s., White, 1876	124

TERTIARY PERIOD.

Mollusca.
Conchifera.

Unio, Retzius	125
U. petrinus, n. s., White, 1876	125
U. propheticus, n. s., White, 1876	125–126
U. brachyopisthus, n. s., White, 1876	126
U. shoshonensis, n. s., White, 1876	126–127
Corbicula, Mühlfeldt	127
C. powelli, n. s., White, 1876	127–128
Pisidium, Pfeiffer	128
P. saginatum, n. s., White, 1876	128
Mesodesma, Deshayes	128
M. bishopi, n. s., White, 1876	128
Corbula, Bruguière	129
C. subundifera, n. s., White, 1876	129

Gasteropoda.

Succinea, Draparnaud	129
S. papillispira, n. s., White, 1876	129–130
Helix, Linnæus	130
H. riparia, n. s., White, 1876	130
H. peripheria, n. s., White, 1876	130
Pupa, Lamarck	130
P. inculata, n. s., White, 1876	130–131
P. arenula, n. s., White, 1876	131
Neritina, Lamarck	131
N. volvilineata, n. s., White, 1876	131
Melania, Lamarck	131
M. larunda, n. s., White, 1876	131–132
Hydrobia, Hartmann	132
H. recta, n. s., White, 1876	132
H. utahensis, n. s., White, 1876	132–133
Viviparus, Montfort	133

	Page.
V. plicapressus, n. s., White, 1876	133
Leioplax, Troschel	133
L. ? turricula, n. s., White, 1876	133–134
Tulotoma, Haldeman	134
T. thompsoni, n. s., White, 1876	134
Phorus, Montfort	134
P. exoneratus, n. s., White, 1876	134–135

50.

WHITE, C. A. In Memoriam: Fielding Bradford Meek. <Am. Jour. Sci., 3d ser., vol. xiii, pp. 169–171. New Haven, 1877.

Same. New Haven, 1877. 8vo, 3 pages. One hundred separates printed with half-title and repaging.

A brief sketch of the life and labors of the paleontologist, Mr. Meek.

51.

WHITE, C. A. Descriptions of new species of fossils from the Paleozoic rocks of Iowa. <Proc. Acad. Nat. Sci. Phila., for 1876, vol. xxviii, pp. 27–34. Philadelphia, 1877.

Same. Philadelphia, 1876. 8vo., pp. 27–31. Fifty separates printed without title-page, covers, or repaging.

Thirteen species are described as new, and the genus *Strobilocystites* is proposed.

Radiata.
  Actinozoa.

	Page.
Chætetes, Fischer	27
C. muscatinensis, n. s., White, 1877	27
Monticulipora, d'Orbigny	27
M. monticula, n. s., White, 1877	27
Lophophyllum, E. & H., 1877	27
L. expansum, n. s., White, 1877	27–28

  Echinodermata.

Strobilocystites, n. g., White, 1877	28
S. calvini, n. s., White, 1877	28–29
Megistocrinus, Owen	29
M. farnsworthi, n. s., White, 1877	29–30

Mollusca.
  Brachiopoda.

Stricklandinia, Billings	30
S. castellana, n. s., White, 1877	30

  Conchifera.

Paracyclas, Hall	31
P. sabini, n. s., White, 1877	31
Allorisma, King	31
A. marionensis, n. s., White, 1877	31–32

  Gasteropoda.

Bellerophon, Montfort	32
B. bowmani, n. s., White, 1877	32
Euomphalus, Sowerby	32
E. springvalensis, n. s., White, 1877	32–33

  Pteropoda.

Conularia, Miller	33
C. molaris, n. s., White, 1877	33

  Cephalopoda.

Cyrtoceras, Goldfuss	33
C. dictyum, n. s., White, 1877	33–34

Articulata.
  Vermes.

Tentaculites, Schlotheim	34
T. hoyti, n. s., White, 1877	34

52.

WHITE, C. A. Paleontological papers, No. 1: Descriptions of Unionidæ and Physidæ, collected by Prof. E. D. Cope, from the Judith River Group of Montana, during the summer of 1876. <Bulletin U. S. Geol. and Geog. Surv. of the Terr., vol. iii, art. xx, pp. 599–602. Washington, 1877.

Same. Washington: Government Printing Office, 1877. 8vo, pp. 599–602. Two hundred separates printed (author's edition) without repaging, but with title-page, and in paper covers, together with Paleontological Papers Nos. 2, 3, 4, and 5.

 Six species are described as new. This is the first of a series of short unillustrated articles which reached eleven in number, and then ceased with the suspension of the survey in charge of Dr. Hayden. This form of title has not since been used by the author.

                        Page.
Unionidæ.
 Unio primævus, n. s., White, 1877 599–600
 Unio cryptorhynchus, n. s., White, 1877 600
 Unio senectus, n. s., White, 1877 600–601
 Anodonta propatoris, n. s., White, 1877 601
Physidæ.
 Bulinus atavus, n. s., White, 1877 601–602
 Physa copei, n. s., White, 1877 602

53.

WHITE, C. A. Paleontological Papers No. 2: Descriptions of new species of Uniones, and a new genus of fresh-water Gasteropoda from the Tertiary strata of Wyoming and Utah. <Bulletin U. S. Geol. and Geog. Surv. of the Terr., vol. iii, art. xxi, pp. 603–606. Washington, 1877.

Same. Washington: Government Printing Office, 1877. 8vo, pp. 603–606. Two hundred separates printed (author's edition) without repaging, but with title-page, and in paper covers, together with Paleontological Papers Nos. 1, 3, 4, and 5.

 Four species are described as new, two preoccupied names changed, and the genus *Cassiopella* proposed.

                        Page.
Unionidæ.
 Unio proavitus, n. s., White, 1877 603–604
 U. holmesianus, n. s., White, 1877 604
 U. endlichi, n. s., White, 1877 604–605
 U. couesi, [n. s.,] White, 1877 605
 U. meeki, [n. s.,] White, 1877 605
 U. mendax, n. s., White, 1877 605–606
Ceriphasiidæ.
 Cassiopella, n. g., White, 1877 606

54.

WHITE, C. A. Paleontological Papers No. 3: Catalogue of the Invertebrate Fossils, hitherto published from the fresh- and brackish water deposits of the western portion of North America. <Bulletin U. S. Geol. and Geog. Surv. of the Terr., vol. iii, art. xxii, pp. 607–614. Washington, 1877.

Same. Washington: Government Printing Office, 1877. 8vo, pp. 607–614. Two hundred separates printed (author's edition) without repaging, but with title-page, and in paper covers, together with Paleontological Papers Nos. 1, 2, 4, and 5.

 This paper contains a list of the fossil non-marine invertebrate forms at that time known. They have since been much increased. (See entry No. 115.)

 The author has somewhat modified certain views expressed in this paper. These later views are published in entry No. 115.

55.

WHITE, C. A. Paleontological Papers No. 4: Comparison of the North American Mesozoic and Cenozoic Unionidæ and associated mollusks with living species. <Bulletin U. S. Geol. and Geog. Surv. of the Terr., vol. iii, art. xxiii, pp. 615-624. Washington, 1877.

Same. Washington: Government Printing Office, 1877. 8vo, pp. 615-624. Two hundred separates printed (author's edition) without repaging, but with title-page, and in paper covers, together with Paleontological Papers Nos. 1, 2, 3, and 5.

> The relations of the living Uniones of the Mississippi River system with the fossil forms from western North America are pointed out.

56.

WHITE, C. A. Paleontological Papers No. 5: Remarks on the Paleontological characteristics of the Cenozoic and Mesozoic groups as developed in the Green River region. <Bulletin U. S. Geol. and Geog. Surv. of the Terr., vol. iii, art. xxiv, pp. 625-629. Washington, 1877.

Same. Washington: Government Printing Office, 1877. 8vo, pp. 625-629. Two hundred separates printed (author's edition) without repaging, but with title-page, and in paper covers, together with Paleontological Papers Nos. 1, 2, 3, and 4.

Same. Washington: Government Printing Office, 1878. 8vo, pp. 721-724. Two hundred separates printed (author's edition) without repaging, but with title-page, and in paper covers, together with Paleontological Papers No. 6.

57.

WHITE, C. A. Circulation of the Blood, subjectively seen. <Am. Journ. of the Med. Sci., vol. lxxiii, n. s., p. 279. Philadelphia, 1877.

> The appearance of multitudes of rapidly moving corpuscles to be observed when the eyes are shut, especially when there is more or less cerebral excitement, the author conceives to be due to the impact of the blood corpuscles behind the layer of rods and cones.

58.

WHITE, C. A. Paleontological Papers No. 6: Descriptions of new species of Invertebrate fossils from the Laramie Group. < Bulletin U. S. Geol. and Geog. Surv. of the Terr., vol. iv., art. xxviii, pp. 707-719. Washington, 1878.

Same. Washington: Government Printing Office, 1878. 8vo, pp. 707-719. Two hundred separates printed (author's edition) without repaging, but with title-page, and in paper covers, together with Paleontological Papers No. 7.

> Nineteen species of mollusks are described as new.

	Page.
Conchifera.	
Volsella, Scopoli	707
Subgenus *Brachydontes*, Swainson	707
V. (Brachydontes) regularis, n. s., White, 1877	707-708
V. (Brachydontes) laticostata, n. s., White, 1877	708
Nuculana, Link	708
N. inclara, n. s., White, 1877	708-709
Anodonta, Cuvier	709
A. parallela, n. s., White, 1877	709
Unio, Retzius	709
U. goniambonatus, n. s., White, 1877	709-710
U. aldrichi, n. s., White, 1877	710-711
Corbicula, Mergele	711
C. cleburni, n. s., White, 1877	711
C. cardiniæformis, n. s., White, 1877	711-712
C. obesa, n. s., White, 1877	712-713

	Page.
Subgenus *Leptesthes*, Meek	713
C. [*Leptesthes*] *macropistha*, n. s., White, 1877	713
Acella, Haldeman	714
A. haldemani, n. s., White, 1877	714
Physa, Draparnaud	714
P. felix, n. s., White, 1877	714
Helix, Linnæus	714
H. evarstonensis, n. s., White, 1877	714
Neritina, Lamarck	715
N. natleiformis, n. s., White, 1877	715
Subgenus *Velatella*, Meek	715
N. (*Velatella*) *baptista*, n. s., White, 1877	715–716
Goniobasis, Lea	716
G. endlichi, n. s., White, 1877	716
Viviparus, Lamarck	716
V. prudentia, n. s., White, 1877	716–717
V. couesi, n. s., White, 1877	717–718
Odontobasis, Meek	718
O. ? formosa, n. s., White, 1877	718–719

59.

WHITE, C. A. Paleontological Papers No. 7: On the distribution of Molluscan species in the Laramie Group. <Bulletin U. S. Geol. and Geog. Surv. of the Terr. vol. iv, art. xxix, pp. 721–724. Washington, 1878.

The unity of the Judith River, Fort Union, Lignitic, and Bitter Creek series, as the Laramie Group is shown in this paper.

60.

WHITE, C. A. Paleontological papers No. 8. Remarks upon the Laramie group. <Bulletin U. S. Geol. and Geogr. Surv. of the Terr., vol. iv, art. xxxvi, pp. 865–876. Washington, 1878.

Same. Washington: Government Printing Office, 1878. 8vo, pp. 865–876. Two hundred separates printed (author's edition), with title-page and paper covers, but without repaging.

The characteristics of the group and its probable geological age are discussed.

61.

WHITE, C. A., *and* NICHOLSON, H. A. Bibliography of North American Invertebrate Paleontology; being a report upon the publications that have hitherto been made upon the invertebrate paleontology of North America, including the West Indies and Greenland. By C. A. White, M. D., Paleontologist of the United States Geological Survey, and H. Alleyne Nicholson, M. D., D. Sc., Professor at the University of St. Andrews, Scotland. <Department of the Interior; United States Geological Survey of the Territories. Miscellaneous publications No. 10. Washington: Government Printing Office, 1878. 8vo, 132 pp.

CONTENTS.

Part I. Embracing titles and abstracts of publications made in the United States. By C. A. White, M. D. pp. 7–69.

Part II. Embracing titles and abstracts of publications made in British North America, in the West Indies, and Europe. By H. Alleyne Nicholson, M. D., D. Sc., pp. 71–132.

62.

WHITE, C. A. Change in the habits of Toads. <Nature. 4to, vol. xvii, p. 242. London and New York, 1878.

Observations made in Colorado along irrigating ditches show that the toads there have formed, or reverted to, the habit of diving to the bottom, when disturbed, and remaining there like frogs.

63.

WHITE, C. A. Note on the Re-establishment of Forests in Iowa, now in progress. <Amer. Jour. Sci., 3d ser., vol. xvi, p. 328. New Haven, 1878.

> This note has reference to an article by Prof. Asa Gray in the same volume of the Journal, entitled, Forestry and Archæology.

64.

WHITE, C. A. Report on the Geology of a portion of Northwestern Colorado. <Tenth Annual Report U. S. Geol. and Geogr. Surv. of the Terr. (for the year 1876), pp. 1–60. One map and one plate. Washington, 1878.

> The structure of the eastern end of the Uinta Mountain range is shown, and its relation to the Park range of the Rocky Mountain system explained. The isolated up-thrust mountains are described, and the term "partiversal," as applied to dips, is first used in this report. The results of this work, as shown on the accompanying map, were subsequently incorporated in sheets iv and xi of the Atlas of Colorado and Portions of the Adjacent Territory, published in 1877 by the same survey.

65.

WHITE, C. A. Note on the Garter Snake. <Amer. Naturalist, vol. xii, p. 53. Philadelphia, 1878.

> This note records a supposed instance of the swallowing of a quantity of air by a garter snake immediately before diving to the bottom of a pool of water.

66.

WHITE, C. A. Contributions to Invertebrate Paleontology, No. 1: Cretaceous Fossils of the Western States and Territories. <Eleventh Annual Report U. S. Geol. and Geogr. Surv. of the Terr. (for the year 1877), pp. 273–319, pls. i–x. Washington, 1879.

Same. Washington: Government Printing Office, 1879. 8vo, pp. 273–319, pls. i–x. Two hundred separates printed without repaging, but with title-page and paper covers. The title upon the title-page differs from that which heads the text by the omission of the word "Invertebrate."

> Fifty-six species are described and figured, a part of which are new, but most of which had been previously described by different authors, without illustrations. This series of "Contributions" was established for the purpose of illustrating species of fossils previously described, as well as new forms. They number eight in all, and that form of title has not been subsequently used by the author in connection with his work for the United States Surveys.

	Page.
Radiata.	
Actinaria.	
Caryophyllia, Lamarck.	
C. johannis, n. s., White, 1879, pl. vi, figs. 6 a, b	274–275
C. egeria, n. s., White, 1879, pl. vi, figs. 7 a, b	275
Mollusca.	
Conchifera.	
Ostrea, Linnæus	275
O. quadriplicata, Shumard, 1860, pl. v, fig. 6 a, pl. viii, figs. 3 a, b	275–276
O. (Alectryonia) bellaplicata, Shumard, 1860, pl. iv, figs. 3 a, b, pl. viii, figs. 2 a, b	276–277
O. (Alectryonia) sannionis, White, 1876, pl. ii, figs. 2 a–e	277–278
Exogyra, Say	278
E. valkeri, n. s., White, 1879, pl. i, figs. 1 a, b	278
Placunopsis, Morris & Lycett	278
P. hilliardensis, n. s., White, 1879, pl. vii, fig. 14 a	278–279
Plicatula, Lamarck	279
P. hydrotheca, White, 1876, pl. vi, figs. 3 a, b	279
Pteria, Scopoli	279
P. parkensis, White, 1876, pl. iii, fig. 3 a	279–280
Subgenus, Oxytoma, Meek	280

	Page.
P. (Oxytoma ?) gastrodes, Meek, 1873, pl. x, fig. 1 *a*	280-281
Subgenus *Pseudoptera*, Meek	281
P. (Pseudoptera) propleura, Meek, 1873, pl. x, figs. 2 *a-c*	281-284
Inoceramus, Sowerby	284
I. howelli, White, 1876, pl. iv, figs. 1 *a-c*	284-285
I. gilberti, White, 1876, pl. iii, figs. 1 *a-c*	285
I. oblongus, Meek, 1871, pl. ii, figs. 1 *a, b*	285-286
Barbatia, Gray	286
B. coalvillensis, White, 1876, pl. vi, figs. 2 *a, b*	286-287
Crassatella, Lamarck	287
C. cimarronensis, n. s., White, 1879, pl. v, figs. 3 *a-c*	287-288
Tancredia, Lycett	288
T. ? cœlionotus, n. s., White, 1879, pl. v, figs. 2 *a-d*	288
Cyrena, Lamarck	289
C. securis, Meek, 1873, pl. iii, figs. 2 *a-c*	289
C. inflexa, Meek, 1871, pl. x, figs. 7 *a, b*	290
Corbula, Bruguière	290
C. nematophora, Meek, 1873, pl. iii, figs. 4 *a-d*	290-291
Cardium, Linnæus	291
C. pauperculum, Meek, 1872, pl. ix, fig. 3 *a*	291
C. trite, n. s., White, 1879, pl. v, figs. 4 *a, b*	291-292
Cardium —— ?, White, 1879, pl. ix, figs. 2 *a-c*	292
Trapezium truncatum, Meek, 1871, pl. x, figs. 6 *a, b*	292-293
Trapezium ? micronema, Meek, 1873, pl. x, figs. 5 *a*	293
Baroda wyomingensis, Meek, 1871, pl. x, figs. 3 *a, b*	293-294
Baroda subelliptica, n. s., White, 1879, pl. x, figs. 4 *a-d*	294-295
Mactra ? holmesi, Meek, 1875, pl. vi, figs. 4 *a-c*	295-297
Mactra ? cañonensis, Meek, 1871, pl. ix, figs. 11 *a-c*	297-298
Pachymya, Sowerby	298
P. austinensis, Shumard, pl. viii, figs. 1 *a, b*, and pl. v, figs. 7 *a, b*	298
P. ? herseyi, n. s., White, 1879, pl. v, figs. 5 *a, b*	298-299
Glycimeris, Lamarck	299
G. berthoudi, n. s., White, 1879, pl. vi, figs. 1 *a, b*	299-300
Parapholas, Conrad	300
P. sphenoideus, White, 1876, pl. v, figs. 1 *a-d*	300-302
Gasteropoda.	
Paliurus, Gabb	302
P. pentangulatus, n. s., White, 1879, pl. iv, figs. 4 *a, b*	302-303
Anisomyon, M. & H	303
A. centrale, Meek, 1872, pl. ix, figs. 1 *a-d*	303-304
Actæon, Montfort	304
A. woosteri, n. s., White, 1879, pl. vii, figs. 9 *a-c*	304
Actæonina, d'Orbigny	305
A. prosocheila, n. s., White, 1879, pl. vii, figs. 10 *a, b*	305
Turbonilla, Risso	305
T. (Chemnitzia) coalvillensis, Meek, 1873, pl. ix, figs. 5 *a, b*	305-306
Physa, Draparnaud	306
P. carletoni, Meek, 1873, pl. vii, fig. 12 *a*	306-307
Physa —— ?, White, 1879, pl. vii, fig. 13 *a*	307-308
Neritina, Lamarck	308
N. pisum, Meek, 1873, pl. vii, figs. 11 *a-c*	308
N. incompta, White, 1876, pl. vii, figs. 6 *a-c*	308-309
N. (Velatella) patelliformis, Meek, 1873, pl. vii, figs. 7 *a-d*	309
N. (Velatella) patelliformis var. *weberensis*, White, 1879, pl. vii, figs. 8 *a* and *b*	309-310
Euspira, Agassiz	310
E. coalvillensis, White, 1876, pl. iv, figs. 2 *a* and *b*	310
Anchura, Conrad	311
A. haydeni, n. s., White, 1879, pl. vii, fig. 1 *a*	311-312
A. (Drepanocheilus) ruida, White, 1876, pl. vii, figs, 4 *a, b*	312
A. (Drepanocheilus) mudgeana, n. s., White, 1879, pl. vii, figs. 3 *a, b*	312-313
A. (Drepanocheilus) prolabiata, White, 1876, pl. vii, fig. 2 *a*	313-314
Turritella, Lamarck	314
T. marnochi, n. s., White, 1879, pl. vii, figs. 5 *a, b*	314-315
T. coalvillensis, Meek, 1873, pl. ix, fig. 4 *a*	315-316
T. (Aclis ?) micronema, Meek, 1873, pl. ix, fig. 8 *a*	316

	Page.
Eulimella, Forbes	316
E ? funicula, Meek, 1873, pl. ix, fig. 10 a	316–317
Fusus, Lamark	317
F. (Neptunea?) gabbi, Meek, 1873, pl. ix, fig. 9 a	317
Admetopsis, Meek	317
A. rhomboides, Meek, 1873, pl. ix, figs. 6 a, b	317–318
A. subfusiformis, Meek, 1873, pl. ix, fig. 7 a	318–319

67.

WHITE, C. A. Descriptions of new species of Invertebrate fossils from the Carboniferous and Upper Silurian rocks of Illinois and Indiana. <Proc. Acad. Nat. Sci., Phila., vol. xxx, pp. 29–37. Philadelphia, 1878.

Same. 8 vo., pp. 29–37. Philadelphia, 1878. Thirty separates printed without title-page, covers, or repaging.

Nine species are described as new.

	Page.
Radiata.	
Actinozoa.	
Baryphyllum, Edwards & Haime	29
B. fungulus, n. s., White, 1878	29–30
Echinodermata.	
Platycrinus, Miller	30
P. bonoensis, n. s., White, 1878	30–31
Scaphiocrinus, Hall	31
S. gibsoni, n. s., White, 1878	31–32
S. gurleyi, n. s., White, 1878	32–33
Lepidesthes, M. & W	33
L. colletti, n. s., White, 1878	33–34
Mollusca.	
Polyzoa.	
Ptilodictia, Lonsdale	35
P. triangulata, n. s., White, 1878	35
Conchifera.	
Astartella, Hall	35
A. gurleyi, n. s., White, 1878	35–36
Cephalopoda.	
Nautilus, Breynius	36
N. danvillensis, n. s., White, 1878	36–37
Articulata.	
Vermes.	
Serpula, Linnæus	37
S. insita, n. s., White, 1878	37

68.

WHITE, C. A. Anecdote of the Great Horned Owl. <American Naturalist, vol. xiii, p. 783. Philadelphia, 1879.

This anecdote is told to illustrate the precision, rapidity, and extent of the action of the muscles of the neck of that owl. Its authorship is indicated only by the initials "C. A. W."

69.

WHITE, C. A. Remarks on the Jura-Trias of Western North America. <Am. Jour. Sci., vol. xvii, 3d ser., pp. 214–218. New Haven, 1879.

Same. New Haven, 1879. 8vo., pp. 214–218. Twenty separates printed without title-page, covers, or repaging.

Reference is made in this article to the Triassic fossils which are described in Paleontological Papers No. 9.

70.

WHITE, C. A. Report on the Paleontological Field-work for the season of 1877. <Eleventh Annual Report U. S. Geol. and Geog. Surv. of the Terr. (for the year 1877), pp. 161–272. Washington, 1879.

Same. Washington: Government Printing Office, 1879. 8vo, pp. 161-272. Two hundred separates printed, with title-page and paper covers, but without re-paging.

This report gives results of observations in Colorado, Utah, and Wyoming. The existence of the Green River and Bridger groups south of the Uinta Mountains is first announced in this report; and descriptions of three species of fossils are given in foot-notes.

	Page.
Bulinus disjunctus, n. s., White, 1879	170
Pinna lakesi, n. s., White, 1879	181
Corbula dubiosa, n. s., White, 1879	249
Crow Creek section	164
List of fossils from the valley of Crow Creek, Colorado	165
Notes on the Laramie fossils obtained in the valley of Crow Creek, Colorado	165-175
List of the fossils collected from Cretaceous strata in the valley of the Cache à la Poudre, from five to twelve miles west of Greeley, Colo	175-176
List of Cretaceous fossils collected at Fossil Ridge, three miles southeastward from Spring cañon, and about six miles south of Fort Collins, Colorado	176-177
List of fossils collected in the valley of Little Thompson Creek	177-178
List of the fossils collected at the mouth of Saint Vrains River, Colorado	178-179
Notes on the fossils of the Fox Hills group as developed in Colorado, east of the Rocky Mountains	179-189
Bijou Creek section	189
List of fossils from the valley of Bijou Creek, Colorado	190
Notes on the Laramie fossils collected in the valley of Bijou Creek, Colorado	190-197
List of Cretaceous fossils from the vicinity of Golden City and Morrison, Colo	197
Notes on the fossils from the vicinity of Golden City and Morrison	179-204
List of Cretaceous fossils collected on Sage Creek, an upper tributary of Yampa River, Colorado	205
Notes on the Cretaceous fossils of Sage Creek	205-207
List of Laramie fossils collected in Yampa valley, near Cañon Park, Northwestern Colorado	207
Notes on the Laramie fossils of Yampa valley	208-211
List of the Laramie fossils found in the Danforth Hills, Northwestern Colorado	211
Notes on the Laramie fossils of Danforth Hills	211-215
List of Laramie fossils collected at Rock Springs, Wyoming	215
List of Laramie fossils from Bitter Creek valley, two miles west of Point of Rocks station, Wyoming	215-216
List of the Laramie fossils collected at Point of Rocks Station, Wyoming	216
List of the Laramie fossils collected at Black Buttes Station, Wyoming	216-217
Notes on the Laramie fossils of Bitter Creek valley	217-222
Section of Laramie strata at Black Butte Station	223
List of Cretaceous fossils from a cañon six miles northwest from White River Indian Agency, Northwestern Colorado	224
Notes on the Cretaceous fossils from near White River Indian Agency	224-226
List of fossils of the Wasatch group collected in White River valley, Colorado	226
Notes on the Wasatch fossils of White River valley	226-229
List of Cretaceous fossils at Dodd's Ranch on Ashley's Fork, Utah	229
Notes on the Cretaceous fossils from Ashley's Fork	229-232
List of fossils from the Cretaceous series at Coalville, Utah	232-233
Notes on the Cretaceous fossils of Coalville, Utah	233-241
List of Laramie fossils from Bear River valley, Wyoming	241-242
Notes on Laramie fossils of Bear River valley, Wyoming	242-248
List of Cretaceous fossils from the valley of Bear River, Wyoming	248
Notes on the Cretaceous fossils of Bear River valley	248-249
List of Cretaceous fossils collected at Hilliard Station, Wyoming	249
Notes on the Cretaceous fossils of Hilliard Station	249-251
General discussion	251-265
Table showing the geographical distribution of the Cretaceous species collected during the season of 1877	252-253
Fox Hills Group	252-253
Colorado Group	253
Table showing the geographical distribution of the fossils of the Laramie Group, collected during the season of 1877	255
Catalogue of fossils	265

	Page.
List of Cretaceous fossils sent by Mr. Arthur Lakes from Bear Creek valley, near Morrison, Colorado	265-266
List of fossils sent by Mr. L. C. Wooster from the vicinity of Greeley, Colo	266
List of Cretaceous fossils sent by Mr. J. C. Hersey from Colorado	266-267
List of Cretaceous fossils sent by Capt. E. L. Berthoud from Colorado	267-268
List of Cretaceous fossils sent by Prof. B. F. Mudge from Dennison, Tex	268
List of Cretaceous fossils sent by Mr. G. W. Marnoch from Helotes, Bexar County, Texas	268-269
List of Cretaceous fossils sent by D. H. Walker to the Smithsonian Institution from near Salado, Bell County, Texas	269-270
Concluding remarks	270-271

71.

WHITE, C. A. Paleontological Papers No. 9: Fossils of the Jura-Trias of Southeastern Idaho. <Bulletin U. S. Geol. and Geog. Surv. of the Terr., vol. v, art. v, pp. 105-117. Washington, 1879.

Same. Washington: Government Printing Office, 1879. 8vo, pp. 105-117. Two hundred separates printed (author's edition) without repaging, but with titlepage, and in paper covers, together with Paleontological Papers No. 10.

Ten species are described, most of which are new; and the genus *Meekoceras*, Hyatt, is described for the first time in this article.

	Page.
Brachiopoda.	
Terebratula, Lihwyd	108
T. semisimplex, n. s., White, 1879	108
T. augusta, Hall & Whitfield	108-109
Conchifera.	
Aviculopecten, McCoy	109
A. pealei, n. s., White, 1879	109-110
A. altus, n. s., White, 1879	110
A. idahoensis, Meek	110-111
Cephalopoda.	
Meekoceras, n. g., Hyatt, 1879	111-112
M. aplanatum, n. s., White, 1879	112-113
M. mushbachanus, n. s., White, 1879	113-114
M. gracilitatis, n. s., White, 1879	114-116
M. gracilitatis var., White, 1879	116
Arcestes, Suess	116
A. ? cirratus, n. s., White, 1879	116-117
Arcestes? ——? White, 1879	117
Arcestes? ——? White, 1879	117

72.

WHITE, C. A. Paleontological Papers No. 10. Conditions of Preservation of Invertebrate Fossils. <Bulletin U. S. Geol. and Geog. Surv. of the Terr. Vol. v, art. viii, pp. 133-141. Washington, 1879.

Same. Washington: Government Printing Office, 1879. 8vo., pp. 133-141. Two hundred separates printed (author's edition) without repaging, but with title page and with paper covers, together with Paleontological Papers No. 9.

The fact is pointed out that shells of different classes and families which have been fossilized under the same conditions are differently preserved.

73.

WHITE, C. A. Paleontological Papers No. 11. Remarks upon Certain Carboniferous Fossils from Colorado, Arizona, Idaho, Utah, and Wyoming, and Certain Cretaceous Corals from Colorado, together with Descriptions of New Forms. <Bulletin U. S. Geol. and Geog. Surv. of the Terr. Vol. v, art. xiv, pp. 209-221. Washington, 1879.

Same. Washington: Government Printing Office, 1879. 8vo., pp. 209-221. Two hundred separates printed (author's edition) without repaging, but with title-page and paper covers. The series of Paleontological Papers closes with this number.

Thirteen species are described, most of which are new. Attention is called to the commingling of Upper and Lower Carboniferous types in the Carboniferous strata of the western portion of North America.

CARBONIFEROUS.

	Page.
Actinozoa.	
Leptopora, Winchell	211
L. winchelli, n. s., White, 1879	211-212
Echinodermata.	
Granatocrinus, Troost	212
G. lotoblastus, White? 1879	213-213
Polyzoa.	
Archimedes, Lesueur	213
Archimedes, ———?, White, 1879	213-214
Ptilodyctia, Lonsdale	214
P. triangulata, White, 1879	214-215
Brachiopoda.	
Retzia, King	215
R. woosteri, n. s., White, 1879	215-216
Conchifera.	
Nuculana, Link	217
N. obesa, n. s., White, 1879	216-216
Nucula, Lamarck	217
N. perumbonata, n. s., White, 1879	217
Allorisma, King	217
A.? gilberti, n. s., White, 1879	217-218
Gasteropoda.	
Bellerophon, Monfort	218
B. subpapillosus, White, 1879	218-219
Murchisonia, D'Archiac	219
M. terebra, n. s., White, 1879	219
Pleurotomaria, Defrance	219
P. grayvillensis, Norwood & Pratten	219-220

CRETACEOUS FORMS.

Actinozoa.	
Chætetes, Fischer	220
C. ?? dimissus, n. s., White, 1879	220-221
Beaumontia, Edwards & Haime	221
B.? solitaria, n. s., White, 1879	221

74.

WHITE, C. A., and NICHOLSON, H. A. Supplement to the Bibliography of North American Invertebrate Paleontology. <Bulletin U. S. Geol. and Geog. Surv. of the Terr., vol. v, pp. 143-152. Washington, 1879.

Same. Washington: Government Printing Office, 1879, pp. 143-152. Two hundred separates printed (author's edition) with title-page and paper covers, but without repaging.

This supplement embraces works which were published after the publication of the Bibliography, besides some that were inadvertently omitted then.

Part I. Publications made in the United States. By C. A. White, pp. 143-149.
Part II. Publications made in British North America, the West Indies, and Europe. By H. Alleyne Nicholson, pp. 150-152.

75.

WHITE, C. A. Paleontology: Fossils of the Indiana Rocks. <State of Indiana. Second Annual Report of the Department of Statistics and Geology, 471-522, pls. i-xi. Indianapolis, 1880.

Same. In Indiana Geological Report, 1879-'80. From the Second Annual Report of the Bureau of Statistics and Geology. John Collett, Chief of Department. Indianapolis, 1881. 8vo. pp. 103-154, and 11 plates.

This book is an extract of the geological and biological matter from the volume of the preceding entry, repaged and bound in cloth, with new title-page as above.

This work is designed for popular use, and, with the exception of one variety, none of the species described and figured are new. The illustrations are also largely copies of formerly published figures. Forty-nine species of characteristic Silurian, Devonian, and Carboniferous forms are given.

LOWER SILURIAN.

Mollusca.
Brachiopoda.

	Page.
Strophomena, Rafinesque	481
S. alternata, Conrad, pl. i, figs. 6 and 7	481–482
S. planumbona, Hall, pl. ii, figs. 13 and 14	483–484
Orthis, Dalman	484
O. subquadrata, Hall, pl. i, figs. 3–5	484–485
O. occidentalis, Hall, pl. ii, figs. 10–12	485–487
O. biforata, Schlotheim var. *acutilirata*, Conrad, pl. ii, figs. 5–9	487–489
Rhynchonella, Fischer	489
R. capax, Conrad, pl. i, figs. 9–11	489–490
R. dentata, Hall, pl. i, figs. 12–14	490–491

Lamellibranchiata.

Megaptera, M. & W.	491
Ambonychia (*Megaptera*) *casei*, M. & W., pl. i, figs. 1 and 2	491–492

Gasteropoda.

Cyclonema, Hall	492
C. bilix, Conrad, pl. ii, figs. 3 and 4	492–493

Crustacea.
Trilobites.

Calymene, Brongniart	493
C. senaria, Conrad, pl. ii, figs. 1 and 2	493–495

UPPER SILURIAN.

Echinodermata.

Eucalyptocrinus, Goldfuss	495
E. crassus, Hall, pl. iii, fig. 1	495–496

Brachiopoda.

Rhynchonella, Fischer	496
R. tennesseensis, Roemer, pl. iii, figs. 2–4	496–497
Spirifer, Sowerby	497
S. radiata, Sowerby, pl. iii, figs. 5 and 6	497

Gasteropoda.

Platyostoma, Conrad	497
P. niagarense, Hall, pl. iii, figs. 7 and 8	497–498

Crustacea.
Trilobites.

Cyphaspis, Burmeister	498
C. christyi, Hall, pl. iii, fig. 9	498–499

DEVONIAN.

Polypi.

Zaphrentis, Rafinesque & Clifford	499
Zaphrentis ———? White, 1880, pl. v, figs. 3 and 4	499
Favosites, Lamarck	499
Favosites ———? White, 1880, pl. v, figs. 1 and 2	499

Brachiopoda.

Strophodonta, Hall	500
S. demissa, Conrad, pl. iv, figs. 6 and 7	500–501
Orthis, Dalman	501
O. iowensis, Hall (?), pl. v, figs. 10–12	501–502
Atrypa, Dalman	502
A. reticularis, Linnæus, pl. v, figs. 7–9	502
Athyris, McCoy	502
A. vittata, Hall, pl. iv, figs. 8 and 9	502–503
Spirifer, Sowerby	503

	Page.
S. acuminata, Conrad, pl. iv, figs. 1-3	503-504
S. euritines, Owen, pl. iv, figs. 4 and 5	504
S. gregaria, Clapp, pl. iv, figs. 10 and 11	504-505

Lamellibranchiata.
 Paracyclas, Hall 505
 P. elliptica var. occidentalis, Hall, pl. v, figs. 5 and 6 505

SUB-CARBONIFEROUS.
Polypi.
 Lithostrotion, Fleming.
  L. mamillare, Castelnau, pl. vi, figs. 1 and 2 506
  (L. canadense, Castelnau), pl. vi, figs. 1 and 2 506
Echinodermata.
 Taxocrinus, Phillips 506
 T. multibrachiatus, Lyon & Cassady var. colletti, White, 1880, pl. vii, fig. 3 506-507
 Scaphiocrinus, Hall 507
 S. gibsoni, White, 1878, pl. vii, fig. 7 507-508
 S. gurleyi, White, 1878, pl. vii, fig. 8 509
 Actinocrinus, Miller 510
 A. wachsmuthi, White, 1878, pl. vii, fig. 6 510
 Pentremites, Say 511
 P. pyriformis, Say, pl. vii, fig. 9 511
 P. godoni, Defrance, pl. vii, figs. 10 and 11 511
 P. conoideus, Hall, pl. vii, fig. 12 512
Brachiopoda.
 Spirifer, Sowerby 512
 S. textus, Hall, pl. vii, figs. 1 and 2 512-513
Pteropoda.
 Conularia, Miller 513
 C. missouriensis, Swallow? pl. vi, fig. 4 513
Gasteropoda.
 Platyceras, Conrad 514
 P. equilatera, Hall, pl. vii, fig. 5 514
Cephalopoda.
 Goniatites, De Haan 514
 G. oweni, Hall, pl. vii, figs. 3 and 4 515
Crustacea.
 Trilobites.
 Phillipsia, Portlock 515
 P. bufo, M. & W., pl. vi, fig. 5 515-516

COAL MEASURES.
Brachiopoda.
 Productus, Sowerby 516
 P. costatus, Sowerby, pl. viii, figs. 7 and 8 516-517
 Spirifer, Sowerby 517
 S. cameratus, Morton, pl. viii, fig. 3 517-518
Lamellibranchiata.
 Allorisma, King 518
 A. subcuneata, M. & H.? pl. viii, figs. 1 and 2 518-519
Gasteropoda.
 Polyphemopsis, Portlock 519
 P. fusiformis, Hall, pl. viii, fig. 6 519
 Pleurotomaria, Defrance 519
 P. tabulata, Hall, pl. viii, figs. 4 and 5 519-520

FOSSIL PLANTS.
Neuropteris 520
 N. hirsuta, Lesqx. pl. ix, figs. 1-3 520
 N. rarinervis, Bunbury, pl. x, figs. 1-3 520-521
Callipteris 521
 C. sullivanti, Lesqx. pl. ix, fig. 4 521
Annularia 521
 A. longifolia, Brongniart? pl. xi, figs. 1 and 2 521-522
Odontopteris 522
 O. subcuneata, Bunbury? pl. xi, fig. 3 522
Sphenopteris 522
 S. acuta, Brongniart? pl. xi, fig. 4 522

76.

WHITE, C. A. Progress of Invertebrate Paleontology in the United States for the year 1879. <Amer. Nat. vol. xiv, pp. 250-260. Philadelphia, 1880.
Same. Philadelphia, 1880, 8vo, pp. 250-260. One hundred separates printed without title-page, covers, or repaging.

> This article, like those for the years 1880, 1881, and 1882, respectively, is a list of the titles of the works published within the year designated, together with remarks upon them.

77.

WHITE, C. A. On the Antiquity of Certain Subordinate Types of Fresh-water and Land Mollusca. <Amer. Jour. Sci., 3d ser., vol. xx, pp. 44-49. New Haven, 1880.
Same. New Haven, 1880, 8vo, pp. 44-49. Twenty separates printed without title-page, covers, or repaging.

> It is shown that not only were many of the genera of living North American Mollusca established as early as the close of the Cretaceous period, but many of the subgenera were established thus early also.

78.

WHITE, C. A. Descriptions of new species of Carboniferous Invertebrate Fossils. <Proc. U. S. Nat. Mus., vol. ii, pp. 252-260, one plate. Washington, 1880.
Same. Washington: Government Printing Office, 1879, 8vo, pp. 252-260, one plate. One hundred and fifty separates printed without title-page, covers, or repaging.

> Six species of Echinoderms are described, and the genus *Lecythiocrinus* is proposed.

	Page.
Actinozoa.	
Acervularia, Schweigger	255
A. adjunctiva, n. s., White, 1879, plate i, figs. 1-3	255
Echinodermata.	
Lecythiocrinus, n. g., White, 1879	256-257
L. olliculæformis, n. s., White, 1879, pl. i, figs. 4 and 5	257
Erisocrinus, M. & W	257
E. planus, n. s., White, 1879, pl. i, figs. 6 and 7	257-258
Cyathocrinus, Miller	258
C. stillativus, n. s., White, 1879, pl. i, figs. 9 and 10	258-259
Rhodocrinus, Miller	259
R. vesperalis, n. s., White, 1879, pl. i, figs. 11 and 12	259-260
Archæocidaris, McCoy	260
A. dininnii, n. s., White, 1879, pl. i, figs. 13 and 14	260

79.

WHITE, C. A. Note on Endothyra ornata. <Proc. U. S. Nat. Mus., vol. ii, p. 291. Washington, 1880.
Same. Washington: Government Printing Office, 1879, 8vo, p. 291. One hundred and fifty separates printed, together with the papers of the two following entries, without title-page, covers, or repaging.

> The discovery of that foraminiferous form in the Carboniferous rocks of Wyoming is noticed. It was previously known only in Europe.

80.

WHITE, C. A. Note on Criocardium and Ethmocardium. < Proc. U. S. Nat. Mus., vol. ii, pp. 291-292. Washington, 1880.
Same. Washington: Government Printing Office, 1879, 8vo, pp. 291-292. One hundred and fifty separates printed, together with the papers of the preceding and the following entry, without title-page, covers, or repaging.

> The new genus *Ethmocardium* is proposed.

---

* Ληκύθιον, a small oil flask.

81.

WHITE, C. A. Descriptions of new Cretaceous Invertebrate Fossils from Kansas and Texas. <Proc. U. S. Nat. Mus., vol. ii, pp. 292-298, pls. i-v. Washington, 1880.
Same. Washington: Government Printing Office, 1879, 8vo, pp. 292-298, and five plates. One hundred and fifty separates printed, together with the papers of the two preceding entries, without title-page, covers, or repaging.

Six species are described as new and illustrated.

	Page.
Mollusca.	
Ostrea, Linnæus	293
Subgenus *Alectryonia*, Fischer	293
O. (Alectryonia) blackii, n. s., White, 1879, pl. iv, figs. 1 and 2	293
Exogyra, Say	293
E. forniculata, n. s., White, 1879, pl. iv, figs. 3 and 4	293-294
E. winchelli, n. s., White, 1879, pl. ii, figs. 1 and 2, and pl. iii, figs. 1 and 2	294-295
Gervillia, Defrance	295
G. mudgeana, n. s., White, 1879, pl. v, figs. 3 and 4	295-296
Pteria, Scopoli	296
Subgenus *Oxytoma*, Meek	296
P. (Oxytoma) salinensis, n. s., White, 1879, pl. v, figs. 1 and 2	296-297
Pachymya, Sowerby	297
P.? compacta, n. s., White, 1879, pl. vi, figs. 3 and 4	297
Thracia, Leach	297
T. myæformis, n. s., White, 1879, pl. vi, figs. 1 and 2	297-298

82.

WHITE, C. A. Report on the Carboniferous Invertebrate Fossils of New Mexico. <Report upon U. S. Geog. and Geol. Surveys West of the 100th Meridian; in charge of Capt. George M. Wheeler, Corps of Engineers, U. S. Army. 4to., vol. iii; Supplement. Geology. Appendix, pp. i-xxxviii, pl. i-ii. Washington, 1881.
Same. Washington: Government Printing Office, 1881, 4to, pp. i-xxxviii, and two plates. Seventy-five separates printed, without title-page, covers, or repaging.

This report embraces an annotated catalogue of the species collected at each locality, and also descriptions of seventeen species, a part of which are new.

DESCRIPTIONS OF SPECIES.

Echinodermata.
Archæocidaris, McCoy ... xxii
A. triplex, n. s., White, 1881, pl. iv, figs. 3 a–e ... xxii–xxiii

Brachiopoda.
Orthis, Dalman ... xxiii
O. resupinoides, Cox ? 1857, pl. iii, figs. 2 a, b ... xxiii–xxiv

Polyzoa.
Ptilodyctia, Lonsdale ... xxiv
P. triangulata, White, 1878, pl. iv, figs. 2 a–e ... xxiv–xxv

Conchifera.
Myalina, de Koninck ... xxv
M. permiana, Swallow, 1[8]58, pl. iii, figs. 1 a–d ... xxv–xxvii
Pleurophorus, King ... xxvii
P. subcostatus, M. & W., 1866, pl. iii, fig. 8 a ... xxvii

Pteropoda.
Conularia, Miller ... xxviii
C. crustula, White, 1880, pl. iii, figs. 4 a, b ... xxviii

Gasteropoda.
Soleniscus, M. & W ... xxviii
S. brevis, n. s., White, 1881, pl. iv, figs. 5 a–e ... xxviii–xxix
S. planus, n. s., White, 1881, pl. iv, figs. 4 a–e ... xxix–xxx
Bellerophon, Montfort ... xxx
B. inspeciosus, n. s., White, 1881, pl. iv, figs. 1 a–d ... xxx–xxxi
Pleurotomaria, Defrance ... xxxi
P. perizomata, n. s., White, 1881, pl. iii, figs. 5 a–e ... xxxi
Murchisonia, d'Archiac ... xxxi
M. copei, n. s., White, 1881, pl. iii, figs. 10 a, b ... xxxi–xxxii

	Page.
Rotella, Lamarck	xxxii
R. verruculifera, n. s., White, 1881, pl. iv, figs. 7 *a–d*	xxxii–xxxiii
Naticopsis, McCoy	xxxiii
N. wheeleri, Swallow, var. White, 1881, pl. iv, figs. 6 *a, b*	xxxiii–xxxiv
N. monilifera, White, 1880, pl. iii, figs. 3 *a–d*	xxxiv–xxxv
N. altonensis, McChesney, pl. iii, fig. 6*a*	xxxv
Loxonema, Phillips	xxxv
L. rugosa, M. & W., pl. iii, fig. 7 *a*	xxxv
Aclis, Loven	xxxv
A. ? stevensoni, n. s., White, 1881, pl. iii, figs. 9 *a, b*	xxxv–xxxvi

83.

WHITE, C. A. Progress of Invertebrate Paleontology in the United States for the year 1880. <Amer. Nat., vol. xvi, pp. 273–279. Philadelphia, 1881.

Same. Philadelphia, 1881, 8vo, pp. 273–279. One hundred separates printed without title-page, covers, or repaging.

> This article is similar to that of entry No. 74.

84.

WHITE, C. A. English Sparrows refusing to eat worms. <Amer. Nat., vol. xv, pp. 671–672. Philadelphia, 1881.

> It is observed that the sparrows refused to eat earth worms, which are favorite food with other birds, when they come to the surface after rains.

85.

WHITE, C. A. Note on the occurrence of Productus giganteus in California. <Proc. U. S. Nat. Mus., vol. iii, pp. 46–47, pl. i. Washington, 1881.

Same. Washington: Government Printing Office, 1880, 8vo, pp. 46–47, and one plate. One hundred and fifty separates printed, together with the papers of the two following entries, without title-page, covers, or repaging.

> The volume was first distributed without the plate; but the latter was afterward printed and distributed.

86.

WHITE, C. A. Note on Acrothele. <Proc. U. S. Nat. Mus., vol. iii, p. 47. Washington, 1881.

Same. Washington: Government Printing Office, 1880, 8vo, p. 47. One hundred and fifty separates printed, together with the last, and the next following entry.

> It is shown that *Aerotreta subsidua*, White, from Southern Utah, really belongs to the genus *Acrothele*, Linnarsson.

87.

WHITE, C. A. Description of a new Cretaceous Pinna from New Mexico. <Proc. U. S. Nat. Mus., vol. iii, pp. 47–48. Washington, 1881.

Same. Washington: Government Printing Office, 1880, 8vo, pp. 47–48. One hundred and fifty separates printed, together with the papers of the two preceding entries, without-title page, covers, or repaging.

> The author now thinks that the proposed new species (*Pinna stevensoni*) is only a variety of *P. petrina*, White.

	Page.
Pinna stevensoni, n. s., White, 1880	47–48

88.

WHITE, C. A. Note on the occurrence of Stricklandinia salteri and S. davidsoni in Georgia. <Proc. U. S. Nat. Mus., vol. iii, pp. 48–49. Washington, 1881.

Same. Washington: Government Printing Office, 1880, 8vo, pp. 48–49. One hundred and fifty separates printed without title-page, covers, or repaging.

89.

WHITE, C. A. Description of a very large fossil Gasteropod from the State of Puebla, Mexico. <Proc. U. S. Nat. Mus., vol. iii, pp. 140-142, and one plate. Washington, 1881.

Same. Washington: Government Printing Office, 1880, 8vo, pp. 140-142, and one plate. One hundred and fifty separates printed without title-page, covers, or repaging.

The species described is from strata believed to be of Cretaceous age.

	Page.
Tylostoma, Sharpe	141
T. princeps, n. s., White, 1880, pl. ii, figs. 1 and 2	141-142

90.

WHITE, C. A. Descriptions of new Invertebrate fossils from the Mesozoic and Cenozoic rocks of Arkansas, Wyoming, Colorado, and Utah. <Proc. U. S. Nat. Mus., vol. iii, pp. 157-162. Washington, 1881.

Same. Washington: Government Printing Office, 1880, 8vo, pp. 157-162. One hundred and fifty separates printed without title-page, covers, or repaging.

Nine species are described as new; but none are illustrated.

	Page.
Mollusca.	
Conchifera.	
Pteria, Scopoli	157
Subgenus *Oxytoma*, Meek	157
P. (Oxytoma) erecta, n. s., White, 1880	157-158
Solemya, Lamarck	158
S. bilix, n. s., White, 1880	158
Lucina, Bruguière	158
L. profunda, n. s., White, 1880	158-159
Gasteropoda.	
Planorbis, Guettard	159
P. æqualis, n. s., White, 1880	159
Subgenus *Gyraulus*, Agassiz	159
Planorbis (Gyraulus) militaris, n. s., White, 1880	159-160
Limnæa, Lamarck	160
Subgenus *Leptolimnea*, Swainson	160
Limnæa (Leptolimnea) minuscula, n. s., White, 1880	160
Helix, Linnæus	160
Subgenus *Patula*, Haldeman	160
H. (Patula) sepulta, n. s., White, 1880	160
Articulata.	
Vermes.	
Spirorbis, Lamarck	161
S.? dickhauti, n. s., White, 1880	161
Crustacea.	
Callianassa, Leach	161
C. ulrichi, n. s., White, 1880	161-162

91.

WHITE, C. A. [Review of] A. G. Wetherby: Description of new fossils from the Lower Silurian and Subcarboniferous rocks of Ohio and Kentucky. (Journ. Cincinnati Soc. Nat. History, vol. iv, no. 1 [April, 1881], pp. 77-85, pl. ii.) <Neues Jahrbuch für Mineralogie, Geologie, und Palæontologie, ii Band, p. 408. Stuttgart, 1881.

92.

WHITE, C. A. [Review of] A. G. Wetherby: Description of new fossils from the Lower Silurian and Subcarboniferous rocks of Kentucky. (Journ. Cincinnati Soc. Nat. History, vol. iv, no. 2 [July, 1881], pp. 177-179, pl. v.) < Neues Jahrbuch für Mineralogie, Geologie, und Palæontologie, ii Band, p. 408. Stuttgart, 1881.

93.

WHITE, C. A. [Review of] S. A. Miller: New species of fossils and remarks upon others from the Niagara Group of Illinois. (Journ. Cincinnati Soc. Nat. History, vol. iv, no. 2 [July, 1881], pp. 166-176, pl. iv.) <Neues Jahrbuch für Mineralogie, Geologie, und Paläontologie. ii Band, p. 404. Stuttgart, 1881.

94.

WHITE, C. A. Fossils of the Indiana rocks, No. 2. < Indiana, Department of Geology and Natural History, Eleventh Annual Report. John Collet, State Geologist, 1881; pp. 347-401, pls. xxxvii–lv. Indianapolis, 1882.

Same. Indianapolis, 1882; 8vo, pp. 347-401, with 19 plates and the index for the whole volume. Fifty separates printed without title-page or repaging, and without date.

The object of his memoir, like that of entry No. 75, is largely that of popular instruction. The greater part of the species here described and figured are republished from the works of other authors, only four of the species being new. Twelve of the plates are devoted to figures of fossil corals, which figures were engraved many years ago by J. W. Van Cleve, to accompany a work on fossil corals by himself, which he did not live to publish or to finish. There are 29 species of these corals here described and figured, besides 20 other species, all from Paleozoic rocks.

Description of fossils.

UPPER SILURIAN.

	Page.
Mollusca.	
Cephalopoda.	
Gyroceras, Meyer	356
G. elrodi, n. s., White, 1882, pl. xxxvii, fig. 1, and pl. xxxviii, figs. 2-4	356-358
Orthoceras, Breynius	358
O. annulatum, Sowerby, 1818, pl. xxxviii, fig. 1	358

SUBCARBONIFEROUS.

Gasteropoda.	
Patella, Linnæus	359
P. leverttei, n. s., White, 1882, pl. xxxix, figs. 4 and 5	359
Bellerophon, Montfort	359
B. sublævis, Hall, 1858, pl. xl, figs. 5-7	359-360
B. gibsoni, n. s., White, 1882, pl. xli, figs. 4-6	360-361
Brachiopoda.	
Terebratula, Llhwyd	361
T. formosa, Hall, 1858, pl. xxxix, figs. 6-8	361
Bryozoa.	
Archimedes, Lesueur	361
A. laxa, Hall, 1857, pl. xli, fig. 7	361-362
Radiata.	
Echinodermata.	
Lepidesthes, M. & W.	362
L. colletti, White, 1878, pl. xli, figs. 2 and 3	362-363
Agaricocrinus, Troost	363
A. springeri, n. s., White, 1882, pl. xl, figs. 2-4	363-364
Onychocrinus, Lyon & Casseday	365
O. exsculptus, Lyon & Casseday, 1860, pl. xl, fig. 1	365-366
O. ramulosus, Lyon & Casseday, 1859, pl. xxxix, figs. 2 and 3	366-367
Cyathocrinus, Miller	367
C. multibrachiatus, Lyon & Casseday, 1859, pl. xxxix, fig. 1	367
Platycrinus, Miller	368
P. hemisphericus, M. & W., 1865, pl. xli, fig. 1	368-369
Protista.	
Porifera.	
Palæacis, Haime	369
P. cuneatus, M. & W., 1860, pl. xli, figs. 8 and 9	369-370

Page.

COAL-MEASURES.

Mollusca.
Gasteropoda.
Polyphemopsis, Portlock 370
P. *nitidula*, M. & W., pl. xlii, figs. 7 and 8 370–371
Conchifera.
Nucula, Lamarck 371
N. *ventricosa*, Hall, 1858, pl. xlii, figs. 9 and 10 371–372
Brachiopoda.
Spirifer, Sowerby 372
Subgenus *Martinia*, McCoy 372
S. (*Martinia*) *lineatus*, Martin, pl. xlii, figs. 4–6 372–373
Productus, Sowerby 373
P. *punctatus*, Martin, pl. xlii, figs. 1–3 373

FOSSIL PLANTS.

Taonurus.
T. *colletti*, Lesquereux, 1870, pl. xliii, fig. 1 374
Sphenophyllum, Brongniart 374
S. *schlotheimi*, Brongniart, pl. xliii, fig. 2 374–375
S. *emarginatum*, Brongniart, pl. xliii, fig. 3 375
Van Cleve's Fossil Corals, identified by C. A. White, M. D 375

LOWER SILURIAN.

Streptelasma, Hall 376
S. *corniculum*, Hall, pl. li, figs. 2–4 376
Palæophyllum, Billings 377
P. *divaricans*, Nicholson, pl. lii, fig. 4 377–378
Favistella, Hall 378
F. *stellata*, Hall, pl. xliv, figs. 1 and 2 378
Protarea, Edwards & Haime 378
P. *vetusta*, Edwards & Haime, pl. xlix, fig. 4 378–379
Constellaria, Dana 379
C. *antheloidea*, Hall, pl. xlvi, figs. 1–3 379–380
Monticulipora, d'Orbigny 380
M. *frondosa*, d'Orbigny, pl. xlviii, figs. 2 and 3 380–381

UPPER SILURIAN.

Lyellia, Edwards & Haime 381
L. *americana*, Edwards & Haime, pl. xlvii, fig. 5 381–382
Halysites, Fischer 382
H. *catenulata*, Linnæus, pl. xlvi, figs. 4–7 382
Heliolites, Dana 383
H. *elegans*, Hall, pl. xlviii, fig. 4 383
Favosites, Lamarck 383
F. *favosus*, Goldfuss, pl. lii, figs. 1 and 2 383–384
Cladopora, Hall 384
C. *reticulata*, Hall, pl. xlvii, fig. 6 384–385
Clathropora, Hall 385
C. *frondosa*, Hall, pl. lv, fig. 3 385–386

DEVONIAN.

Acervularia, Schweigger 386
A. *davidsoni*, Edwards & Haime, pl. xlix, fig. 2 386
Diphyphyllum, Lonsdale 387
D. *archiaci*, Billings, pl. l, fig. 1 387
D. *stramineum*, Billings, pl. xlviii, fig. 1 388
D. *arundinaceum*, Billings, pl. li, fig. 1 389–390
Eridophyllum, Edwards & Haime 390
E. *strictum*, Edwards & Haime, pl. xlix, fig. 1 390–391
Cystiphyllum, Lonsdale 391
C. *vesiculosum*, Goldfuss, pl. lv, figs. 1 and 2 391–392
Zaphrentis, Rafinesque 393
Z. *rajinesquii*, Edwards & Haime, pl. xlv, figs. 3–5 393
Amplexus, Sowerby 393

	Page
A. yandelli, Edwards & Haime, pl. xlv, figs. 1 and 2	393-394
Favosites, Lamarck	394
F. basaltica, Goldfuss, pl. liv, fig. 1	394-395
F. polymorpha, Goldfuss, pl. l, fig. 2, pl. liii, figs. 1 and 2	395
F. polymorpha var. *dubia*, Nicholson, 1874, pl. liii, fig. 3	396
Favosites —— (?) White, 1882, pl. liv, fig. 2	396
Fistulipora, McCoy	396
F. canadensis, Billings, pl. xlvii, figs. 1 and 2	396-397
Alveolites.	
A. goldfussi, Billings, pl. liv, fig. 3	397-398
Striatopora, Hall	398
S. linnæana, Billings, pl. xlvii, fig. 7	398
Syringopora, Goldfuss	398
S. perelegans, Billings, pl. xlix, fig. 3	398-399
S. maclurei, Billings, pl. xlvii, fig. 3	399-400
Stromatopora, Blainville	400
S. pustulifera, Winchell ? pl. liii, figs. 4 and 5	400

SUBCARBONIFEROUS.

Lithostrotion, Fleming	401
L. mamillare, Castelnau, pl. lii, fig. 3	401

95.

WHITE, C. A. Progress of Invertebrate Paleontology in the United States for the year 1881. <American Naturalist, vol. xvi, pp. 887-891. Philadelphia, 1882.

Same. Philadelphia, 1882. 8vo, pp. 887-891. One hundred separates printed without title page, covers, or repaging.

This article is similar to those of entries 74 and 81.

96.

WHITE, C. A. Artesian Wells upon the Great Plains. <North American Review, vol. 135, pp. 187-195. New York, 1882.

Same. New York, 1882. 8vo, 9 pages. Fifty separates printed without title-page or covers, but repaged.

This article gives some results of the examination of the region of Eastern Colorado by a commission appointed by the U. S. Commissioner of Agriculture, of which the author was chief.

97.

WHITE, C. A. Tanganyika Shells. <Nature, 4to, vol. xxv, pp. 101-102. London and New York, 1882.

This note suggests the probable identity of *Paramelania* Smith, a proposed molluscan subgenus now living in Lake Tanganyika, Africa, with the fossil genus *Pyrgulifera* Meek, from the Laramie Group of Southwestern Wyoming.

98.

WHITE, C. A. On certain conditions attending the geological descent of some North American types of fresh-water gill-bearing Mollusks. <Amer. Jour. Sci., 3d ser., vol. xxiii, pp. 382-386. New Haven, 1882.

Same. New Haven, 1882. 8vo, pp. 382-386. Twenty separates printed without title-page, covers, or repaging.

The substance, not the words, of this article is an extract from parts of an illustrated work, the title of which is given in entry No. 105. The opinion is advanced that the present gill-bearing fauna of the Mississippi River system has been derived in part from the Laramie sea and the fresh-water lakes which succeeded it, by means of the persistence of their outlets as rivers, down to the present time.

99.

WHITE, C. A. Artesian Wells upon the Great Plains; being the report of a geological commission appointed to examine a portion of the Great Plains east of the Rocky Mountains, and report upon the localities deemed most favorable for making experimental borings. Department of Agriculture, pp. 1-38, pl. i. Washington, 1882.

The map is a copy of part of sheet xi of the atlas of Colorado, published by the United States Geological Survey of the Territories.

100.

WHITE, C. A. On certain Cretaceous fossils from Arkansas and Colorado. <Proc. U. S. National Museum, vol. iv, pp. 136-139, pl. i. Washington, 1882.

Same. Washington, Government Printing Office, 1882. 8vo, pp. 136-139, and one plate. One hundred and fifty separates printed without title-page, covers, or repaging.

Six species are described and figured, two of which are new.

	Page.
Callianassa ulrichi, White, 1880, pl. i, figs. 10 and 11	137
Tubulostium dickhauti, White, 1880, pl. i, figs. 12 and 13	138
Cantharus? julesburgensis, n. s., White, 1882, pl. i, figs. 1 and 2	138
Lucina profunda, White, 1880, pl. i, figs. 5 and 6	138
Lucina cleburni, n. s., White, 1882, pl. i, figs. 3 and 4	139
Solemya bilix, White, 1880, pl. i, fig. 9	139
Pteria (Oxytoma) erecta, White, 1880, pl. i, figs. 7 and 8	139

101.

WHITE, C. A. [Review of] S. A. Miller: Description of some new and remarkable crinoids and other fossils of the Hudson River Group, and notice of Strotocrinus bloomfieldensis. (Journ. Cincinnati Soc. Nat. History. Vol. iv, No. 1 [April, 1881], pp. 69-77, pl. i.) <Neues Jahrbuch für Mineralogie, Geologie, und Paläontologie. 1 Band, p. 307. Stuttgart, 1881.

102.

WHITE, C. A. [Review of] Henry Newton, E. M., and Walter P. Jenney, E. M.: Report on the Geology and resources of the Black Hills of Dakota. 4to, pp. 1-555, with Atlas. Washington, 1880. <Neues Jahrbuch für Mineralogie, Geologie, und Paläontologie. II Band, pp. 216-218. Stuttgart, 1882.

103.

WHITE, C. A. [Review of] United States Geological Survey: Annual Report for 1881 of J. W. Powell, Director. Imperial 8vo, pp. 1-558, lxi plates, and 32 woodcuts. Washington, 1882. <Neues Jahrbuch für Mineralogie, Geologie, und Paläontologie. II Band, pp. 365-366. Stuttgart, 1882.

104.

WHITE, C. A. Contributions to Invertebrate Paleontology, No. 2: Cretaceous fossils from the Western States and Territories. <Twelfth Annual Report of the United States Geological and Geographical Survey of the Territories (for the year 1878). pp. 1-39; appendix, pp. 38, 39, pls. xi-xviii. Washington, 1883.

Same. Washington: Government Printing Office, 1880. 8vo, pp. 1-39, and plates 11-18. Two hundred separates printed (author's edition) without repaging, but with title-page and covers, together with Contributions to Invertebrate Paleontology, Nos. 3, 4, 5, 6, 7, and 8. In 1883 100 additional copies of the same were printed in the same form. The title-page in both these cases differs from the title which heads the text by the omission of the word "Invertebrate."

Thirty-seven species are described and figured, five of which are new. On plate xviii ten species are also illustrated which were originally published by Dr. Shumard, but were never before illustrated.

CRETACEOUS.

	Page.
Actinozoa.	
Chætetes Fischer	7
C.? ? dimissus, White, 1879, pl. xii, fig. 14 *a*	7
Beaumontia, Edwards & Haime	8
B. ? solitaria, n. s., White, 1880, pl. xii, figs. 13 *a–c*	8
Echinodermata.	
Ophioderma	8
O.? bridgerensis, Meek, 1873, pl. xii, fig. 12 *a*	8–9
Conchifera.	
Ostrea, Linnæus	9
O. soleniscus, Meek, 1873, pl. xi, figs. 2 *a, b*	9–10
O. anomioides, Meek, 1873, pl. xi, figs. 4 *a, b*	10–11
Subgenus *Alectryonia*, Fischer	11
O. (Alectryonia) blackii, White, 1880, pl. xiv, figs. 1 *a, b ;* pl. xvii, fig. 4 *a*	11–12
Exogyra, Say	12
E. winchelli, White, 1880, pl. xiii, figs. 1 *a–d*	12
E. forniculata, White, 1880, pl. xiv, figs. 2 *a, b*	13–14
Anomia, Linnæus	14
A. propatoris, n. s., White, 1880, pl. xii, figs. 15 *a, b*	14–15
Pteria, Scopoli	15
P.? ? stabilitatis, n. s., White, 1880, pl. xvii, fig. 3 *a*	15
Subgenus *Oxytoma*, Meek	15
P. (Oxytoma) salinensis, White, 1880, pl. xvi, figs. 2 *a, b*	15–16
Gervillia, Defrance	16
G. mudgeana, White, 1880, pl. xiv, figs. 3 *a, b*	16–17
Pinna, Linnæus	17
P. lakesii, White, 1879, pl. xi, figs. 1 *a, b*	17–18
Volsella, Scopoli	18
Subgenus *Brachydontes*, Swainson	18
V. (Brachydontes) multilinigera, Meek, 1873, pl. xi, fig. 3 *a*	18–19
Barbatia, Gray	19
B. barbulata, n. s., White, 1880, pl. xi, fig. 5 *a*	19
Cyrena, Lamarck	20
C. carletoni, Meek, 1873, pl. xii, figs. 16 *a, b*	20–21
Pharella, Gray	21
P.? pealei, Meek, 1873, pl. xi, figs. 6 *a, b*	21–22
Tapes, Mühlfeldt	22
T. hilgardi, Shumard, 1860, pl. xvi, figs. 3 *a–c*	22
Pachymya, Sowerby	22
P.? compacta, White, 1880, pl. xvii, figs. 4 *a, b*	22–23
Thracia, Leach	23
T. myæformis, White, 1880, pl. xvii, figs. 2 *a, b*	23
Gasteropoda.	
Melampus, Montfort	23
M.? antiquus, Meek, 1873, pl. xii, figs. 11 *a–d*	23–25
Melampus ——— ? Meek, 1873, pl. xii, fig. 6 *a*	25
Neritina, Lamarck	25
N. bannisteri, Meek, 1873, pl. xii, figs. 10 *a–c*	25–26
N. pisiformis, Meek, 1873, pl. xii, figs. 9 *a–c*	26–27
Subgenus *Velatella*, Meek	27
N. (Velatella) bellatula, Meek, 1873, pl. xii, figs. 8 *a, b*	27–28
N. (Velatella) carditoides, Meek, 1873, pl. xii, fig. 7 *a*	28–29
Euspira, Agassiz	29
E. utahensis, White	29
Tessarolax, Gabb	29
T. kitzii, n. s., White, 1880, pl. xv, fig. 2 *a*	29–30
Lispodesthes, White	30
L.? obscurata, n. s., White, 1880, pl. xi, figs. 7 *a, b*	30–31
Turritella, Lamarck	31
T. spironema, Meek, 1873, pl. xii, fig. 3 *a*	31–32
Eulimella, Forbes	32
E.? chrysallis, Meek, 1873, pl. xii, fig. 4 *a*	32–33
E.? inconspicua, Meek, 1873, pl. xii, fig. 5 *a*	33
Valvata Müller	33
V. nana, Meek, 1873, pl. xii, figs. 17 *a, b*	33–34

	Page.
Fusus, Lamarck	34
F. ? utahensis, Meek, 1873, pl. xii, fig. 2 *a*	34
Fasciolaria, Lamarck	34
Subgenus *Piestocheilus*, Meek	34
F. (Piestocheilus) alleni, n. s., White, 1880, pl. xii, fig. 1 *a*	34–35
Cephalopoda.	
Prionocyclus, Meek	35
P. wyomingensis, Meek, 1870, pl. xv, figs. 1 *a–e*	35–36
Articulata.	
Crustacea.	
Paramithrax, Milne-Edwards	37
P. ? walkeri, n. s., 1880, Whitfield, pl. xvi, fig. 1 *a*, and pl. xvii, fig. 1 *a*	37–38

Appendix to Contributions to Invertebrate Paleontology No. 2.

Cidaris hemigranosus, Shumard, pl. xviii, figs. 2 *a, b*	38
Gervillia gregaria, Shumard, pl. xviii, fig. 3 *a*	38
Nucula bellastriata, Shumard, pl. xviii, figs. 5 *a–c*	38
Nucula haydeni, Shumard, pl. xviii, figs. 6 *a, b*	38
Cardium choctawense, Shumard, pl. xviii, figs. 7 *a–c*	38
Cytheria lamarensis, Shumard, pl. xviii, figs. 4 *a, b*	39
Ancyloceras annulatum, Shumard, pl. xviii, figs. 10 *a, b*	39
Scaphites vermiculus, Shumard, pl. xviii, fig. 8 *a*	39
Ammonites graysonensis, Shumard, pl. xviii, figs. 9 *a, b*	39
Ammonites swallovii, Shumard, pl. xviii, fig. 1 *a*	39

105.

WHITE, C. A. Contributions to Invertebrate Paleontology No. 3.—Certain Tertiary Mollusca from Colorado, Utah, and Wyoming. <Twelfth Annual Report of the U. S. Geol. and Geog. Surv. of the Terr. (for the year 1878), pp. 41–48, and pl. xix. Washington, 1883.

Same. Washington, Government Printing Office, 1880. 8vo, pp. 41–48, and pl. xix. Three hundred separates printed. See remarks following entry No. 104.

Ten species are described and figured, one of which is new.

	Page.
Conchifera.	
Unio, Retzius	41
U. shoshonensis, White, 1876, pl. xix, figs. 2 *a, b*	41–42
U. washakiensis, Meek, 1871, pl. xix, figs. 3 *a, b*	42–43
U. meeki, White, 1877, pl. xix, fig. 1 *a*	43–44
Gasteropoda.	
Planorbis, Müller	44
P. cirratus, White, 1879, pl. xix, figs. 5 *a–c*	44–45
Physa, Draparnaud	45
P. bridgerensis, Meek, 1873, pl. xix, figs. 10 *a, b*	45
Succinea, Draparnaud	45
Subgenus *Brachyspira*, Pfeiffer	45
S. (Brachyspira) papillispira, White, 1876, pl. xix, fig. 4 *a*	45–46
Pupa, Lamarck	46
P. arenula, White, 1876, pl. xix, figs. 8 *a, b*	46
P. atavuncula, n. s., White, 1880, pl. xix, fig. 9 *a*	46–47
Subgenus *Leucocheila*, Albers	47
P. (Leucocheila) inculata, White, 1876, pl. xix, figs. 7 *a–c*	47
Bythinella, Moquin-Tandon	48
B. gregaria, Meek, 1871, pl. xix, figs. 6 *a, b*	48

106.

WHITE, C. A. Contributions to Invertebrate Paleontology No. 4.—Fossils of the Laramie Group. <Twelfth Annual Report of the U. S. Geol. and Geog. Surv. of the Terr. (for the year 1878), pp. 49–103, and pls. xx–xxx. Washington, 1883.

Same. Washington, Government Printing Office, 1880. 8vo, pp. 49–103, and pls. xx–xxx. Three hundred separates printed. See remarks following entry No. 104.

The object of this article is to give a list of all the mollusca that were known from the Laramie Group up to the time of publication, and to illustrate those which had not before been illustrated.

	Page.
Conchifera.	
Ostrea, Linnæus	56
O. glabra, M. & H., 1857	56
Anomia, Linnæus	57
A. gryphorhynchus, Meek, 1872, pl. xxv, figs. 1 c–c	57
A. micronema, Meek, 1875, pl. xxv, figs. 2 a–d	57–58
Volsella, Scopoli	58
Subgenus *Brachydontes*, Swainson	58
V. (Brachydontes) regularis, White, 1878, pl. xxv, fig. 3 a	58–59
V. (Brachydontes) laticostata, White, 1878, pl. xxv, fig. 4 a	59
Axinæa, Poli	59
A. holmesiana, n. s., White, 1880, pl. xx, figs. 2 a, b	59–60
Nuculana, Link	60
N. inclara, White, 1878, pl. xxv, fig. 7 a	60, 61
Anodonta, Cuvier	61
A. propatoris, White, 1877, pl. xxiv, figs. 2 a–d	61–62
A. parallela, White, 1878, pl. xxiv, fig. 3 a	62
Unio, Retzius	62
U. aldrichi, White, 1878, pl. xxix, figs. 2 a, b	62–63
U. goniambonatus, White, 1878, pl. xxix, figs. 1 a, b	63–64
U. brachyopisthus, White, 1876, pl. xxii, figs. 2 a, b	64
U. conesi, White, 1877, pl. xxvii, fig. 1 a	64, 65
U. propheticus, White, 1876, pl. xxii, fig. 5 a	65
U. proavitus, White, 1877, pl. xxii, figs. 3 a–d	65–66
U. endlichi, White, 1877, pl. xxvi, figs. 1 a, b	66–67
U. holmesianus, White, 1877, pl. xxii, figs. 4 a–e	67–68
U. danæ, M. & H.? pl. xxvii, figs. 2 a, b	68
U. cryptorhynchus, White, 1877, pl. xxiv, figs. 1 a, b	68–69
U. senectus, White, 1877, pl. xxviii, figs. 1 a–c	69
U. primœvus, White, 1877, pl. xxix, figs. 3 a, b	70
U. priscus, M. & H	70
U. subspatulatus, M. & H	71
U. deweyanus, M. & H	71
U. retustus, Meek	71
U. belliplicatus, Meek	71
U. gonionotus, White, 1876, pl. xxvi, figs. 2 a–e	71–72
Sphærium, Scopoli	72
S. planum, M. & H	72
S. recticardinale, M. & H	72
S. formosum, M. & H	72
S. subellipticum, M. & H	72
Corbicula, Megerle	72
C. obesa, White, 1878, pl. xxiii, figs. 3 a–e	72–73
C. cardiniæformis, White, 1878, pl. xxv, figs. 5 a, b	73
C. cleburni, White, 1878, pl. xxiii, figs. 1 a–c	73–74
C. cytheriformis, M. & H., pl. xxi, figs. 4 a–d	74
C. nebrascensis, M. & H	74
C. occidentalis, M. & H., 1856, pl. xxi, figs. 3 a–c	75
Subgenus *Leptesthes*, Meek	75
C. (Leptesthes) fracta, Meek, 1871, pl. xxiii, figs. 2 a–e and pl. xxi, fig. 5 a	75–77
C. (Leptesthes) planumbona, Meek, 1875, pl. xxi, figs. 2 a–d	77–78
C. (Leptesthes) macropistha, White, 1878, pl. xxiii, figs. 4 a–f	78–79
C. (Leptesthes) subelliptica, M. & H	79
Subgenus *Veloritina*, Meek	79
C. (Veloritina) durkeei, Meek	79–80
Corbula, Bruguière	80
C. subtrigonalis, M. & H	80
C. perundata, M. & H	80
C. crassatelliformis, Meek	80
C. tropidophora, Meek	80

	Page
O. mactriformis, M. & H	80
C. undifera, Meek, 1873, pl. xxix, figs. 4 *a–f*	80–81
O. undifera var. *subundifera*, White, 1880, pl. xxix, figs. 5 *a–c*	81–82
C. pyriformis, Meek	82
Gasteropoda.	
Rhytophorus, Meek	82
R. priscus, Meek	82
R. meekii, n. s., White, 1880, pl. xxx, figs. 8 *a*, *b*	82–83
Acroloxus, Beck	83
A. minutus, M. & H	83
Planorbis, Müller	83
P. convolutus, M. & H	83
Subgenus *Bathyomphalus*, Agassiz	83
P. (Bathyomphalus) amplexus, M. & H	83
P. (Bathyomphalus) planoconvexus, M. & H	83
Limnæa, Lamarck	84
L. nitidula, Meek	84
Subgenus *Pleurolimnæa*, Meek	84
L. (Pleurolimnæa) tenuicostata, M. & H	84
Acella, Haldeman	84
A. haldemani, White, 1878, pl. xxx, figs. 9 *a*, *b*	84
Physa, Draparnaud	84
P. felix, White, 1878, pl. xxii, fig. 1 *a*	84–85
P. copei, White, 1877, pl. xxiv, figs. 4 *a*, *b*	85
Physa ——— ? White, 1880, pl. xxx, fig. 11 *a*	85
Bulinus, Adanson	86
B. atavus, White, 1877, pl. xxiv, figs. 5 *a*, *b*	86
B. disjunctus, White, 1879, pl. xxiv, figs. 6 *a*, *b*	86–87
B. longiusculus, M. & H	87
B. ? rhomboideus, M. & H	87
B. subelongatus M. & H	87
Vitrina, Draparnaud	87
V. ? obliqua, M. & H	87
Hyalina, Férussac	87
H. ? occidentalis, M. & H	87
H. ? evansi, M. & H	87–88
Helix, Linnæus	88
H. retusta, M. & H	88
Thaumastus, Albers	88
T. limnæformis, M. & H	88
Columna, Perry	88
C. teres, M. & H	88
C. vermicula, M. & H	88
Neritina, Lamarck	88
N. volvilineata, White, 1876, pl. xxi, figs. 6 *a*, *b*	88–89
N. naticiformis, White, 1878, pl. xxx, figs. 3 *a*, *b*	89
Subgenus *Velatella*, Meek	89
N. (Velatella) baptista, White, 1878, pl. xxix, figs. 6 *a*, *b*	89–90
Cerithidea, Swainson	90
Subgenus *Pirenella*, Gray	90
C. (Pirenella) nebrascensis, M. & H	90
Goniobasis, Lea	90
G. cleburni, White, 1876, pl. xxx, figs. 4 *a–d*	91
G. chrysallis, Meek, 1871, pl. xxx, figs. 6 *a*, *b*	91–92
G. chrysalloidea, White, 1876, pl. xxx, figs. 5 *a*, *b*	92
G. endlichi, White, 1878, pl. xxx, figs. 7 *a–c*	92–93
G. macilenta, White, 1879, pl. xxx, fig. 10 *a*	93
G. gracilenta, M. & H	94
G. convexa, M. & H	94
G. invenusta, M. & H	94
G. sublævis, M. & H	94
G. ? omitta, M. & H	94
G. ? subturtuosa, M. & H	94
G. nebrascensis, M. & H	94

	Page.
G. tenuicarinata, M. & H	94
Melania, Lamarck	94
M. ? insculpta, Meek, 1873, pl. xx, fig. 4 a	94–95
M. wyomingensis, Meek, 1873, pl. xxviii, figs. 6 a, b	95–96
Pyrgulifera, Meek	96
P. humerosa, Meek	96
Cassiopella, White	96–97
C. turricula, White, 1876, pl. xxvii, figs. 3 a–g	97
Hydrobia, Hartmann	97
H. anthonyi, M. & H	97
H. warrenana, M. & H	97
H. subconica, M. & H	97
H. ? culinoides, M. & H	97
Micropyrgus, Meek	98
M. minutulus, M. & H	98
Viviparus, Montfort	98
V. plicapressus, White, 1876, pl. xxviii, figs. 3 a, b	98
V. prudentius, White, 1878, pl. xxviii, figs. 5 a, b	98–99
V. couesi, White, 1878, pl. xxx, fig. 1 a	99
V. leai, M. & H	100
V. retusus, M. & H	100
V. peculiaris, M. & H	100
V. trochiformis, M. & H	100
V. reynoldsianus, M. & H	100
V. leidyi, M. & H	100
V. conradi, M. & H	100
Tulotoma, Haldeman	100
T. thompsoni, White, 1876, pl. xxviii, figs. 2 a–h	100–101
Campeloma, Rafinesque	101
C. vetula, M. & H	101
C. multistriata, M. & H	101
C. multilineata, M. & H	101
C. macrospira, Meek, pl. xxx, fig. 2 a	102
Valvata, Müller	102
V. subumbilicata, M. & H	102
V. parvula, M. & H	102
V. ? montanaensis, Meek	102
Odontobasis, Meek	102
O. buccinoides, White, 1876, pl. xx, figs. 3 a, b	102–103
O. ? formosa, White, 1878, pl. xxviii, fig. 7 a	103

107.

WHITE, C. A. Contributions to Invertebrate Paleontology, No. 5. Triassic Fossils of Southeastern Idaho. <Twelfth Annual Report U. S. Geol. and Geog. Surv of the Terr. (for the year 1878), pp. 105–118, pls. xxxi–xxxii. Washington, 1883. Same. Washington: Government Printing Office, 1880, 8vo, pp. 105–118, pls. xxxi–xxxii. Three hundred separates printed. See remarks following entry No. 93.

This article is essentially a republication of Paleontological paper No. 9. (See entry No. 71.)

	Page.
Brachiopoda.	
Terebratula, Llhwyd	108
T. semisimplex, White, 1879, pl. xxxi, figs. 3 a–c	108–109
T. augusta, H. & Whitf. ?	109
Conchifera.	
Aviculopecten, McCoy	109
A. ? pealei, White, 1879, pl. xxxii, fig. 4 a	109–110
A. ? altus, White, 1879, pl. xxxii, fig. 3 a	110
A. ? idahoensis, Meek, 1872, pl. xxxii, fig. 2 a	110–111
Cephalopoda.	
Meekoceras, Hyatt	112
M. aplana[a]tum, White, 1879, pl. xxxi, figs. 1 a–d	112–113
M. mushbachianum, White, 1879, pl. xxxii, figs. 1 a–d	114

WRITINGS OF C. A. WHITE. 163

	Page.
M. gracilitatis, White, 1879, pl. xxxi, figs. 2 a-d	115-116
M. gracilitatis var., White, 1880	116
Arcestes, Suess	116
A. ? cirratus, White, 1879	116-117
Arcestes ——? White? 1880	117-118

108.

WHITE, C. A. Contributions to Invertebrate Paleontology, No. 6.—Certain Carboniferous Fossils from the Western States and Territories. <Twelfth Annual Report U. S. Geol. and Geog. Surv. of the Terr. (for the year 1878), pp. 119-141, and pls. xxxiii-xxxvi. Washington, 1883.

Same. Washington: Government Printing Office, 1883. 8vo, pp. 119-141, and pls. xxxiii-xxxvi. Three hundred separates printed. See remarks following entry No. 104.

Twenty-seven species of fossils are described and illustrated, all of which had been previously published.

	Page.
Actinozoa.	
Amplexus, Sowerby	120
A. zaphrentiformis, White, 1876, pl. xxxiii, figs. 1 a-d	120
Acervularia, Schweigger	120
A. adjunctiva, White, 1880, pl. xxxv, figs. 1 a-d	120-121
Leptopora, Winchell	121
L. winchelli, White, 1879, pl. xxxiv, fig. 11 a	121-122
Echinodermata.	
Platycrinus, Miller	122
P. haydeni, Meek, 1873, pl. xxxiii, fig. 7 a	122-123
Lecythiocrinus, White	123
L. olliculæformis, White, 1880, pl. xxxv, figs. 2 a, b	124
Eupachycrinus, M. & W.	124
E. platybasis, White, 1876, pl. xxxiii, fig. 8 a	124-125
Cyathocrinus, Miller	125
C. stillatives, White, 1880, pl. xxxv, figs. 3 a, b	125
Erisocrinus, M. & W	126
E. typus, M. & W., 1866, pl. xxxiii, fig. 5 a	126-127
E. (Ceriocrinus) planus, White, 1880, pl. xxxv, figs. 5 a, b	127-128
E. (Ceriocrinus) inflexus, Geinitz, 1866, pl. xxxiv, figs. 9 a, b	128
Poteriocrinus, Miller	128
P. montanaensis, Meek, 1873, pl. xxxiii, fig. 6 a	128-129
Rhodocrinus, Miller	129
R. vesperalis, White, 1880, pl. xxxv, figs. 4 a, b	129-130
Archæocidaris, McCoy	130
A. cratis, White, 1876, pl. xxxiii, fig. 2 a	130
A. dininnii, White, 1880, pl. xxxv, figs. 6 a-c	131
Polyzoa.	
Ptilodyctia, Lonsdale	131
P. triangulata, White, 1878, pl. xxxiii, figs. 3 a-c	131-132
Brachiopoda.	
Productus, Sowerby	132
P. giganteus, Martin, pl. xxxvi, figs. 1 a-c	132
Rhynchonella, Fischer	133
R. endlichi, Meek, 1875, pl. xxxiii, figs. 4 a, b, and pl. xxxvi, figs. 2 a, b	133-134
Retzia, King	134
R. woosteri, White, 1879, pl. xxxiv, figs. 8 a, b	134
Spirifer, Sowerby	135
S. agelaius, Meek, 1873, pl. xxxiv, figs. 10 a, b	135
Conchifera.	
Nucula, Lamarck	136
N. perumbonata, White, 1879, pl. xxxiv, figs. 7 a, b	136
Nuculana, Link	136
N. obesa, White, 1879, pl. xxxiv, figs. 2 a-c	136-137
Allorisma, King	137
A. ? gilberti, White, 1879, pl. xxxiii, figs. 9 a, b	137-138

Gastropoda. | Page.
- *Bellerophon*, Montfort...... 138
- *B. subpapillosus*, White, 1879, pl. xxxiv, fig. 3 a...... 138
- *Naticopsis*, McCoy...... 139
- *N. remex*, White, 1876, pl. xxxiv, fig. 6 a...... 139
- *Murchisonia*, d'Archiac...... 139
- *M. terebra*, White, 1879, pl. xxxiv, fig. 4 a...... 139-140
- *Pleurotomaria*, Defrance...... 140
- *P. taggerti*, Meek, 1874, pl. xxxiv, figs. 1 a, b...... 140
- *P. grayvillensis*, Norwood & Pratten, 1855, pl. xxxiv, fig. 5 a...... 140-141

109.

WHITE, C. A. Contributions to Invertebrate Paleontology No. 7. Jurassic Fossils from the Western Territories. <Twelfth Annual Report of the U. S. Geol. and Geog. Surv. of the Terr. (for the year 1878), pp. 143-153, pls. xxxvii-xxxviii. Washington, 1883.

Same. Washington: Government Printing Office, 1880. 8vo, pp. 143-153, and plates xxxvii-xxxviii. Three hundred separates printed. See remarks following entry No. 93.

Sixteen species are described and figured, part of which are new, and the genus Lyosoma is diagnosed.

	Page.
Camptonectes (Agassiz), Meek......	143
C. platessiformis, White, 1876, pl. xxxvii, fig. 5 a......	143-144
Arieulopecten, McCoy......	144
A.? superstrictus, n. s., White, 1880, pl. xxxvii, figs. 4 a, b......	144
Gervillia, Defrance......	145
G. montanaensis, Meek, 1873, pl. xxxvii, figs. 1 a, b......	145
Volsella, Scopoli......	145
V. subimbricata, Meek, 1873, pl. xxxvii, figs. 2 a-c......	145-146
V. (Modiolina) platynota, n. s., White, 1880, pl. xxxvii, figs. 3 a, b......	146-147
Mytilus, Linnæus......	147
M. whitei, Whitfield, 1877, pl. xxxvii, fig. 9 a......	147
Trigonia, Bruguière......	147
T. montanaensis, Meek, 1873, pl. xxxviii, fig. 2 a......	147-148
T. americana, Meek, 1873, pl. xxxviii, figs. 1 a, b......	148
Astarte, Sowerby......	149
A. packardi, n. s., White, 1880, pl. xxxvii, figs. 6 a, b......	149
Cardinia, Agassiz......	149
C. præcisa, n. s., White, 1880, pl. xxxvii, figs. 7 a, b......	149-150
Tancredia, Lycett......	150
T. extensa, White, 1880, pl. xxxviii, fig. 4 a......	150
Pholadomya, Sowerby......	150
P. kingii, Meek, 1873, pl. xxxvii, figs. 3 a, b......	150-151
Goniomya, Agassiz......	151
G. montanaensis, Meek, 1873, pl. xxxvii, fig. 8 a......	151
Myacites (Schlotheim), Munster......	151
M. subcompressus, Meek, 1873, pl. xxxviii, figs. 5 a-c......	151-152
*Lyosoma, n. g., White, 1880......	152-153
L. powelli, White, 1876, pl. xxxviii, figs. 6 a-d......	153

110.

WHITE, C. A. Contributions to Invertebrate Paleontology No. 8. Fossils from the Carboniferous Rocks of the Interior States. <Twelfth Annual Report of the U. S. Geol. and Geog. Surv. of the Terr. (for the year 1878). pp. 155-171, pls. xxxix-xlii. Washington, 1883.

Same. Washington: Government Printing Office, 1880. 8vo, pp. 155-171, and plates xxxix-xlii. Three hundred separates printed. See remarks following entry No. 93. The series of contributions closes with No. 8, and it is not again resumed in that form.

Twenty-eight species are described and figured, part of which are new.

---

* λυω, to loosen, and σωμα, the body.

	Page.
Radiata.	
Actinaria.	
Zaphrentis, Rafinesque	155
Z. elliptica, White, 1862, pl. xxxix, figs. 4 a, b	155–156
Z. calceola, White & Whitf., 1862, pl. xxxix, figs. 6 a–d	156
Hadrophyllum, Edwards & Haine	156
H. glans, White, 1862, pl. xxxix, figs. 5 a. b	156–157
Lophophyllum, Edwards & Haine	157
L. expansum, White, 1876, pl. xxxix, figs. 4 a, b	157
Chonophyllum, Edwards & Haine	157
C. sedaliense, n. s., White, 1880, pl. xxxix, fig. 1 a	157
Michilinia, De Koninck	157
M. ? placenta, n. s., White, 1880. pl. xxxix, figs. 1 a–d	157–158
M. expansa, n. s., White, 1880, pl. xxxix, figs. 2 a, b	158
Lithostrotion, Fleming	159
L. microstylum, n. s., White, 1880, pl. xl, fig. 7 a	159
L. mamillare, Castelnau, pl. xl, figs. 6 a, b	159–160
Echinodermata.	
Platycrinus, Miller	160
P. bonoensis, White, 1878, pl. xl, fig. 5 a	160–161
Scaphiocrinus, Hall	161
S. gibsoni, White, 1878, pl. xl, fig. 4 a	161–162
S. gurleyi, White, 1878, pl. xl, fig. 3 a	162
Actinocrinus, Miller	162
A. wachsmuthi, n. s., White, 1880, pl. xl, figs. 1 a, b	162–163
Lepidesthes, M. & W	163
L. colleti, White, 1878, pl. xl, figs. 2 a, b	163–164
Mollusca.	
Molluscoidea.	
Brachiopoda.	
Orthis, Dalman	164
O. thiemei, White, 1860, pl. xli, figs. 4 a–d	164–165
Rhynchonella, Fischer	165
R. ottumwa, White, 1862, pl. xli, figs. 5 a–c	165
Spirifer, Sowerby	165
S. subcardiiformis, Hall, 1858, pl. xli, figs. 2 a–c	165–166
Mollusca, vera.	
Conchifera.	
Anthracoptera, Salter	166
A. polita, n. s., White, 1880, pl. xlii, figs. 5 a, b	166
Astartella, Hall	166
A. gurleyi, White, 1878, pl. xlii, figs. 6 a, b	166–167
Allorisma, King	167
A. marionensis, White, 1876, pl. xli, figs. 1 a, b	167–168
Gasteropoda.	
Euomphalus, Sowerby	167
E. springvalensis, White, 1876, pl. xli, figs. 1 a, b	167–168
Platyceras, Conrad	168
P. tribulosum, n. s., White, 1880, pl. xli, figs. 6 a, b	168
Naticopsis, McCoy	168
N. monilifera, n. s., White, 1880, pl. xlii, figs. 3 a–c	168
Pleurotomaria, Defrance	169
P. broadheadi, n. s , White, 1880, pl. xlii, figs. 1 a, b	169
P. newportensis, n. s., White, 1880, pl. xlii, figs. 2 a, b	169
Pteropoda.	
Conularia, Miller	170
C. crustula, n. s., White, 1880, pl. xlii, fig. 4 a	170
Cephalopoda.	
Nautilus, Breynius	170
N. danvillensis, White, 1878, pl. xlii, fig. 7 a	170–171
Articulata.	
Vermes.	
Serpula, Linnæus	171
S. insita, White, 1878, pl. xlii, fig. 8 a	171

111.

WHITE, C. A. Forestry in the Great Prairie Region. <American Journal of Forestry, vol. i (May No.), pp. 366–370. Cincinnati, 1883.
Same. Cincinnati, 1883. 8vo, pp. 1–6. Forty separates printed without title-page, or covers, but repaged.

> This Journal was started at the beginning of 1883, with Dr. F. B. Hough as editor and Robert Clarke & Co. as publishers. It was published one year and then discontinued.

112.

WHITE, C. A. The reversion of Sunflowers at night. <Nature. 4to, vol. [xxvii?] p. 241. London and New York, 1883.

> Sunflowers are observed to turn to the eastward immediately after dark.

113.

WHITE, C. A. New Molluscan forms from the Laramie and Green River groups; with discussion of some associated forms heretofore known. <Proc. U. S. National Museum, vol. v, pp. 94–99, pls. i–ii. Washington, 1883.
Same. Washington: Government Printing Office, 1882. 8vo, pp. 94–99, and two plates. One hundred and fifty separates printed, together with the paper of the next entry, without title-page, covers, or repaging.

> Six species are described and figured as new; and *Paramelania*, Smith, is figured and compared with *Pyrgulifera*, Meek. Also the discovery of the under valve of *Anomia micronema* is announced and figured; and the prismatic structure of that valve suggested as the cause that it has so generally been destroyed.

	Page.
Unio, Retzius	94
U. clinopisthus, n. s., White, 1882, pl. iii, figs. 1 and 2	94
Corbicula, Mühlfeldt	94
C. berthoudi, n. s., White, 1882, pl. iv, figs. 1–3	94–95
C. augheyi, n. s., White, 1882, pl. iv, figs. 4–6	95
Neritina, Lamarck	95
N. bruneri, n. s., White, 1882, pl. iv, figs. 7 and 8	95–96
Melanopsis, Lamarck	96
M. americana, n. s., White, 1882, pl. iv, figs. 9 and 10	96
Campeloma, Rafinesque	97
C. producta, n. s., White, 1882, pl. iii, figs. 7–9	97

114.

WHITE, C. A. The Molluscan Fauna of the Truckee Group, including a new form. <Proc. U. S. National Museum, vol. v, pp. 99–101, pl. i. Washington, 1883.
Same. Washington: Government Printing Office, 1882. 8vo, pp. 99–101, and one plate. One hundred and fifty separates printed, together with the paper of the last entry, without title-page, covers, or repaging.

> Nine species are figured, which constitute the entire molluscan fauna of the Truckee Group as it was then known to the author. A species of *Latia* is described, which is the only one of that genus yet known in North America.

	Page.
Melania sculptilis, Meek, pl. v, fig. 1	100
Melania subsculptilis, Meek, pl. v, fig. 2	100
Melania taylori, Gabb, pl. v, fig. 3	100
Lithasia antiqua, Gabb, pl. v, fig. 4	100
Carinifex (Vorticifex) tryoni, Meek, pl. v, figs. 5–7	100
Carinifex (Vorticifex) binneyi, Meek, pl. v, figs. 8 and 9	100
Ancylus undulatus, Meek, pl. v, figs. 10 and 11	100
Sphærium rugosum, Meek, pl. v, figs. 14–16	100
Sphærium? idahoense, Meek, pl. v, figs. 12 and 13	100
Latia dallii, n. s., White, 1882, pl. v, figs. 17–20	100–101

115.

WHITE, C. A. A Review of the Non-Marine Fossil Mollusca of North America. <Third Annual Report of the Director of the United States Geological Survey to the Secretary of the Interior, 1881–'82. By J. W. Powell, Director, pp. 403–550 and pls. p. i–xxxii. Washington, 1883.

Same. Washington: Government Printing Office, 1883. Imp. 8vo, pp. 1–80 and pls. 1–32. One hundred copies printed early in 1883 with paging, numbering of the plates, and title-page separate from that of the volume.

Same. One hundred more copies printed at the time of the issuance of the volume in 1884.

> This memoir contains a discussion of the families which are represented in the brackish and fresh water deposits, an annotated catalogue of all the known species in the order in which the formations occur, general discussion of the conditions which prevailed when the mollusks lived, and the manner in which the lines of descent of a part of them have been continued to the present time. Two hundred and twenty-seven species are noticed, all of which are illustrated. Although the volume of which this work is an extract, bears the date 1883 upon its title-page, the volume, as a whole, was not issued until near the middle of 1884.

[The pagination in the separates is different from that in the volume. Both are referred to.]

DEVONIAN.

	Page.
Strophites grandaeva, Dawson, pl. i, fig. 1	49, 455
Anodonta? angustata (Vanuxem), Hall, pl. i, figs. 2 and 3	18, 424
Anodonta? cattskillensis (Vanuxem), Hall, pl. i, fig. 4	18, 424

CARBONIFEROUS.

Pupa vetusta, Dawson, pl. ii, figs. 1 and 2	50, 456
Dawsonella meeki, Bradley, pl. ii, figs. 3 and 4	47, 453
Anthracopupa ohioensis, Whitfield, pl. ii, figs. 5–8	50, 456
Pupa bigsbyi, Dawson, pl. ii, figs. 9 and 10	50, 456
Pupa vermillionensis, Bradley, pl. ii, figs. 13 and 14	50, 456
Zonites priscus, Dawson, pl. ii, figs. 11 and 12	47, 453
Naiadites carbonaria, Dawson, pl. ii, fig. 15	19, 425
Naiadites elongata, Dawson, pl. ii, fig. 16	19, 425
Naiadites laevis, Dawson, pl. ii, fig. 17	19, 425

JURASSIC AND TRIASSIC. (?)

Unio stewardii, White, pl. iii, fig. 1	20, 426
Unio nucalis, M. & H., pl. iii, figs. 2–4	20, 426
Unio cristonensis, Meek, pl. iii, fig. 5	19, 425
Planorbis veternus, M. & H., pl. iii, fig. 6	40, 446
Valvata scabrida, M. & H., pl. iii, fig. 7	64, 470
Viviparus gillianus, M. & H., pl. iii, fig. 8	60, 470
Lioplacodes veternus, M. & H., pl. iii, fig. 9	60, 470
Neritina nebrascensis, M. & H., pl. iii, figs. 10, 11	51, 457

CRETACEOUS.

Margaritana nebrascensis, M. & H., pl. iv, figs. 1 and 2	21, 427
Cyrena dakotensis, M. & H., pl. iv, figs. 3 and 4	30, 436
Physa ———? White, 1879, pl. iv, fig. 5	38, 444
Unio penultimus, Gabb, pl. v, fig. 1	21, 427
Unio hubbardii, Gabb, pl. v, figs. 2 and 3	21, 477
Cyrena carletoni, Meek, pl. v, figs. 4 and 5	30, 436
Anomia propatoris, White, pl. v, figs. 6 and 7	16, 422
Neritina (Velatella) bellatula, Meek, pl. v, figs. 8 and 9	52, 458
Neritina (Velatella) carditoides, Meek, pl. v, fig. 10	52, 458
Neritina banistri, Meek, pl. v, figs. 11 and 12	52, 458
Melampus? antiquus, Meek, pl. v, figs. 13–16	38, 444
Melampus? ———, Meek, 1873, pl. v, fig. 17	38, 444
Physa carletoni, Meek, pl. v, fig. 18	43, 449
Valvata nana, Meek, pl. v, figs. 19 and 20	64, 470

BEAR RIVER, LARAMIE.

	Page.
Unio belliplicatus, Meek, pl. vi, figs. 1-3	24, 430
Pyrgulifera humerosa, Meek, pl. vi, figs. 4-6	54, 460
Goniobasis cleburni, White, pl. vi, figs. 7-9	56, 462
Goniobasis chrysalloidea, White, pl. vi, figs. 10 and 11	56, 462
Goniobasis macilenta, White, pl. vi, fig. 12	56, 462
Goniobasis chrysalis, Meek, pl. vi, figs. 13 and 14	56, 462
Limnæa (Limnophysa) nitidula, Meek, pl. vi, figs. 15 and 16	39, 445
Physa ——? White, 1880, pl. vi, fig. 17	43, 449
Limnæa (Acella) haldemani, White, pl. vi, figs. 19 and 20	39, 445
Unio retustus, Meek, pl. vii, figs. 1-4	24, 430
Neritina naticiformis, White, pl. vii, figs. 5 and 6	52, 458
Goniobasis endlichi, White, pl. vii, figs. 7-9	57, 463
Viviparus couesii, White, pl. viii, fig. 1	61, 467
Rhytophorus priscus, Meek, pl. viii, figs. 2 and 3	38, 444
Rhytophorus meekii, White, pl. viii, figs. 4 and 5	38, 444
Campeloma macrospira, Meek, pl. viii, figs. 6 and 7	63, 469
Corbicula (Veloritina) durkeei, Meek, pl. viii, figs. 8-11	31, 437
Corbicula pyriformis, Meek, pl. viii, figs. 12-16	35, 441

LARAMIE.

Ostrea glabra, M. & H., pls. ix-xi, and pl. xii, fig. 1	15, 421
Ostrea subtrigonalis, Evans & Shumard, pl. xii, figs. 2-5	15, 421
Anomia micronema, Meek, pl. xii, figs. 6-11	16, 422
Anomia gryphorhynchus, Meek, pl. xii, figs. 12-15	16, 422
Volsella (Brachydontes) regularis, White, pl. xiii, fig. 1	17, 423
Volsella (Brachydontes) laticostata, White, pl. xiii, fig. 2	17, 423
Unio proavitus, White, pl. xiii, figs. 3-6	27, 433
Unio gonionotus, White, pl. xiii, figs. 7-10	27, 433
Unio priscus, M. & H., pl. xiv, fig. 1	26, 432
Unio subspatulatus, Meek, pl. xiv, figs. 2 and 3	25, 431
Unio primævus, White, pl. xiv, figs. 4 and 5	26, 432
Unio cryptorhynchus, White, pl. xiv, figs. 6 and 7	25, 431
Unio endlichi, White, pl. xv, figs. 1 and 2	26, 432
Unio propheticus, White, pl. xv, fig. 3	27, 433
Unio aldrichi, White, pl. xv, figs. 4, 5	27, 433
Unio couesii, White, pl. xvi, fig. 1	26, 432
Unio holmesianus, White, pl. xvi, figs. 2-6	27, 433
Unio brachyopisthus, White, pl. xvi, figs. 7 and 8	27, 433
Unio danæ, M. & H., pl. xvii, figs. 1-3	25, 431
Unio deweyanus, M. & H., pl. xvii, figs. 4 and 5	25, 431
Corbicula occidentalis, M. & H., pl. xvii, figs. 6 and 7, pl. xxiii, figs. 1-6	31, 437
Sphærium planum, M. & H., pl. xvii, fig. 8	33, 439
Sphærium recticardinale, M. & H., pl. xvii, fig. 9	33, 439
S. subellipticum, M. & H., pl. xvii, fig. 10	33, 439
S. formosum, M. & H., pl. xvii, fig. 11	33, 439
Unio mendax, White, pl. xviii, figs. 3-5	27, 433
Unio danæ, M. & H..? pl. xviii, figs. 1 and 2	27, 433
Corbula undifera, Meek, pl. xviii, figs. 6-9	36, 440
Corbula undifera var. *subundifera*, White, pl. xviii, figs. 10 and 11	36, 440
Corbula mactriformis, M. & H., pl. xviii, figs. 12-15	36, 440
Unio senectus, White, pl. xix, figs. 1 and 2	26, 432
Unio goniambonatus, White, pl. xix, figs. 3 and 4	27, 433
Anodonta parallela, White, pl. xix, fig. 5	23, 429
Anodonta propatoris, White, pl. xix, figs. 6-9	23, 429
Corbula subtrigonalis, M. & H., pl. xix, figs. 10-17	36, 442
Corbicula (Leptesthes) fracta, Meek, pl. xx, figs. 1-6	33, 439
Corbicula cleburni, White, pl. xx, figs. 7-9	31, 437
Corbicula subelliptica, M. & H., pl. xx, figs. 10 and 11	31, 437
Corbicula nebrascensis, M. & H., pl. xx, figs. 12 and 13	31, 437
Pisidium saginatum, White, pl. xx, figs. 14 and 15	34, 440
Corbicula berthoudi, White, pl. xxi, figs. 1-3	32, 438
Corbicula maykepi, n. s., White, 1883, pl. xxi, figs. 4-6	32, 438
Corbicula umbonella, Meek, pl. xxi, figs. 7-10	32, 438
Corbicula (Leptesthes) macropistha, White, pl. xxi, figs. 11-14	31, 437

	Page.
Corbicula cytheriformis, M. & H., pl. xxii, figs. 1–6	31,437
Corbicula (Leptesthes) planumbona, Meek, pl. xxii, figs. 7–9	31,437
Corbicula (Leptesthes) cardiniæformis, White, pl. xxii, figs. 10–15	31,437
Corbicula obesa, White, pl. xxiii, figs. 7–11	31,437
Neritina volvilineata, White, pl. xxiii, figs. 12 and 13	52,458
Neritina bruneri, White, pl. xxiii, figs. 14 and 15	53,459
Neritina (Velatella) baptista, White, pl. xxiii, figs. 16–20	52,458
Melanopsis americana, n. g., White, 1883, pl. xxiii, figs. 21–23	55,461
Limnæa (Pleurolimnæa) tenuicostata, M. & H., pl. xxiii, fig. 24	39,445
Cassiopella turricula, White, pl. xxiii, figs. 25–29	58,464
Viviparus retusus, M. & H., pl. xxiv, figs. 1–3	61,467
Viviparus conradi, M. & H., pl. xxiv, figs. 4–6	61,467
Viviparus leidyi, M. & H., pl. xxiv, fig. 7	61,467
Viviparus leidyi var. *formosus*, M. & H., pl. xxiv, figs. 8 and 9	61,467
Viviparus trochiformis, M. & H., pl. xxiv, figs. 10–16	61–467
Tulotoma thompsoni, White, pl. xxiv, figs. 17–22	61,467
Viviparus peculiaris, M. & H., pl. xxiv, figs. 23 and 24	61,467
Viviparus plicapressus, White, pl. xxiv, figs. 25 and 26	61,467
Acroloxus minutus, M. & H., pl. xxiv, fig. 27	45,451
Physa copei, White, pl. xxv, figs. 1 and 2	44,450
Physa felix, White, pl. xxv, fig. 3	44,450
Bulinus disjunctus, White, pl. xxv, figs. 4 and 5	45,451
Bulinus atavus, White, pl. xxv, figs. 6 and 7	44,450
Bulinus longiusculus, M. & H., pl. xxv, fig. 8	45,451
Bulinus rhomboideus, M. & H., pl. xxv, fig. 9	45,451
Bulinus subelongatus, M. & H., pl. xxv, figs. 10 and 11	44,450
Helix kanabensis, White, pl. xxv, figs. 12–14	48,454
Columna teres, M. & H., pl. xxv, fig. 15	48,454
Columna vermicula, M. & H., pl. xxv, fig. 16	48,454
Viviparus prudentius, White, pl. xxv, figs. 17 and 18	61,467
Viviparus panguitchensis, White, pl. xxv, figs. 19–21	61,467
Viviparus reynoldsianus, M. & H., pl. xxv, figs. 22 and 23	61,467
Thaumastus limnæiformis, M. & H., pl. xxv, fig. 24	48,454
Melania wyomingensis, Meek, pl. xxvi, figs. 1–3	54,460
Melania insculpta, Meek, pl. xxvi, figs. 4 and 5	54,460
Goniobasis convexa, M. & H., pl. xxvi, figs. 6 and 7	57,463
Goniobasis convexa var. *impressa*, M. & H., pl. xxvi, figs. 8 and 9	57,463
Goniobasis? omitta, M. & H., pl. xxvi, fig. 10	57,463
Goniobasis tenuicarinata, M. & H., pl. xxvi, fig. 11	57,463
Goniobasis gracilenta, Meek, pl. xxvi, figs. 12 and 13	57,463
Goniobasis nebrascensis, M. & H., pl. xxvi, figs. 15 and 16	57,463
Goniobasis invenusta, M. & H., pl. xxvi, fig. 17	57,463
Goniobasis sublævis, M. & H., pl. xxvi, fig. 18	57,463
Limnæa? compactilis, Meek, pl. xxvi, fig. 14	39,445
Cerithidea? nebrascensis, M. & H., pl. xxvi, fig. 19	57,463
Micropyrgus minutulus, M. & H., pl. xxvi, fig. 20	59,465
Campeloma producta, White, pl. xxvi, figs. 21–27	63,469
Campeloma multilineata, M. & H., pl. xxvii, figs. 1–7	63,469
Campeloma vetula, M. & H., pl. xxvii, figs. 8 and 9	63,469
Campeloma multistriata, M. & H., pl. xxvii, fig. 15	63,469
Viviparus leai, M. & H., pl. xxvii, figs. 10–14	61,467
Planorbis convolutus, M. & H., pl. xxvii, fig. 16	41,447
Planorbis (Bathyomphalus) planoconvexus, M. & H., pl. xxvii, figs. 17 and 18	41,447
Planorbis (Bathyomphalus) amplexus, M. & H., pl. xxvii, figs. 19 and 20	41,447
Planorbis (Bathyomphalus) kanabensis, M. & H., pl. xxvii, figs. 21–23	41,447
Valvata? montanaensis, Meek, pl. xxvii, fig. 24	64,470
Valvata subumbilicata, M. & H., pl. xxvii, fig. 25	64,470
Hyalina? evansi, M. & H., pl. xxvii, fig. 26	46,452
Hyalina? occidentalis, M. & H., pl. xxvii, fig. 27	46,452
Helix? vetusta, M. & H., pl. xxvii, fig. 28	48,454
Helix evanstonensis, White, pl. xxvii, figs. 29–31	48,454
Helix sepulta, White	48,454
Vitrina obliqua, M. & H., pl. xxvii, figs. 32 and 33	46,452
Goniobasis? subtortuosa, M. & H., pl. xxvii, fig. 34	57,463
Hydrobia utahensis, White, pl. xxvii, fig. 35	60,466

	Page.
Hydrobia subconica, Meek, pl. xxvii, fig. 36	59, 465
Hydrobia? cutimoides, Meek, pl. xxvii, fig. 37	59, 465
Hydrobia recta, White, pl. xxvii, fig. 38	60, 466
Hydrobia anthonyi, M. & H., pl. xxvii, fig. 39	59, 465
Hydrobia warrenana, M. & H., pl. xxvii, fig. 40	59, 465

EOCENE.

Unio clinopisthus, White, pl. xxviii, figs. 1 and 2	28, 434
Unio shoshonensis, White, pl. xxviii, fig. 3	29, 435
Unio haydeni, Meek, pl. xxviii, figs. 4 and 5	29, 435
Unio washakiensis, Meek, pl. xxviii, fig. 6-8	29, 435
Unio tellinoides, Hall, pl. xxviii. fig. 9	29, 435
Planorbis (Gyraulus) militaris, White, pl. xxviii, figs. 10 and 11	41, 447
Bythinella gregaria, Meek, pl. xxviii, figs. 12 and 13	60, 466
Bulimus floridanus, Conrad, pl. xxviii, fig 14	48, 454
Melania claibornensis, Heilprin, pl. xxviii, fig. 15	54, 460
Planorbis utahensis, Meek, pl. xxix, figs. 1-3	41, 447
Planorbis utahensis var. *spectabilis*, Meek, pl. xxix, figs. 4-6	41, 447
Planorbis circulus, White, pl. xxix, fig. 7	42, 448
Planorbis æqualis, White, pl. xxix, figs. 8-10	42, 448
Helix peripheria, White, pl. xvix, figs. 11 and 12	49, 455
Helix riparia, White, pl. xxix, figs. 13 and 14	49, 455
Pupa incolata, White, pl. xxix, figs. 15-17	50, 456
Pupa atavuncula, White, pl. xxix, fig. 18	50, 456
Pupa arenula, White, pl. xxix, fig. 19	50, 456
Limnæa similis, Meek, pl. xxix, figs. 20 and 21	39, 445
Limnæa vetusta, Meek, pl. xxix, figs. 22 and 23	39, 445
Limnæa miniuscula, White, pl. xxix, figs. 24 and 25	40, 446
Succinea (Brachyspira) papillispira, White, pl. xxix, fig. 26	51, 457
Anodonta decurtata, Conrad, pl. xxix, figs. 27 and 28	73, 479
Macrocyclis spatiosa, M. & H., pl. xxx, figs. 1-3	46, 452
Helix? veterna, M. & H., pl. xxx, figs. 4 and 5	48, 454
Physa pleromatis, White, pl. xxx, figs. 6-8	44, 450
Physa bridgerensis, Meek, pl. xxx, figs. 9 and 10	44, 450
Viviparus paludinæformis, Hall, pl. xxx, figs. 11 and 12	62, 468
Viviparus wyomingensis, Meek, pl. xxx, figs. 13 and 14	62, 468
Goniobasis tenera, Hall, and varieties, pl. xxxi	58, 464

MIOCENE AND PLIOCENE?

Melania sculptilis, Meek, pl. xxxii, fig. 1	55, 461
Melania subsculptilis, Meek, pl. xxxii, fig. 2	55, 461
Melania taylori, Gabb, pl. xxxii, fig. 3	55, 461
Lithasia antiqua, Gabb, pl. xxxii, fig. 4	59, 465
Carinifex (Vorticifex) binneyi, Meek, pl. xxxii, figs. 5 and 6	42, 448
Carinifex (Vorticifex) tryoni, Meek, pl. xxxii, figs. 7-9	42, 448
Ancylus undulatus, Meek, pl. xxxii, fig. 10	45, 451
Sphærium rugosum, Meek, pl. xxxii, figs. 11-13	34, 440
Sphærium idahoense, Meek, pl. xxxii, figs. 14 and 15	34, 440
Planorbis retustus, M. & H., pl. xxxii, figs. 16-18	42, 448
Planorbis leidyi, M. & H., pl. xxxii, figs. 19-21	42, 448
Planorbis nebrascensis, Evans & Shumard, pl. xxxii, figs. 22 and 23	42, 448
Planorbis lunata, Conrad, pl. xxxii, figs. 24 and 25	42, 448
Limnæa meekii, Evans & Shumard, pl. xxxii, figs. 26 and 27	40, 446
Limnæa shumardi, M. & H., pl. xxii, figs. 28 and 29	40, 446
Limnæa (Polyrhytis) kingii, Meek, pl. xxxii, figs. 30 and 31	40, 446
Helix leidyi, Hall & Meek, pl. xxxii, figs. 32 and 33	49, 455
Helix (Zonites) marginicola, Conrad, pl. xxxii, fig. 34	47, 453
Physa secalina, Evans & Shumard, pl. xxxii, figs. 35 and 36	44, 450
Latia dallii, n. s. White, 1883, pl. xxxii, figs. 37-40	45, 451
Unio meekii, White	28, 434
Unio leai, Meek	28, 434
Viviparus ionicus, White	62, 468

116.

WHITE, C. A. Progress of Invertebrate Paleontology in the United States for the year 1882. < Amer. Nat., vol. xvii, pp. 598–603. Philadelphia, 1883.

Same. Philadelphia, 1883. 8vo, pp. 598–603. One hundred separates printed without title-page, covers, or repaging.

> This series of articles, which was begun for the year 1879, was discontinued with this article; but the series is continued by Mr. J. B. Marcou.

117.

WHITE, C. A. Glacial Drift in the Upper Missouri River Region. < Amer. Jour. Sci., 3d ser., vol. xxv, p. 206. New Haven, 1883.

Same. New Haven, 1883. 8vo, p. 206. Twenty separates printed without title-page, covers, or repaging.

> The existence of true northern glacial drift in the region of the mouth of the Yellowstone River is announced.

118.

WHITE, C. A. Late observations concerning the Molluscan Fauna and the Geographical extent of the Laramie Group. < Amer. Jour. Sci., 3d ser., vol. xxv, pp. 207–209. New Haven, 1883.

Same. New Haven, 1883. 8vo, pp. 207–209. Twenty separates printed without title-page, covers, or repaging.

> The discovery of Laramie fossils in the State of Nuevo Leon, Mexico, and in the Saskatchewan Valley, British America, is announced.

119.

WHITE, C. A. On the existence of a deposit in Northeastern Montana and Northwestern Dakota, that is possibly equivalent with the Green River Group. < Amer. Jour. Sci., 3d ser., vol. xxv, pp. 411–416. New Haven, 1883.

Same. New Haven, 1883. 8vo, pp. 411–416. Twenty separates printed without title-page, covers, or repaging.

> Some teleost fish remains were obtained from certain layers which rest comformably upon the top of the Laramie Group. On pages 414-416 of this article Prof. E. D. Cope describes a new genus of fishes and two species.

120.

WHITE, C. A. The burning of Lignite in situ. < Amer. Jour. Sci., 3d ser., vol. xxvi, pp. 24–26. New Haven, 1883.

Same. New Haven, 1883. 8vo, pp. 24–26. Twenty separates printed without title-page, covers, or repaging.

> The opinion is advanced that the burning of the beds of lignite of the Laramie Group has been mainly the result of spontaneous ignition, and that these fires probably began as early as the later Tertiary and before the advent of man.

121.

WHITE, C. A. On the commingling of ancient faunal and modern floral types in the Laramie Group. < Amer. Journ. Sci., 3d ser., vol. xxvi, pp. 120–123. New Haven, 1883.

Same. New Haven, 1883. 8vo, pp. 120–123. Twenty separates printed without title-page, covers, or repaging.

> It is shown that well-known species of Miocene plants are found associated with Dinosaurs and characteristic Laramie mollusks.

172   BULLETIN NO. 30, UNITED STATES NATIONAL MUSEUM.

122.

WHITE, C. A. [Administrative report for the year 1882–'83.] < Fourth Annual Report of the Director of the United States Geological Survey, pp. 42–44. Washington, 1883.

123.

WHITE, C. A. A review of the Fossil Ostreidæ of North America; and a comparison of the fossil with the living forms. < Fourth Annual Report of the Director of the United States Geological Survey. pp. 281–333, and pls. xxxiv–lxxvi. Washington, 1883.

Same. Washington: Government Printing Office, 1883. Imp. 8vo, pp. 281–333, and pls. xxxiv–lxxvi. One hundred separates printed with paper covers and title-page, but without repaging.

This work is an annotated catalogue of the species. It contains two appendices, by Prof. A. Heilprin and Mr John A. Ryder, respectively. The former is on the North American Tertiary *Ostreidæ*. pp. 309–316, and pls. lxii–lxxii. The latter is a sketch of the life-history of the oyster. pp. 317–333, and pls. lxxiii–lxxv.

CARBONIFEROUS.

	Page.
Ostrea, Linnæus	288
O. patercula, Winchell, pl. xxxiv, figs. 1 and 2	288

JURASSIC.

Ostrea, Linnæus	289
O. engelmanni, Meek, pl. xxxiv, figs. 3 and 4	289
O. strigilecula, White, pl. xxxv, figs. 9–11	289–290
O. (Alectryonia) procumbens, White, pl. xxxv, figs. 6–8	290
Gryphœa, Lamarck	290
G. calceola, Quenstedt, var. *nebrascensis*, M. & H., pl. xxxv, figs. 1–5	290

CRETACEOUS.

Ostrea, Linnæus	291
O. americana, Deshayes, pl. lvi, figs. 1 and 2, pl. lvii, figs. 1 and 2	291
O. anomiæformis, Roemer	291
O. anomioides, Meek, pl. xxxix	291
O. appressa, Gabb, pl. xxxix, fig. 9	291–292
O. bella, Conrad, pl. xxxix, fig. 6	292
O. bellorugosa, Shumard	292
O. belliplicata, Shumard, pl. lxxviii, figs. 1–3	292
O. blackii, White, pl. xlv, fig. 1, and pl. xlvi, fig. 2	292
O. barrandei, Coquand, pl. xliv, figs. 1 and 2, pl. xlv, fig. 2, and pl. xlvi, fig. 1	292–293
O. breweri, Gabb	293
O. bryani, Gabb	293
O. carinata (Lamarck), Roemer, pl. xliii, figs. 1–4	293
O. coalvillensis, Meek, pl. xxxvi, figs. 1–4	293
O. confragosa, Conrad	293
O. congesta, Conrad, pl. xxxix, figs. 11–13	294
O. convexo, Say, pl. xlviii, figs. 1–5	294
O. cortez, Conrad, pl. xxxvii, figs. 3 and 4	294
O. crenulata, Tuomey	294
O. crenulimargo, Roemer, pl. xliii, figs. 8, 9	294
O. crenulimarginata, Gabb, pl. xl, fig. 2	294
O. cretacea, Morton, Owen, pl. xxxix, figs. 1–3	294–295
O. denticulifera, Conrad	295
O. diluriana, Linnæus, pl. xl, fig. 1, pl. xli, figs. 1 and 2	295
O. elegantula, Newberry, pl. xxxvi, figs. 5–7	295
O. exogyrella, Gabb	296
O. falcata, Morton, pl. xlii, figs. 2–9	296
O. franklini, Coquand, pl. xxxix, figs. 1–3	296
O. gabbana, M. & H.	296

	Page.
O. inornata, Meek	296
O. idriænsis, Gabb, pl. xxxiv, figs. 7 and 8	296
O. (Alectryonia) larva, Lamarck, pl. xlii, figs. 2-9	296
O. lateralis, Nilsson	297
O. littlei, Gabb	297
O. lugubris, Conrad, pl. xli, fig. 3	297
O. lyoni, Shumard	297
O. malleiformis, Gabb, pl. l, fig. 8	297
O. mesenterica, Morton, pl. xlii. figs. 2-9	297
O. mortoni, Gabb	297
O. multilirata, Conrad, pl. xxxviii, figs. 1 and 2	298
O. nasuta, Morton, pl. xlii, figs. 2-9	298
O. owenana, Shumard	298
O. panda, Morton	298
O. pandæformis, Gabb	298
O. patina, M. & H., pl. xlvii, figs. 4-6	298
O. peculiaris, Conrad	298
O. pellucida, M. & H., pl. l, figs. 6 and 7	299
O. planovata, Shumard	299
O. plumosa, Morton, pl. xxxvii, figs. 5 and 6	299
O. prudentia, White, pl. xl, figs. 5 and 6	299
O. quadriplicata, Shumard, pl. xliii, figs. 5-7	299-300
O. robusta, Conrad, pl. xl, figs. 3 and 4	300
O. sannionis, White, pl. xlv, figs. 3-7	300
O. soleniscus, Meek, pl. xlii, fig. 1	300
O. subalata, Meek, pl. xxxix, fig. 10	300
O. subovata, Shumard	301
O. subspatulata, Forbes, pl. xxxvii. figs. 1 and 2	301
O. tecticostata, Gabb, pl. l, figs. 3 and 4	301
O. torosa, Morton	301
O. translucida, M. & H	301
O. tuomeyi, Coquand	301-302
O. uniformis, Meek, pl. xlviii, figs. 6 and 7	302
O. velicata, Conrad	302
O. vomer, Morton, pl. xlviii, figs. 8-10	302
Gryphæa, Lamarck	302
G. mucronata, Gabb	302
G. mutabilis, Morton, pl. xlviii, figs. 1-5	302
G. navia, Conrad	302
G. pitcheri, Morton, pl. xlix, figs. 1-6	302-303
G. thirsæ, Gabb	303
G. vesicularis, Lamarck, pl. xlviii, figs. 1-5	303
G. vomer, Morton	303
Exogyra, Say	303
E. arietina, Roemer, pl. lvi, figs. 3-5	303-304
E. aquila, Goldfuss, pl. liii, figs. 1 and 2	304
E. columbella, Meek, pl. lv, figs. 5 and 6	304
E. costata, Say, pl. lvi, figs. 1 and 2, and pl. lvii, figs. 1 and 2	304
E. fimbriata, Conrad	305
E. forniculata, White, pl. lii, figs. 1 and 2	305
E. fragosa, Conrad	305
E. interrupta, Conrad	305
E. læviuscula, Roemer, pl. lii, figs. 3-5	305-306
E. matheroniana, d'Orbigny	306
E. plicata, Lamarck	306
E. ponderosa, Roemer, pl. l, figs. 1 and 2	306
E. parasitica, Gabb, pl. lv, figs. 3 and 4	306
E. texana, Roemer, pl. li, figs. 1-5	306
E. walkeri, White, pl. liv, figs. 1 and 2	307
E. winchelli, White, pl. lv, figs. 6 and 7; pl. lvi, figs. 1 and 2	307

LARAMIE GROUP.

Ostrea Linnæus.

O. glabra, M. & H., pls. lviii, lix, lx, lxi	307-308
O. subtrigonalis, Evans & Shumard, pl. lxi, figs. 4-7	308

124.

WHITE, C. A. [Review of] E. O. Ulrich. Description of two new species of Crinoids. (Journ. Cincinnati Soc. Nat. History, vol. v, No. 3, p. 118, pl. v.) <Neues Jahrbuch für Mineralogie, Geologie und Paläontologie. 2. Band, p. 118. Stuttgart, 1883.

125.

WHITE, C. A. [Review of] P. de Loriol. Description of a new species of Bourgueticrinus. (Journ. Cincinnati Soc. Nat. History, vol. v, No. 3, p. 118, pl. v.) <Neues Jahrbuch für Mineralogie, Geologie und Paläontologie. 2. Band, p. 118. Stuttgart, 1883.

126.

WHITE, C. A. [Review of] S. A. Miller. Description of three new orders and four new families in the class Echinodermata, and eight new species from the Silurian and Devonian formations. (Journ. Cincinnati Soc. Nat. History, vol. v, No. 4, pp. 221-231, pl. ix. <Neues Jahrbuch für Mineralogie, Geologie und Paläontologie. 2. Band, p. 117. Stuttgart, 1883.

127.

WHITE, C. A. [Review of] S. A. Miller. Description of three new species and remarks upon others. (Journ. Cincinnati Soc. Nat. History, vol. v, No. 3, pp. 116-117, pl. v.) <Neues Jahrbuch für Mineralogie, Geologie und Paläontologie. 2. Band, p. 98. Stuttgart, 1883.

128.

WHITE, C. A. [Review of] C. Schlumberger. Remarks upon a species of Cristellaria. (Journ. Cincinnati Soc. Nat. History, vol. v, No. 3, p. 119, pl. v.) <Neues Jahrbuch für Mineralogie, Geologie und Paläontologie. 2. Band, p. 409. Stuttgart, 1883.

129.

WHITE, C. A. [Review of] vol. 3, Supplement. Geology. U. S. Geographical Surveys West of the one hundreth Meridian. <Neues Jahrbuch für Mineralogie, Geologie und Paläontologie. 1. Band, pp. 232-233. Stuttgart, 1883.

130.

WHITE, C. A. [Review of] Geology of Wisconsin. Final reports of the State Geological Survey. 4 volumes, royal 8vo. Published under the direction of the Chief Geologist by the Commissioners of Public Printing. <Neues Jahrbuch für Mineralogie, Geologie und Paläontologie. 2. Band, pp. 341-349. Stuttgart, 1883.

131.

WHITE, C. A. [Review of] Clarence E. Dutton, Captain of Ordnance, U. S. Army. The Tertiary History of the Grand Cañon District. 4to, pp. i-xiv and 1-264; pls. i-xlii, with folio atlas containing 23 plates. Washington: Government Printing Office. 1882. <Neues Jahrbuch für Mineralogie, Geologie und Paläontologie. 2. Band, pp. 190-191. Stuttgart, 1883.

132.

WHITE, C. A. [Review of] John Collett. Department of Geology and Natural History. (Eleventh Annual Report, 1881. Indianapolis, 1882. 8vo, pp. 414, pls. lv, and three small maps.) <Neues Jahrbuch für Mineralogie, Geologie und Paläontologie. 2. Band, pp. 189-190. Stuttgart, 1883.

133.

WHITE, C. A. [Review of] S. A. Miller. Description of ten new species of fossils. (Journ. Cincinnati Soc. Nat. Hist., vol. v, No. 2, pp. 79-88, pls. iii-iv.) <Neues Jahrbuch für Mineralogie, Geologie und Paläontologie. 2. Band, p. 98. Stuttgart, 1883.

134.

WHITE, C. A. [Review of] S. A. Miller. Description of two new genera and eight new species of fossils from the Hudson River Group, with remarks upon others. (Journ. Cincinnati Soc. Nat. Hist., vol. v, No. 1, pp. 34-44, pls. i and ii.) <Neues Jahrbuch für Mineralogie, Geologie und Paläontologie. 2. Band, pp. 97-98. Stuttgart, 1883.

135.

WHITE, C. A. [Annual report to the director of the National Museum.] <Report of the Assistant Director of the U. S. National Museum, G. Brown Goode, for the year 1882. pp. 31-32. Washington, 1883.

A portion of Dr. White's statements and recommendations are quoted.

136.

WHITE, C. A. On the Macrocheilus of Phillips, Plectostylus of Conrad, and Soleniscus of Meek & Worthen. <Proc. U. S. National Museum, vol. vi, pp. 184-187, pl. viii. Washington, 1884.

Same. Washington: Government Printing Office, 1884, 8vo, pp. 184-187, pl. viii. One hundred and fifty separates printed without title-page, covers, or repaging.

Many of the Carboniferous species which have hitherto been referred by authors to *Macrocheilus* are herein referred to *Soleniscus*.

	Page.
Soleniscus ? (Macrocheilus) ponderosus, Swallow ? pl. viii, figs. 1 and 2	187
Soleniscus ? (Macrocheilus) primigenius, Conrad, pl. viii, fig. 3	187
Soleniscus (Macrocheilus) fusiformis, Hall, pl. viii, figs. 4-6	187
Soleniscus (Macrocheilus) newberryi, Hall, pl. viii, figs. 7 and 8	187
Soleniscus planus, White, pl. viii, figs. 9, 10	187
Soleniscus (Macrocheilus) ventricosus, Hall, pl. viii, figs. 11 and 12	187
Soleniscus (Macrocheilus) texanus, Shumard, pl viii, figs. 13 and 14	187
Soleniscus ? (Macrocheilus) medialis M. & W., pl. viii, figs. 15 and 16	187
Soleniscus (Macrocheilus) paludineaformis, Hall, pl. viii, fig. 17	187
Soleniscus typicus, M. & W., pl. viii, figs. 18 and 19	187

137

WHITE, C. A. On the character and function of the epiglottis in the Bull snake (Pityophis). <Amer. Nat., vol. xviii, pp. 19-21. Philadelphia, 1884.

Same. Philadelphia, 1884, 8vo, pp. 19-21. Fifty separates, printed without title-page, covers, or repaging.

It is shown that the hoarse hiss of this snake is produced by the fluttering of the thin erect epiglottis in the current of air expelled from the rima glottidis.

138.

WHITE, C. A. The permanence of the domestic instinct in the cat. <Amer. Nat., vol. xviii, pp. 213-214. Philadelphia, 1884.

Same. Philadelphia, 1884, 8vo, pp. 213-214. Fifty separates printed without title-page, covers, or repaging.

A story of a cat which had spent a year alone in the wilderness of the Upper Missouri.

139.

WHITE, C. A. Glacial drift in Montana and Dakota. <Amer. Jour. Sci. 3d ser., vol. xxvii, pp. 112-113. New Haven, 1884.

Same. New Haven, 1884, 8vo 3d ser., vol. xxvii, pp. 112-113. Twenty separates printed without title-page, covers, or repaging.

> The presence of true northern glacial drift is observed along the Upper Missouri River, from the Great Falls to Bismarck.

140.

WHITE, C. A. Description of certain aberrant forms of the Chamidæ from the Cretaceous rocks of Texas. <Bull. of the U. S. Geol. Surv. No. 4. On Mesozoic fossils. pp. 5 (93)-9 (94), pls. i-v. Washington, 1884.

	Page
Requienia, Matheron	6
R. patagiata, n. s., White, 1884, pl. i, figs. 1-8, and pl. ii, figs. 1-4	6-7
R. texana, Roemer, 1852, pl. ii, figs. 5-7	7
Monopleura, Matheron	8
M. marcida, n. s., White, 1884, pls. iii and iv	8
M. pinguiscula, n. s., White, 1884, pl. v	8-9

141.

WHITE, C. A. On a small collection of Mesozoic fossils, obtained in Alaska by Mr. W. H. Dall, of the United States Coast Survey. <Bull. of the U. S. Geol. Surv. No. 4. On Mesozoic fossils. pp. 10 (98)-15 (103), pl. vi. Washington, 1884.

Mollusca.

	Page.
Aucella, Keyserling	13
A. concentrica, Fischer var. White, 1884, p!. vi, figs. 2-12	13-14
Cyprina, Lamarck	14
C. ? dallii, n. s., White, 1884, pl. vi, fig. 1	14
Belemnites, Lamarck	14
B. macritatis, n. s., White, 1884, pl. vi, figs. 13-14	14-15

142.

WHITE, C. A. On the Nautiloid genus Enclimatoceras Hyatt, and a description of the type species. <Bull. of the U. S. Geol. Surv. No. 4. On Mesozoic fossils. pp. 16 (104)-17 (105), pls. vii-ix. Washington, 1884.

	Page.
Enclimatoceras, n. g., Hyatt, 1884	16-17
E. (Nautilus) ulrichi, n. s., White, 1884, pls. vii, viii, and ix	17

143.

WHITE, C. A. On the adaptability of the prairies for artificial forestry. <Science, 4°, vol. iii, pp. 438-443. Cambridge, 1884.

> The view is held that the prairie soil is well adapted to the growth of forest trees; and that the prairies are such only because their occupation by forests has not been accomplished by the natural distribution of trees; also that such distribution has long been retarded by prairie fires.

144.

WHITE, C. A. Enemies and parasites of the oyster, past and present. <Science, 4to, vol. iii, p. 618. Cambridge, 1884.

> It is shown that *Cliona* or a similar burrowing sponge infested certain *Brachiopod* shells as early as the Devonian, and that they were as common upon the fossil *Ostreidæ* as upon the living. Also that remains of star fishes are rarely found with fossil *Ostreidæ*, although they are so common an enemy to living oysters.

145.

WHITE, C. A. Certain phases in the geological history of the North American Continent, biologically considered. <Proceedings of the Washington Biological Society. vol. ii, pp. 41–66. Washington, 1884.

Same. One hundred separates printed with paper covers, title-page, and repaging.

Address as retiring president of the Biological Society of Washington.

146.

WHITE, C. A. Fossils of the Indiana rocks, No. 3. <Indiana, Department of Geology and Natural History. Thirteenth annual report, John Collett, State geologist. pp. 105–180, pls. xxiii–xxxix. Indianapolis, 1884.

This paper contains some elementary remarks and description of three new species, but it is mainly devoted to the republication of well-known forms.

Same. Indianapolis, 1884, pp. —. Fifteen separates printed.

FAUNA OF THE COAL-MEASURES.

Description of species.

	Page.
Protozoa.	
Foraminifera.	
Fusulina, Fischer	116
F. cylindrica, Fischer, pl. xxiii, figs. 1–3	116–117
Coelenterata.	
Polypi.	
Zaphrentis, Rafinesque	117
Z. gibsoni, n. s., White, 1884, pl. xxiii, figs. 4 and 5	117–118
Lophophyllum, Edwards & Haime	118
L. proliferum, McChesney, pl. xxiii, figs. 6 and 7	118
Axophyllum, Edwards & Haime	118
A. rudis, White & St. John, 1867, pl. xxiii, figs. 8 and 9	118–119
Campophyllum, Edwards & Haime	119
C. torquim, Owen, 1852, pl. xxiii, figs. 10–13	119
Michelinia, de Koninck	119
M. eugeneae, n. s., White, 1884, pl. xxiii, figs. 14–16	119–120
Brachiopoda.	
Lingula, Bruguière	120
L. umbonata, Cox, 1857, pl. xxv, fig. 14	120
Discina, Lamarck	121
D. nitida, Phillips, pl. xxv, fig. 10	121
D. convexa, Shumard, 1858, pl. xxv, fig. 9	121
Crania, Retzius	121
C. modesta, White & St. John, 1867, pl. xxxv, fig. 9, and pl. xxxvi, fig. 5	121–122
Productus, Sowerby	122
P. nebrascensis, Owen, 1852, pl. xxiv, figs. 7–9	122–123
P. symmetricus, McChesney, 1866, pl. xxv, figs. 1 and 2	123
P. punctatus, Martin, pl. xxvii, figs. 1–3	124
P. costatus, Sowerby, pl. xxiv, figs. 4–6, and pl. xxv, figs. 3–5	124–125
P. semireticulatus, Martin, pl. xxiv, figs. 1–3	125–126
P. cora, d'Orbigny pl. xxvi, figs. 1–3	126–127
P. longispinus, Sowerby, pl. xxiv, figs. 10 and 11	127–128
Chonetes, Fischer	128
C. verneuiliana, Norwood & Pratten, 1854, pl. xxv, figs. 7 and 8	128
Orthis, Dalman	129
O. pecosi, Marcou, 1858, pl. xxxii, figs. 20–22	129
Hemipronites, Pander	129
Streptorhynchus, King	129
Hemipronites crassus, M. & H., pl. xxvi, figs. 4–11	129–130
Meekella, White & St. John	130
M. striatocostata, Cox, 1857, pl. xxvi, figs. 12–14	130–131
Syntrielasma, M. & W	131
S. hemiplicata, Hall, 1852, pl. xxvi, figs. 15–18	131–132
Rhynchonella, Fischer	132
R. uta, Marcou, 1858, pl. xxv, fig. 6	132

	Page
Spirifer, Sowerby	132
S. cameratus, Morton, 1836, pl. xxxv, figs. 3–5	132–133
S. (Martinia) lineatus, Martin, pl. xxvii, figs. 4–6	133–134
S. (Martinia) planoconvexa, Shumard, 1855, pl. xxxii, figs. 23 and 24	134–135
Spiriferina, d'Orbigny	135
S. kentuckensis, Shumard, 1855, pl. xxxv, figs. 13 and 14	135
Athyris, McCoy	136
A. subtilita, Hall, pl. xxxv, figs. 6–9	136
Retzia, King	136
R. mormonii, Marcou, 1858, pl. xxxv, figs. 10–12	136–137
Terebratula, Lhwyd	137
T. bovidens, Morton, 1836, pl. xxxii, figs. 17–19	137
Polyzoa.	
Synocladia, King	138
S. biserialis, Swallow, 1858, pl. xxv, figs. 11–13	138
Conchifera.	
Lima, Bruguière	138
L. retifera, Shumard, 1858, pl. xxviii, fig. 4	138–139
Monopteria, M. & W.	139
M. gibbosa, M. & W., 1866, pl. xxx, figs. 11 and 12	139
Myalina, de Koninck	140
M. subquadrata, Shumard, 1855, pl. xxix, figs. 1 and 2, and pl. xxx, figs. 1 and 2	140
M. recurvirostris, M. & W., 1866, pl. xxix, figs. 3 and 4	140–141
M.? swallovi, McChesney, 1860, pl. xxx, figs. 6–8	141
Entolium, Meek	142
E. aviculatum, Swallow, 1858, pl. xxviii, figs. 7 and 8	142
Eumicrotis, Meek	142
E. hawni, M. & H., 1866, pl. xxx, fig. 10	142–143
Aviculopecten, McCoy	143
A. occidentalis, Shumard, 1855, pl. xxviii, fig. 3	143
A. carboniferus, Stevens, 1858, pl. xxviii, figs. 5 and 6	144
A.? interlineatus, M. & W., pl. xxx, fig. 9	145
Pinna, Linnæus	145
P. peracuta, Shumard, 1858, pl. xxviii, figs. 1 and 2	145–146
Nuculana, Link	146
N. bellistriata, Stevens, 1858, pl. xxxi, figs. 8 and 9	146
Nucula, Lamarck	146
N. ventricosa, Hall, 1858, pl. xxvii, figs. 9 and 10	146–147
Schizodus, King	147
S. wheeleri, Swallow, 1862, pl. xxx, figs. 3–5	147
Clinopistha, M. & W	147
C. radiata, Hall, 1858, pl. xxxi, figs. 6 and 7	147–148
Edmondia, de Koninck	148
E. aspinwallensis, Meek, 1872, pl. xxxi, figs. 4 and 5	148
Alloriama, King	148
A. subcuneata, M. & H., 1864, pl. xxxi, figs. 1–3	148–149
Gasteropoda.	
The genera *Macrocheilus* and *Soleniscus*	149–152
Soleniscus, M. & W	152
S. typicus, M. & W., 1866, pl. xxxiv, figs. 18, 19	152
S. (Macrocheilus) newberryi, Stevens, 1858, pl. xxxiv, figs. 7 and 8	153
S. planus, White, 1881, pl. xxxiv, figs. 9 and 10	153–154
S. (Macrocheilus) fusiformis, Hall, 1858, pl. xxxiv, figs. 4–6	154
S. (Macrocheilus) paludinæformis, Hall, 1858, pl. xxxiv, fig. 17	154–155
S. (Macrocheilus) ventricosus, Hall, 1858, pl. xxxiv, figs. 11 and 12	155
S. (Macrocheilus) texanus, Shumard? 1859, pl. xxxiv, figs. 13 and 14	155–156
S. (Macrocheilus) medialis, M. & W., 1866, pl. xxxiv, figs. 15 and 16	156
S. (Macrocheilus) ponderosus, Swallow?, 1858, pl. xxxiv, figs. 1 and 2	156
S. (Macrocheilus) primigenius, Conrad, 1835, pl. xxxiv, fig. 3	157
Bellerophon, Montfort	157
B. crassus, M. & W., 1866, pl. xxxiii, figs. 1 and 2	157
B. percarinatus, Conrad, 1842, pl. xxxiii, figs. 9–14	15*
B. carbonarius, Cox, 1857, pl. xxxiii, figs. 6–8	158–15*
B. nodocarinatus, Hall, 1858, pl. xxxiii, figs. 3–5	159
Platyceras, Conrad	159

	Page
P. nebrascense, Meek, 1872, pl. xxxii, figs. 15 and 16	159-160
Pleurotomaria, Defrance	160
P. turbiniformis, M. & W., 1866, pl. xxxii, figs. 7 and 8	160
P. tabulata, Hall, pl. xxxii, figs. 4 and 5	160-161
P. sphærulata, Conrad, 1842, pl. xxxii, figs. 1-3	161
Euomphalus, Sowerby	161
E. rugosus, Hall, 1858, pl. xxxii, figs. 11 and 12	161-162
Naticopsis, McCoy	162
N. nana, M. & W., 1866, pl. xxxvi, figs. 6 and 7	162
N. wheeleri, Swallow, 1860, pl. xxxii, figs. 13 and 14	162-163
Polyphemopsis, Portlock	163
P. peracuta, M. & W., pl. xxxii, figs. 9 and 10	163
P. nitidula, M. & W., 1866, pl. xxvii, figs. 7 and 8	163
Polyphemopsis? —— (?) White, 1884, pl. xxxii, fig. 6	164
Cephalopoda.	
Orthoceras, Breynius	164
O. rushensis, McChesney, pl. xxxvi, fig. 5	164
Nautilus, Breynius	165
N. winslowi, M. & W., 1873, pl. xxxvi, figs. 1 and 2	165
N. forbesianus, McChesney, pl. xxxvi, figs. 3 and 4	165
N. missouriensis, Swallow ? 1857, pl. xxxv, figs. 1 and 2	166
Crustacea.	
Gnathostomata.	
Lenia, Jones	167
L. tricarinata, M. & W., 1868, pl. xxxix, figs. 10-13	167-168
Merostomata.	
Eurypterus, DeKay	
E. (Anthraconectes) mazonensis, M. & W., 1868, pl. xxxvii, figs. 1-3	168-170
Euproops, M. & W	170
E. danæ, M. & W., 1866, pl. xxxix, fig. 1	170-172
E. colletti, n. s., White, 1884, pl. xxxix, fig. 2	172
Trilobita.	
Phillipsia, Portlock	173
P. (Griffithides?) scitula, M. & W., 1873, pl. xxxix, figs. 6-9	173-174
P. (Griffithides?) sangamonensis, M. & W., 1873, pl. xxxix, figs. 4 and 5	174-176
Isopoda.	
Acanthotelson, M. & W	176
A. stimpsoni, M. & W., 1866, pl. xxxvii, figs. 4 and 5	176-177
A. eveni, M. & W., 1868, pl. xxxviii, figs. 4-7	177-178
Dithyrocaris, Scouler	178
D. carbonarius, M. & W., 1873, pl. xxxix, fig. 3	178
Macroura.	
Palæocaris, M. & W	179
P. typus, M. & W., pl. xxxviii, figs. 1-3	179-180
Anthrapalæmon, Salter	180
A. gracilis, M. & W., 1865, pl. xxxviii, figs. 8 and 9	180

147.

WHITE, C. A. [Review of] Geological Survey of Illinois; A. H. Worthen, Director. Vol. vii, Geology and Paleontology. Springfield. 1883. pp. 1-373; pls. i-xxxi. <Science. 4to. Vol. iii, pp. 332-333. Cambridge, 1884.

148.

WHITE, C. A. Notes on the Jurassic strata of North America. <Amer. Jour. Sci. 3d ser., vol. xxix, pp. 228-232. New Haven, 1885.

A criticism of the views published by Mr. J. F. Whiteaves in Geol. Surv. Canada, Mesozoic Fossils, vol. 1, part ii. Montreal, 1884. The identification of the following species is considered very doubtful:

	Page
Belemnites densus, M. & H	229
Lyosoma Powelli, White	230
Myacites (Pleuromya) subcompressa, Meek	230

	Page.
Astarte Packardi, White	230
Arca (Cucullœa) inornata, M. & H	230
Modiola (Volsella) subimbricata, M. & H	230
Pteria (Oxytoma) mucronata, M. & H	230
Camptonectes extenuatus, M. & H	230–231
Gryphœa nebrascensis, M. & H	231

149.

WHITE, C. A. The Genus Pyrgulifera, Meek, and its Associates and Congeners. <Amer. Jour. Sci., 3d ser., vol. xxix, pp. 277–280. New Haven, 1885.

A summary of the occurrence, fossil and recent, of the genus *Pyrgulifera*.

150.

WHITE, C. A. On Marine Eocene, Fresh-water Miocene, and other Fossil Mollusca of Western North America. <Bull. U. S. Geol. Surv., No. 18, pp. 1–19, pls. 1–iii. Washington, 1885.

This paper is divided into three parts, the first on "The occurrence of *Cardita planicosta*, Lamarck, in Western Oregon," pl. i, pp. 7–9.

The second on "Fossil Mollusca, from the John Day group in Eastern Oregon," pp. 10–16.

The following new species are described:

	Page.
Unio condoni, n. s., White, 1885, pl. ii, figs. 1–3	13–14
Helicidœ	14
Helix (Aglaia) fidelis, Gray, pl. iii, figs. 1–3	14
Helix (Patula) perspectiva, Say, pl. iii, fig. 7	14
Helix (Monodon?) dallii, Stearns, MS., 1885, pl. iii, figs. 4–6	14–15
Gonostoma yatesii, Cooper, pl. iii, figs. 8–12	16

The third part contains "Supplementary notes on the non-marine fossil mollusca of North America."

	Page.
Ampullaria powelli	18
Cerithidea nebrascensis	19
Dreissena leucophœata	19
Physa prisca	18
Unio martini	18
Zaptychius carbonaria	18

151.

WHITE, C. A. On new Cretaceous Fossils from California. <Bull. U. S. Geol. Surv., No. 22, pp. 1 (349);–15 (361), pls. i–v. Washington, 1885.

	Page.
Chamidœ.	
Coralliochama, n. g., White, 1885	9–10
C. orcutti, n. s., White, 1885, pls. i–iv	10–12
Trochidœ.	
Trochus, Linnæus	12
Subgenus *Oxystele*, Philippi	12
T. (Oxystele) euryostomus, n. s., White, 1885, pl. v, figs. 9–11	12
Neritidœ.	
Nerita, Linnæus	12
Nerita, ——— ? White, 1885	12
Cerithiidœ.	
Cerithium, Bruguière	13
C. pillingi, n. s., White, 1885, pl. v, figs. 3–6	13
C. totium sanctorum, n. s., White, 1885, pl. v, figs. 12 and 13	13
Solariidœ.	
Solarium, Lamarck	14
S. wallalense, n. s., White, 1885, pl. v, figs. 1 and 2	14

Besides the foregoing, Dr. White has in press an important work on the Cretaceous invertebrates of Brazil, which were collected by the Imperial Geological Commission under the direction of the late Prof. Ch. Fred. Hartt. The work is in process of publication at Rio de Janeiro by the Brazilian National Museum. It is to appear in Volume VII of the "Archives" of that museum, in both the Portuguese and English languages, and will be illustrated with 28 lithographic plates of figures.

The work consists of five parts, which are as follows in the original Portuguese edition—

Contribuiçoes para a Palæontologia do Brazil:
    No. 1. Conchiferos Cretaceos.
    No. 2. Gasteropodes Cretaceos.
    No. 3. Cephalopodes Cretaceos.
    No. 4. Molluscos Cretaceos de Agua doce do Grupo da Bahia.
    No. 5. Echinodermes Cretaceos.

Two hundred and fourteen species in all are published and figured in this work, of which 116 species are diagnosed as new. Four new genera are proposed, three of gasteropoda and one of echinoids. The former are *Orvillia*, *Cylindritella*, and *Cypræactæon*. The latter is *Heteropoda*, the generic diagnosis of which was furnished to the author by Prof. P. de Loriol, of Geneva.

PART III.

THE PUBLISHED WRITINGS OF CHARLES DOOLITTLE WALCOTT.

III.—PUBLISHED WRITINGS OF CHARLES DOOLITTLE WALCOTT.

1.

WALCOTT, C. D.   Description of a New Species of Trilobite.   <Cincinnati Quarterly Journal of Science, vol. ii, pp. 273–274, figs. 18 a, b.   July.   Cincinnati, 1875.

Page.
Spherocoryphe, Angelin.
S. robustus, n. s., Walcott, 1875, p. 274, figs. 18 a, b.................................... 273–274

2.

WALCOTT, C. D.   New Species of Trilobite from the Trenton Limestone at Trenton Falls, N. Y.   <Cincinnati Quarterly Journal of Science, vol. ii, pp. 347–349, fig. 27.   October.   Cincinnati, 1875.

Page.
Remopleurides, Portlock.
R. striatulus, n. s., Walcott, 1875, fig. 27 a, b, and A................................ 347–349

3.

WALCOTT, C. D.   Notes on Ceraurus pleurexanthemus, Green.   <Annals Lyc. Nat. Hist. N. Y., vol. xi, pp. 155–162, pl. xi.   November.   New York, 1875.

4.

WALCOTT, C. D.   Preliminary Notice of the Discovery of the Remains of the Natatory and Branchial appendages of Trilobites.   <28th Regent's Report N. Y. State Mus. Nat. Hist., pp. 89–92.   1877.   Albany, 1879.

5.

WALCOTT, C. D.   Descriptions of New Species of Fossils from the Trenton limestone. <28th Regent's Report N. Y. State Mus. Nat. Hist., pp. 93–97.   1877.   Albany, 1879.

Page.
Conularia, Miller, MS., 1818... 93
C. quadrata, n. s., Walcott, 1877... 93
Conchopeltis, n. g., Walcott, 1877... 93
C. alternata, n. s., Walcott, 1877.. 93–94
C. minnesotensis, n. s., Walcott, 1877... 94
Bathyurus, Billings... 94
B. longispinus, n. s., Walcott, 1877.. 94–96
Asaphus, Brongniart, 1822... 96
A. romingeri, n. s., Walcott, 1877... 96–97
A. wisconsensis, n. s., Walcott, 1877.. 97

6.

WALCOTT, C. D.   Notes on some Sections of Trilobites from the Trenton limestone, [and] Note upon the Legs of Trilobites.   <31st Regent's Report N. Y. State Mus. Nat. Hist., pp. 61–65, pl. i.   Albany, 1879.

Published in advance, September 20, 1877, pp. 1–17, pl. i.
Extracts from the 31st Regent's report, published in March, 1879.

---

* κογχη, shell; πελτη, shield.

7.

WALCOTT, C. D. Note upon the Eggs of the Trilobite. <31st Regent's Report N. Y. State Mus. Nat. Hist., pp. 66–67? Albany, 1879.

Published in advance, September 20, 1877, pp. 11-13.
Extracts from the 31st Regent's report, published in March, 1879.

8.

WALCOTT, C. D. Descriptions of New Species of Fossils from the Chazy and Trenton limestones. <31st Regent's Report N. Y. State Mus. Nat. Hist., pp. 68–71. Albany, 1879.

Published in advance, September 20, 1877, pp. 15-21.
Extract from the 31st Regent's report, published in March, 1879.

	Page.
Arionellus, Barrande, 1846	15, 68
A. pustulatus, n. s., Walcott, 1877	15, 68
Ceraurus, Green, 1832	15, 68
C. rarus, n. s., Walcott, 1877	15-16, 68
Encrinurus, Emmerich, 1845	16, 68
E. trentonensis, n. s., Walcott, 1877	16, 68
E. varicostatus, n. s., Walcott, 1877	16, 69
Acidaspis, Murchison, 1839	16, 69
A. parvula, n. s., Walcott, 1877	16, 17, 69
Dalmanites, Barrande, 1852	17, 69
D. intermedius, n. s., Walcott, 1877	17-18, 69-70
Illænus, Dalman, 1826	19, 70
I. indeterminatus, n. s., Walcott, 1877	19-20, 70-71
I. milleri, Billings	20, 71
Asaphus, Brongniart, 1822	20, 71
A. homalonotoides, n. s., Walcott, 1877	20-21, 71

9.

WALCOTT, C. D. Descriptions of New Species of Fossils from the Calciferous formation. <32d Regent's Report N. Y. State Mus. Nat. Hist., pp. —. Albany.

Published in advance, January 3, 1879, pp. 1-4.

	Page.
Platyceras, Conrad, 1840	1
P. minutissimum, n. s., Walcott, 1879	1
Metoptoma, Phillips, 1836	1
M. cornutaforme, n. s., Walcott, 1879	1
Conocephalites, Adams, 1848	1
C. calciferus, n. s., Walcott, 1879	1, 2
C. hartii, n. s., Walcott, 1879	2, 3
Ptychaspis, Hall, 1863	3
P. speciosus, n. s., Walcott, 1879	3
Bathyurus armatus, Billings	3, 4

10.

WALCOTT, C. D. The Utica slate and Related formations. Fossils of the Utica slate and Metamorphoses of Triarthrus Becki. <Trans. Albany Institute, vol. x, pp. 1-38, pls. i-ii. Albany, 1879.

Published in advance, June, 1879.

	Page.
*Cyathophycus, n. g., Walcott, 1879	18
C. reticulatus, n. s., Walcott, 1879, pl. ii, figs. 16, 16 *a–d*	19
C. subsphericus, n. s., Walcott, 1879, pl. ii. fig. 17	19
†*Discophycus*, n. g., Walcott, 1879	19
D. typicalis, n. s., Walcott, 1879, pl. ii, figs. 18, 18*a*	19

* Κύαθος, a cup; φῦκος, a weed.   † Δίσκος, a disk; φῦκος, a sea-weed.

	Page.
Graptolithus, Linnæus, 1736	20
G. annectans, n. s., Walcott, 1879	20
Dendrograptus, Hall, 1865	20
D. simplex, n. s., Walcott, 1879, pl. i, figs. 5, 5 *a*, *b*, and 6	20
D. tenuiramosus, n. s., Walcott, 1879, pl. i, fig. 4	21
D. compactus, n. s., Walcott, 1879, pl. i, fig. 1	21
Sagenella, Hall, 1852	22
S. ambigua, n. s., Walcott, 1879, pl. i, figs. 3, 3 *a*	22
Modiolopsis, Hall, 1847	22
M. cancellata, n. s., Walcott, 1879, pl. i, figs. 8, 8 *a*	22
Orthoceras, Breynius, 1732	22
O. oneidaense, n. s., Walcott, 1879, pl. i, figs. 7, 7 *a*	22–23
Beyrichia cincinnatiensis, Miller	23
Triarthrus becki, Green, 1832	23–24
Metamorphoses of *Triarthrus becki*	24–26
Degree of development	26–29
Periods of development	29–30
Comparison of parts during development	30–31
Table of development	31–32
Ornamentation	32–33
Catalogue of fossils occurring in the Utica slate	34–38

11.

WALCOTT, C. D. The Permian and other Paleozoic Groups of the Kanab Valley, Arizona. <Amer. Jour. Sci., 3d ser., vol. xx, pp. 221-225. September. New Haven, 1880.

12.

WALCOTT, C. D. The Trilobite: New and Old Evidence Relating to its Organization, <Bull. Mus. Comp. Zool., at Harvard College, vol. viii, No. 10, pp. 191-224, pls. i-vi. March. Cambridge, 1881.

13.

WALCOTT, C. D. On the nature of Cyathophycus. <Amer. Jour. Sci., 3d ser., vol. xxii, pp. 394-395. November. New Haven, 1881.

14.

WALCOTT, C. D. Description of a New Genus of the The Order Euripterida from the Utica slate. <Amer. Jour. Sci., 3d ser., vol. xxiii, pp. 213-216. March. New Haven, 1882.

	Page.
Echinognathus, n. g., Walcott, 1882.	
E. clevelandi, n. s., Walcott, fig. 1, p. 213; fig. 2, p. 214	213–216

15.

WALCOTT, C. D. Injury sustained by the Eye of a Trilobite at the time of Moulting of the shell. <Amer. Jour. Sci., 3d ser., vol. xxvi, p. 302. October. New Haven, 1883.

16.

WALCOTT, C. D. Descriptions of new species of Fossils from the Trenton group of New York. <35th Regents Rep., N. Y. State Mus. Nat. Hist., pp. 207-214, pl. xvii. Albany, 1884.

Published in advance. October 15, 1883, pp. 1-8, pl. xvii.

	Page.
Glyptocrinus, Hall	1, 207
G. argutus, n. s., Walcott, 1883, pl. xvii, fig. 9	1, 207
G.? subnodosus, n. s., Walcott, 1883, pl. xvii, fig. 3	2, 208
Merocrinus, n. g., Walcott, 1883	2–3, 208–209
M. typus, n. s., Walcott, 1883, pl. xvii, fig. 5	3, 209

	Page.
M. corroboratus, n. s., Walcott, 1883, pl. xvii, fig. 6	4, 210
Iocrinus, Hall	4, 210
I. trentonensis, n. s., Walcott, 1883, pl. xvii, figs. 7, 8	4–5, 210–211
Dendrocrinus, Hall	5, 211
D. retractilis, n. s., Walcott, 1883, pl. xvii, fig. 4	5, 211
Calceocrinus, Hall	6, 212
C. barrandii, n. s., Walcott, 1883, pl. xvii, figs. 1, 2	6, 212
Metoptoma, Phillips	6, 212
M. billingsi, n. s., Walcott, 1883, pl. xvii, figs. 12, 12 a	6–7, 212–213
Beyrichia, McCoy	7, 213
B. bella, n. s., Walcott, 1883, pl. xvii, figs. 11, 11 a	7, 213
Leperditia, Rouault	7, 213
Subgenus *Isochilina*, Jones	7, 213
S. (Isochilina) armeta, n. s., Walcott, 1883, pl. xvii, fig. 10	7–8, 213–214

17.

WALCOTT, C. D. Cambrian system of the United States and Canada. <Bull. Philosophical Soc., Washington, vol. vi, pp. 97–102. Washington. 1884.

Separates were published in [December, 1883].

18.

WALCOTT, C. D. Pre-carboniferous strata in the Grand Cañon of the Colorado, Arizona. <Amer. Jour. Sci., 3d ser., vol. xxvi, pp. 437–442, and p. 484. December. New Haven, 1883.

19.

WALCOTT, C. D. Fresh-water shells from the Paleozoic Rocks of Nevada. <Science, vol. ii, pp. 808–809. December. Cambridge, 1883.

	Page.
Zaptychius, n. g., Walcott, 1883	808
Zaptychius carbonaria, n. s, Walcott, 1883, fig.1	808
Physa prisca, n. s., Walcott, 1883, fig. 2	809
Ampullaria powelli, n. s., Walcott, 1883, fig. 3	809

20.

WALCOTT, C. D. Appendages of the Trilobite. Notes on the original specimen described by Prof. Mickleborough (Journ. Cincinnati Soc. Nat. Hist., vol. vi, p. 200, 1883). <Science, vol. iii, pp. 279–281, figs. 3. March. Cambridge, 1884.

21.

WALCOTT, C. D. Note on Paleozoic Rocks of Central Texas. <Amer. Jour. Sci., 3d ser., vol. xxvii, pp. 431–433. December. New Haven, 1884.

22.

WALCOTT, C. D. Deer Creek Coal-field, White Mountain Indian Reservation, Arizona. Report and Appendix. <U. S. Senate Ex. Doc. No. 20, 48th Congress, 2d session, pp. 1–7. December. Washington, 1884.

23.

WALCOTT, C. D. Paleontology of the Eureka District. <Monograph viii, U. S. Geol. Surv., pp. i–xiii, 1–298, pls. i–xxiv, figs. 1–7 in text. Washington, 1884.

CONTENTS.

	Page.
Letter of transmittal to Mr. Arnold Hague, by the author	v
Letter of transmittal to the director, by Mr. Arnold Hague	vii
Preface	ix

	Page.
Summary of results	1-9
Fossils of the Cambrian	11-64
Observations on *Olenellus howelli*	32-39
Fossils of the Lower Silurian	65-98
Fossils of the Devonian	99-211
Fossils of the Carboniferous	212-267
Systematic list of species	268-281
Paleozoic section in Central Nevada	283-285
Index	287-296

FOSSILS OF THE CAMBRIAN.

Prospect Mountain Group.

Porifera.

Protaspongia, Salter	11
P. fenestrata, Salter, 1864, pl. ix, figs. 5, 5 a, b	11-12

Brachiopoda.

Lingulepis, Hall	12
L. mœra, H. & W., 1877	12-13
L.? minuta, H. & W., 1877	13
Lingula, Bruguière	13
L.? manticula, White, 1874, pl. ix, figs. 3 and pl. xi, fig. 2	13-14
Obolella, Billings	14
O. discoidea, H. & W., 1877	14
Acrothele, Linnarsson	14
A.? dichotoma, n. s., Walcott, 1884, pl. ix, fig. 11	14-15
Scenella, Billings	15
S.? conula, n. s., Walcott, 1884, pl. ix, fig. 6	15-16
Acrotreta, Kutorga	16-17
A. gemma, Billings, 1865, pl. i, figs. 1 a, 1 b, 1 d-f; pl. ix, figs. 9, 9 a	17-18
Kutorgina, Billings	18
K. whitfieldi, n. s., Walcott, 1884, pl. ix, figs. 4, 4 a	18-19
K. prospectensis, n. s., Walcott, 1884, pl. ix, figs. 1 a, b	19
K. sculptilis, Meek sp., 1873, pl. i, figs. 7, 7 a, b; pl. ix, fig. 7	20-21
Leptæna, Dalman	22
L. melita, H. & W., 1877	22
Orthis, Dalman	22
O. eurekensis, n. s., Walcott, 1884, pl. ix, figs. 8, 8 a	22-23

Pteropoda.

Stenotheca, Salter	23
S. elongata, n. s., Walcott, 1884, pl. ix, figs. 2, 2 a	23
Hyolithes, Eichwald	23
H. primordialis, Hall sp., 1861	23-24

Pœcilopoda.

Agnostus, Brongniart	24
A. richmondensis, n. s., Walcott, 1884, pl. ix, fig. 10	24-25
A. seclusus, n. s., Walcott, 1884, pl. ix, fig. 14	25
A. bidens, Meek, 1873, pl. ix, figs. 13, 13 a	26, 27
A. communis, H. & W., 1877	27
A. neon, H. & W., 1877	27
A. prolongus, H. & W., 1877	28
Olenellus, Hall	28
O. iddingsi, n. s., Walcott, 1884, pl. ix, fig. 12	28
O. gilberti, Meek, MSS., 1874, pl. ix, figs. 16, 16 a ; pl. xxi, fig. 13	29
O. howelli, Meek, MSS., 1874, pl. ix, figs. 15, 15 a, b, and pl. xxi, figs. 1-9	30-31
Observations on *Olenellus howelli*, pl. xxi, figs. 1-7	32-39
Dicellocephalus, Owen	40
D. bilobatus, H. & W., 1877	40
D. oscrola, Hall, 1863, pl. ix, fig. 25	40
D. nasutus, n. s., Walcott, 1884, pl. x, fig. 15	40-41
D. richmondensis, n. s., Walcott, 1884, pl. x, fig. 7	41-42
D.? angustifrons, n. s., Walcott, 1884, pl. x, figs. 1, 1 a, b	42-43
D. iole, n. s., Walcott, 1884, pl. x, fig. 19	43-44
D. marica, n. s., Walcott, 1884, pl. x, fig. 13	44-45
D.? quadriceps, H. & W., 1879, pl. ix, fig. 24	45

	Page.
D. ? expansus, n. s., Walcott, 1884, pl. ix, fig. 19	45-46
Ptychoparia, Corda	46
P. ? prospectensis, n. s., Walcott, 1884, pl. ix, fig. 20	46-47
P. ? linnarssoni, n. s., Walcott, 1884, pl. ix, figs. 18, 18 a	47-48
P. (Solenopleura?) breviceps, n. s., Walcott, 1884, pl. x, fig. 9	49
P. ? pernasutus, n. s., Walcott, 1884, pl. x, figs. 8, 8 a, b	49-50
P. (Euloma?) dissimilis, n. s., Walcott, 1884, pl. ix, fig. 28	51
P. occidentalis, n. s., Walcott, 1884, pl. x, fig. 5	51-52
P. similis, n. s., Walcott, 1884, pl. x, fig. 10	52-53
P. similis var. *robustus*, n. var., Walcott, 1884, pl. i, figs. 9, 9 a	53
P. (Euloma?) affinis, n. s., Walcott, 1884, pl. x, fig. 12	54
P. læviceps, n. s., Walcott, 1884, pl. x, figs. 17, 18	54-55
P. oweni, M. & H., 1861, pl. x, figs. 3, 3 a	55-56
P. anytus, H. & W., 1877, pl. ix, fig. 26	56
P. granulosus, H. & W., 1877	57
P. hagueí, H. & W., 1877	57
P. nitidus, H. & W., 1877	57-58
P. unisulcatus, H. & W., 1877	58
Subgenus *Pterocephalus*, Roemer	58
P. (Pterocephalus) occidens, n. s., Walcott, 1884, pl. ix, fig. 21	58-59
P. (Pterocephalus) laticeps, H. & W., 1877	59
Anomocare, Angelin	59
A. ? parvum, n. s., Walcott, 1884, pl. ix, fig. 17	59-60
Ptychaspis, Hall	60
P. minuta, Whitfield ? 1878, pl. x, fig. 23	60-61
Chariocephalus, Hall	61
C. ? tumifrons, H. & W., 1877, pl. x, fig. 16	61
Agraulos, Corda	61
A. ? globosus, n. s., Walcott, 1884, pl. ix, fig. 23	61-62
Arethusiana, Barrande	62
A. americana, n. s., Walcott, 1884, pl. ix, fig. 27	62-63
Ogygia, Brongniart	63
O. ? spinosa, n. s., Walcott, 1884, pl. ix, fig. 22	63
O. ? problematica, n. s., Walcott, 1884, pl. x, figs. 2 a, b, and 4	63-64

FOSSILS OF THE LOWER SILURIAN.

Pogonip Group.

Rhizopoda.

Receptaculites Defrance	65
R. mammillaris, MSS., Newberry, 1880, pl. xi, fig. 11	65-66
R. elongatus, n. s., Walcott, 1884	66-67
R. ellipticus, n. s., Walcott, 1884, pl. xi, fig. 12	67

Brachiopoda.

Obolella, Billings	67
O. ? ambigua, n. s., Walcott, 1884, pl. i, figs. 2 a-c	67-68
Schizambon, n. g., Walcott, 1884	69-70
S. typicalis, n. s., Walcott, 1884, pl. i, figs. 3 a-d	70-71
Strophomena, Rafinesque	71
S. nemea, H. & W., 1877	71
Orthis, Dalman	72
O. perveta, Conrad, 1843, pl. xi, figs. 3 a, b	72
O. testudinaria, Dalman, 1827, pl. xi, figs. 10, 10 a	72-73
O. hamburgensis, n. s., Walcott, 1884, pl. xi, figs. 5, 5 a	73
O. iowensis, n. s., Walcott, 1884, pl. xi, figs. 6, 6 a	74
O. tricenaria, Conrad, 1843, pl. xi, figs. 4, 4 a	74-75
Streptorhynchus, King	75
S. minor, n. s., Walcott, 1884, pl. xi, fig. 9	75
Triplesia, Hall	75
T. calcifera, Billings, 1861, pl. xi, figs. 7, 8	75-76

Lamellibranchiata.

Tellinomya, Hall	76
T. contracta, Salter? 1859, pl. xi, figs. 15, 15 a	76
T. ? hamburgensis, n. s., Walcott, 1884, pl. xi, figs. 1, 1 a	76-77
Modiolopsis, Hall	77

	Page.
M. occidens, n. s., Walcott, 1884, pl. i, fig. 5, and pl. xi, figs. 14, 14 *a*	77-78
M. pogonipensis, n. s., Walcott, 1884, pl. i, fig. 6, and pl. xi, fig. 13	78
Gasteropoda.	
Raphistoma, Hall	78
R. nasoni, Hall, 1861, pl. xi, figs. 21, 21 *a*	78-79
Murchisonia, D'Archiac and De Verneuil	79
M. milleri, Hall? 1877, pl. i, figs. 12, 12 *a*, *b*	79-80
Pleurotomaria, Defrance	80
P. ionensis, n. s., Walcott, 1884, pl. xi, fig. 22	80
Helicotoma, Salter	81
Helicotoma, sp. ? Walcott, 1884	81
Maclurea, Le Sueur	81
M. annulata, n. s., Walcott, 1884, pl. xi, figs. 19, 19 *a*	81-82
M. subannulata, n. s., Walcott, 1884, pl. xi, figs. 18, 18 *a*, *b*	82
M. carinata, n. s., Walcott, 1884, pl. xi, figs. 20, 20 *a*	82-83
Maclurea, sp. ? Walcott, 1884	83
Metoptoma, Phillips	83
M. phillipsi, n. s., Walcott, 1884, pl. i, figs. 4, 4 *a*	83-84
M. ? analoga, n. s., Walcott, 1884, pl. i, figs. 11, 11 *a*	84
Cyrtolites, Conrad	84
C. sinuatus, H. & W., 1877	84
Pteropoda.	
Coleoprion, Sandberger	85
C. minuta, n. s., Walcott, 1884, pl. xi, figs. 17, 17 *a*, and pl. xii, fig. 21	85
Hyolithes, Eichwald	85
H. vanuxemi, n. s., Walcott, 1884, pl. xi, figs. 16, 16 *a*, *b*	85-86
Cephalopoda.	
Orthocerata, pl. xii, figs. 1, 1 *a–c*, 2, 3	86
Endoceras protciforme, Hall? 1847, pl. xii, figs. 1, 1 *a–c*	86
Orthoceras multicameratum, Hall? 1847, pl. xii, fig. 3	86
Orthoceras, sp. ? Walcott, 1884, pl. xii, fig. 2	86
Endoceras multitubulatum, Hall? 1847	87
Orthoceras, Walcott, 1884, pl. xii, fig. 1 *b* and figs. 1, 2, p. 87	87
Crustacea.	
Leperditia, Rouault	88
L. bivia, White, 1874	88
Beyrichia, McCoy	88
Beyrichia, sp. ? Walcott, 1884	88
Plumulites, Barrando	88
Pœcilopoda	89
Dicellocephalus, Owen	89
D. finalis, n. s., Walcott, 1884, pl. xii, figs. 12, 12 *a*, *b*	89-90
D. inexpectans, n. s., Walcott, 1884, pl. i, fig. 10	90
Ptychoparia, Corda	91
P. ? annectans, n. s., Walcott, 1884, pl. xii, fig. 18	91
Bathyurus, Billings	91
B. ? tuburculatus, n. s., Walcott, 1884, pl. xii, fig. 9	91-92
B. ? congeneris, n. s., Walcott, 1884, pl. xii, fig. 8	92-93
B. ? simillimus, n. s., Walcott, 1884, pl. xii, fig. 11	93
Cyphaspis, Burmeister	93
C.? brevimarginatus, n. s., Walcott, 1884, pl. xii, fig. 10	93-94
Amphion, Pander	94
A. nevadensis, n. s., Walcott, pl. xii, fig. 13	94
Ceraurus, Green	95
Ceraurus ——? Walcott, 1884, pl. xii, fig. 17	95
Symphysurus, Goldfuss	95
S.? goldfussi, n. s., Walcott, 1884, pl. xii, fig. 16	95
Barrandia, McCoy	96
B. mecovi, n. s., Walcott, 1884, pl. xii, fig. 5	96
Barrandia? sp. ? Walcott, 1884, pl. xii, fig. 6	96-97
Illænurus, Hall	97
I. eurekensis, n. s., Walcott, 1884, pl. xii, figs. 4, 4 *a*	97-98
Asaphus, Brongniart	98
A. caribouensis, n. s., Walcott, 1884, pl. xii, figs. 7, 7 *a*, *b*	98
A. ? curiosa, Billings, 1865, pl. xii, fig. 15	98

FOSSILS OF THE DEVONIAN.

	Page.
Porifera.	
Palæomanon, Roemer	99
P. roemeri, n. s., Walcott, 1884, pl. xiii, fig. 12	99
Astylospongia, Roemer	99
Astylospongia, sp. ? Walcott, 1884	99
Stromatopora, Goldfuss	100
Actinozoa.	
Favosites hemisphericа, Yandell & Shumard, 1876	100
Favosites basaltica, Goldfuss, 1829	100–101
Favosites, n. sp., Walcott, 1884	101
Fistulipora, sp. ? Walcott, 1884	101
Alveolites rockfordensis, Hall ? 1864	102
Cladopora pulchra, Rominger? 1876	102
Cladopora, sp. nndt., Walcott, 1884	102
Thecia ramosa, Rominger ? 1876	102–103
Syringopora hisingeri, Billings, 1850	103
Syringopora pereleganѕ, Billings, 1859	103
Aulopora serpens, Goldfuss?	103
Cyathophyllum corniculum, Milne-Edwards ?	104
Cyathophyllum rugosum, Edwards and Haime, 1876	104
Cyathophyllum davidsoni, Milne-Edwards, 1876	105
Cyathophyllum, n. s., Walcott, 1884	104–105
Cyathophyllum, n. s., Walcott, 1884	105
Acervularia pentagona, Goldfuss, 1877	105
Pachyphyllum woodmani (White), II. & W., 1864	105
Diphyphyllum simcoense, Billings, 1876	105–106
Cystiphyllum americanum, Milne-Edwards, 1876	106
Cystiphyllum, n. s., Walcott, 1884	106
Brachiopoda.	
Lingula, Bruguière	106
L. lena, Hall, 1867, pl. xiii, fig. 2	106–107
L. ligea, Hall, 1860, pl. ii, fig. 2	107
L. ligea, var. *nevadensis*, n. var., Walcott, 1884, pl. ii, fig. 3	107
L. alba-pinensis, n. s., Walcott, 1884, pl. ii, figs. 1, 1 a	108
L. lonensis, n. s., Walcott, 1884, pl. xiii, figs. 1, 1 a	108–109
L. whitei, n. s., Walcott, 1884, pl. xiii, fig. 3	109–111
Discina, Lamarck	112
D. minuta, Hall, 1843, pl. xiii, fig. 5	112
Discina, sp. ? Walcott, 1884	112
D. lodensis, Hall, 1843, pl. ii, figs. 5, 5 a	112–113
Pholidops, Hall	113
P. bellula, n. s., Walcott, 1884, pl. ii, figs. 6, 6 a, b	113–114
P. quadrangularis, n. s., Walcott, 1884, pl. ii, fig. 7	114
Orthis, Dalman	114
O. mcfarlanei, Meek, 1868	114
O. impressa, Hall, 1843, pl. xiii, fig. 13	115
O. tulliensis, Vanuxem, 1842, pl. ii, figs. 12, 12 a	115–116
Skenidium, Hall	116
S. devonicum, n. s., Walcott, 1884, pl. xiii, figs. 4, 4 a	116
Streptorhynchus, King	117
S. chemungensis, Conrad sp., 1842, pl. xiii, figs. 7 and 16	117–118
Strophomena, Blainville	118
S. rhomboidalis, Wilckens, sp	118
Strophodonta, Hall	118
S. demissa, Conrad, sp., 1842, pl. ii, figs. 9, 9 a, b	118–119
S. patersoni, Hall, 1857	119
S. inequiradiata, Hall, 1857, pl. xi, figs. 11, 11 a	120
S. perplana, Conrad, sp., 1842, pl. xiii, fig. 11	120–121
S. punctulifera, Conrad, sp., pl. xiii, fig. 10	121
S. arcuata, Hall ? 1858	121
S. calvini, Miller, 1883, pl. xiii, fig. 6	122
Chonetes, Fischer	122–123
C. hemispherica, Hall, 1857	123
C. deflecta, Hall, 1857, pl. ii, figs. 8, 8 a, b	124

	Page.
C. mucronata, Hall?, 1843	124-125
C. setigera, Hall, 1843	125
C. macrostriata, n. s., Walcott, 1884, pl. ii, fig. 13, pl. xiii, figs. 14, 14 a-c	126-127
C. filistriata, n. s., Walcott, 1884, pl. xiii, figs. 15, 15 a	127-128
Productus, Sowerby	128
Subgenus Productella, Hall, 1867	128
P. (Productella) subaculeatus, Murch, pl. vii, fig. 2, pl. xiii, figs. 19, 19 a, 20, 20 a	128-129
P. (Productella) shumardianus, Hall, 1858, pl. xiv, fig, 1	129-130
P. (Productella) hallanus, n. s., Walcott, 1884, pl. xiii, figs. 17, 17 a	130-131
P. (Productella) navicella, Hall, 1857, pl. xiii, fig. 9	131
P. (Productella) truncatus, Hall, 1857, pl. xiv, fig. 2	131-132
P. (Productella) lachrymosus var. linus, Conrad, sp., 1842, pl. xiii, figs. 18, 18 a	132
P. (Productella) lachrymosus var. stigmatus, Hall, 1867	132-133
P. (Productella) speciosus, Hall, 1857, pl. xiii, fig. 8	133
P. hirsutiforme, n. s., Walcott, 1884, pl. ii, figs. 10, 10 a	133-134
Spirifera, Sowerby	134
S. disjuncta, Sowerby, 1840	134-135
S. raricosta, Conrad, sp., 1842, pl. iv, figs. 2, 2 a, pl. xiv, fig. 12	135
S. varicosa, Hall, 1857	136
S. parryana, Hall ?, 1858, pl. xiv, fig 10	137
Spirifera (sp. undt.), Walcott, 1884	137
S. englemanni, Meek, 1860	138
S. pinonensis, Meek, 1870, pl. iv, figs. 1, 1 a-f	138
Subgenus Martinia, McCoy	139
S. (Martinia) glabra, Martin	139
S. (Martinia) glabra var. nevadensis, n. var. Walcott, 1884, pl. iii, fig. 5, pl. xiv, figs. 14, 14 a, b	139-140
S. (Martinia) maia, Billings, 1860, pl. iii, figs. 1, 1 a-e, pl. xiv. figs. 13, 13 a	141-142
S. (Martinia) undifera, Roemer, 1844, pl. iii, figs. 3, 3 a, b; 6, 6 a, pl. xiv, figs. 11, 11 a, b.	143-145
Cyrtina, Davidson	146
C. davidsoni, n. s., Walcott, 1884, pl. iii, figs. 2, 2 a-e	146-147
C. hamiltonensis, Hall, 1857	147
Nucleospira, Hall	147
N. concinni, Hall, 1843	147
Athyris, McCoy	148
A. angelica, Hall, 1861	148
Athyris (sp. undt.), Walcott, 1884	148
Meristella, Hall	148
Whitfieldia, Davidson	148
M. (Whitfieldia) nasuta, Conrad, sp., 1842, pl. iii, fig. 8, 8 a, b	148-149
Atrypa, Dalman	150
A. reticularis (Linnæus, sp.), Dalman, pl. xiv, figs. 6, 6 a, b	150
A. desquamata, Sowerby, pl. xiv, figs. 4, 4 a	150-151
Trematospira, Hall	151
T. infrequens, Hall, n. s., Walcott, 1884, pl. iv, figs. 3, 3 a, b	151
Rhynchonella, Fischer	152
R. horsfordi, Hall, 1860, pl. xiv, fig. 3; pl. xv, fig. 6	152
R. tethys, Billings, 1860	152
R.? occidens, n. s., Walcott, 1884, pl. xv, figs. 3, 3 a, b	152-153
R. castanea, Meek, 1868, pl. xv, figs. 1, 1 a; 4, 4 a	153-155
R. duplicata, Hall, 1843, pl. xiv, fig. 8	155
R. pugnus, Martin, 1809, pl. xiv, figs. 7, 7 a	155-157
R. emmonsi, H. & W., 1877	157
Subgenus Leiorhynchus, Hall	157
R. (Leiorhynchus) nevadensis, n. s., Walcott, 1884, pl. xiv, figs. 9, 9 a, b	157-158
R. (Leiorhynchus) sinuatus, Hall, 1867, pl. xiv. fig. 5	158-159
R. (Leiorhynchus) laura, Billings, 1860	159
Pentamerus, Sowerby	159
Subgenus Gypidula Hall	159
P. comis, Owen, 1852, pl. iii, figs. 4, 7; pl. xiv, figs. 15, 15 a, b; pl. xv, figs. 5, 5 a, b...	159-161
P. lotis, n. s., Walcott, 1884, pl. iii, figs. 9, 9 a-c	161-162
Cryptonella, Hall	163
C.? circula, n. s., Walcott, 1884, pl. xv, figs. 2, 2 a, b	163
C. pinonensis, n. s., Walcott, 1884, pl. iv, figs. 4, 4 a, b	163-164

Lamellibranchiata.

	Page.
Pterinea, Goldfuss	165
P. flabella, Conrad, sp., 1842, pl. xv, fig. 12; pl. v, fig. 6	165
P. newarkensis, n. s., Walcott, 1884, pl. v, fig. 12	165–166
Actinopteria, Hall	166
A. boydi, Conrad, sp., 1842, pl. v, fig. 2	166
Leiopteria, Hall, 1883, pl. v, figs. 10, 10 a	166
L. rafinesquei, Hall, 1883, pl. v, figs. 10, 10 a	166
Leptodesma, Hall	167
L. transversa, n. s., Walcott, 1884, pl. v, fig. 13	167
Limoptera, Hall	167
L. sarmenticia, n. s., Walcott, 1884, pl. v, figs. 3, 3 a, b	167–168
Mytilarca, Hall	168
M. dubia, n. s., Walcott, 1884, pl. iv, fig. 5	168
M. chemungensis, Conrad, 1842, pl. iv, fig. 9	168–169
Subgenus *Plethomytilus*, Hall	169
M. (Plethomytilus) oviformis, Conrad, 1842, pl. v, fig. 11	169
Modiomorpha, Hall	169
M. altiforme, n. s., Walcott, 1884, pl. v, fig. 9	169–170
M. oblonga, n. s., Walcott, 1884, pl. v, fig. 7	170
M. obtusa, n. s., Walcott, 1884, pl. iv, figs. 8, 8 a	171
Goniophora, Phillips	171
G. perangulata, Hall, 1870, pl. xv, fig. 10	171
Nucula, Lamarck	172
N. rescuensis, n. s., Walcott, 1884, pl. xv, fig. 9	172
Nucula, sp. ? Walcott, 1884	172
Dystactella, Hall	172
D. insularis, n. s., Walcott, 1884, pl. xv, fig. 8	172, 173
Megambonia, Hall	173
M. occidualis, n. s., Walcott, 1884, pl. v, fig. 1	173
Nyassa, Hall	173
N. parva, n. s., Walcott, 1884, pl. xv, figs. 14, 14 a	173, 174
Grammysia, De Verneuil	174
G. minor, n. s., Walcott, 1884, pl. xv, figs. 15, 15 a	174–175
Sanguinolites, McCoy	175
S. ? combensis, n. s., Walcott, 1884, pl. xv, fig. 16	175
S. ? gracilis, n. s., Walcott, 1884, pl. iv, fig. 10	175–176
S. rigidus, White & Whitfield, 1862, pl. xvi, fig. 6	176
S. ? sanduskyensis, Meek, 1871, pl. v, fig. 4	176–177
S. ventricosus, White & Whitfield, sp., 1862, pl. xv, fig. 13	177
Conocardium, Brown	177
C. neradensis, n. s., Walcott, 1884, pl. xvi, figs. 4, 4 a	177–178
Paracyclas, Hall	178
P. occidentalis, H. & W., 1872	178
Posidonomya, Bronn	178
P. levis, n. s., Walcott, 1884, pl. iv, fig. 6	178–179
P. devonica, n. s., Walcott, 1884, pl. iv, fig. 7	179–180
Microdon, Conrad	180
Subgenus *Cypricardella*, Hall	180
M. (Cypricardella) macrostriatus, n. s., Walcott, 1884, pl. v, fig. 5	180
Anodontopsis, McCoy	180
A. amygdaleformis, n. s., Walcott, 1884, pl. xv, figs. 7, 7 a, b	180–181
Schizodus, King	181
Subgenus *Cytherodon*, Hall	181
S. (Cytherodon) orbicularis, n. s., Walcott, 1884, pl. v, figs. 8, 8 a	181
Cypricardinia, Hall	182
C. indenta, Conrad, sp., 1842, pl. v, fig. 14; pl. xv, fig. 11	182
Gasteropoda.	
Platyceras, Conrad	182
P. conradi, n. s., Walcott, 1884, pl. xvi, figs. 1, 1 a	182–183
P. nodosum, Conrad, 1841, pl. vi, figs. 5, 5 a, b	183
P. undulatum, n. s., Walcott, 1884, pl. vi, figs. 2, 2 a	184
P. thetiforme, n. s., Walcott, 1884, pl. vi, figs. 4, 4 a, b	184
Platyostoma, Conrad	185
P. lineatum, Conrad, 1842	185

	Page.
Euomphalus, Sowerby	185
E. eurekensis, n. s., Walcott, 1884, pl. xvi, figs. 2, 2 a	185-186
E. (Phanerotinus) laxus, Hall, 1861, pl. vi, fig. 3	186
Ecculiomphalus, Portlock	187
E. devonicus, n. s., Walcott, 1884, pl. vi, figs. 6, 6 a	187
Straparollus, Montfort	187
S. newarkensis, n. s., Walcott, 1884, pl. xvi, figs. 7, 7 a	187-188
Platyschisma, McCoy	188
P. ? meropi, n. s , Walcott, 1884, pl. xvii, figs. 1, 1 a-c	188
P. ? ambiguum, n. s., Walcott, 1884, pl. xvii, figs. 3, 3 a	188-189
Callonema, Hall	189
C. occidentalis, n. s., Walcott, 1884, pl. xvi, figs. 3, 3 a	189
Loxonema, Phillips	190
L. eurekensis, n. s., Walcott, 1884, pl. xvi, fig. 8	190
L. nobile, n. s., Walcott, 1884, pl. xvi, fig. 9	190-191
L. ? subattenuatum, Hall ? 1861	191
L. approximatum, n. s., Walcott, 1884, pl. vi, fig. 7	191-192
Loxonema ? sp. undt., Walcott, 1884	192
Loxonema, sp. undt., Walcott, 1884	192
Bellerophon, Montfort	192
B. perplexa, n. s., Walcott, 1884, pl. xvii, figs. 6, 6 a, b	193
B. combsi, n. s., Walcott, 1884, pl. xvii, figs. 9, 9 a, b	193-194
B. lyra, Hall	194
B. leda, Hall	194
B. mera, Hall ?	194
B. pelops, Hall ?	194
Scoliostoma, Braun	195
S. americana, n. s., Walcott, 1884, pl. vi, figs. 1, 1 a-e	195
Metoptoma, Phillips	195
M. ? devonica, n. s., Walcott, 1884, pl. xvii, figs. 2, 2 a	195-196
Pteropoda.	
Tentaculites, Schlotheim	196
T. gracilistriatus, Hall	196-197
T. scalariformis, Hall	197
T. attenuatus, Hall	197
T. bellulus, Hall ?	197
Styliola, Le Sueur	197
S. fissurella, Hall	197
S. fissurella var. *intermittens*, Hall	197-198
Conularia, Miller	198
Conularia, sp. undt., Walcott, 1884	198
Hyolithes, Eichwald	199
Hyolithes, sp. (?), Walcott, 1884, pl. vi, figs. 8, 8 a	199
Coleolus, Hall	199
C. lævis, n. s., Walcott, 1884, pl. vi, fig. 9	199-200
Cephalopoda.	
Orthoceras, Breynius	200-202
Gomphoceras, Sowerby	202
G. suboriforme, n. s., Walcott, 1884, pl. xvii, figs. 8, 8 a	202-203
Cyrtoceras, Goldfuss	203
C. nevadense, n. s., Walcott, 1884, pl. xvii, figs. 7, 7 a	203
Goniatites, De Haan	203
G. desideratus, n. s., Walcott, 1884, pl. xvii, fig. 10	203-204
Crustacea.	
Beyrichia, McCoy	204
B. (Primitia) occidentalis, n. s , Walcott, 1884, pl. xvii, figs. 4, 4 a	204-206
Leperditia, Rouault	206
L. rotundata, n. s., Walcott, 1884, pl. xvi, fig. 5	206
Pœcilopoda.	
Phacops, Emmerich	207
P. rana, Green, sp., 1832	207
Dalmanites, Emmerich	207
D. merki, n. s., Walcott, 1884, pl. xvii, figs. 5, 5 a-e	207-209
Dalmanites, undt. sp., Walcott, 1884	210
Proetus, Steininger	210

	Page.
P. haldemani, Hall, 1861	210
P. marginalis, Conrad, sp., 1839	210-211
Phillipsia, Portlock	211
P. coronata, Hall ? 1876	211
Supposed eggs of the Trilobite	211

FOSSILS OF THE CARBONIFEROUS.

Echinodermata.

Archæocidaris, McCoy	212
Archæocidaris, sp., ? Walcott, 1884	212-213

Brachiopoda.

Discina, Lamarck	213
D. newberryi, Hall, 1863, pl. xviii, figs. 3, 2, 2 a ?	213
D. nitida, Phillips, 1836, pl. vii, figs. 4, 4 a	213-214
D. connata, n. s., Walcott, 1884, pl. vii, fig. 3, 3 a	214
Productus, Sowerby	214
P. subaculeatus, Murch., pl. vii, fig. 2	214-215
Spirifera, Sowerby	215
S. trigonalis, Martin, sp., 1809, pl. xviii, fig. 11	215-216
S. leidyi, N. & P., 1855, pl. xviii, figs. 4, 4 a	216
S. annectans, n. s., Walcott, 1884, pl. xviii, figs. 7, 7 a	216-217
S. neglecta, Hall, 1858, pl. xviii, fig. 10	217
S. desiderata, n. s., Walcott, 1884, pl. vii, fig. 8	217-218
Subgenus *Spiriferina*, D'Orbigny	218
S. (Spiriferina) cristata, Schlotheim, 1816, pl. xviii, figs. 12, 13	218-219
Syringothyris, Winchell	219
S. cuspidata, Martin, sp., 1796	219-220
Retzia, King	220
R. radialis, Phillips, sp., 1836, pl. vii, figs. 5, 5 a–h	220-222
Athyris, McCoy	222
A. hirsuta, Hall, 1857, pl. xviii, fig. 5	222
Rhynchonella, Fischer	223
R. eurekensis, n. s., Walcott, 1884, pl. xviii, figs. 8, 8 a–c	223
R. thera, n. s., Walcott, 1884, pl. vii, figs. 6, 6 a–c	223-224
Camarophoria, King	224
C. cooperensis, Shumard, 1855, pl. xviii, fig. 6	224
Terebratula, Llhwyd	224
T. hastata, Sowerby	224

Lamellibranchiata.

Aviculopecten, McCoy	226
A. haguei, n. s., Walcott, 1884, pl. xix, fig. 4	226-227
A. eurekensis, n. s., Walcott, 1884, pl. xix, figs. 2, 3	227
A. peroccidens, n. s., Walcott, 1884, pl. viii, fig. 8	227-228
A. pintöensis, n. s., Walcott, 1884, pl. viii, fig. 6	228
A. affinis, n. s., Walcott, 1884, pl. xix, figs. 1, 1 a	229-230
Aviculopecten, sp. ? Walcott, 1884	230
Streblopteria, McCoy	230
S. similis, n. s., Walcott, 1884, pl. viii, figs. 4, 4 a–d, and pl. xix, fig. 7	230-231
Crenipecten, Hall	231
C. hallanus, n. s., Walcott, 1884, pl. viii, figs. 7, 7 a–c	231-232
Pterinopecten, Hall	232
P. hoosacensis, n. s., Walcott, 1884, pl. viii, fig. 9	232-233
P. spio, n. s., Walcott, 1884, pl. viii, figs. 1, 1 a	233
Pterinea, Goldfuss	234
P. pintöensis, n. s., Walcott, 1884, pl. xix, fig. 10	234
Leptodesma, Hall	234
Leptodesma, sp. ? Walcott, 1884	234
Ptychopteria, Hall	235
P. protuforme, n. s., Walcott, 1884, pl. viii, fig. 5	235
Pinna, Linnæus	235
P. inexpectans, n. s., Walcott, 1884, pl. xix, fig. 11	235-236
P. consimilis, n. s., Walcott, 1884, pl. xx, fig. 13	236
Myalina, De Koninck	237
M. congeneris, n. s., Walcott, 1884, pl. xix, fig. 6, and pl. xxii, fig. 10	237
M. nemesis, n. s., Walcott, 1884, pl. xix, fig. 5, and pl. xxii, fig. 7	237-238
M. nessus, n. s., Walcott, 1884, pl. xxii, figs. 8, 8 a	238

	Page.
Modiola, Lamarck	239
M. ? nevadensis, n. s., Walcott, 1884, pl. xix, fig. 8	239
Modiomorpha, Hall	239
M. ambigua, n. s., Walcott, 1884, pl. xx, fig. 1	239–240
M. ? desiderata, n. s., Walcott, 1884, pl. xx, fig. 3	240
M. ? pintoensis, n. s., Walcott, 1884, pl. xx, fig. 2	240–241
Nucula, Lamarck	241
N. insularis, n. s., Walcott, 1884, pl. xx, fig. 14	240
N. levatiforme, n. s., Walcott, 1884, pl. xxii, figs. 1, 1 *a*	241–242
Solenomya, Lamarck	242
S. curta, n. s., Walcott, 1884, pl. xxii, figs. 6, 11	242
Macrodon, Lycett	243
M. hamiltonæ, Hall, 1870, pl. xxiii, figs. 5, 5 *a–c*	243
M. truncatus, n. s., Walcott, 1884, pl. viii, fig. 2	243–244
Grammysia, De Verneuil	244
G. hannibalensis, Shumard, sp., 1855, pl. xx, fig. 4	244
G. arcuata, Conrad, sp., 1841, pl. xx, fig. 5	245
Edmondia, De Koninck	245
E. medon, n. s., Walcott, 1884, pl. xxiii, fig. 6	245
E. ? circularis, n. s., Walcott, 1884, pl. xxii, fig. 9	246
Pleurophorus, King	246
P. meeki, n. s., Walcott, 1884, pl. viii, fig. 3	246
Sanguinolites, McCoy	247
S. æolus, H. & W., 1870, pl. xx, figs. 6, 7, 9	247
S. retusus, n. s., Walcott, 1884, pl. xx, fig. 10	247–248
S. simplex, n. s., Walcott, 1884, pl. xx, fig. 11	248
S. salteri, n. s., Walcott, 1884, pl. xx, fig. 12	248–249
S. ? nuenia, n. s., Walcott, 1884, pl. xix, fig. 9	249
S. striatus, n. s., Walcott, 1884, pl. xxiii, fig. 7	249–250
Microdon, Conrad	250
Subgenus *Cypricardella*, Hall	250
M. (Cypricardella) connatus, n. s., Walcott, 1884, pl. xxiv, figs. 5, 5 *a*	250–251
Cardiola, Broderip	251
C. ? filicostata, n. s., Walcott, 1884, pl. xxii, figs. 4, 4 *a*	251
Schizodus, King	252
S. cuneatus, Meek, 1875, pl. xx, fig. 8	252
S. deparcus, n. s., Walcott, 1884, pl. xxii, fig. 5	252
S. curtiforme, n. s., Walcott, 1884, pl. xxii, figs 3, 3 *a*	253
S. pintoensis, n. s., Walcott, 1884, pl. xxii, figs. 2, 2 *a*	253–254
Gasteropoda.	
Platyceras, Conrad	254
P. occidens, n. s., Walcott, 1884, pl. xxiv, figs. 9, 9 *a*	254
P. piso, n. s., Walcott, 1884, pl. xxiv, figs. 7, 7 *a, b*	254–255
Platyostoma, Conrad	255
P. inornatum, n. s., Walcott, 1884, pl. xxiv, figs. 3, 3 *a*	255
Euomphalus, Sowerby	255
E. (Straparollus) subrugosus, M. & W., 1873, pl. xviii, fig. 19	255–256
Bellerophon, Montfort	256
B. majusculus, n. s., Walcott, 1884, pl. xxiii, figs. 1, 1 *a*; pl. xxiv, fig. 6, and fig. 3, p. 257.	256–257
B. textilis, Hall, ? 1877, pl. xviii, fig. 18	257–258
Loxonema, Phillips	258
L. bella, n. s., Walcott, 1884, pl. xxiv, figs. 1, 1 *a*	258–259
Pleurotomaria, Defrance	259
P. nodomarginata, McChesney, 1860, pl. xviii, fig. 15	259
P. nevadensis, n. s., Walcott, 1884, pl. xxiv, figs. 2, 2 *a*	259–260
Macrocheilus, Phillips	260
Macrocheilus, sp., ? Walcott, 1884, pl. xxiv, fig. 8	260
Metoptoma, Phillips	260
M. peroccidens, n. s., Walcott, 1844, pl. xviii, fig. 16	260
Ampullaria, Lamarck	261
A. ? powelli, Walcott, 1883, figs. 4, 5, p. 261	261
Pulmonifera.	
Physa, Draparnaud	262
P. prisca, Walcott, 1883, fig. 6, p. 262	262
Zaptychius, Walcott, 1883	263
Z. carbonaria, Walcott, 1883, fig. 7, p. 263	263

Pteropoda.   Page.
   Conularia, Miller 264
   C. missouriensis, Swallow, ? 1860, pl. xxiii, fig. 4 264
   Hyolithes, Eichwald 264
   H. carbonaria, n. s., Walcott, 1884, pl. xxiii, fig. 3 264
Cephalopoda.
   Orthoceras, Breynius 265
   O. randolphensis, Worthen, ? 1882, pl. xviii. fig. 17 265
   O. eurekensis, n. s., Walcott, 1884, pl. xxiii, figs 2, 2 *a* 265–266
   Orthoceras, sp., ? Walcott, 1884 266
Pœcilopoda.
   Griffithides, Portlock 266
   G. portlocki, M. & W., 1865, pl. xxiv, figs 4, 4 *a*, *b* 266–267
   Systematic list of fossils of each geologic formation 268–281
   Palæozoic section in Central Nevada 281–283

24.

WALCOTT, C. D.—On the Cambrian Faunas of North America; Preliminary studies. <Bull. U. S. Geol. Surv. No. 10, pp. 1-74, pl. i-x. Washington, 1884.

This contains three parts. The first is a " Review of the fauna of the Saint John formation, contained in the Hartt collection." Mr. Matthew proposed the specific names for the new species excepting for one, *Harttia matthewi*, the type of the n. g. *Harttia*, Walcott. The author does not accept the genus *Conocephalites* and refers its different species to some of *Ptychoparia* and one of *Conocoryphe*. The following species are mentioned:

Page.
Eocystites, Billings 14, 294
E. primævus, Billings, 1868, pl. i, fig. 2 .... 14–15, 294–295
Lingula, Bruguière 15, 295
L. ? dawsoni, n. s., Matthew, MSS., 1884, pl. v, fig. 8 .... 15, 295
Acrothele, Linnarson 15, 295
A. matthewi, Hartt, sp., 1868, pl. i, figs. 4, 4 *a* 15–16, 295–296
Obolella, Billings 16, 296
O. transversa, Hartt, 1868, pl. i, figs. 5, 5 *a* 16, 299
Obolella, sp., undt., Walcott, 1884 .... 16–17, 296–297
Orthis, Dalman 17, 297
O. billingsi, Hartt, 1868, pl. i, figs 1, 1 *b–d* 17–18, 297–298
Orthis, ? sp., Walcott, 1884, pl. i, fig. 1 *a* 18, 298
Harttia, n. g., Walcott, 1884 .... 18–19, 298–299
H. matthewi, n. s., Walcott, 1884, pl. i, fig. 3 .... 19, 299
Palæacmea, H. and W. (*Stenotheca*) 19, 299
Stenotheca acadica, Hartt, sp., 1868, pl. i, fig. 6 .... 19, 299
Hyolithes, Eichwald 20, 300
H. acadica, n. s., Hartt, sp., MSS., 1884, pl. ii, fig. 5 .... 20, 300
H. danianus, n. s., Matthew, MSS., 1884, pl. ii, figs. 7 *a*, *b* 20–21, 300–301
H. micmac, n. s., Matthew, MSS., 1884, pl. ii, fig. 6 .... 21–22, 301–302
Agnostus, Brongniart 22, 302
A. acadicus, Hartt, 1868, pl. ii, figs. 2 2 *a–c* 22–23, 302–303
Microdiscus, Emmons 23, 303
M. dawsoni, Hartt, 1868, pl. ii, figs. 3, 3 *a* 23, 303
M. punctatus, Salter, 1864, pl. ii, figs. 1, 1 *a–c* 24–25, 303–304
Paradoxides, Brongniart 25, 305
P. lamellatus, Hartt, 1868, pl. iii, figs. 2, 2 *a* 25, 305
P. acadicus, Matthew, 1882, pl. iii, figs. 3, 3 *a* 25–27, 305–307
P. eteminicus, Matthew, 1883, pl. iii, figs. 1, 1 *a–g* 27, 307
Conocoryphe, Corda 28, 308
C. (Subgenus ?), matthewi, Hartt, sp., 1868, pl. iv, figs. 1, 1 *a*, *b* 28–30, 308–310
C. walcotti, Matthew (in lit.), 1884 .... 30–31, 310–311
Bailiella (new subgenus), Matthew, 1884 .... 31–32, 311–312
C. (Bailiella), baileyi, Hartt, sp., 1868, pl. iv, figs. 3, 3 *a*; pl. v, figs. 7, 7 *a* ... 32–33, 312–313
C. elegans, Hartt, sp., 1868, pl. iv, figs. 2, 2 *a*, *b* 33–34, 313–314
Ptychoparia, Corda 34–36, 314–316
P. robbi, Hartt, sp., 1868, pl. vi, figs 1, 1 *a* 36–37, 316–317
P. ouangondiana, Hartt, sp., 1868, pl. v, figs. 4, 4 *a–f* 37–38, 317–318

	Page.
P. ouangondiana var. *aurora*, Hartt, sp., 1868, pl. v, fig. 5	38–39, 318–319
P. quadrata, Hartt, sp., 1868, pl. v, fig. 1	39, 319
P. orestes, Hartt, sp., 1868, pl. v, figs. 3, 3 *a*	39–40, 319–320
P. orestes var. *thersites*, Hartt, sp., 1868, pl. v, fig. 2	40–41, 320–321
P. tener, Hartt, sp., 1868, pl. v, figs. 6, 6 *a*, *b*	41–42, 321–322

The second part is on the "Fauna of the Braintree argillites." The author doubts the specific difference between *Paradoxides harlani* and *P. benetti*. He describes the following forms:

	Page.
Hyolithes, Eichwald	44, 324
H. shaleri, n. s., Walcott, 1884, pl. vii, figs. 4, 4 *a–c*	44–45, 324–325
Paradoxides, Brongniart	45, 325
P. harlani, Green, 1834, pl. vii, fig. 3; pl. viii, figs. 1, 1 *a–e*; pl. ix, fig. 1	45–47, 325–327
Ptychoparia, Corda	47, 327
P. rogersi, n. s., Walcott, 1884, pl. vii, fig. 2	47–48, 327–328
Agraulos, Corda	48, 328
A. quadrangularis, Whitfield, sp., 1884, pl. vii, fig. 1	48–49, 328–329

The third part is "On a new genus and species of Phyllopoda from the middle Cambrian."

	Page.
Protocaris, n. g., Walcott, 1884	50, 330
P. marshi, n. s., Walcott, 1884, pl. x, fig. 1	50–51, 350–351

25.

WALCOTT, C. D. Paleontologic Notes. <Amer. Jour. Sci., 3d ser., vol. xxix, February, pp. 114–117, pl. on p. 116. New Haven, 1885.

OBOLIDÆ.

	Page
Linnarssonia, n. g., Walcott, 1885	115
Obolella chromatica, figs. 1, 2, p. 116	116
Linnarssonia transversa, Hartt, sp., figs. 3, 4, p. 116	116
L. sagittatis, figs. 5–8, p. 116	116
L. transversa, figs. 6, 7, p. 116	116

26.

WALCOTT, C. D. Paleozoic Notes; New Genus of Cambrian Trilobites, Mesonacis. <Amer. Jour. Sci., 3d ser., vol. xxix, April, pp. 328–330, figs. 2. New Haven, 1885.

	Page.
Mesonacis, n. g., Walcott, 1885, figs. 1, 2, p. 329	328–330

27.

WALCOTT, C. D. Note on some Paleozoic Pteropods. <Amer. Journ. Sci., 3d ser., vol. xxx, July, pp. 17–21, figs. 1–6. New Haven, 1885.

	Page.
Matthevia, n. g., Walcott, 1885	17–18
M. variabilis, n. s., Walcott, 1885, figs. 1–6, p. 20	18–19
Note on Hyolithes (Camarotheca) emmonsi, Ford	19–21

PART IV.

PUBLICATIONS BASED UPON THE PALEONTOLOGICAL COLLECTIONS OF THE UNITED STATES GOVERNMENT

BY

JACOB WHITMAN BAILEY,
TIMOTHY ABBOTT CONRAD,
JAMES DWIGHT DANA,
CHRISTIAN GOTTFRIED EHRENBERG,
JAMES HALL,
ANGELO HEILPRIN,
ALPHEUS HYATT,
JULES MARCOU,

JOHN STRONG NEWBERRY,
I. N. NICOLLET,
DAVID DALE OWEN,
HIRAM A. PROUT,
JAMES SCHIEL,
BENJAMIN F. SHUMARD,
ROBERT PARR WHITFIELD.

I.—THE WRITINGS OF JACOB WHITMAN BAILEY.

1.

BAILEY, J. W. [Descriptions of fossil fresh-water infusoria from Oregon.] <Rep. Expl. Exp. to the Rocky Mountains and to Oregon and North California, by J. C. Frémont. Appendix A. Geological formations by James Hall, p. 302, pl. v. Washington, 1845.

	Page.
Eunotia librile, Ehrenberg, pl. v, figs. 1, 2, 3	302
Eunotia gibba, Ehr., pl. v, figs. 4, 5	302
Pinnularia pachyptera?, Ehr., pl. v, figs. 6	302
Cocconema cymbiforme?, Ehr., pl. v, figs. 7, 8, 9	302
Gomphonema clavatum?, Ehr., pl. v, figs. 10, 11	302
Gomphonema minutissimum, Ehr., pl. v, fig. 12	302
Gallionella ——, n. s., J. W. Bailey, 1845, pl. v, figs. 13, 14, 15	302
Gallionella distans?, J. W. Bailey, pl. v, fig. 16	302
Cocconeis prætexta, Ehr., pl. v, figs. 17, 18	302
Fragillaria, J. W. Bailey, 1845, pl. v, fig. 19	302
Surirella, J. W. Bailey, 1845, pl. v, fig. 20	302
Fragillaria rhabdosoma?, J. W. Bailey, 1845, pl. v, fig. 21	302
Spiculæ of fresh-water sponges, pl. v, figs. 22, 23	302
——?, pl. v, fig. 24	302
Scale = 10-100ths of millimeter magnified equally with drawings, pl. v, fig. 25	302

2.

BAILEY, J. W. Letter upon Infusorial Fossils submitted to him by Dr. Schiel. <Rep. Expls. and Survs. from the Mississippi River to the Pacific Ocean. Report of Expls. for a route for the Pacific Railroad of the line of the forty-first parallel of north latitude, by Lieut. E. G. Beckwith, 1854. Vol. ii, chap. x, pp. 111, 112, pl. iii (pars.). Washington, 1855.

	Page.
Epithemia, pl. iii, figs. 5, 6	111
Cocconema asperum, Ehr	111
Cocconema cymbiforme, Ehr	111
Discoplea atmospherica, Ehr	111
Surirella campylodiscus?, Ehr., pl. iii, fig. 4	111
Cocconeis, pl. iii, fig. 3	111
Cymbella gibba, n. s., Bailey, 1855, pl. iii, fig. 1	111
Cymatopleura? campylodiscus, n. s., Bailey, 1855, pl. iii, figs. 2 *a*, *b*	111
Galleinella, pl. iii, figs. 7 *a*, *b*	112
Galleinella varians	112
Stephanodiscus ——?	112
Pennularia nobilis, Ehr.	112
Pennularia viridis	112
Epithemia	112
Surirella splendida (? Ehr.)	112
Pollen of pine	112
Spongiolites	112

II.—THE WRITINGS OF TIMOTHY ABBOTT CONRAD.

1.

CONRAD, T. A. Observations on a portion of the Atlantic Tertiary region, with a description of New Species of organic remains. <2d Bull. Proc. Natl. Institution, pp. 171-194, pls. i and ii. Washington, 1842.

	Page.
Section of the cliff at Claiborne	174
Classification of Tertiary formations	176
Table of Atlantic supracretaceous deposits	177
Geographical range of Lower Tertiary	178
Geographical range of Medial Tertiary	179
Localities of the Upper Tertiary	176
Medial Tertiary period	180
Section at Fair Haven	181
List of species obtained	181-182
(Section) Cliff near Beckett's	182
(List of fossils)	185
(Section) Saint Mary's River	185
Organic remains found on Saint Mary's River	186-187
Post-Pliocene period	187
Upper Tertiary formation	187
(Sections near Patuxent River)	188
Section near the mouth of Potomac	189
(List of species)	190
List of Fossil shells at Benner's	191-192
Descriptions of new Tertiary fossils	192
Lower Tertiary or Eocene fossils	192
Ostrea, Linnæus	192
O. sellæformis, n. s., Conrad, 1842, pl. i, fig. 1	192-193
Pholadomya	193
P. marylandica, n. s., Conrad, 1842, pl. i, fig. 3	193
Pholas	193
P. petrosa, n. s., Conrad, 1842, pl. ii, fig. 4	193
Isocardia, Lamarck	193
I. markoëi, n. s., Conrad, 1842, pl. ii, fig. 1	193
Pecten, Lamarck	194
P. humphreysii, n. s., Conrad, 1842, pl. ii, fig. 2	194
Dispotæa, Say	194
D. constricta, n. s., Conrad, 1842, pl. i, fig. 2	194
Scalaria, Lamarck	194
S. expansa, n. s., Conrad, 1842, pl. ii, fig. 3	194
Buccinum, Lamarck	194
B. integrum, n. s., Conrad, 1842, pl. ii, fig. 5	194
Scutella, Lamarck	194
S. aberti, n. s., Conrad, 1842	194

2.

CONRAD, T. A. [Descriptions of the fossil shells of **Astoria, Oregon.**] <U. S. Expl. Exp. under the command of Charles Wilkes, vol. x, Geology, Appendix i, pp. 723-728, pls. xvii-xxi. Philadelphia, 1849.

	Page.
Mya abrupta, n. s., Conrad, 1849, pl. xvii, figs. 5, 5 a	723
Thracia trapezoides, n. s., Conrad, 1849, pl. xvii, figs. 6 a, b	723
Solemya ventricosa, n. s., Conrad, 1849, pl. xvii, figs. 7, 8	723

	Page.
Donax? protexta, n. s., Conrad, 1849, pl. xvii, fig. 9	723–724
Venus bisecta, n. s., Conrad, 1849, pl. xvii, figs. 10, 10 a	724
Venus angustifrons, n. s., Conrad, 1849, pl. xvii, fig. 11	724
Venus lamellifera, n. s., Conrad, 1849, pl. xvii, figs. 12, 12 a	724
Venus brevilineata, n. s., Conrad, 1849, pl. xvii, fig. 13	724
Venus ——, sp., Conrad, 1849, pl. xviii, figs. 1, 1 a	724
Lucina acutilineata, n. s., Conrad, 1849, pl. xviii, figs. 2, 2 a, b	725
Tellina arctata, n. s., Conrad, 1849, pl. xviii, figs. 3, 3 a	725
Tellina emacerata, n. s., Conrad, 1849, pl. xviii, fig. 4	725
Tellina albaria, n. s., Conrad, 1849, pl. xviii, fig. 5	725
Tellina nasuta, n. s., Conrad, 1849	725
Tellina bitruncata, n. s., Conrad, 1849	725
Nucula dicaricata, n. s., Conrad, 1849, pl. xviii, figs. 6, 6 a	725–726
Nucula impressa, n. s., Conrad, 1849, pl. xviii, figs. 7 a–e	726
Pectunculus patulus, n. s., Conrad, 1849, pl. xviii, figs. 8, 8 a	726
Pectunculus nitens, n. s., Conrad, 1849, pl. xviii, figs. 9 a, b	726
Arca devincta, n. s., Conrad, 1849, pl. xviii, figs. 10 a	726
Arca ——, Conrad, 1849, pl. xviii, figs. 11 a, b	726
Cardita subtenta, n. s., Conrad, 1849, pl. xviii, figs. 12, 12 a	726
Pecten propatulus, n. s., Conrad, 1849, pl. xviii, figs. 13, 13 a	726
Terebratula nitens, n. s., Conrad, 1849, pl. xix, figs. 1, 1 a	726–727
—— ? Conrad, 1849, pl. xix, fig. 2	727
Dolium petrosum, n. s., Conrad, 1849, pl. xix, figs. 3 a, b, 4 a, b, and 5 a, b	727
Sigaretus scopulosus, n. s., Conrad, 1849, pl. xix, figs. 6, 6 a–d	727
Natica saxea, n. s., Conrad, 1849, pl. xix, figs. 7 a, b	727
Bulla petrosa, n. s., Conrad, 1849, pl. xix, fig. 8	727
Crepidula prærupta, n. s., Conrad, 1849, pl. xix, figs. 9 a, and 10 a, b	727
Crepidula —— ? Conrad, 1849, pl. xix, figs. 11 a, b	727
Rostellaria indurata, n. s., Conrad, 1849, pl. xix, fig. 12	727–728
Cerithium mediale, n. s., Conrad, 1849, pl. xx, figs. 1 a	728
Buccinum? derinctum, n. s., Conrad, 1849, pl. xx, figs. 2, 2 a	728
Fusus geniculus, n. s., Conrad, 1849, pl. xx, fig. 3	728
Fusus corpulentus, n. s., Conrad, 1849, pl. xx, fig. 4	728
Nautilus angustatus, n. s., Conrad, 1849, pl. xx, figs. 5, 6	728
Teredo substriata, pl. xx, figs. 7 a, b	728

The figures from 8 to 13, inclusive, on pl. xx, representing species from Astoria, are given of natural size, without names.

The plates were destroyed by fire, but about a dozen photographic copies of them have been taken by the U. S. Geological Survey.

3.

CONRAD, T. A. Description of the Fossils of Syria, collected in the Palestine expedition. <Official Rep. of the U. S. Exp. to explore the Dead Sea and the river Jordan. Sec. vi, Paleontological Report, pp. 209–235 [pls. i–xvi]. Baltimore, 1852.

The plates are numbered in a very irregular manner.

	Page.
Echinodermata.	
Echinus, Lin., Lam	212
E. syriacus, n. s., Conrad, 1852, pl. i, fig. 1, and pl. xxii, fig. 127	212
Holaster, Agassiz	212
H. syriacus, n. s., Conrad, 1852, pl. i, fig. 2	212
Cidaris, Lam	212
Spines of *Cidaris*, Conrad, 1852, pl. i, figs. 3–5	212
Testacea.	
Bivalves.	
Ostrea, Lin., Lam	212
O. virgata, Goldfuss, Nyst., pl. i, figs. 6–8	212
O. syriaca, n. s., Conrad, 1852, pl. ii, fig. 12	212
O. linguloides, n. s., Conrad, 1852, pl. ii, fig. 13	213
O. scapha, Rœmer, pl. xv, figs. 78, 79	213
Exogyra, Say	213
Ostrea bousingaultii, d'Orbigny	213
E. bousingaultii, d'Orbiguy, pl. i, fig. 9, and pl. ii, figs. 10, 11	213

	Page.
Pecten, Gault., Lam.	213
Pecten, ——, Conrad, 1852	213
Nucula, Lam.	213
N. submucronata, n. s., Conrad, 1852, pl. ii, fig. 14	213
N. parallela, n. s., Conrad, 1852, pl. ii, fig. 15	214
N. syriaca, n. s., Conrad, 1852, pl. iii, fig. 16	214
N. myiformis, n. s., Conrad, 1852, pl. iii, fig. 17	214
N. perobliqua, n. s., Conrad, pl. iii, fig. 18	214
Trigonia, Brug	214
T. syriaca, n. s., Conrad, 1852, pl. iii, figs. 19–23	214
T. alta, n. s., Conrad, 1852, pl. iv, fig. 24	214
T. cuneiformis, n. s., Conrad, 1852, pl. iii, fig. 22	214–215
Isocardia, Lam.	215
I. crenulata, n. s., Conrad, 1852, pl. iv, fig. 26	215
Astarte, Sow.	215
A. syriaca, n. s., Conrad, 1852, pl. iv, fig. 25	215
A. orientalis, n. s., Conrad, 1852, pl. iv, fig. 27	215
A. pervetus, n. s., Conrad, 1852, pl. iv, fig. 28	215
A. engonata, n. s., Conrad, 1852, pl. iv, fig. 29	215
A. arctata, n. s., Conrad, 1852, pl. xx, fig. 119	215
Arca, Lin.	215
A. syriaca, n. s., Conrad, 1852, pl. v, fig. 30	215
A. brevifrons, n. s., Conrad, 1852, pl. v, fig. 31	215
A. indurata, n. s., Conrad, 1852, pl. v, fig. 33	216
A. orientalis, n. s., Conrad, 1852, pl. v, fig. 36	216
A. declivis, n. s., Conrad, 1852, pl. v, fig. 32	216
A. subrotunda, n. s., Conrad, 1852, pl. v, fig. 34	216
A. acclivis, n. s., Conrad, 1852, pl. v, fig. 35	216
Corbula, Brug.	216
C. congesta, n. s., Conrad, 1852, pl. v, fig. 37, and pl. xxii, fig. 130	216
Cardium, Lin., Lam.	216
C. biseriatum, n. s., Conrad, 1852, pl. vi, figs. 38–40	216–217
C. crebriechinatum, n. s., Conrad, 1852, pl. vi, figs. 41–43, and pl. xv, fig. 77 ; Appendix, pl. ii, fig. 16	217
C. syriacum, n. s., Conrad, 1852, pl. vii, fig. 45	217
C. hermonense, n. s., Conrad, 1852, pl. xxii, fig. 129	217
Cardium ? Conrad, 1852, pl. xv, fig. 76	217
Pholadomya, Sow	217
P. decisa, n. s., Conrad, 1852, pl. vii, fig. 44	217
Panopœa, Ménard	217
P. pecterosa, n. s., Conrad, 1852, pl. vii, fig. 46	217
Inoceramus, Sow	218
I. lynchii, n. s., Conrad, 1852, pl. viii, fig. 47	218
Mactra, Lin., Lam.	218
M. petrosa, n. s., Conrad, 1852, pl. viii, fig. 48	218
M. pervetus, n. s., Conrad, 1852, pl. viii, fig. 49	218
M. arciformis, n. s., Conrad, 1852, pl. viii, fig. 50	218
M. syriaca, n. s., Conrad, pl. viii, fig. 51	218
Venus, Lin., Lam.	218
V. syriaca, n. s., Conrad, 1852, pl. ix, fig. 52	218
V. indurata, n. s., Conrad, 1852, pl. ix, fig. 53	219
Cytherea, Lam	219
C. syriaca, n. s., Conrad, 1852, pl. ix, figs. 54–56	219
Lucina, Lam.	219
L. syriaca, n. s., Conrad, 1852, pl. x, fig. 57	219
L. ? subtruncata, n. s., Conrad, 1852, pl. xv, fig. 76	219
Tellina, Lin., Lam.	219
T. syriaca, n. s., Conrad, 1852, pl. x, figs. 59–61	219
T. obruta, n. s., Conrad, 1852, pl. x, fig. 58	219
Orbicula ? Lam.	219
O. subobliqua, n. s., Conrad, 1852, pl. x, fig. 61½	219
Univalves:	
Chenopus, Phill.	220
C. turriculoides, n. s., Conrad, 1852, pl. x, fig. 62	220
C. induratus, n. s., Conrad, 1852, pl. xi, fig. 69	220

	Page.
O. syrincus, n. s., Conrad, pl. xii, fig. 71	220
Natica, Lam	220
N. indurata, n. s., Conrad, 1852, pl. xi, figs. 65 and 68	220
N. syriaca, n. s., Conrad, 1852, pl. xii, fig. 70	220
Phorus, Montf.	220
P. syriacus, n. s., Conrad, 1852, pl. xi, fig. 60	220
Turritella, Lam	220
T. syriaca, n. s., Conrad, 1852, pl. xv, fig. 75	220–221
T. magnicostata, n. s., Conrad, 1852, pl. x, fig. 63–64	221
T. peralveata, n. s., Conrad, 1852, pl. xx, fig. 120	221
Nerinea, Defr.	221
N. syriaca, n. s., Conrad, 1852, pl. xii, fig. 72; pl. xi, fig. 67	221
N. rhamdunensis, n. s., Conrad, 1852, pl. xxii, fig. 132	221
Strombus, Lin., Lam.	221
S. pervetus, n. s., Conrad, 1852, pl. xiii, fig. 73	221
Ammonites	221
A. syriacus, n. s., Conrad, 1852, pl. xiv, fig. 74	221

ORGANIC REMAINS OF THE CHALK.

Astarte, Sow	222
A. undulosa, n. s., Conrad, 1852, pl. xvi, figs. 81, and 86; pl. xvii, figs. 89, 90, 99	222
A. mucronata, n. s., Conrad, pl. xvii, fig. 88	222
Corbula, Lam.	222
C. sublineolata, n. s., Conrad, 1852, pl. xvi, fig. 83	222
C. syriaca, n. s., Conrad, 1852, pl. xxi, fig. 125	222
Opis, Defr.	222
O. undatus, n. s., Conrad, 1852, pl. xvii, fig. 87	222
Nucula, Lam.	222
N. perovata, n. s., Conrad, 1852, pl. xvii, fig. 91	222
N. crebrilineata, n. s., Conrad, 1852, pl. xvii, fig. 92–93	223
N. ———, Conrad, 1852, Appendix, pl. i, fig. 5	223
N. perdita, n. s., Conrad, 1852, pl. xvii, fig. 96	223
N. ———, Conrad, 1852, pl. xix, fig. 111	223
Cucullæa, Lam.	223
C. subrotunda, n. s., Conrad, 1852, pl. xvii, fig. 94	223
C. lintea, n. s., Conrad, 1852, pl. xvii, fig. 95	223
C. parallela, n. s., Conrad, 1852, pl. xvii, fig. 98	223
Arca, Lin.	223
A. fabiformis, n. s., Conrad, 1852, pl. xvii, fig. 97	223
Crassatella, Lam.	223
C. syriaca, n. s., Conrad, 1852, pl. xvii, fig. 100	223–224
Lithodomus, Cuv.	224
L. cretaceus, n. s., Conrad, 1852, pl. xvii, fig. 101	224
Gryphæa, Lam.	224
G. capuloides, n. s., Conrad, 1852, pl. xviii, figs. 103 and 104	224
G. vesicularis, Brown, pl. xviii, fig. 105	224
Exogyra, Say	224
E. densata, n. s., Conrad, pl. xviii, fig. 102	224
E. densata, var., Conrad, 1852, pl. xviii, fig. 106	224
Avicula, Lam.	225
A. samariensis, n. s., Conrad, 1852, pl. xix, fig. 107	225
Pecten, Lin., Lam.	225
P. delumbis, n. s., Conrad, 1852, pl. xix, fig. 110, and Appendix, pl. i, fig. 4	225
P. obrutus, n. s., Conrad, 1852, pl. xix, fig. 114	225
Cardium, Lin., Lam.	225
C. bellum, n. s., Conrad, 1852, Appendix, pl. i, fig. 3	225
C. ovulum, n. s., Conrad, 1852, pl. xix, fig. 108	225
Astarte, Sow.	225
A. lintea, n. s., Conrad, 1852, pl. xix, fig. 109	225
A. sublineolata, n. s., Conrad, pl. xix, fig. 112	225
Venus, Lin., Lam.	225
V. perovalis, n. s., Conrad, Appendix, pl. i, fig. 2	225
Inoceramus, Sow.	226
I. aratus, n. s., Conrad, 1852, pl. xix, fig. 113	226
Lucina, Lam.	226

WRITINGS OF T. A. CONRAD. 209

	Page.
L. safedensis, n. s., Conrad, 1852, pl. ix, fig. 115	226
Terebratula, Lam.	226
T. hermonensis, n. s., Conrad, 1852, pl. xx, fig. 123	226
Univalves:	
Fusus, Lam.	226
F. ellerii, n. s., Conrad, 1852, pl. xvi, fig. 82	226
Chenopus, Phillipi	226
Chenopus, Conrad, 1852	226
Hippurites, Lam.	226
H. syriacus, n. s., Conrad, 1852, pl. xvi, fig. 84	226
Nerinea, Defr.	227
N. cretacea, n. s., Conrad, 1852, pl. xvi, fig. 85	227
Ancyloceras, d'Orbigny	227
A. safedensis, n. s., Conrad, 1852, pl. xx, figs. 117–118	227
Baculites, Lam.	227
B. syriacus, n. s., Conrad, 1852, pl. xx, fig. 121	227
B. ——, Conrad, 1852, pl. xx, fig. 122	227
Ammonites, Brug.	227
A. safedensis, n. s., Conrad, 1852, pl. xxi, fig. 124	227
Nummulites	227
N. arbiensis, n. s., Conrad, 1852, pl. xxii, fig. 127	227
Dentalium, Lin.	228
D. cretaceum, n. s., Conrad, 1852, Appendix, pl. i, fig. 1	228
Echinodermata	228
Echinus, Lin.	228
E. kerakensis, n. s., Conrad, 1852, pl. xix, fig. 116	228

APPENDIX.

JURASSIC FORMS.

Bivalves:	
Janira, Shum.	230
J. syriaca, n. s., Conrad, 1852, pl. i, fig. 6	230
Ostræa, Lin., Lam.	230
O. corticosa, n. s., Conrad, 1852, pl. i, fig. 7	230
O. virgata? n. s., Conrad, pl. i, fig. 8	230
Opis, Defr.	231
O. equalis, n. s., Conrad, 1852, pl. ii, fig. 9	231
O. orientalis, n. s., Conrad, 1852, pl. ii, fig. 10	231
O. obrutus, n. s., Conrad, 1852, pl. ii, fig. 12	231
Astarte, Sow.	231
A. lucinoides, n. s., Conrad, 1852, pl. ii, fig. 11	231
A. subcordata, n. s., Conrad, 1852, pl. ii, fig. 13	231
Inoceramus, Sow.	231
I. syriacus, n. s., Conrad, 1852, pl. ii, fig. 14	231
I. elevatus, n. s., Conrad, 1852, pl. ii, fig. 15	231
Pholadomya, Sow.	231
P. syriaca, n. s., Conrad, 1852, pl. ii, fig. 17	231
Cardium, Lin.	231
C. crebricchinatum, n. s., Conrad, 1852, pl. ii, fig. 16	231
Arca, Lin.	231
A. longa, n. s., Conrad, 1852, pl. iii, fig. 18	231
A. bhamdunensis, n. s., Conrad, 1852, pl. iii, fig. 19	232
A. cuneus, n. s., Conrad, 1852, pl. iii, fig. 22	232
Cucullæa, Lam.	232
C. opiformis, n. s., Conrad, 1852, pl. iii, fig. 21	232
Nucula, Lam.	232
N. abrupta, n. s., Conrad, 1852, pl. iii, fig. 20	232
N.? obtenta, n. s., Conrad, 1852, pl. iii, fig. 23	232
Tellina, Lin.	232
T. syriaca?, Conrad, 1852, pl. iii, fig. 25, and pl. x, figs. 59–61	232
Orbicula, Cuv.	232
O.? syriaca, n. s., Conrad, 1852, pl. iii, fig. 24	232
Trigonia, Brug.	232
T. syriaca, n. s., Conrad, 1852, pl. iv, fig. 26	232

	Page.
T. distans, n. s., Conrad, 1852, pl. iv, fig. 27	232
Panopæa, Mén.	232
P. orientalis, n. s., Conrad, 1852, pl. iv, fig. 28	232–233
Univalves:	
Nerinea, Defrance	233
N. ? cochleæformis, n. s., Conrad, 1852, pl. iv, fig. 29	233
N. ———, Conrad, 1852, pl. iv, figs. 30, 31	233
N. ? orientalis, n. s., Conrad, 1852, pl. v, fig. 32	233
N. syriaca, n. s., Conrad, 1852, pl. v, figs. 33, 34, 35, 37, 38	233
N. abbreviata, n. s., Conrad, 1852, pl. v, fig. 36	233
N. ———, Conrad, 1852	233
Actæonella, d'Orb.	233
A. syriaca, n. s., Conrad, 1852, pl. v, fig. 40	233
Cerithium, Adans., Lam.	233
C. bilineatum, n. s., Conrad, 1852, pl. v, fig. 39	233
Natica, Lam.	233
N. orientalis, n. s., Conrad, 1852, pl. v, fig. 41	233
Turritella, Lam.	234
T. syriaca, n. s., Conrad, 1852, pl. v, fig. 42	234
Cancellaria, Lam.	234
C. petrosa, n. s., Conrad, 1852, pl. v, fig. 43	234
Bivalves:	
Lithodomus, Cuv.	234
L. stamineus, n. s., Conrad, 1852, pl. v, fig. 44	234
Cardium, Lin.	234
C. biseriatum, n. s., Conrad, 1852, pl. v, fig. 45	234
Univalves:	
Ammonites	234
A. libanensis, n. s., Conrad, 1852, pl. vi, fig. 46	234
Hippurites, Lam.	234
H. liratus, n. s., Conrad, 1852, pl. vii, figs. 47, 48	234
H. plicatus, n. s., Conrad, 1852, pl. vii, fig. 49	234
Natica, Lam.	234
N. ? scalaris, n. s., Conrad, 1852, pl. vii, fig. 50	234
Chenopus	235
C. ———, Conrad, pl. viii, figs. 51, 52	235
Bivalves:	
Corbula, Lam.	235
C. alethensis, n. s., Conrad, 1852, pl. viii, fig. 53	235
Orbicula, Lam.	235
Orbicula ? ———, Conrad, 1852, pl. viii, fig. 55	235
Echinodermata	235
Echinus, Lin.	235
E. libanensis, n. s., Conrad, 1852, pl. viii, fig. 54	235
E. bullatus, n. s., Conrad, 1852, pl. viii, fig. 56	235

4.

CONRAD, T. A. Report on the fossil shells collected in California by William P. Blake, geologist of the expedition under the command of Lieut. R. S. Williamson, United States Topographical Engineers. <Reports of Explorations and Surveys from the Mississippi River to the Pacific Ocean. Appendix to the Preliminary Geological Report of William P. Blake. Paleontology article I, pp. 5–20, 8vo, House Document 129, Washington (1855).

CATALOGUE.

	Page.
Eocene	7
Miocene and recent formations	7–8

DESCRIPTIONS OF FOSSIL SHELLS FROM THE EOCENE AND MIOCENE FORMATIONS OF CALIFORNIA.

Eocene.

Cardium, Lin.	9
C. linteum, n. s., Conrad, 1855, pl. I, fig. 1	9

	Page.
Dosinia, Scopoli	9
D. alta, n. s., Conrad, 1855, pl. i, fig. 2	9
Meretrix, Lam.—*Cytherea*, Lam.	9
M. uvasana, n. s., Conrad, 1855, pl. i, fig. 3	9
M. californiana, n. s., Conrad, 1855, pl. i, fig. 4	9
Crassatella, Lam.	9
C. uvasana, n. s., Conrad, 1855, pl. i, fig. 5	9
C. alta, n. s., Conrad, 1855	9
Mytilus, Lin	10
M. humerus, n. s., Conrad, 1855, pl. i, fig. 10	10
Cardita, Brug.	10
C. planicosta, n. s., Conrad, pl. i, fig. 6	10
Natica, Adanson	10
N. œtites ?, Conrad, 1833, pl. i, fig. 7	10
N. gibbosa and *semilunata*, Lea	10
N. alveata, u. s., Conrad, 1855, pl. i, figs. 8, 8 *a*	10
Turritella, Lam.	10
T. uvasana, n. s., Conrad, 1855, pl. i, fig. 12	10-11
Voluta[l]ithes, Swains. [e]	11
V. californiana. n. s., Conrad, 1855, pl. i, fig. 9	11
Busycon, ?	11
B. ? blakei, n. s., Conrad, 1855, pl. i, fig. 13	11
Clavatula, ? Swains.	11
C. ? californica, n. s., Conrad, 1855, pl. i, fig. 11	11

FOSSILS OF THE MIOCENE AND RECENT FORMATIONS OF CALIFORNIA.

Cardium, Lin.	11
C. modestum, n. s., Conrad, 1855, pl. ii, fig. 15	11
Nuculana, Lam.	11
N. decisa, n. s., Conrad, 1855, pl. ii, fig. 19	11-12
Corbula	12
C. diegoana, n. s., Conrad, 1855, pl. ii, fig. 16	12
Meretrix, Lam.	12
M. uniomeris, n. s., Conrad, 1855, pl. ii, fig. 20	12
M. decisa, n. s., Conrad, 1855, pl. ii, fig. 27	12
M. tularana, n. s., Conrad, 1855, pl. ii, figs. 22, 22 *a*	12
Tellina, Lin	12
T. diegoana, n. s., Conrad, 1855, pl. ii, fig. 28	12
T. congesta, n. s., Conrad, 1855, pl. ii, figs. 14, 18, 21	12-13
T. pedroana, n. s., Conrad, 1855, pl. ii, fig. 17	13
Arca, Lin	13
A. microdonta, n. s., Conrad, 1855, pl. ii, fig. 29	12
Tapes	13
T. diversum, Sow., pl. ii, figs. 24, *a*, and 26	13
Saxicava Fleur de Bell	13
S. abrupta, n. s., Conrad, 1855, pl. ii, figs. 25, 25 *a*	13
Petricola, Lam.	
P. pedroana, n. s., Conrad, pl. ii, fig. 23	13-14
Schizothœrus, Conrad.	14
S. nuttalli, n. s., Conrad, 1855, pl. iii, figs. 33, 33 *a*	14
Lutraria ?, Lam.	14
L. traskei, n. s., Conrad, 1855, pl. iii, fig. 30	14
Mactra, Lin.	14
M. diegoana, n. s., Conrad, 1855, pl. iv, fig. 35	14
Modiola, Lam.	14
M. contracta, n. s., Conrad, 1855, pl. iv, fig. 35	14
Mytilus, Lin.	15
M. pedroanus, n. s., Conrad, 1855, pl. —, fig. 40	15
Pecten, Lin	15
P. deserti, n. s., Conrad, 1855, pl. —, fig. 41	15
Anomia, Lin.	15
A. subcostata, n. s., Conrad, 1855, pl. —, fig. 34	15
Ostrea, Lin.	15
O. vespertina, Conrad, 1855, pl. —, fig. 36-38	15

	Page.
O. heermani, Conrad	15-16
Penitella	16
P. spelæum, n. s., Conrad, 1855, pl. —, figs. 43, 43 *a*, *b*	16
Fissurella, Lam.	16
F. crenulata, Sow., pl. —, fig. 44	16
Crepidula, Lam., *Crypta*, Humph.	16
C. princeps, n. s., Conrad, 1855, pl. —, fig. 52	16
Natica	16
N. diegoana, n. s., Conrad, 1855, pl. —, fig. 39	16
Trochita, Schum.	17
T. diegoana, n. s., Conrad, 1855, pl. —, fig. 42	17
Crucibulum, Shum	17
C. spinosum, n. s., Conrad, 1855, pl. —, fig. 46	17
Nassa, Lam.	17
N. int[er]striata, n. s., Conrad, 1855, pl. —, fig. 49	17
N. pedroana, n. s., Conrad, 1855, pl. —, fig. 48	17
Strephona, Browne, *Oliva*, Lam.	17
S. pedroana, n. s., Conrad, 1855, pl. —, fig. 51	17
Littorina, Ferr.	17
L. pedroana, n. s., Conrad, 1855, pl. —, fig. 50	17
Stramonita, Shum., *Purpura*, Lam.	17
S. petrosa, n. s., Conrad, 1855, pl. —, fig. 47, 47 *a*	17-18

TERTIARY SHELLS OF THE ISTHMUS OF DARIEN.

Miocene?

Gratelupia? Desmoulins	18
G. mactropsis, n. s., Conrad, 1855, pl. —, fig. 54	18
Meretrix	18
M. dariena, n. s., Conrad, 1855, pl. —, fig. 55	18
Tellina, Lin.	18
T. dariena, n. s., Conrad, 1855, pl. —, fig. 53	19

MIOCENE FOSSILS FROM OCOYA CREEK.

Natica	18
N. ocoyana, n. s., Conrad, 1855, pl. vi, fig. 57	18
N. geniculata, n. s., Conrad, 1855, pl. vi, fig. 67	18-19
Bulla	19
B. jugularis, n. s., Conrad, 1855, pl. vi, figs. 62, 62 *a*, *b*	19
Pleurotoma	19
P. transmontana, n. s., Conrad, 1855, pl. vi, fig. 60	19
P. ocoyana, n. s., Conrad, 1855, pl. vi, fig. 71	19
Sycotopus	19
S. ocoyanus, n. s., Conrad, 1855, pl. vi, fig. 72	19
Turritella	19
T. ocoyana, n. s., Conrad, 1855, pl. vii, figs. 73, 73 *a*, 73 *b*	19
Colus	19
C. arctatus, n. s., Conrad, 1855, pl. vii, fig. 76	19
Tellina	19
T. ocoyana, n. s., Conrad, 1855, pl. vii, fig. 75	19
Pecten	19
P. nevadanus, n. s., Conrad, 1855, pl. vii, fig. 77	19
P. catillifornis, n. s., Conrad, 1855, pl. viii, fig. 83	20

5.

CONRAD, T. A. Descriptions of the Fossil Shells. <Reports of Expls. & Survs. from the Mississippi River to the Pacific Ocean. Vol. v, part ii, appendix, article ii, pp. 317-329, pls. ii-ix. Washington, 1856.

	Page.
Catalogue	318-320

DESCRIPTIONS OF FOSSIL SHELLS FROM THE EOCENE AND MIOCENE FORMATIONS OF CALIFORNIA.

Eocene.

Cardium, Lin	320
C. linteum, Conrad, 1855, pl. ii, fig. 1	320
Dosinia, Scopoli	320

	Page.
D. alta, Conrad, 1855, pl. ii, fig. 2	320
Meretrix, Lam.—*Cytherea*, Lam	320
M. uvasana, Conrad, 1855, pl. ii, fig. 3	320
M. californiana, Conrad, 1855, pl. ii, fig. 4	320
Crassatella, Lam.	320
C. uvasana, Conrad, 1855, pl. ii, fig. 5	320–321
C. alta, Conrad, 1855	321
Mytilus, Lin.	321
M. humerus, Conrad, 1855, pl. ii, fig. 10	321
Cardita, Brug.	321
C. planicosta, Conrad, 1855, pl. ii, fig. 6	321
Natica, Adamson	321
N. œtites? Conrad, 1833, pl. ii, fig. 7	321
N. gibbosa and *semilunata*, Lea, 1833	321
N. alveata, Conrad, pl. ii, figs. 8, 8 *a*	321
Turritella, Lam.	321
T. uvasana, Conrad, 1855, pl. ii, fig. 12	321–322
Voluta[l]*ithes*, Swains.	322
V. californiana, Conrad, 1855, pl. ii, fig. 9	322
Busycon?	322
B. ? blakei, Conrad, 1855, pl. ii, fig. 13	322
Clavatula? Swains.	322
C.? californica, Conrad, 1855, pl. ii, fig. 11	322

FOSSILS OF THE MIOCENE AND RECENT FORMATIONS OF CALIFORNIA.

Cardium, Lin.	322
C. modestum, Conrad, 1855, pl. iii, fig. 15	322
Nucula, Lam.	322
N. decisa, Conrad, 1855, pl. iii, fig. 19	322
Corbula	322
C. diegoana, Conrad, 1855, pl. iii, fig. 16	322–323
Meretrix, Lam.	323
M. uniomeris, Conrad, 1855, pl. iii, fig. 20	323
M. decisa, Conrad, 1855, pl. iii, fig. 27	323
M. tularana, Conrad, 1855. pl. iii, fig. 22, 22 *a*	323
Tellina, Lin.	323
T. diegoana, Conrad, 1855, pl. iii, fig. 28	323
T. congesta, Conrad, 1855, pl. iii, figs. 14, 18, 21, 21 *a*	323
T. pedroana, Conrad, 1855, pl. iii, fig. 17	323
Arca, Lin.	323
A. microdonta, Conrad, 1855, pl. iii, fig. 29	323–324
Tapes	324
T. diversum, Sow., pl. iv, figs. 31, 32 *a*, *b*	324
Saxicava, Fleur de Bell	324
S. abrupta, Conrad, 1855, pl. iii, figs. 25, 25 *a*	324
Petricola, Lam.	324
P. pedroana, Conrad, 1855, pl. iii, fig. 24	324
Schizothærus, Conrad	324
S. nuttalli, Conrad, 1855, pl. iv, figs. 23, 33 *a*	324
Lutraria? Lam	324
L. traskei, Conrad, 1855, pl. iii, fig. 23	324–325
Mactra, Lin.	325
M. diegoana, Conrad, 1855, pl. v, fig. 45	325
Modiola, Lam.	325
M. contracta, Conrad, pl. v, fig. 35	325
Mytilus, Lin.	325
M. pedroanus, Conrad, 1855, pl. v, fig. 40	325
Pecten, Lin.	325
P. deserti, Conrad, 1855, pl. v, fig. 41	325
Anomia	325
A. subcostata, Conrad, 1855, pl. v, fig. 34	325
Ostrea, Lin.	325
O. vespertina, Conrad, 1855, pl. v, figs. 36–38	325–326
O. heermanni, Conrad, 1855, pl. —, figs. —	326

	Page.
Penitella	326
P. spelæa, Conrad, 1855, pl. v, figs. 43, 43 *a, b*	326
Fissurella, Lam.	326
F. creanlata, Sow., pl. v, fig. 44	326
Crepidula, Lam., *Crypta*, Humph.	326
C. princeps, Conrad, 1855, pl. vi, figs. 52, 52 *a*	326
Narica	326
N. diegoana, Conrad, 1855, pl. v, fig. 39	326
Trochita, Shum.	327
T. diegoana, Conrad, 1855, pl. v, fig. 42	327
Crucibulum, Shum.	327
C. spinosum, Conrad, 1855, pl. v, figs. 46, 46 *a*	327
Nassa, Lam.	327
N. interstriata, Conrad, 1855, pl. vi, fig. 49	327
N. pedroana, Conrad, 1855, pl. vi, fig. 48	327
Strephona, Browne, *Oliva*, Lam.	327
S. pedroana, Conrad, 1855, pl. vi, fig. 51	327
Littorina, Ferr.	327
L. pedroana, Conrad, 1855, pl. vi, fig. 50	327
Stramonita, Shum., *Purpura*, Lam	327
S. petrosa, Conrad, 1855, pl. vi, figs. 47, 47 *a*	327

TERTIARY SHELLS OF THE ISTHMUS OF DARIEN.

Miocene.

Gratelupia? Desmoulins.	
G. ? manetropsis, Conrad, 1855, pl. vi, fig. 54	328
Meretrix	328
M. dariena, Conrad, 1855, pl. vi, fig. 55	328
Tellina, Lin.	328
T. dariena, Conrad, 1855, pl. vi, fig. 53	328

MIOCENE FOSSILS FROM OCOYA CREEK.

Natica	328
N. ocoyana, Conrad, 1855, pl. vii, figs. 57, 57 *a*	328
N. geniculata, Conrad, 1855, pl. vii, fig. 67	328
Bulla	328
B. jugularis, Conrad, 1856, pl. vii, figs. 62, 62 *a, b*	328
Pleurotoma	328
P. transmontana, Conrad, 1855, pl. vi, fig. 69	328–329
Syctopus	329
S. ocoyanus, Conrad, 1855, pl. vii, figs. 72, 72 *a*	329
Turritella	329
T. ocoyana, Conrad, 1855, pl. viii, figs. 73, 73 *a, b*	329
Colus	329
C. arctatus, Conrad, 1855, pl. viii, fig. 76	329
Tellina	329
T. ocoyana, Conrad, 1855, pl. viii, figs. 75, 75 *a*	329
Pecten	329
P. nevadanus, Conrad, 1855, pl. viii, fig. 77	329
P. catilliformis, Conrad, 1855, pl. ix, fig. 83	329

6.

CONRAD, T. A. Descriptions of one Tertiary and eight New Cretaceous Fossils from Texas, in the Collection of Major Emory. <Proc. Acad. Nat. Sci. Philadelphia, vol. vii, pp. 268–269, February, 1855. Philadelphia, 1856.

	Page.
Rostellites, Conrad	268
R. texanus, n. s., Conrad, 1856	268
Turritella, Lam.	268

	Page.
T. irrorata, Conrad, n. s., 1856	268
Caprina	268
C. planata, n. s., Conrad, 1856	268
C. occidentalis, n. s., Conrad, 1856	268
Neithea, Drouet	269
N. occidentalis, n. s., Conrad, 1856	269
Mactra, Lin.	269
M. texana, n. s. Conrad, 1856	269
Exogyra, Say	269
E. fragosa, n. s., Conrad, 1856	269
E. fimbriata, n. s., Conrad, 1856	296

Tertiary species:

Ostrea contracta, n. s., Conrad, 1856	269

7.

CONRAD, T. A. Descriptions of three new genera; twenty-three new species of Middle Tertiary Fossils from California, and one from Texas. <Proc. Acad. Nat. Sci. Philadelphia, vol. viii, pp. 312–316, 1856. Philadelphia, 1857.

	Page.
Janira, Shum.	312
J. bella, n. s., Conrad, 1857	312–313
Pallium, Klein	313
P. estrellanum, n. s., Conrad, 1857	313
P. crassicardo, n. s., Conrad, 1857	313
Pecten, Lin.	313
P. mekii, n. s., Conrad, 1857	313
P. altiplectus, n. s., Conrad, 1857	313
Pachydesma, Conrad	313
P. inezana, n. s., Conrad, 1857	313
Mulinia, Gray	313
M. densata, n. s., Conrad, 1857	313
Thracia, Leach	313
T. mactropis, n. s., Conrad, 1857	313
Mya, Lin.	313
M. montereyana, n. s., Conrad, 1857	313–314
Arca, Lin.	314
A. canalis, n. s., Conrad, 1857	314
A. trilineata, n. s., Conrad, 1857	314
A. congesta, n. s., Conrad, 1857	314
Axinæa, Poli, *Pectunculus*, Lam	314
Axinæa barbarensis, n. s., Conrad, 1857	314
Arcopagia	314
A. medialis, n. s., Conrad, 1857	314
Tapes, Sowerby	314
T. linteatum, n. s., Conrad, 1857	314
Cryptomya, Conrad	314
C. ovalis, n. s., Conrad, 1857	314
Cyclas, Klein, *Lucina*, Lam.	314
Cyclas tetrica, n. s., Conrad, 1857	314
Spondylus estrallensis, n. s., Conrad, 1857	315
Dosinia, Scopoli	315
D. longula, n. s., Conrad, 1857	315
D. alta, n. s., Conrad, 1857	315
Lutraria	315
L. transmontana, n. s., Conrad, 1857	315
Schizopyga, n. g., Conrad, 1857	315
S. californiana, n. s., Conrad, 1857	315
Tamiosoma, n. g., Conrad, 1857	315
T. gregaria, n. s., Conrad, 1857	315

ECHINODERMS.

	Page.
Astrodapsis, n. g., Conrad, 1857	315
A. antiselli, n. s., Conrad, 1857	315
Mellita texana, n. s., Conrad, 1857	316

8.

CONRAD, T. A. Description of the Tertiary Fossils collected on the Survey. <Reports of Explorations and Surveys from the Mississippi River to the Pacific Ocean. Vol. vi, part ii, No. 2, pp. 69-73, pls. ii-v. Washington, 1857.

CALIFORNIA FOSSILS.

Univalves.	Page.
Schizopyga, n. s., Conrad, 1857	69
S. californiana, n. s., Conrad, 1857, pl. ii, fig. 1	69
Bivalves.	
Cryptomya, Conrad, 1857	69
C. ovalis, Conrad, 1856, pl. ii, fig. 2	69
Thracia, Leach	69
T. maetropis, Conrad, 1856, pl. ii, fig. 3	69-70
Mya, Lin.	70
M. montereyana, Conrad, 1856, pl. ii, fig. 4	70
M. ? subsinuata, n. s., Conrad, 1857, pl. ii, fig. 5	70
Arcopagia, Leach	70
A. medialis, Conrad, 1856, pl. ii, fig. 6	70
Tapes, Sowerby	70
T. linteatum, Conrad, 1856, pl. ii, fig. 7	70
Arca, Lin.	70
A. canalis, Conrad, 1856, pl. ii, fig. 8	70
A. trilineata, Conrad, 1856, pl. ii, fig. 9	70
A. congesta, Conrad, 1856, pl. ii, fig. 10	70-71
Axinaea, Poli, *Pectunculus*, Lam	71
A. barbarensis, n. s., Conrad, 1857, pl. iii, fig. 11	71
Mulinia, Gray	71
M. densata, Conrad, 1856, pl. iii, fig. 12	71
Dosinia, Scopoli	71
D. longula, Conrad, 1856	71
D. alta, Conrad, 1856, pl. iii, figs. 13 *a, b*	71
Pecten, Lin.	71
P. pabloensis, n. s., Conrad, 1857, pl. iii, fig. 14	71
Pallium, Klein	71
P. estrellanum, Conrad, 1856, pl. iii, fig. 15	71
Janira, Shum.	71
J. bella, Conrad, 1856, pl. ii, fig. 16	71-72
Ostrea, Linn.	72
O. titan, Conrad, 1855, pl. iv, fig. 17, and pl. v, fig. 17 *a*	72

FOSSILS OF GATUN, ISTHMUS OF DARIEN.

Malea, Valenc.	72
M. ringens, n. s., Conrad, 1857, pl. v, fig. 22	72
Turritella, Lam.	72
T. altilira, n. s., Conrad, 1857, pl. v, fig. 19	72
T. gatunensis, n. s., Conrad, 1857, pl. v, fig. 20	72
Triton, Lam.	72
Cytherea ? Lam.	72
C. ! (Meretrix) dariena, ? n. s., Conrad, 1857, pl. v, fig. 21	72
Taminoma, n. g., Conrad, 1857	72
T. gregaria, n. s., Conrad, 1857, pl. iv, fig. 18	72-73
Pandora, Lam.	73
P. bilirata, Conrad, 1855, pl. v, fig. 25	73
Cardita, Brug.	73
C. occidentalis, Conrad, 1855, pl. v, fig. 24	73
Diadora, Gray	73
D. crucibuliformis, Conrad, 1855, pl. v, fig. 23	73

9.

CONRAD, T. A. Report on the Palæontology of the Survey. <Reports of Explorations and Surveys from the Mississippi River to the Pacific Ocean. Vol. vii, part iii, pp. 189-196, plates i-x. Washington, 1857.

	Page.
Hinnites, Defrance	190
H. crassa, n. s., Conrad, 1857, pl. ii, figs. 1, 2	190
Pecten, Lin.	190
P. meekii, n. s., Conrad, 1857, pl. i, fig. 1	190
P. deserti, Conrad	190
P. discus, n. s., Conrad, 1857, pl. iii, fig. 1	190-191
P. magnolia, n. s., Conrad, 1857, pl. i, fig. 2	191
P. altiplicatus, n. s., Conrad, 1857, pl. iii, fig. 2	191
Pallium, Conrad	191
P. estrellanum, n. s., Conrad, 1857, pl. iii, figs. 3, 4	191
Spondylus, Bond., Lam.	191
S. estrellanus, n. s., Conrad, 1857, pl. i, fig. 3	191
Tapes ? Mühlf.	192
T. montana, n. s., Conrad, 1857, pl. v, figs. 3, 5	192
T. inezensis, n. s., Conrad, 1857, pl. vii, fig. 1	192
Venus, Lin.*	192
V. pajaroana, n. s., Conrad, 1857, pl. iv, figs. 1, 2	192
Arcopagia, Brown	192
A. unda, n. s., Conrad, 1857, pl. iv, figs. 3, 4	192
Cyclas, Klein, *Lucina*, Lam.	192
C. permacra, n. s., Conrad, 1857, pl. vii, fig. 4	192
C. estrellana, n. s., Conrad, 1857, pl. vi, fig. 6	192
Area, Lin.	192
A. obispoana, n. s., Conrad, 1857, pl. v, fig. 1	192
Pachydesma, Conrad	193
P. inezana, n. s., Conrad, 1857, pl. v, figs. 2, 4	193
Crassatella, Lam.	193
C. collina, n. s., Conrad, 1857, pl. vi, figs. 1, 2	193
Ostrea, Lin.	193
O. subjecta, n. s., Conrad, 1857, pl. ii, fig. 3	193
O. panzana, n. s., Conrad, 1857, pl. ii, fig. 4	193
Dosinia Scopoli, *Azthemis* Poli	193
D. alta, n. s., Conrad, 1857, pl. —, fig. —	193
D. longula, n. s., Conrad, 1857, pl. vii, fig. 2	193-194
D. montana, n. s., Conrad, 1857, pl. vi, fig. 4	194
D. subobliqua, n. s., Conrad, 1857, pl. vi, fig. 5	194
Mytilus, Lin.	194
M. inezensis, n. s., Conrad, 1857, pl. viii, figs. 2, 3	194
Lutraria, Lam.	194
L. transmontana, n. s., Conrad, 1857, pl. v, fig. 6	194
Axinea, Sow., *Pectunculus*, Lam.	194
A. barbarensis, n. s., Conrad, 1857, pl. vi, fig. 3	194
Mactra ?	194
M. ? yubiotensis, n. s., Conrad, 1857, pl. vii, fig. 4	194
Glycimeris, Lam., *Panopœa*, Menard	194
G. estrellanus, n. s., Conrad, 1857, pl. vii, fig. 5	194
Perna, Lam.	195
P. montana, n. s., Conrad, 1857, pl. —, fig. —	195
Univalves.	
Trochita	195
T. costellata, n. s., Conrad, 1857, pl. vii, fig. 3	195
Turritella, Lam.	195
T. inezana, n. s., Conrad, 1857, pl. viii, fig. 4	195
T. variata, n. s., Conrad, 1857, pl. viii, fig. 5	195
Natica	195
N. inezana, n. s., Conrad, 1857, pl. x, figs. 5, 6	195
Multivalves.	
Balanus estrellanus, n. s., Conrad, 1857, pl. viii, fig. 1	195
Echinoderm.	
Astrodapsis, Conrad	196
A. antiselli, n. s., Conrad, 1857, pl. x, figs. 1, 2	196

10.

CONRAD, T. A. Descriptions of Cretaceous and Tertiary fossils. <United States and Mexican Boundary Survey. Report of William H. Emory. Vol. I, part ii, pp. 141-174, pls. i-xxi. Washington, 1857.

	Page.
Eocene species	141
Cretaceous fossils from Oak Creek, Texas	141
Cretaceous fossils from between Rio San Pedro and Rio Puercos	142
Cretaceous fossils from between El Paso and Frontera	142
Cretaceous fossils from Leon Springs	142
Cretaceous fossils from Jacun, 3 miles below Laredo	143
Cretaceous fossils from Lepan Hills	143
Cretaceous ? fossils from Dry Creek, Mexico	143
Cretaceous fossils from various localities	143

Polypi:
Turbinolia, Lam	144
T. texana, n. s., Conrad, 1857, pl. ii, figs. 3 a, b	144

Bivalves:
Caprina, D'Orbigny	147
C. occidentalis, Conrad, pl. ii, figs. 1 a-c	147
C. planata, Conrad, pl. ii, figs. 2 a, b	147
Terebratula, Lhwyd., Lam	147
T. wacoensis, Roemer, pl. iii, fig. 1	147
T. choctawensis, Shumard	147
Trigonia, Lam	147
T. emoryi, Conrad, pl. iii, fig. 2 a-c	148
T. texana, n. s., Conrad, 1857, pl. iii, figs. 3 a-c	148
Mactra, Lin., Lam	148
M. texana, Conrad, pl. iv, figs. 1 a, b	148
Cucullaea, Lam	148
C. terminalis, n. s., Conrad, 1857, pl. iv, figs. 2 a, b	148
Arca, Lin	148
A. subelongata, n. s., Conrad, 1857, pl. vi, figs. 3 a, b	148
Areopagia	149
A. texana, Roemer, pl. iv, figs. 3 a, b	149
Cardium, Lin., Lam	149
C. mediale, n. s., Conrad, 1857, pl. iv, figs. 4 a, b	149
C. congestum, n. s., Conrad, 1857, pl. vi, figs. 5 a-d	149
Subgenus Protocardia, Beyrich	149
C. (Protocardia) multistriatum, Shumard, pl. vi, figs. 4 a-c	149
C. (Protocardia) texanum, n. s., Conrad, 1857, pl. v, figs. 6 a-c	150
C. (Protocardia) filosum, n. s., Conrad, 1857, pl. vi, figs. 7 a, b	150
Cardita, Lam., Blainville	150
C. eminula, n. s., Conrad, 1857, pl. vi, fig. 8	150
Corbula	150
C. occidentalis, n. s., Conrad, 1857, pl. vi, fig. 9	150
Neithea, Drouet	150
N. occidentalis, Conrad, pl. v, figs. 1 a, b	150-151
N. texana, Roemer, sp., pl. v, figs. 2 a, b	151
Lima	151
L. wacoensis, Roemer, sp., pl. v, figs. 4 a, b	151
L. leonensis, n. s., Conrad, 1857, pl. v, figs. 3 a-c	151
Inoceramus, Sowerby	151
I. confertim-annulatus, Roemer, sp., pl. v, fig. 5	151
I. naytilopsis, n. s., Conrad, 1857, pl. v, figs. 6 a, b	152
I. texanus, n. s., Conrad, 1857, pl. v, fig. 7	152
I. crispii, Mantell, pl. v, fig. 8	152
Pholadomya, Sowerby	152
P. texana, n. s., Conrad, 1857, pl. xix, fig. 3	152
Astarte, Sowerby	152
A. texana, n. s., Conrad, 1857, pl. v, fig. 9	152
Cytherea, Lam	153
C. leonensis, n. s., Conrad, 1857, pl. vi, fig. 1	153
C. texana, n. s., Conrad, 1857, pl. vi, fig. 2	153
Plicatula, Lam	153

	Page.
P. incongrua, n. s., Conrad, 1857, pl. vi, figs. 10 a, b...	153
Exogyra, Say	153
E. arietina, Rœmer, pl. vii, figs. 1 a–e	153
E. fimbriata, Conrad, pl. vii, figs. 2 a, b	154
E. læviuscula, Rœmer, pl. vii, figs. 4 a, b	154
E. matheroniana, d'Orbigny, pl. viii, figs. 1 a–c, and pl. xi, figs. 1 a, b	154
E. costata var. Conrad, 1857, pl. viii, fig. 2	154
E. costata, Say, pl. ix, figs. 1 and 2, and pl. x, fig. 1, and pl. viii, fig. 3	154–155
E. fragosa, Conrad, pl. viii, figs. 2 a, b	155
Gryphœa, Lam	155
G. pitcheri, Morton, pl. vii, fig. 3, and pl. x, figs. 2 a, b	155
Ostrea, Linn	150
O. subspatulata, Lyell & Sowerby, pl. x, figs. 3 a, b	155–156
O. bella, n. s., Conrad, 1857, pl. x, figs. 4 a, b	156
O. lugubris, n. s., Conrad, 1857, pl. x, figs. 5 a, b	156
O. carinata, Lam., pl. x, fig. 6	156
O. vellicata, n. s., Conrad, 1857, pl. xi, figs. 2 a, b	156
O. robusta, n. s., Conrad, 1857, pl. xi, figs. 3 a, b	156–157
O. cortex, n. s., Conrad, 1857, pl. xi, figs. 4 a–d	157
O. multiliratu, n. s , Conrad, 1857, pl. xii, figs. 1 a–d	157

Univalves :

Natica, Lam	157
N. texana, n. s., Conrad, 1857, pl. xiii, figs. 1 a, b	157
N. collina, n. s., Conrad, 1857, pl. xiii, figs. 2 a, b	157
Rostellaria ? Lam	157
R. ? collina, n. s., Conrad, 1857, pl. xiii, figs. 3 a, b	157
R. ? collina [*texana*], n. s., Conrad, 1857, pl. xiii, figs. 4 a, b	158
Buccinopsis	158
B. parryi, n. s., Conrad, 1857, pl. xiii, figs. 5 a, b	158
Turritella, Lam	158
T. planilateris, n. s., Conrad, 1857, pl. xiv, figs. 1 a, b	158
Rostellites, Conrad	158
R. texana, Conrad, pl. xiv, figs. 2 a, b	158
Nerinea, Defrance	158
N. schottii, n. s., Conrad, 1857, pl. xiv, figs. 3 a, b	158–159
Nodosaria, Lam	159
N. texana, n. s., Conrad, 1857, pl. xiv, figs. 4 a–c	159
Ammonites, Lam	159
A. pleurisepta, n. s., Conrad, 1857, pl. xv, figs. 1 a–c	159
A. geniculatus, n. s., Conrad, 1857, pl. xv, figs. 2 a, b	159
A. texanus, Rœmer, pl. xvi, figs. 1 a–d	159–160
A. leonensis, n. s., Conrad, 1857, pl. xvi, figs. 2 a, b	160

TERTIARY FOSSILS.

Ostrea, Lin	160
O. vespertina, Conrad, pl. xvii, figs. 1 a–d	180
O. veleniana, n. s., Conrad, 1857, pl. xvii, figs. 2 a, b	160
O. contracta, Conrad, pl. xviii, a–d	160–161
Anomia, Lin	161
A. subcostata, Conrad, pl. xix, figs. 1 a, b	161
Cardita, (Lam.) Blain	161
C. planicosta, Lam., pl. xix, figs. 2, a, b	101
Corbula, Lam	161
C. nasuta, Conrad, pl. xix, fig. 4	161
Venus vespertina, n. s., Conrad, 1857, pl. xix, figs. 5 a, b	162
Cytherea, Lam	162
C. nuttali, Conrad, pl. iv, fig. 5	162
Volutalithes, Swainson	162
V. sayana, Conrad, pl. xix, fig. 6	162
Natica, Lam	162
N. limula, Conrad, pl. xix, fig. 7	162
Turritella, Lam	163
Turritella, ——, Conrad, 1857, pl. xix, fig. 8	163
Cassidula, Humphreys	163
Subgenus *Lacinia*, Conrad	163
C. (Lacinia) alveata, Conrad, pl. xix, fig. 9	163

APPENDIX.

Cretaceous fossils.

	Page.
Cardita subtetrica, n. s., Conrad, 1857, pl. xxi, fig. 5	164
Pholadomya sancti-sabæ, Rœmer, pl. xxi, fig. 4	164
Capsa, Lam.	164
C. texana, n. s., Conrad, 1857, pl. xxi, fig. 6	164
Terebratula leonensis, n. s., Conrad, 1857, pl. xxi, fig. 2	164
Turritella leonensis, n. s., Conrad, 1857, pl. xxi, figs. 7 a, b	165
Hamites larvatus, Conrad, pl. xxi, fig. 8	165

11.

Conrad, T. A. Check list of the Invertebrate Fossils of North America. Eocene and Oligocene. <Smithsonian miscellaneous collections, 200 pp., i–iv, and 1–41. Washington, 1866.

NOTES AND EXPLANATIONS.

	Page.
Nummulites floridana, Conrad=*Nemophora floridana*, Con.	33
Turbinolia goldfussii, Lea=*Platytrochus goldfussii* (Lea), Edwards	33
Turbinolia stokesii, Lea=*Platytrochus stokesii* (Lea), Edwards	33
Anthophyllum cuneiforme, Conrad=*Flabellum cuneiforme* (Con.), Lonsdale	33
Turbonillia maclurii, Lea=*Endopachys maclurii* (Lea), Conrad	33
Scutella lyelli, Conrad=*Mortonia (Periarchus) lyelli*, Conrad	33
Orbitolites interstitia (Lea), *Lunulites* Gabb & Horn=*Lunulites interstitia*, Lea	33
Lunulites bouei, Lea=*Discoflustrellaria bouei* (Lea), Gabb & Horn	33
Orbitolites discoidea Lea=*Capularia discoidea* (Lea), Gabb & Horn	33
Lunulites duclosii, Lea=*Heteractis duclosii* (Lea), Gabb & Horn	33
Terebratula wilmingtonensis, Lyell & Sowerby=*Rhynchonella wilmingtonensis* (Lyell & Sowerby), Conrad	33
Ostrea eversa, Deshayes=*Gryphostrea eversa* (Deshayes), Conrad	33
Pecten calvatus, Morton=*Camptonectes calvatus* (Morton), Conrad	33
Leda compsa, Gabb=*Nuculana compsa* (Gabb), Conrad	33
Nucula cultelliformis, Rogers=*Nuculana cultelliformis* (Rogers), Conrad	33
Nucula magna, Lea=*Nuculana magna* (Lea), Conrad	33
Nucula media, Lea=*Nuculana media* (Lea), Conrad	33
Leda Oregona, Shumard=*Nuculana Oregona* (Shumard), Conrad	33
Nucula ovula, Lea=*Nuculana ovula* (Lea), Conrad	33
Nucula parva, Rogers=*Nuculana parva* (Rogers), Conrad	33
Nucula plana, Lea=*Nuculana plana* (Lea), Conrad	33
Nucula plicata, Lea=*Nuculana plicata* (Lea), Conrad	33
Nucula pulcherrima, Lea=*Nuculana pulcherrima* (Lea), Conrad	33
Nucula semen, Lea=*Nuculana semen*, (Lea) Conrad	33
Cucullæa onoucheila, Rogers=*Latiarca onoucheila*, (Rogers) Conrad	33
Cucullæa transversa, Rogers=*Latiarca transversa*, (Rogers) Conrad	33
Arca rhomboidella, Lea=*Anomolocardia rhomboidella*, (Lea) Conrad	33
Pectunculus ellipsis, Lea=*Limopsis ellipsis*, (Lea) Conrad	33
Nucula pectuncularis, Lea=*Limopsis pectuncularis*, (Lea) Conrad	33
Crenella concentrica, Gabb=*Stalignium concentricum*, (Gabb) Conrad	33
Astarte minutissima, Lea=*Micromeris minutissima*, (Lea) Conrad	34
Astarte parva, Lea=*Micromeris parva*, (Lea) Conrad	34
Egeria inflata, Lea=*Sphærella inflata*, (Lea) Conrad	34
Dosinia gyrata, Gabb=*Lucina gyrata*, (Gabb) Conrad	34
Corbis lamellosa, Conrad (not Lam.)=*Gafrarium liratum*, Conrad	34
Cytherea lenticularis, Rogers=*Dosiniopsis lenticularis*, (Rogers) Conrad	34
Venus floridana, Conrad=*Cryptogramma floridana*, Conrad	34
Venus penita, Conrad=*Cryptogramma ? penita*, Conrad	34
? Cytherea ovata, Rogers=*Dione ovata*, (Rogers) Conrad	34
Egeria plana, Lea=*Tellina plana*, (Lea) Conrad	34
Anatina claibornensis, Lea=*Periploma claibornensis*, (Lea) Conrad	34
Solen diegoensis, Gabb=*Plectosolen ? diegoensis*, (Gabb) Conrad	34
Solen parallelus, Gabb=*Plectosolen parallelus*, (Gabb) Conrad	34
Bulla dekayi, Lea=*Cylichna dekayi*, (Lea) Conrad	34
Actæon impressa, Gabb=*Tornatellæa impressa*, (Gabb) Conrad	34
Pasithea striata, Lea=*Actæonema striata*, (Lea) Conrad	34
Pasithea sulcata, Lea=*Actæonema sulcata*, (Lea) Conrad	34

	Page.
Marginella biplicata, Lea = *Ringicula biplicata*, (Lea) Conrad	34
Melinia nitidula, Meek = *Limnæa nitidula*, Meek	34
Limnæa tenuicostata, M. & H. = *Limnæa (Pleurolimnæa) tenuicostata*, M. & H.	34
Bulimus limnæiformis, M. & H. = *Spiraxis hay*..., Meek	34
Helix spatiosa, Meek = *Macrocyclis spatiosa*, (M. & H.) Meek	34
Bulimus perversus, M. & H. = *Clausilia contraria*, (M. & H.) Meek	34
Bulimus teres, M. & H. = *Clausilia teres*, (M. & H.) Meek	34
Bulimus vermiculus, M. & H. = *Clausilia vermicula*, (M. & H.) Meek	34
Dentalium (Ditrupa?) pusillum, Gabb = *Gadus pusillus*, (Gabb) Conrad	34
Ditrupa subcoarctata, Gabb = *Gadus subcoarctatus*, (Gabb) Conrad	34
Rotella nana, Lea = *Umbonium nana*, (Lea) Conrad	34
Narica diegoana, Conrad = *Vanikoro diegoana*, (Conrad) Meek	34
Hipponyx pygmæa, Lea = *Concholepas pygmæa*, (Lea) Conrad	34
Calyptræa trochiformis, Lea = *Trochita trochiformis*, (Lea) Conrad	34
Galerus excentricus, Gabb = *Galeropsis excentricus*, (Gabb) Conrad	34
Turritella striata, Lea = *Mesalia striata*, (Lea) Conrad	34
Melania? multistriata, M. & H. = *Campeloma multistriata*, (M. & H.) Meek	34
Paludina multilineata, M. & H. = *Campeloma multilineatum*, (M. & H.) Meek	35
Paludina vetula, M. & H. = *Campeloma vetulum*, (M. & H.) Meek	35
Melania anthonii, M. & H. = *Hydrobia anthonii*, M. & H.	35
Melania minutula, M. & H. = *Micropurgus minutulus*, M. & H.	35
Natica alabamiensis, Whitfield = *Lacunaria alabamiensis*, (Whitfield) Conrad	35
Natica erecta, Whitfield = *Lacunaria erecta*, (Whitfield) Conrad	35
Melania humerosa, M. & H. = *Tiara humerosa*, Meek	35
Cerithium nodulosum, Hall = *Goniobasis? nodulosa*, (Hall) Meek	35
Melania arcta, M. & H. = *Goniobasis? arcta*, Meek	35
Cerithium fremontii, Hall = *Goniobasis? fremontii*, (Hall) Meek	35
Melania simpsoni, Meek = *Goniobasis? simpsoni*, Meek	35
Melania sublævis, M. & H. = *Goniobasis? sublævis*, (M. & H.) Meek	35
Melania subtortuosa, M. & H. = *Goniobasis? subtortuosa*, (M. & H.) Meek	35
Melania tenuicarinata, M. & H. = *Goniobasis? tenuicarinata*, (M. & H.) Meek	35
Cerithium tenerum, Hall = *Goniobasis? tenera*, (Hall) Meek	35
Solarium henrici, Lea = *Architectonica henrici*, (Lea) Conrad	35
Solarium ornatum, Lea = *Architectonica ornata*, (Lea) Conrad	35
Delphinula plana, Lea = *Architectonica plana*, (Lea) Conrad	35
Solarium pseudogranulatum, d'Orbigny = *Architectonica pseudogranulata*, (d'Orbigny) Conrad	35
Delphinula depressa, Lea = *Solariorbis depressus*, (Lea) Conrad	35
Turbo lineatus, Lea = *Solariorbis lineatus*, (Lea) Conrad	35
Planaria nitens, Lea = *Solariorbis nitens*, (Lea) Conrad	35
Pasithea accisculata, Lea = *Eulima acicsulata*, (Lea) Conrad	35
Pasithea lugubris, Lea = *Eulima lugubris*, (Lea) Conrad	35
Pasithea notata, Lea = *Eulima notata*, (Lea) Conrad	35
Pasithea scale, Lea = *Eulima scale*, (Lea) Conrad	35
Actæon melanellus, Lea = *Obeliscus melanellus*, (Lea) Conrad	35
Actæon pygmæus, Lea = *Obeliscus pygmæus*, (Lea) Conrad	35
Actæon striatus, Lea = *Obeliscus striatus*, (Lea) Conrad	35
Mitra costata, Lea = *Pyramimitra costata*, (Lea) Conrad	35
Natica minima, Lea = *Lunatia minima*, (Lea) Conrad	35
Natica alveata, Conrad = *Ampullina alveata*, Conrad	35
Natica gibbosa, Lea = *Neverita gibbosa*, (Lea) Conrad	35
Naticina obliqua, Gabb = *Catinus obliquus*, (Gabb) Conrad	35
Sycotypus penitus, Conrad = *Ficopsis penitus*, Conrad	35
Buccinum sowerbii, Lea = *Semicassis sowerbii*, (Lea) Conrad	35
Fusus remondii, Gabb = *Ficopsis remondii*, (Gabb) Conrad	36
Fusus cooperi, Gabb = *Ficopsis cooperi*, (Gabb) Conrad	36
Ficus mammillatus, Gabb = *Ficopsis mammillatus*, (Gabb) Conrad	36
Hemifusus hornii, Gabb = *Priscoficus hornii*, (Gabb) Conrad	36
Mitra flemingii, Lea = *Caricella flemingii*, (Lea) Conrad	36
Mitra fusoides, Lea = *Conomitra fusoides*, (Lea) Conrad	36
Mitra mooreana, Gabb = *Lapparia mooreana*, (Gabb) Conrad	36
Mitra lineata, Lea = *Fusimitra? lineata*, (Lea) Conrad	36
Mitra minima, Lea = *Fusimitra? minima*, (Lea) Conrad	36
Fasciolaria moorei, Gabb = *Cordieria moorei*, (Gabb) Conrad	36
Fasciolaria plicata, Lea = *Latirus (Peristernia) plicatus*, (Lea) Conrad	36

	Page.
Ancillaria elongata, Gabb = *Lamprodoma elongata*, (Gabb) Conrad	36
Oliva gracilis, Lea = *Lamprodoma gracilis*, (Lea) Conrad	36
Olivia phillipsii, Lea = *Lamprodoma phillipsii*, (Lea)Conrad	36
Anolax giganteа, Lea = *Ancillopsis altile*, Conrad	36
Anolax plicata, Lea = *Olivula? plicata*, (Lea) Conrad	36
Agaronia punctulifera, Gabb = *Olivula punctulifera*, (Gabb) Conrad	36
Fusus taitii, Lea = *Cornulina armigera*, Conrad	36
Monoceras sulcatum, Lea = *Pseudoliva sulcata*, (Lea) Conrad	36
Tritonium diegoensis, Gabb = *Buccinofusus diegoensis*, (Gabb) Conrad	36
Pleurotoma beaumontii, Lea = *Surcula beaumontii*, (Lea) Conrad	36
Pleurotoma cœlata, Lea = *Surcula cœlata*, (Lea) Conrad	36
Pleurotoma childreni, Lea = *Surcula childreni*, (Lea) Conrad	36
Pleurotoma desnoyersii, Lea = *Surcula desnoyersii*, (Lea) Conrad	36
Pleurotoma kellogii, Gabb = *Surcula kellogii*, (Gabb) Conrad	36
Pleurotoma monilifera, Lea = *Surcula monilifera*, (Lea) Conrad	36
Pleurotoma nodocarinata, Gabb = *Surcula nodocarinata*, (Gabb) Conrad	36
Pleurotoma obliqua, Lea = *Surcula obliqua*, (Lea) Conrad	36
Pleurotoma rugosa, Lea = *Surcula rugosa*, (Lea) Conrad	36
Pleurotoma sayi, Lea = *Surcula sayi*, (Lea) Conrad	36
Pleurotoma varicostata, Gabb = *Surcula varicostata*, (Gabb) Conrad	36
Pleurotoma lonsdalii, Lea = *Drillia lonsdalii*, (Lea) Conrad	36
Pleurotoma texana, Gabb = *Drillia texana*, (Gabb) Conrad	36
Papillina altilis, Conrad = *Clavifusus altile*, Conrad	36
Fusus cooperi, Conrad = *Clavifusus cooperi*, Conrad	36
Fusus conybearni, Lea = *Strepsidura conybearii*, (Lea) Conrad	36
Pelagus vanuxemi, Conrad = *Aturia vanuxemi*, Conrad	36
Nautilus lamarckii, Deshayes = *Cymomia lamarckii*, (Deshayes) Conrad	37
Clavella Vicksburgensis, Conrad = *Piestocheilus Vicksburgensis*, (Con.) Meek	37
Nummulites mantelli, Morton = *Orbitolites (Orbitoides) mantelli*, Morton	37
Scutella crustuloides, Morton = *Mortonia (Periarchus) crustuloides*, (Mort.) Conrad	37
Scutella pileus-sinensis, Ravenel = *Mortonia (Periarchus) pileus-sinensis*, (Rav.) Conrad	37
Echinus infulatus, Morton = *Cœlopleurus inflatus*, (Morton) Desor	37
Scutella jonesii, Forbes = *Clypeaster jonesii*, (Forbes) Desor	37
Scutella rogersi, Morton = *Clypeaster rogersi*, (Morton) Conrad	37
Mortonia tumida, Conrad = *Clypeaster tumidus*, Conrad	37
Pyrgorhynchus mortonis, Mich. = *Echinianthus mortonis*, (Mich.) Desor	37
Catopygus patelleformis, Bouvé = *Cassidulus patelliformis*, (Bouvé) Desor	37
Cellepora tubulata, Lonsdale = *Eschara tubulata*, (Lonsd.) Gabb & Horn	37
Terebratula lachryma, Morton = *Terebratulina lachryma*, (Mort.) Conrad	37
Plagiostoma dumosa, Morton = *Spondylus dumosus*, (Morton) Conrad	37
Umbrella planulata, Conrad = *Operculatum planulatum*, Conrad	37
Doliopsis, n. g., Conrad = *Galeodea (Galeodaria) quinquecostata*, Conrad	37

12.

CONRAD, T. A. [Description of a new genus and subgenus.] <Rep. U. S. Geol. and Geogr. Surv. Terr., F. V. Hayden, vol. ii. The vertebrata of the Cretaceous formations of the West by E. D. Cope, pp. 23–24. Washington, 1875.

	Page
Haploscapha, n. g., Conrad, 1875	23
H. grandis, n. s., Conrad, 1875	23–24
Cucullifera, n. s. g., Conrad, 1875	24
H. (Cucullifera) eccentrica, n. s., Conrad, 1875	24

III.—THE WRITINGS OF JAMES DWIGHT DANA.

1.

DANA, J. D. Zoophytes. <U. S. Expl. Exp. during the years 1838–1842 under the command of Charles Wilkes, U. S. N., vol. vii, pp. 1–741, Atlas pls. i–lxi. Philadelphia, 1846.

Many fossil genera are mentioned and discussed in this work.

2.

DANA, J. D. Genera of Fossil Corals of the family Cyathophyllidæ. <Amer. Journ. Sci., 2d ser., vol. i, pp. 178–189, figs. 1–5. New Haven, 1846.

This article is extracted from the report of the United States Exploring Expedition during the years 1838–'42, under the command of Charles Wilkes, U. S. N. Zoophytes, by James D. Dana, Geologist of the Expedition, pp. 1–741. 4to. See entry No. 1.

	Page.
Family *Cyathophyllidæ*	179–182
Cyathophyllum	182–183
Calophyllum, n. g., Dana, 1846	183–184
Amplexus, Sowerby	184
Caninia, Michelin	184
Acervularia, Schweigger	184–186
Arachnophyllum, n. g., Dana, 1846, fig. 1, p. 186	186
Cystiophyllum, Lonsdale	186
Clisiophyllum, n. g., Dana, 1846, figs. 2 and 3, p. 187	187
Michelinia, Koninck	187
Columnaria, Goldfuss, fig. 4, p. 188	188
Sarcinula, Lamarck	188–189

3.

DANA, J. D. Description of Fossil Shells of the collections of the Exploring Expedition under the command of Charles Wilkes, U. S. N., obtained in Australia, from the lower layers of the coal formation in Illawara, and from a deposit probably of nearly the same age at Harper's Hill, valley of the Hunter. <Amer. Journ. Sci., 2d ser., vol. iv. Appendix, pp. 151–160. New Haven, 1847.

	Page.
Bellerophon undulatus, n. s., Dana, 1847	151
Bellerophon strictus, n. s., Dana, 1847	151
Platyschisma ? depressum, n. s., Dana, 1847	151
Pleurotomaria tri-filata, n. s., Dana, 1847	151
Pleurotomaria nuda, n. s., Dana, 1847	151
Natica ——— *?* Dana, 1847	151
Patella tenella, n. s., Dana, 1847	151–152
Pentadia, n. g., Dana, 1847	152
P. spatangus, n. s., Dana, 1847	152
P. reniformis, n. s., Dana, 1847	152
P. trigonia, n. s., Dana, 1847	152
Lingula ovata, n. s., Dana, 1847	152
Terebratula amygdala. n. s., Dana, 1847	152
Terebratula elongata, n. s., Dana, 1847	152
Productus fragilis, n. s., Dana, 1847	153
Solen (Solecurtus?) ellipticus, n. s., Dana, 1847	153
Solen (Solecurtus?) planulatus, n. s., Dana, 1847	153
Pholadomya undata, n. s., Dana, 1847	153
Allorisma audax, n. s., Dana, 1847	153
Cleobis, n. g., Dana, 1847	154
C. grandis, n. s., Dana, 1847	154

	Page.
C. gracilis, n. s., Dana, 1847	154
C. ? recta, n. s., Dana, 1847	154
Astarte gemma, n. s., Dana, 1847	154-155
Astartila, n. g., Dana, 1847	155
A. intrepida, n. s., Dana, 1847	155
A. cyprina, n. s., Dana, 1847	155
A. cytherea, n. s., Dana, 1847	155
A. polita, n. s., Dana, 1847	155
A. cyclas, n. s., Dana, 1847	155
A. transversa, n. s., Dana, 1847	155-156
Cardinia, Agassiz	156
C. recta, n. s., Dana, 1847	156
C. cuneata, n. s., Dana, 1847	156
Pyramus, n. g., Dana, 1847	156-157
P. ellipticus, n. s., Dana, 1847	157
P. myiformis, n. s., Dana, 1847	157
Nucula abrupta, n. s., Dana, 1847	157
Nucula ——— ? Dana, 1847	157
Cypricardia rugulosa, n. s., Dana, 1847	157
Cypricardia sinuosa, n. s., Dana, 1847	157-158
Myenia, n. g., Dana, 1847	158
M. elongata, n. s., Dana, 1847	158
M. valida, n. s., Dana, 1847	158
Eurydesma elliptica, n. s., Dana, 1847	158
Eurydesma globosa, n. s., Dana, 1847	158
Modiolopsis simplex, n. s., Dana, 1847	158
Modiolopsis siliqua, n. s., Dana, 1847	159
Modiolopsis prœrupta, n. s., Dana, 1847	159
Modiolopsis imbricata, n. s., Dana, 1847	159
Modiolopsis arcodes, u. s., Dana, 1847	159
Modiolopsis acutifrons, n. s., Dana, 1847	159-160
Avicula ——— ? Dana, 1847	160
Pecten comptus, n. s., Dana, 1847	160
Pecten tenuicollis, n. s., Dana, 1847	160
Pecten lenniseulus, n. s., Dana, 1847	160

4.

DANA, J. D. Fossils of the Exploring Expedition under the command of Charles Wilkes, U. S. N., a fossil fish from Australia, and a Belemnite from Terra Del Fuego. <Am. Journ. Sci., vol. v, 2d ser., pp. 433-435. New Haven, 1848.

	Page.
Urosthenes, n. g., Dana, 1848	433-434
U. australis, n. s., Dana, 1848	434
Helicerus, n. g., Dana, 1848	434
H. fuegiensis, n. s., Dana, 1848	434
Cardinia ? exilis, McCoy=*Cardinia recta*, Dana	434
Pleurotomaria morrisiana, McCoy=*Pleurotomaria triflata*, Dana	434
Pachydomus ovalis, McCoy	434
Pachydomus pusillus, McCoy	434
Pachydomus sacculus, McCoy	434
Eurydesma cordata ? Morris	434
Notomya, McCoy=*Pyramus*, Dana	434

5.

DANA, J. D. Descriptions of fossils. <U. S. Expl. Exp., 1838-'42, under the command of Charles Wilkes, U. S. N. Geology. By James D. Dana. Vol. x, Appendix, pp. 681-720, pars. Atlas, xxi, pls. pars. Philadelphia, 1849.

APPENDIX I.

FOSSILS OF NEW SOUTH WALES.

	Page.
Pisces	681
Urosthenes, Dana, 1848	681
Urosthenes australis, Dana, 1848	681-682

	Page.
Mollusca	682
Brachiopoda	682
Terebratula amygdala, Dana, 1847	682
Terebratula elongata, Dana, 1847, pl. i, figs. 3 a, b	682–683
Terebratula ——?. Dana, 1849, pl. i, figs. 4 a, b	683
Terebratula ? ——, Dana, 1849, pl. i, fig. 5	683
Spirifer glaber, pl. i, fig. 6 a, b	683
Spirifer darwinii (J. Morris), pl. i, fig. 7 a	684
Spirifer duodecicostatus (M'Coy), pl. ii, figs. 1 a, 1 b	684
Spirifer ——, Dana, 1849, pl. ii, fig. 2	684
Spirifer vespertilio (G. Sowerby), pl. ii, figs. 3 a–c	685
Spirifer phalæna, n. s., Dana, 1849, pl. ii, fig. 4	685
Siphonotreta ? curta, n. s., Dana, 1849, pl. ii, figs. 5 a, b	685
Lingula ovata, n. s., Dana, 1849, pl. ii, figs. 6 a, b	685–686
Productus fragilis, Dana, pl. vii, figs. 7 a–c	686
Productus brachythærus (G. Sowerby), pl. ii, fig. 8	686
Acephala	686
Solecurtus ? ellipticus, Dana, pl. ii, fig. 9	686
Solecurtus (Psammobia ?) planulatus, Dana, pl. ii, fig. 10	686–687
Pholadomya (Platymya) undata, Dana, pl. ii, figs. 11 a, b	687
Pholadomya (Homomya) glendonensis, n. s., Dana, 1849, pl. ii, fig. 12	687
Pholadomya (Homomya) audiæ, Dana, pl. iii, figs. 1 a–c	687
Pholadomya (Homomya) curvata (?) (J. Morris). Dana, pl. iii, figs. 2 a, b	687
Astarte gemma, Dana, pl. iii, figs. 4 a, b	686
Astartila. Dana, 1847	688
A. intrepida, Dana, 1847, pl. iii, figs. 5, 5 a	689
A. cyprina, Dana, 1847, pl. iii, figs. 6, 6 a	686
A. cytherea, Dana, 1847, pl. iv, figs 7, 1 a	689
A. polita, Dana, 1847, pl. iv, figs. 2 a–c	690
A. cyclas, Dana, 1847, pl. iv, figs. 3, 3 a	690
A. transversa Dana, 1847, pl. iv, figs. 4 a, b	690
A. ? corpulenta, n. s., Dana, 1849, pl. iii, figs. 3 a–c	691
Cardinia recta, Dana, pl. iv, figs 5 a, b	691
Cardinia ? cuneata, Dana, pl. iv, figs. 6, 6 a–c	692
Cardinia ? costata (J. Morris), Dana, pl. iv, figs. 8 a–e	692
Pachydomus	692–693
P. cuneatus (J. D. Sowerby), Morris, pl. v, figs. 1, 1 a	693
P. antiquatus. (J. D. Sowerby), Morris, pl. v, fig. 2	693–694
P. levis (J. D. Sowerby), Morris	694
Mæonia, Dana	694
Mæonia, Dana	694
Pyramis, Dana	695
Cleobis, Dana	695
Mæonia elongata, Dana, 1847, pl. v, figs. 3 a–c	695
M. valida, Dana, 1847, pl. v, figs. 4 a–b	695
M. axinia, Dana, 1847, pl. v, figs. 5 a, b	696
M.? carinata (J. Morris), Dana, pl. vi, figs. 1 a, b	696
M. fragilis, n. s., Dana, 1849, pl. vi, figs. 2, 3	696–697
M. mytiformis, Dana, 1847, pl. vi, fig. 4 a	697
? M. elliptica, Dana, 1849, pl. vi, fig. 5 a, c	697
M. gigas (M'Coy), Dana	697
M. grandis, Dana, 1847, pl. vi, fig. 7	697–698
M. gracilis, Dana, 1847, pl. vii, figs. 1 a, c	698
M.? recta, Dana, 1847, pl. vii, fig. 2	698
Nucula abrupta, Dana, 1847, pl. vii, fig. 3	698
Nucula concinna, n. s., Dana, 1849, pl. vii, fig. 4	699
Nucula glendonensis, n. s., Dana, 1849, pl. vii, fig 5	699
Eurydesma, Morris	699
E. elliptica, Dana, 1847, pl. vii, figs. 6 a–d ?	700
E. globosa, Dana, 1847, pl. vii, figs. 7, 7 a	700
E. sacculus (M'Coy), Dana, pl. vii, figs. 8 a, b	700
E. cordata (Morris)	700–701
Cardium australe (M'Coy), Dana, pl. viii fig. 2	701
Cardium ferox, n. s., Dana, 1849, pl. viii, fig. 3	701
Cypricardia	701–702

	Page.
C. acutifrons, Dana, 1847, pl. viii, figs. 4 *a, b*	702
C. imbricata, Dana, 1847, pl. viii, fig. 5	702
C. arcodes, Dana, 1847, pl. viii, figs. 8 *a, b*	702–703
C. prærupta, Dana, 1847, pl. viii, fig. 10	703
C. siliqua, Dana, 1847, pl. ix, figs. 1 *a, b*	703
C. simplex. Dana, 1847, pl. ix, fig. 2	703–704
C. (Aricula?) veneris, Dana, 1847, pl. ix, figs. 3 *a, b*	704
Aricula ralgensis? (Verneuil), pl. ix, fig. 4	704
Pterinea macroptera (J. Morris)	704
Pecten comptus, Dana, 1847, pl. ix, fig. 5	704
Pecten tenuisculus, Dana, 1847, pl. ix, figs. 6 *a*, 6*b*	704–705
Pecten tenuicollis, Dana, 1847, pl. ix, fig. 7	705
Pecten mitis, n. s., Dana, 1849, pl. ix, figs. 8 *a, b*	705
Pecten illawarrensis (J. Morris), pl. ix, fig. 9	705
Pecten squamuliferus [?], Morris	705
——— ?, pl. ix, fig. 10, Dana, 1849	705–706

Gasteropoda.

Pileopsis tenella, Dana, 1847, pl. ix, figs. 13 *a*, b	706
Pileopsis alta, n. s., Dana, 1849, pl. ix, fig. 14 *a*	706
Pleurotomaria morrisiana, McCoy, pl. ix, figs. 15, 15 *a*, 16	706
Pleurotomaria nuda. Dana, 1847, pl. ix, figs. 17 *a–c*	706
Pleurotomaria strzeleckiana, Morris	707
Platychisma oculus (Morris), pl. x, fig. 1	707
Platychisma rotundatum, Morris	707
Platychisma depressum, Dana, 1847, pl. x, figs. 2 *a, b*	707
Natica ——— ? Dana, 1847, pl. x, figs. 2 *a, b*	707
Bellerophon undulatus Dana, 1847, pl. x, figs. 4 *a, b*	707
Bellerophon strictus, Dana, 1847, pl. x, figs. 5, *a, b*	707–708
Bellerophon micromphalus, Morris, pl. x, figs. 6 *a, b*	708

Cephalopoda.

Theca lanceolata (Morris), pl. x, figs. 7 *a, b*	708
Conularia	708–709
C. inornata, n. s., Dana, 1849, pl. x, fig. 8	709–710
C. lerigata. Morris, pl. x, fig. 9	710
C. tenuistriata? (M'Coy)	710

Radiata.

Fenestella internata (Lonsdale), pl. x, fig. 13	710
Fenestella media, n. s., Dana, 1849, pl. x, figs. 14, 14 *a*, and fig. 15	710
Fenestella ampla (Lonsdale), pl. xi, figs. 1, 1 *a*; 2, 2 *a*	710
Fenestella fossula (Lonsdale), pl. xi, figs. 3 *a, b*	710–711
Fenestella gracilis, n. s. (Dana), 1849, pl. xi, fig. 4	711
Fenestella, pl. xi, figs. 5, 5 *a*	711
Chetetes crinita (Lonsdale), Dana, pl. xi, figs. 6, 6 *a–c*	711
Chetetes tasmaniensis (Lonsdale), Dana, pl. xi, figs. 7, 7 *a*, 8, 8 *a*	711
Chetetes ovata (Lonsdale), Dana, pl. xi, figs. 9, 9 *a, b*	712
Chetetes gracilis. n. s., Dana, 1849, pl. xi, figs. 10, 10 *a–c*	712
Hemitrypa?, pl. xv, fig. 10	712
Encrinital remains. pl. xi, figs. 12 *a, b*. and pl. xi. figs. 13, 14, and fig. 15	712
Pentadia, Dana, 1847	712–713
Pentadia corona, pl. x, figs. 10, 10 *a–c*, 11 and 12	713

FOSSIL PLANTS.

Coniferæ.

Fruit scales, pl. xii, figs. 1, 2, 3, 4, 5, 5 *a, b*, 6, 7, 8, 8 *a–d*	714
Nœggerathia, n. g., Dana, 1849	715
N. spatulata, n. s., Dana, 1849, pl. xii, fig. 9	715
N. media, n. s., Dana, 1849, pl. xii, fig. 10	715
N. elongata (J. Morris), Dana, 1849, pl. xii, fig. 11	715
Sphenopteris lobifolia (Morris), pl. xii, fig. 12	715–716
Glossopteris browniana (Brongniart), pl. xii, fig. 13, 13 *a–c*, 14 (Young?)	716–717
Glossopteris ampla, n. s., Dana, 1849, pl. xiii, fig. 13, *a, b*	717
Glossopteris reticulum, n. s., Dana, 1849, pl. xiii, figs. 2, 3	717–718
Glossopteris elongata, n. s., Dana, 1849, pl. xiii, fig. 4	718
Glossopteris? cordata, n. s., Dana, 1849, pl. xiii, fig. 5	718
Glossopteris linearis (McCoy)	718

	Page.
Phyllotheca australis, pl. xiii, fig. 6, pl. xiv, figs. 1 [2?]	718–719
Clasteria, n. g., Dana, 1849	719
C. australis, n. s., Dana, 1849	719–720
Anarthrocanna australis, n. s., Dana, 1849, pl. xiv, fig. 6 a	720
Cystoseirites? pl. xiv, fig. 6, b	720
Austrella rigida, n. s., Dana, 1849, pl. xiv, figs. 7, 8	720
Conferrites? tenella, n. s., Dana, 1849, pl. xiv, fig. 9	720

FOSSILS FROM TIERRA DEL FUEGO.

Helcarus, Dana, 1848	720
H. fuegiensis, Dana, pl. xv, fig. 1 a–c	720–721

FOSSILS FROM SAN LORENZO, PERU.

Trigonia Lorentii, n. s. (Dana), 1849, pl. xv, fig. 2 a–c	721
Turbo ——, Dana, 1849, pl. xv, fig. 3 a, b	721
Nautilus tenui—planatus, n. s., Dana, 1849, pl. xv, fig. 4	721

FOSSIL AMMONITE FROM THE ANDES.

Ammonites pickeringi, Dana, pl. xv, fig. 5	721
Ammonites, pl. xv, fig. 6	721
Ostraca pl. xv, fig. 7	722

FOSSILS FROM NORTHWESTERN AMERICA.

Cetacean, pl. xvi, fig. 1	722
Fishes, pl. xvi, fig. 2, pl. xvi, fig. 3, pl. xvii, figs. 1, 2 a, 2 b	722
Crustacea.	
Callianassa oregonensis, n. s., Dana, 1849	722–723
Balanus ——? Dana, pl. xvii, fig. 4	723
Radiata.	
Galerites oregonensis, n. s., Dana, 1849, figs. 5, 6, 6 a, pl. xxi, figs. 7, 8	729

IV.—THE WRITINGS OF CHRISTIAN GOTTFRIED EHRENBERG.

1.

EHRENBERG, C. G. On Infusorial Deposits on the River Chutes in Oregon. Amer. Jour. Sci., 2d ser., Vol. IX, p. 140. New Haven, 1850.

A brief notice of the author's work in the Monatsb. Akad., Berlin, February, 1849, p. 76.

2.

EHRENBERG, C. G. Über das mächtigste bis jetzt bekannt gewordene (angeblich 500 Fuss mächtige) Lager von mikroskopischen reinen kieselschaligen Süsswasser-Formen am Wasserfall-Flusse im Oregon. <Bericht über die, zur Bekanntmachung geeigneten Verhandlungen der Königl. Preuss. Akademie der Wissenschaften zu Berlin aus dem Jahre 1849. February. pp. 76–87. Berlin, 1850.

Amphora libyca.
Campylodiscus americ.?
Cocconeis finnica.
C. concentrica.
C. gemmata.
C. lineata.
C. oblonga.
C. protexta.
C. punctata.
Cocconema asperum.
C. cistula.
C. gibbum.
C. gracile.
C. lanceolatum.
C. lunula.
Discoplea oregonica.
Eunotia amphioxys.
E. argus.
E. gibba.
E. gibberula.
E. granulata.
E. librile.
E. subulata.
E. textricula.
E. uncinata.
E. westermanni.
E. zebra.
E. zebrina.
Fragilaria acuta.
F. amphicephala.
F. rhabdosoma.
Gloeonema paradoxum.?
Gomphonema gracile.
G. herculaneum.
G. longicolle.
G. mammilla.
G. minutissimum.
G. olar.
G. oregonicum.
Gallionella crenata.
G. distans.
G. granulata.
G. laevis.
G. punctata.

G. undulata.
Himantidium arcus.
Navicula sigma.
N. bacillum.
N. scalprum.
N. semen.
N. sificula.
Pinnularia affinis.
P. amphioxys.
P. digitus.
P. gastrum.
P. macilenta.
P. mesogongyla.
P. pachyptera.
P. placentula.
P. oregonica.
P. viridis.
P. viridula.
Podosphenia pupula.
Rhaphoneis foliacea.
R. lanceolata.
R. oregonica.
Surirella bifrons.
S. plicata.
Stauroneis baileyi.
S. semen.
Synedra ulna.
S. splendida.
Amphidiscus armatus.
Lithodontium furcatum.
L. nasutum.
L. scorpius.
Lithostylidium amphiodon.
L. crenulatum.
L. laeve.
L. quadratum.
L. rude.
L. trabecula.
Lithostylidium.?
Spongadithis acicularis.
S. aspera.
S. fustis.
S. inflexa.
S. mesogongyla.

V—THE WRITINGS OF JAMES HALL.

1.

HALL, JAMES. Organic remains. Descriptions of organic remains collected by Captain J. C. Frémont, in the geographical survey of Oregon and North California. <Rep. Expl. Exp. to the Rocky Mountains and to Oregon and North California, by J. C. Frémont. Appendix B, pp. 304–310, pls. i–iv. Washington, 1845.

	Page.
Sphenopteris fremonti. n. s., Hall, 1845, pl ii, figs. 3, 3 a	304
S. triloba, n. s., Hall, 1845, pl. i, fig. 8	304
S. (?) paucifolia, n. s., Hall, 1845, pl. ii, figs. 1 a–d	304–305
S. (?) trifoliata, n. s., Hall, 1845, pl. ii, figs. 2, 2 a	305
Glossopteris phillipsii. n. s., Hall, 1845, pl. ii, figs. 5, 5 a–c	305–306
Pecopteris undulata. n. s., Hall, 1845, pl. i, figs. 1, 1 a	306
Pecopteris undulata var., Hall, 1845, pl. i, figs. 2, 2 a, b	306
Pecopteris (?) odontopteroides, n. s., Hall, 1845, pl. i, figs. 3, 4	306
Trichopteris, n. g., Hall, 1845	306
T. filamentosa, n. s., Hall, 1845, pl. ii, fig. 6	306–307
T. gracilis, n. s., Hall, 1845, pl. i, fig. 5	307
Stems of ferns, pl. i. fig. 7	307
Leaf of a Dycotyledonous plant (?), pl. ii, fig. 4	307
Mya tellinoides, n. s., Hall, 1845, pl. iii, figs. 1, 2	307
Nucula impressa, (?) n. s., Hall, 1845, pl. iii, fig. 3	308
Cytherea parvula, n. s., Hall, 1845, pl. iii, figs. 10, 10 a	308
Pleurotomaria uniangulata, n. s., Hall, 1845, pl. iii, figs. 4, 5	308
Cerithium tenerum, u. s., Hall, 1845, pl. iii, figs. 6, 6 a	308
Cerithium fremonti, n. s., Hall, 1845, pl. iii, figs. 7, 7 a	308
Natica (?) occidentalis, n. s., Hall, 1845, pl. iii, figs. 8, 8 a	308–309
Turritella bilineata, n. s , Hall, 1845, pl. iii, fig. 9	309
Cerithium nodulosum, n. s., Hall, 1845, pl. iii, figs. 11, 12	309
Turbo paludinaeformis, n. s., Hall, 1845, pl. iii, fig. 13	309
Leaves of Dicotyledonous plants, pl. iii, figs. 14, 15	309
Inoceramus ———?, Hall, 1845, pl. iv, figs. 1, 1 a	309–310
Inoceramus ———?, Hall, 1845, pl. iv, fig. 2	310

2.

HALL, JAMES. Description of new or rare species of fossils, from the Palæozoic series. <Rep. on the Geology of the Lake Superior Land district, by J. W. Foster and J. D. Whitney. Part ii, chapter xiii, pp. 203–231, pls. xxiii a–xxxv. Washington, 1851.

POTSDAM AND CALCIFEROUS SANDSTONES.

	Page.
Lingula prima, Conrad, pl. xxiii, figs. 1 a–g	204
Lingula antiqua, Hall, pl. xxiii, figs. 2 a–e	204–205
Trilobites of the Potsdam Sandstone	205
Dikell[o]ocephalus, D. D. Owen, pl. xxiii, figs. 3 a–e, and fig. 4	205–206

FOSSILS FROM THE CHAZY, BIRD'S EYE, BLACK RIVER, AND TRENTON LIMESTONES AND HUDSON RIVER GROUP.

Phænopora multipora. n. s., Hall, 1851, pl. xxiv, figs 1 a, b	206–207
Clathropora flabellata, n. s., Hall, 1851, pl. xxiv, figs. 2 a, b	207
Chætetes lycoperdon. Say, pl. xv, figs. 1 a–d	207–208
Schizocrinus nodosus? pl. xxv, figs. 2 a–c	208

231

	Page
Echinosphaerites ?, n. s., Hall, 1851, pl. xxv, figs. 3 a, b	208-209
Crinoidew, or *Cystidew*	209
—— ?, pl. xxv, figs. 4 a-c	209
Murchisonia major, n. s., Hall, 1851, pl. xxvi, figs. 1 a-c	209-210
Asaph is *barrandi*, n. s., Hall, 1851, pl. xxvii, figs. 1 a-d, and pl. xxviii	210-211
Harpes escanabiw, n. s., Hall, 1851, pl. xxvii, fig. 2 a	211-212
Phacops callicephalus, pl. xxvii, figs. 3 a, b	212
Calemporn gracilis, n. s., Hall, 1851, pl. xxix, figs. 1 a, b	212-213
Sarcinula ? obsoleta, n. s., Hall, 1851, pl. xxix, fig. 2 a, b	213
Modiolopsis pholadiformis, n. s., Hall, 1851, pl. xxx, figs. 1 a-c, and pl. xxxi, fig. 1	213-214
Modiolopsis modiolaris, Hall, pl. xxxi, figs. 2 a-d	214-215
Ambonychia carinata, Hall, pl. xxxi, fig. 3	215

CLINTON GROUP.

Tracks and trails of vertebrates?	215-218
Tracks of Crustaceans?	219-220

NIAGARA GROUP.

Huronia vertebralis, Stokes, pl. xxxiv, fig. 1	221
Huronia annulata, n. s., Hall, 1851, pl. xxxiv, fig. 4	221-222
Discosorus conoideus, pl. xxxiv, figs. 2 and 3	222-223

UPPER HELDERBERG LIMESTONES.

Dictyonema fenestrata, pl. xxxv, figs. 1 a, b	223-224
Proëtus —— ?, Hall, 1851, pl. xxxv, fig. 2	224
Phacops anchiops, pl. xxxv, figs. 3 a, b	224-225

LIST OF FOSSILS.

General remarks on the above list	229
Table of the number of species of fossils found in the State of New York and the Lake Superior district	230

3.

HALL, JAMES. Letter from Professor James Hall, of New York, containing observations on the Geology and Palæontology of the country traversed by the expedition, and notes upon some of the Fossils collected on the route. <Exploration and Survey of the Valley of the Great Salt Lake of Utah, including a reconnaissance of a new route through the Rocky Mountains, by Howard Stansbury. Appendix E, pp. 398-414, pls. i-iv. Philadelphia, 1852.

CORALS.

Cyathophyllidew.	Page.
Cariphyllum ? rugosum, n. s., Hall, 1852, pl. i, figs. 1 a, b	407-408
Zaphrentis ? multilamella, n. s., Hall, 1852, pl. i, fig. 2	408
Zaphrentis stansburii, Hall, n. s., 1852, pl. i, figs. 3 a, b	407
Lithostro[?]tion —— (sp. indet.), Hall, 1852, pl. i, fig. 4 a, b	408
Brachiopoda.	
Terebratula subtilita, n. s., Hall, 1852, pl. iv, figs. 1 a, b, 2 a, b	409
Spirifer hemiplicata, n. s., Hall, 1852, pl. iv, figs. 3 a, b	409
S. octoplicata ? pl. iv, fig. 4 a, b	409-410
S. triplicata, Hall, n. s, 1852, pl. iv, figs. 5 a-c	410
Chonetes variolata, (D'Orb. sp.) DeKoninck, pl. iii, figs. 1 a, b	410
Productus costatus, DeKoninck, pl. iii, fig. 2	411
P. semireticulatus, DeKoninck pl. iii, figs. 3-5 a, b	411
Productus —— (sp. indet.), Hall, 1852, pl. iii, fig. 4	411
Orthis umbraculum ? pl. iii, fig. 6	412
Acephala.	
Avicula ? cesta, pl. ii, figs. 1 a, b	412
Tellinomya producta, Hall, n. s., 1852, pl. ii, fig. 3	412
Cypricardia occidentalis, n. s., Hall, 1852, pl. iv, fig. 2	412
Allorisma terminalis, n. s., Hall, 1852, pl. ii, figs. 4 a, b	413
A. ovata arata, n. s., Hall, 1852, pl. ii, figs. 5 a, b	413
Gasteropoda.	
Pleurotomaria coronata, n. s., Hall, 1852, pl. ii, figs. 6 a-d	413-414
Euomphalus subplanus, n. s., Hall, 1852, pl. ii, figs. 7 a, b	414

4.

HALL, JAMES. Descriptions and Notices of the Fossils collected upon the route. <Rep. Expls. and Survs. from the Mississippi River to the Pacific Ocean, vol. iii. No. 1, general report upon the geological collections, chapter ix, pp. 99–105, pls. i and ii. Washington, 1856.

CRETACEOUS SPECIES.

	Page.
Gryphœa, Lamarck	99
G. pitcheri, Morton, 1834, pl. i, figs. 1–6	99–100
G. pitcheri, var. *navia*, Hall, 1856, (pl. i, figs. 7–10) *	100
Ostrea, Linnæus	100
O. congesta, Conrad, 1843, pl. i, fig. 11	100–101

CARBONIFEROUS SPECIES.

Terebratula [L]hwyd	101
T. millepunctata, n. s., Hall, 1856, pl. ii, figs. 1, 2	101
T. subtilita, Hall, pl. ii, figs. 3–5	101
Spirifer, Sowerby	101
S. lineatus, pl. ii, figs. 6–8	101–102
S. kentuckensis, Shumard, 1855, pl. ii, figs. 10, 11	102
S. cameratus, Morton, 1836, pl. ii, figs. 9, 12, 13	102–103
Productus, Sowerby	103
P. semireticulatus, Martin, 1809, pl. ii, figs. 16, 17	103
P. rogersi, Norwood & Pratten, 1854, pl. ii, figs. 14, 15	104
Imperfect specimens, the specific characters of which are obscure or indeterminable	104–105

5.

HALL, JAMES. Geology and Palæontolgy of the Boundary. <U. S. and Mex. Boundary Survey. Report of William H. Emory, vol. i, part ii, pp. 101–140, and 144–146, pls. i and xx and xxi pars. Washington, 1857.

The fossils figured on pl. xx were apparently named by James Hall, but no descriptions accompany them.

Columnaria thomii, n. s., Hall, 1857, pl. xx, figs, 1 *a–d*.
Terebratula mexicana, n. s., Hall, 1857, pl. xx, figs. 2 *a–c*.
Orthis arachnoides, n. s., Hall, 1857, pl. xx, figs. 3 *a, b*.
Euomphalus michleranus, n. s., Hall, 1857, pl. xx, fig. 4.
Asaphus emoryi, n. s., Hall, 1857, pl. xx, fig. 5.

The Echinoderms in this report are described by Professor Hall.

	Page
Pyrina parryi, n. s., Hall, 1857, pl. i, figs. 1 *a–d*	144–145
Toxaster texanus, Rœmer, pl. i, figs. 2 *a–c*	145
Cyphosoma texanum, Rœmer, pl. i, figs. 3 *a–c*	145
Holectypus planatus, n. s., ? Hall, 1857, pl. i, figs. 4 *a–f*	145–146
Toxaster elegans, Shumard, sp. pl. xxi, figs. 1 *a–e*	146

Gryphea Pitcheri, (Mort.). Hall is figured on pl. xxi, figs. 3 *a–c*, but its locality is not mentioned and it is not described.

6.

HALL, JAMES, *and* WHITFIELD, R. P. Palæontology. <Rep. Geol. Expl. 40th parallel by Clarence King, vol. iv, part ii, pp. 198–302, pls. i–vii. Washington, 1877.

FOSSILS OF THE POTSDAM GROUP.

Brachiopoda.

	Page.
Obolella, Bill.	205
O. discoida, n. s., H. & W., 1877, pl. i, figs. 1, 2	205
Lingulepis, Hall	206
L. mera, n. s., H. & W., 1877, pl. i, figs. 5–7	206

---

* These figures are copied from Marcou's figures of *Gryphœa Pitcheri*, Mort., except fig. 10, which is copied from a figure of Marcou's *Gryphœa dilatata*, Sow., Bull. Geol. Soc. France, 2d ser. vol. xii, pl. xxi, and Geology of North America, pl. iv.

	Page.
L. ? minutus, n. s., H. & W., 1877, pl. i, figs. 3, 4	206–207
Kutorgina, Billings	207
K minutissima, n. s., H. & W., 1877, pl. i, figs. 11, 12	207–208
Leptæna, Dalman	208
L. melita, n. s., H. & W., 1877, pl. i, figs. 13, 14	208–209
Crustacea.	
Conocephalites, Zenker, = *Conocoryphe* Corda	209
Subgenus *Crepicephalus*, Owen? — *Loganellus*, Devine	209
Crepicephalus (*Loganellus*) *hagueii*, n. s., H. & W., 1877, pl. ii, figs. 14, 15	210–212
C. (*Loganellus*) *nitidus*, n. s., H. & W., 1877, pl. figs. 8–10	213–214
C. (*Loganellus*) *granulosus*, n. s., H. & W., 1877, pl. ii, figs. 2, 3	214–215
C. (*Loganellus*) *maculosus*, n. s., H. & W., 1877, pl. ii, figs. 24, 25, and 26?	215–216
C. (*Loganellus*) *unisulcatus*, n. s., H. & W., 1877, pl. ii, figs. 22, 23	216–217
C. (*Loganellus*) *simulator*, n. s., H. & W., 1877, pl. ii, figs. 16–18	218
C. (*Loganellus*) *anytus*, n. s., H. & W., 1877, pl. ii, figs. 19–21	219–220
C. (*Bathyurus?*) *angulatus*, n. s., H. & W., 1877, pl. ii, fig. 28	220–221
Pterocephalus, Roemer	221
Conocephalites (*Pterocephalus*) *laticeps*, n. s. H. & W., 1877, pl. ii, figs. 4–7	221–223
Ptychaspis, Hall	223
P. pusulosa, n. s., H. & W., 1877, pl. ii, fig. 27	223–224
Chariocephalus, Hall	224
C. tumifrons, n. s., H. & W., 1877, pl. ii, figs. 38, 39	224–225
Dikellocephalus, Owen	225–226
D. bilobatus, n. s., H. & W., 1877, pl. ii, fig. 36	226
D. multicinctus, n. s., H. & W., 1877, pl. ii, fig. 37	226–227
D. flabellifer, n. s., H. & W., 1877, pl. ii, figs. 29, 30	227–228
Agnostus, Brongniart	228
A. communis, n. s., H. & W., 1877, pl. i, figs. 28, 29	228–229
A. neon, n. s., H. & W., 1877, pl. i, figs. 26, 27	229–230
A. prolongus, n. s., H. & W., 1877, pl. i, figs. 30, 31	230–231
A. tumidosus, n. s., H. & W., 1877, pl i. fig. 32	231

FOSSILS OF THE LOWER SILURIAN.

Brachiopoda.	
Lingulepsis, Hall	232
L. ella, n. s., H. & W., 1877, pl. i, fig. 8	232
Orthis, Dalman	232
O. pogonipensis, n. s., H. & W., 1877, pl. i, figs. 9, 10	232–233
Strophomena, Rafinesque	233
S. nemea, n. s., H. & W., 1877, pl. i, fig. 15	233–234
Porambonites, Pander	234
P. obscurus, n. s., H. & W., 1877, pl. i, fig. 16	234
Gasteropoda.	
Raphistoma, Hall	235
R. acuta, n. s., H. & W., 1877, pl. i, figs. 20–22	235
Maclurea, Lesueur	235
M. minima, n. s., H. & W., 1877, pl. i, figs. 17–19	235–236
Fusispira, Hall	236
F. compacta, n. s., H. & W., 1877, pl. i, fig. 25	236–237
Cyrtolites, Conrad	237
C. sinuatus, n. s., H. & W., 1877, pl. i, figs. 23, 24	237
Crustacea.	
Conocephalites, Zenker	237
C. subcoronatus, n. s., H. & W., 1877, pl. ii, fig. 1	237–238
Crepicephalus, Owen != *Loganellus*, Devine	238
C.? (Loganellus) quadrans, n. s., H. & W., 1877, pl. ii, figs. 11–13	238–240
Dikellocephalus, Owen	240
D. quadriceps, n. s., H. & W., 1877, pl. i, figs. 37–40	240–241
D. wahsatchensis, n. s., H. & W., 1877, pl. i, fig. 35	241–242
D.? gothicus, n. s., H. & W., 1877, pl. i, fig. 36	242–243
Bathyurus, Billings	243
B. pogonipensis, n. s., H. & W., 1877, pl. i, figs. 33, 34	243–244
Ogygia, Brongn.	244
O. producta, n. s., H. & W., 1877, pl. ii, figs. 31–34	244–245
O. parabola, n. s., H. & W., 1877, pl. ii, fig. 35	245–246

Brachiopoda.

FOSSILS OF THE DEVONIAN.

	Page.
Strophodonta, Hall	246
S. cannce, H. & W., pl. iii, figs. 1-3	246-247
Rhynchonella, Fischer	247
R. emmonsi, n. s., H. & W., 1877, pl. iii, figs. 4-6	247-248

Lamellibranchiata.

Paracyclas, Hall	248
P. peroccidens, n. s., H. & W., 1877, pl. iii, figs. 14-17	248
Nuculites, Conrad	248
N. triangulus, n. s., H. & W., 1877, pl. iii, figs. 12, 13	248-249
Lunulicardium, Munster	249
L. fragosum, Meek, pl. ii, figs. 9, 11	249-250

Gasteropoda.

Bellerophon, Montf.	250
B. neleus, n. s., H. & W., 1877, pl. iii, figs. 18-20	250-251

Radiata.

FOSSILS OF THE WAVERLY GROUP.

Michelina, D'Kon.	251
Michelina —— sp. ? H. & W., 1877, pl. iv, fig. 19	251-252

Brachiopoda.

Streptorhynchus, King	252
S. equivalvis, Hall, pl. iv, figs. 1, 2	252
S. inflatus, White & Whitf., 1862, pl. iv, fig. 3	252-253
Strophomena, Rafinesque	253
S. rhomboidalis, Wilckens, pl. iv, fig. 4	253
Chonetes, Fischer	253
C. loganensis, n. s., H. & W., 1877, pl. iv, fig. 9	253-254
Spirifera, Sow	254
S. centronata, Winchell, 1865, pl. iv, figs. 5, 6	254-255
S. alba-pinensis, n. s., H. & W., 1877, pl. iv, figs. 7, 8	255-256
Athyris, McCoy	256
A. claytoni, u. s., H. & W., 1877, pl. iv, figs. 15-17	256-257
A. planosulcata? n. s., H. & W., 1877, pl. iv, figs. 10, 11	257
Rhynchonella, Fischer	257
R. pustulosa, White ? pl. iv, figs. 12-14	257-258
Terebratula (Llhwyd), Brug.	258
T. utah, n. s., H. & W., 1877, pl. iv, fig. 18	258

Gasteropoda.

Euomphalus, Sow.	259
E. (Straparollus) utahensis, n. s., H. & W., 1877, pl. iv, figs. 20-23	259-260
E. laxus, White, MSS., pl. iv, figs. 24, 25	260-261
E. (Straparollus) ophirensis, n. s., H. & W., 1877, pl. iv, figs. 26, 27	261

Crustacea.

Proetus, Stein.	262
P. peroccidens, n. s, H. & W., 1877, pl. iv, figs. 28-32	262-264
P. loganensis, n. s., H. & W., 1877, pl. iv, fig. 33	264

Brachiopoda.

FOSSILS OF THE LOWER CARBONIFEROUS.

Orthis, Dalman	265
O. resupinata? Martin, sp., pl. v., figs. 1, 2	265
Productus, Sowerby	265
P. flemingi var. *burlingtonensis*, Hall, pl. v, figs. 9-12	265-266
P. lævicostus? White? 1860, pl. v, figs. 7, 8	266-267
P. semireticulatus, Martin, pl. v, figs. 5, 6	267-268
P. elegans, n. s., H. & W., 1877, pl. v, figs. 3, 4	268-269
Spirifera, Sowerby	269
S. striata, pl v. figs. 13-15	269-270
S. setigera, Hall, pl. v, figs. 17, 18	270-271
Spirifera —— sp. ? H. & W., 1877, pl. v, fig. 16	271
Athyris, McCoy	271
A. subquadrata ? Hall, pl. v, figs. 19, 20	271-272

Lamellibranchiata.

FOSSILS OF THE COAL-MEASURES AND PERMO-CARBONIFEROUS.

Aviculopecten, McCoy	273
A. weberensis, n. s., H. & W., 1877, pl. vi, fig. 5	273

	Page
A. certo cardinalis, n. s., H. & W., 1877, pl. vi, fig. 4	273-274
A. par loi, n. s., H. & W., 1877, pl. vi, fig. 6	274-275
Myalina, De Koninck	275
M. avieuloides, M. & H. 1860, pl. vi, fig. 8	275-276
M. permiana, Swallow 1858, pl. vi, fig. 7	276
Sedgwickia, McCoy	276
S. ? concava, M. & H., 1858, pl. vi, fig. 3	276-277
Cardiomorpha, De Koninck	277
C. missouriensis, Swallow, 1858, pl. vi, figs. 1, 2	277
Cephalopoda	
Cyrtoceras, Goldf	278
C. cessator, n. s., H. & W., 1877, pl. vi, fig. 15	278
Goniatites, De Haan	279
G. kingii, n. s., H. & W., 1877, pl. vi, figs. 9-14	279-280

FOSSILS OF THE TRIASSIC FORMATION.

Radiata
Echinodermata.

Pentacrinites, Miller	280
P. asteriscus ? M. & H., 1858, pl. vi, fig. 16	280-281
Brachiopoda.	
Spirifera, D'Orb	281
S. homfrayi ? Gabb, pl. vi, fig. 18	281
Spirifera (Spiriferina ?) alia, n. s., H. & W., 1877, pl. vi. fig. 17	281-282
Terebratula (Llhwyd.), Brug.	282
T. humboldtensis, Gabb, pl. vi, figs. 22-24	282-283
Lamellibranchiata.	
Edmondia, De Koninck	283
E. myrina, n. s., H. & W., 1877, pl. vi, fig. 19	283

FOSSILS OF THE JURASSIC PERIOD.

Brachiopoda.

Rhynchonella, Fischer	284
R. myrina, n. s., H. & W, 1877, pl. vii, figs. 1-5	284
R. gnathophora ? Meek, pl. vii, fig. 6	284-285
Terebratula (Llhwyd.), Brug.	285
T. augusta, n. s., H. & W., 1877, pl. vii, figs. 7-10	285
Lamellibranchiata.	
Ostrea, Linn.	285
Ostrea —— sp. ? H. & W., 1877, pl. vii, fig. 12	285-286
Gryphaea, Lam.	286
G. calceola var. *nebrascensis*, H. & W., 1877, M. & H., 1861, pl. vii, fig. 11	286-287
Aviculopecten, McCoy	288
A. (Eumicrotis ?) augustensis, n. s, H. & W., 1877, pl. vii, figs. 14-16	288
Eumicrotis, Meek	289
E. curta, Hall, pl. vii, fig. 24	289
Camptonectes, Agassiz	289
C. bellistriatus, Meek, pl. vii, fig. 13	289-290
C. extenuatus, Meek, pl. vii, fig. 18	290-291
C. pertenuistriatus, n. s., H. & W., 1877, pl. vii, fig. 17	291-292
Lima, Brug.	292
L. (Plagiostoma) occidentalis, n. s., H. & W., 1877, pl. vii, fig. 23	292-293
Trigonia, Brug	293
T. quadrangularis, n. s., H. & W., 1877, pl. vii, fig. 22	293-294
Leptocardia, (n. g.), H. & W., 1877	294
L. typica, n. s., H. & W., 1877, pl. vii, figs. 26-29	295-296
L. cardioidea, n. s., H. & W., 1877, pl. vii, fig. 25	296-297
Astarte, Sowerby	297
A. ? arenosa, n. s., H. & W., 1877, pl. vi, figs. 20-21	297-298
Gasteropoda.	
Natica, Lam.	298
N. ? lelia, n. s., H & W., 1877, vii, figs. 19-21	298-299
Classified list of the fossils described in this report	301-302

7.

HALL, JAMES, and MEEK, F. B. (see MEEK, F. B., and HALL, JAMES).

VI.—THE WRITINGS OF ANGELO HEILPRIN.

1.

HEILPRIN, ANGELO. On some new species of Eocene mollusca from the southern United States. <Proc. U. S. National Museum, vol. iii, pp. 149-152, pl. 1, 1880. Washington, 1881.

	Page.
Pleurotoma, Lamarck	149
P. pagoda, n. s., Heilprin, 1880, pl. —, fig. 1	149-150
P. venusta, n. s., Heilprin, 1880, pl. —, fig. 2	150
P. platysoma, n. s., Heilprin, 1880, pl. —, fig. 3	150
Eucheilodon, Gabb	150
E. creno-carinata, n. s., Heilprin, 1880, pl. —, fig. 4	150
Scalaria, Lamarck	150
S. unilineata, n. s., Heilprin, 1880, pl. —, fig. 5	150-151
Fusus, Lamarck	151
Subgenus *Strepsidura*, Swainson	151
F. (Strepsidura) marnochi, n. s., Heilprin, pl. —, fig. 6	151
Terebra, Lamarck	151
T. plicifera, n. s., Heilprin, pl. —, fig. 8	151
Crassatella, Lamarck	151
C. declivis, n. s., Heilprin, pl. —, fig. 9	151-152

2.

HEILPRIN, ANGELO. North American tertiary Ostreidæ. <4th Annual Rep. of the director of the U. S. Geological Survey. A review of the fossil ostreidæ of North America; and a comparison of the fossil with the living forms by C. A. White. Appendix I, pp. 309-316, pls. lxiv-lxxii. Washington, 1883.

	Page.
Ostrea, Linnæus	309

EOCENE.

Ostrea alabamensis, Lea, pl. lxiv, figs. 2-4	309
Ostrea carolinensis, Conrad	309
Ostrea compressirostra, Say, pl. lxv, figs. 1-2	309
Ostrea cretacea, Morton	310
Ostrea divaricata, Lea, pl. lxiv, fig. 1	310
Ostrea everna, Mellville, sp., pl. lxiv, figs. 5-8	310
Ostrea falciformis, Conrad	311
Ostrea mortoni, Gabb	311
Ostrea sellæformis, Conrad, pl. lxii, figs. 1,2; pl. lxiii, fig. 1	311
Ostrea thirsæ, Gabb, pl. lxiii, figs. 4-6	311
Ostrea trigonalis, Conrad	311
? *Ostrea twomeyi*, Conrad	311

OLIGOCENE.

Ostrea georgiana, Conrad	311-312
Ostrea vicksburgensis, Conrad, pl. lxiii, figs. 2,3	312

MIOCENE.

Ostrea atwoodi, Gabb, pl. lxviii, figs. 4,5	312
Ostrea borealis, Lamarck	312
Ostrea contracta, Conrad, pl. lxix, figs. 1,2	312

	Page.
Ostrea disparilis, Conrad, pl. lxvi, figs. 1, 2	312
Ostrea panzana, Conrad	313
Ostrea percrassa, Conrad, pl. lxvii, fig. 3	313
Ostrea sculpturata, Conrad, pl. lxx, fig. 2	313
Ostrea subfalcata, Conrad, pl. lxviii, figs. 1-3	313
Ostrea subjecta, Conrad	313
Ostrea tayloriana, Gabb, pl. lxvii, figs. 1, 2	313
Ostrea titan, Conrad	313–314
Ostrea celeriana, Conrad, pl. lxx, fig. 1	314
Ostrea virginica, Gmelin (= *O. virginiana*, Lamarck)	314

PLIOCENE.

Ostrea atwoodi, Gabb	314
Ostrea bourgeoisii, Rémond, pl. lxxi, fig. 1	214
Ostrea heermanni, Conrad	314
Ostrea vespertina, Conrad, pl. lxxi, figs. 2–4	315

POST-PLIOCENE.

Ostrea conchaphila, Carpenter	315
Ostrea fundata (Say?), F. S. Holmes	315
Ostrea gallus, Valenciennes	315
Ostrea lurida, Carpenter, pl. lxxii, figs. 2, 3	316
Ostrea veatchii, Gabb, pl. lxxii, fig. 1	316

VII.—THE WRITINGS OF ALPHEUS HYATT.

1.

HYATT, ALPHEUS. [Descriptions of new genera and remarks on new species of Triassic fossils.] < Rep. Geol. Expl. 40th Parallel, by Clarence King. Vol. iv, part i, pp. 107-128 pars. Washington, 1877.

	Page.
Clydonitidæ, n. f., Hyatt, 1877	107
* *Coroceras*, n. g., Hyatt, 1877	107-108
Clydonites, Hauer	109
C. lævidorsatus, Hauer, sp., 1860, pl. x, fig. 7	109-110
Trachyceratidæ, n. f., Hyatt, 1877	110
† *Gymnotoceras*, n. g., Hyatt, 1877	110-111
G. rotelliforme, Meek, 1877, pl. x, figs. 9, 9 a	112-113
G. blakei, Gabb, sp., 1864, pl. x, figs. 10, a–c, and pl. xi, figs. 6, 6 a	115-116
Trachyceras, Laube	116
T. whitneyi, Gabb, sp., 1864, pl. xi, figs. 3, 3 a	117-118
T. judicaricum, Mojsisovics, 1869, pl. xi, figs. 1, 1 a	118
Arcestidæ	
Arcestes, Suess, 1865	119-1–9
A. ? perplanus, Meek, 1877, pl. xi, figs. 7, 7 a	121
A. gabbi, Meek, 1877, pl. x, figs. 6, 6 a, b	123
Physanoidæ, Hyatt, 1877	124
‡ *Acrochordiceras*, n. g., Hyatt, 1877	124
A. hyatti, Meek, 1877, pl. xi, figs. 5, 5 a	125-126
§ *Eutomoceras*, n. g., Hyatt, 1877	126
‖ *Eudiscoceras*, n. g., Hyatt, 1877	128

2.

HYATT, ALPHEUS. [Description of the new genus Meekoceras, and remarks on the different species of the same.] < Contributions to Invertebrate Paleontology No. 5, Triassic fossils of Southeastern Idaho, by C. A. White. Twelfth Ann. Rep. of the U. S. Geol. and Geogr. Surv. of the Terr., by F. V. Hayden. pp. 112-116, pls. xxxi-xxxii. Washington, 1883.

An author's edition of these contributions was published in 1880.

	Page.
Meekoceras, n. g., Hyatt, 1880	112
M. aplanatum, n. s., White, 1880	113
M. mushbachianum, n. s., White, 1880	114
M. gracilitatis, n. s., White, 1880	115-116
Arcestes? cirratus, n. s., White, 1880	117

3.

HYATT, ALPHEUS. [Description of the new genus Enclimatoceras.] < On the Nautiloid genus Enclimatoceras Hyatt, and a description of the type species. < On Mesozoic fossils, by C. A. White, Bull. U. S. Geol. Surv. No. 4. Vol. i. pp. 16-17 of No. 4, or pp. 104-105 of vol. i. Washington, 1884.

Enclimatoceras, n. g., Hyatt, 1884	104-105

---

* κόρυς, a helmet; κέρας, a horn.
† γυμνός, naked; νῶτος, back; κέρας, a horn.
‡ ἀκροχορδών, a wart; κέρας, a horn.
§ εὖ (augm. part.); τομος, sharp; κέρας, a horn.
‖ εὖ (augm. part.); δίσκος, a quoit; κέρας, a horn.

VIII.—THE WRITINGS OF JULES MARCOU.

1.

MARCOU, JULES. Sur la géologie des Montagnes Rocheuses, entre le Fort Smith (Arkansas) et Albuquerque (Nouveau Mexique). <Bull. Geol. Soc. France, 2d ser., vol. xi, pp. 156–160. Paris, 1854.

> Mentions the existence, at Tucumcari, of the Jurassic formation, with *Gryphæa dilatata* and *Ostrea Marshii*.

2.

MARCOU, JULES. Résumé of a geological reconnaissance extending from Napoleon, at the junction of the Arkansas with the Mississippi, to the Pueblo de los Angeles, in California. <U. S. Pacific R. R. Expl., 1853–'54, vol. iii, 8vo. Report of explorations for a railway route near the 35th parallel of latitude, from the Mississippi River to the Pacific Ocean, by Lieut. A. W. Whipple, Corps of Topographical Engineers, chap. vi, pp. 40–48. H. Doc. 129. [Washington, 1855.]

> Mentions the occurrence of a number of fossils and employs for the first time the name *Gryphæa tucumcarii*.

3.

MARCOU, JULES. Résumé d'une section géologique des Montagnes Rocheuses à San Pedro, sur la côte de l'Océan Pacifique. <Bull. Geol. Soc. France, 2d ser., vol. xi, pp. 474–478. Paris, 1855.

> Mentions the occurrence of a *Gryphæa* at Laguna (New Mexico), and of Carboniferous fossils near the San Francisco Mountains (Arizona).

4.

MARCOU, JULES. Geological notes of a survey of the country comprised between Preston, Red River, and El Paso, Rio Grande del Norte. <U. S. Pacific R. R. Expl., vol. iv, 8vo. Report of explorations of a route for the Pacific Railroad, near the 32d parallel of latitude from the Red River to the Rio Grande, by Brevet Capt. John Pope, Corps of Topographical Engineers. Chap. xiii, pp. 125–128. H. Doc. 126. [Washington, 1855.]

> Mentions the occurrence of a number of fossils.

5.

MARCOU, JULES. Résumé explicatif d'une carte géologique des États-Unis et des provinces anglaises de l'Amérique du Nord, avec un profil géologique allant de la vallée du Mississippi aux côtes du Pacifique, et une planche de fossiles. <Bull. Geol. Soc. France, 2d ser., vol. xii, pp. 813–936, pls. xx–xxi. Paris, 1855.

> The occurrence of many fossils is mentioned in this paper.
>
> *Gryphæa dilatata*, Sowerby, pl. xxi, figs. 1 a, b, and 2.
> *G. dilatata* var. *tucumcarii*, Marcou, 1855, pl. xxi, fig. 3.
> *Ostrea marshii*, Sowerby, pl. xxi, fig. 4.
> *Gryphæa pitcheri*, Morton, pl. xxi, figs. 5 a, b, and 6.

6.

MARCOU, JULES. Notes géologiques sur le pays compris entre Preston, sur la rivière Rouge, et el Paso, sur le rio Grande del Norte. <Bull. Geol. Soc. France, 2d series, vol. xii, pp. 808-813. Paris, 1855.

> In this paper the author mentions the occurrence of many fossils. The notes are made up from collections and observations brought back by Captain Pope, U. S. Top. Engrs., from his survey from the Red River to the Rio Grande, in 1854.

7.

MARCOU, JULES. Résumé and field-notes by Jules Marcou, geologist and mining engineer of the Expedition, with a translation by William P. Blake. <Rep. Expls. and Survs. from the Mississippi River to the Pacific Ocean, vol. iii, part. iv, pp. 121-164. Washington, 1856.

> The occurrence of many fossils is mentioned in these field-notes, originally written in French.

8.

MARCOU, JULES. Résumé of a Geological reconnaissance extending from Napoleon, at the junction of the Arkansas with the Mississippi, to the Pueblo de los Angeles, in California. <Rep. Expls. and Survs. from the Mississippi River to the Pacific Ocean, vol. iii. Rep. of Lieut. A. W. Whipple, part iv, pp. 165-171. Washington, 1856.

> This résumé is reprinted from the preliminary or first report of Lieutenant Whipple, in 8vo. chap. vi, p. 40, Reports of Pacific Railroad Surveys, House Doc. 129, Washington, 1855. A few notes are added.

	Page
Gryphœa Pitcheri, Morton	167
G. dilatata, Sowerby	168
O. Marshii, Sowerby	168

9.

MARCOU, JULES. Geology of North America, with two reports on the Prairies of Arkansas and Texas, the Rocky mountains of New Mexico, and the Sierra Nevada of California, originally made for the United States Government, pp. 1-144, pls. i-ix. Zurich, 1858.

CONTENTS.

	Page
Introduction	1-8
Résumé of a geological reconnaissance extending from Napoleon, at the junction of the Arkansas with the Mississippi, to the pueblo de los Angeles in California. (Extract from Report of Exploration for a railway route, near the thirty-fifth parallel of latitude, from the Mississippi River to the Pacific Ocean, by Lieutenant, now Captain, A. W. Whipple, Corps of Topographical Engineers, chapter vi, page 40, &c., Washington, 1855, House of Representatives, Document No. 129)	9-25
Geological notes of a survey of the country comprised between Preston, Red River, and El Paso, Rio Grande Del Norte. (Extract from Report of Exploration of a route for the Pacific Railroad, near the thirty-second parallel of latitude, from the Red River to the Rio Grande, by Brevet Capt. John Pope, Corps of Topographical Engineers, chapter xiii, Geological Report, p. 125, &c. Washington, 1855)	26-31
Paleontology	32-53
Geology of New Mexico	54-57
On the geology of the United States and the British Provinces of North America. (Extract from Dr. A. Petermann's Geographischen Mittheilungen, Heft 6, in 4to. Gotha, 1855)	58-70
Sketch of a geological classification of the mountains of a part of North America. (Extract from the Annales des Mines, 5me série, tome vii, page 329, &c. Paris, 1855)	71-80
On the Gold of California. (Extract from the Bibothèque Universelle de Genève, février, 1855)	81-84

	Page
Construction of the Geological map of the United States and the British Provinces; Criticisms of the American Journal of Science and Arts, conducted by Professors B. Silliman, B. Silliman, jr., and James D. Dana	85–98
A synopsis of the history of the progress and discoveries of Geology in North America	99–121
List of maps and memoirs on the Geology of North America	122–143
Note	144

PALEONTOLOGY.

FOSSILS OF THE TERTIARY ROCKS.

Ostrea virginica var. *californica*, Marcou, 1858, pl. v, figs. 2, 2 a	32–33

FOSSILS OF THE CRETACEOUS ROCKS.

Ptychodus whipplei, n. s., Marcou, 1858, pl. i, figs. 4, 4 a	33
Ammonites shumardi, n. s., Marcou, 1858, pl. i, figs. 1, 1 a	33–34
Ammonites belknapii, n. s., Marcou, 1858, pl. ii, figs. 1 a, 1 b	34
Ammonites peruvianus, de Buch., pl. v, figs. 1, 1 a, 1 b	34–35
Ammonites gibbonianus, Lea, pl. ii, fig. 2 a, 2 b	35
Ammonites novi-mexicana, n. s., Marcou, 1858, pl. i, figs. 2, 2 a	35–36
Hamites fremonti, n s., Marcou, 1858, pl. i, fig. 3	36
Inoceramus lerouxi, n. s., Marcou, 1858, pl. ii, fig. 3	36–37
Isocardia washita, n. s., Marcou, 1858, pl. iii, figs. 2, 2 a, 2 b	37
Gryphæa sinuata var. *americana*, Marcou, 1858, pl. iii, fig. 1	37–38
Gryphæa pitcheri, Morton, pl. iv, figs. 5, 5 a, b, and 6	38–40
Holaster comanchesi, n. s., Marcou, 1858, pl. iii, figs. 3, 3 a	40–41
Exogyra flabellata, Goldf.	41–42
Cytherea missouriana, Mort	42
Tellina occidentalis, Mort	42
Caprotina texana, Roem	42

FOSSILS OF THE JURASSIC ROCKS.

Gryphæa dilatata var. *tucumcarii*, Marcou, 1858, pl. iv, figs. 1, 1 a, 1, 2, 3	43
Ostrea marshii, Sow., pl iv, fig. 4	43–44

FOSSILS OF THE MOUNTAIN LIMESTONE OR LOWER CARBONIFEROUS ROCKS.

Orthoceras novo-mexicana, n. s., Marcou, 1858, pl. vii, fig. 1	44
Myalina apachesi, n. s., Marcou, 1858, pl. vii, figs. 6, 6 a	44–45
Productus delawarii, n s., Marcou, 1858, pl. v, fig. 3	45
Productus cora, d'Orb., pl. vi, figs. 4, 4 a	45
Productus cora var. *mogayoni*. Marcou, 1858, pl. vi, fig. 5	45–46
Productus semi-recticulatus, Mart., pl. v, figs. 4, 4 a, pl. vi, fig. 6	46
Productus costatus, Sow., pl. v, fig. 5	46–47
Productus flemingii, Sow., pl. vi, fig. 7	47
Productus scabriculus, Mart., pl. v, figs. 6, 6 a	47–48
Productus pyxidiformis, de Kon., pl. vi, figs. 3, 3 a	48
Productus pustulosus, Phill., pl. vi, fig. 1	48
Productus punctatus, Mart., pl. vi, fig. 2	48
Orthis pecosii, n. s., Marcou, 1858, pl. vi, figs. 14, 14 a, b	48–49
Orthis crenistria, Phill	49
Spirifer striatus, Mart., pl. vii, figs. 2, 2 a	49
Spirifer striatus var. *triplicatus*, Marcou, 1858, pl. vii, fig. 3	49, 50
Spirifer rocky-montani, n. s., Marcou, 1858, pl, vii, figs. 4, 4 a–e	50
Spirifer lineatus, Mart., pl. vii, figs. 5, 5 a–c	50
Terebratula rocky-montana, n. s., Marcou, 1858, pl. vi, figs. 13, 13 a–c	50–51
Terebratula mormonii, n. s., Marcou, 1858, pl. vi, figs. 11, 11 a–c	51
Terebratula uta, n. s., Marcou, 1858, pl. vi, figs. 12, 12 a–c	51
Terebratula roysii, Leveillé, pl. vi, figs. 10, 10 a, b	51–52
Terebratula plano-sulcata, Phill., pl. vi, figs. 8, 8 a–b	52
Terebratula subtilita, Hall, pl. vi, figs. 9, 9 a–f	52
Zaphrentis stansburyi, Hall, pl. vii, fig. 7	52–53
Zaphrentis cylindrica, Milne-Edwards & Jules Haime, pl. vii, fig. 8	53
Amplexus coralloides? Sow	53

10.

MARCOU, JULES. Report on the Geology of a portion of Southern California. <Annual Report upon the Geographical Surveys West of the One Hundredth Meridian, in California, Nevada, Utah, Colorado, Wyoming, New Mexico, Arizona, and Montana, by George M. Wheeler, First Lieutenant of Engineers, U. S. A., being Appendix J J of the Annual Report of the Chief of Engineers for 1876. Appendix H, pp. 158–172. Washington, 1876.

	Page.
Pliocene rocks of Los Angeles	158–159
The Sierra of Santa Monica	159–160
Sierra Madre, Pacoña or Pacorina Cañon	160–161
Geology of the vicinity of the San Fernando Mission	161
The San Fernando sierra	161–164
Asphaltum and mineral oil near San Francisquito ranch	164–166
Sierra liebre and California desert	166
Cañada de las Uvas	166–167
Tertiary rocks in the vicinity of Fort Tejon	167–169
Tertiary rocks of California	169
Glacial rocks of Southern California and Pike's Peak	169–170
Mountain chains and their ages	170
The Sierra Madre	170–171
Coast Range	171–172
Sierras of San Fernando and Santa Monica	172
Hills of Los Angeles	172

The occurrence of many fossils is mentioned in this report.

IX.—THE WRITINGS OF JOHN STRONG NEWBERRY.

1.

NEWBERRY, J. S. Palæontology. <Report upon the Colorado River of the West, by Lieut. Joseph C. Ives. Part iii, chapter xi, pp. 116-132, pls. i-iii. Washington. 1861.

	Page.
Archæocidaris, McCoy	116
A. longispinus, n. s., Newberry, 1861, pl. i, figs. 1 and 1 *a*	116
A. ornatus, n. s., Newberry, 1861, pl. i, figs. 2, 3, 3 *a*	116-117
A. gracilis, n. s., Newberry, 1861, pl. i, figs. 4, 4 *a*	117
Ammonites, Brug	117
A. percarinatus, H. & M., 1856	117-118
Nautilus, Breynius	118
Nautilus, sp., Newberry, 1861	118
Bellerophon, Montfort	118
Inoceramus, Sowerby	119
I. problematicus, D'Orb.	119
I. crispii? Mantell	119
Pinna, Linn.	119
P. ? lingula, n. s., Newberry, 1861	119-120
Gryphæa, Lamarck	120
G. pitcheri, Morton	120
G. pitcheri var. *navia*, Hall	120
Allorisma, King	120
A. capax, n. s., Newberry, 1861, pl. i, figs. 9, 9 *a*	120-121
Productus rogersi, Norwood & Pratten	121
P. iresi, n. s., Newberry, 1861, pl. ii, figs. 1-8	122
P. occidentalis, n. s., Newberry, 1861, pl. ii, figs. 9, 10	122
P. calhounianus, Swallow	123
P. costatus, Sowerby	123
P. costatoides, Swallow	123-124
P. semireticulatus, Martin, De Koninck	124
P. nodosus, n. s., Newberry, 1861, pl. i, figs. 7, 7 *b*	124
P. splendens? Norwood & Pratten, 1855	124-125
P. scabriculus, Martin	125
Streptorhynchus, King	125
S. umbraculum, Von Buch	125-126
S. pyramidalis, n. s., Newberry, 1861, pl. ii, figs. 11-13	126
S. occidentalis, n. s., Newberry, 1861, pl. i, figs. 5, 5 *a*	126
Athyris	126
A. subtilita, Hall	126-127
Spirifer, Sowerby	127
S. cameratus, Mort.	127
S. rockymontani, Marcou	127
S. lineatus, De Koninck	127-128
Chonetes, Fisher	128
C. verneuiliana, Norwood & Pratten, pl. ii, fig. 6	128
Rhynconella, Fisher	128
R. uta, Marcou	128
Pecten, Linn.	128
P. occidentalis, Shumard	128
P. (Monotis ?) coloradensis, n. s., Newberry, 1861, pl. i, figs. 6, 6 *a*	129
Fus[u]lina, Fisher	129
F. cylindrica, Fisher	129

FOSSIL PLANTS.	Page.
Cyclopteris, Brong.	129
C. mosquensis, n. s., Newberry, 1861, pl. vi, figs. 1, 2	129–130
Pecopteris, Brong.	130
P. cyclolobu, n. s., Newberry, 1861, pl. iii, figs. 3, 4, 4 *a*	130–131
Neuropteris, Brong.	131
N. angulata, n. s., Newberry, 1861, pl. iii, fig. 5	131
Sphenopteris, Brong.	131
Subspecies, Newberry, 1861	131
Phyllites, Sternb.	131
P. renosissimus, n. s., Newberry, 1861, pl. iii, fig. 6	131
P. coriaceous, n. s., Newberry, 1861, pl. iii, figs. 7, 7 *a*	132
Clathropteris, Brong.	132

2.

NEWBERRY, J. S. Descriptions of the Carboniferous and Triassic Fossils collected on the San Juan Exploring Expedition under Capt. J. N. Macomb, U. S. Engineers. <Rep. Expl. Exp. from Santa Fé, New Mexico, to the junction of the Grand and Green rivers of the Great Colorado of the West in 1859, under the command of Capt. J. N. Macomb, 135–148, pls. iii–viii. Washington, 1876.

	Page.
Deltodus mercurei, n. s., Newberry, 1876, pl. iii, figs. 1, 1 *a*	137
Ptychodus whipplei, Marcou, pl. iii, figs. 2, 2 *a–f*	137–138
Athyris subtilita, Hall, sp	138
Spirifer cameratus, Morton	138
Spirifer (Trigonotreta?) texanus, Meek, 1871, pl. iii, figs. 5, 5 *a*, *b*	139–140
Productus nodosus, Newberry, pl. iii, figs. 3, 3 *d*	140
Pleurotomaria excelsa, n. s., Newberry, 1876, pl. iii, figs. 4, 4 *a*	140–141
Lamna texana, Roemer	141
Oxyrhina mantelli, Agass.	141
Otozamites macombii, n. s., Newberry, 1876, pl. iv, figs. 1, 2, pl. vi, figs. 5, 5 *a*	141–142
Zamites occidentalis, n. s., Newberry, 1876, pl. v, figs. 1, 1 *a*, and 2	142–143
Pecopteris bullatus, Bunbury, pl. vi, figs. 1, 1 *a*	143
Pecopteris mexicana, n. s., Newberry, 1876, pl. vi, figs. 2, 2 *a*	143–144
Pecopteris falcatus, Emmons, pl. vi, fig. 3	144
Pterophyllum fragile, n. s., 1876, Newberry, pl. vi, figs. 6, 6 *a*	144
Pterophyllum robustum, n. s., Newberry, 1876, pl. vi, fig. 7	145
Podozamites crassifolia, n. s., Newberry, 1876, pl. vi, fig. 10	145
Alethopteris whitnei, n. s., Newberry, 1876, pl. vii, figs. 1, 1 *a*, *b*	145–146
Camptopteris remondi, n. s., Newberry, 1876, pl. vii, figs. 2, 2 *a*	146–147
Tæniopteris elegans, n. s., Newberry, 1876, pl. viii, fig. 1	147
Tæniopteris glossopteroides, n. s., Newberry, 1876, pl. viii, figs. 2, 2 *a*	147
Tæniopteris magnifolia, Rogers, pl. viii, figs. 3, 4	147–148
Jeanpaulia radiata, n. s., Newberry, 1876, pl. viii, fig. 6	148

X.—THE WRITINGS OF DAVID DALE OWEN.

1.

OWEN, D. D. Description of some organic remains figured in this work, supposed to be new. <Rep. Geol. Expl. of part of Iowa, Wisconsin, and Illinois, by David Dale Owen. Appendix, pp. 69–86, pls. xi–xviii and p. 33, pl. vii. Washington, 1844.

	Page.
Catenipora escharoides, pl. vii, fig. 2	33
Pentamerus oblongus, pl. vii, fig. 3	33
Pentamerus huspodus, pl. vii, fig. 3	33

APPENDIX.

Cyathopora iowensis, n. s., D. D. Owen, 1844, pl. xi	69
Orthoceras undulatum, n. s., D. D. Owen, 1844, pl. xii. fig. 6	69
Gyroceras cornutus, n. s., D. D. Owen, 1844, pl. xii, fig. 8	69
Delthyris curuleines, n. s., D. D. Owen, 1844, pl. xii, fig. 9	69
Anthophyllum expansum, n. s., D. D. Owen, 1844, pl. xiii, fig. 3	69
Lunulites? dactioloides, n. s., D. D. Owen, 1844, pl. xiii, fig. 4	69
Cyathophyllum caliculare, n. s., D. D. Owen, 1844, pl. xiii, fig. 5	69
Cyathophyllum corinthium, n. s., D. D. Owen, 1844, pl. xiii, fig. 6	69
Cyathophyllum undulatum et multiplicatum, n. s., D. D. Owen, 1844, pl. xiii, fig. 10	69
Sarcinula (Porites?) glabra, n. s., D. D. Owen, 1844, pl. xiii, fig. 11	70
Lamellopora	70
L. infundibularia, n. s., D. D. Owen, 1844, pl. xiv, fig. 1	70
Astrea mamillaris, n. s., D. D. Owen, 1844, pl. xiv, fig. 3	70
Astrea? gygas, n. s., D. D. Owen, 1844, pl. xiv, fig. 7	70
Lingula iowensis, n. s., D. D. Owen, 1844, pl. xv, fig. 1	70
Orthoceras marginale, n. s., D. D. Owen, 1844, pl. xvi, fig. 6	70
Curtoceras conicum, n. s., D. D. Owen, 1844, pl. xvi, fig. 9	70
Strophomena concexa, n. s., D. D. Owen, 1844, pl. xvii, fig. 2	70
Orbitulites? reticulata, n. s., D. D. Owen, 1884, pl. xviii, fig. 7	70
Atrypa limitaris,? Vanuxem, pl. xii, fig. 1	74
Atrypa prisca, Hengevien, pl. xii, fig. 2	74
Calymene bufo, Green, pl. xii, fig. 3	74
Atrypa ? ———, pl. xii, fig. 4	74
Delthyris ? ———, pl. xii, fig. 5	74
Favosites polymorpha ramosa, Goldfuss, pl. xii, fig. 7	74
Atrypa prisca ———, pl. xii, fig. 10	74
Bellerophon ———, pl. xii, fig. 11	74
Nucula? ———, pl. xii, fig. 12	74
Favosites polymorpha, Goldfuss, pl. xii, fig. 13	74
Cyathophyllum helianthoides, Goldfuss, pl. xiii, fig. 1	76
Syringopora (lineata?) ———, pl. xiii, fig. 2	76
Favosites maxima,? Troost, pl. xiii, fig. 7	76
Porites? astraformis, Owen, pl. xiii, fig. 8	76
Phragmoceras centricosum,? ———, pl. xiii, fig. 9	76
Aulopora serpens, Goldfuss, pl. xiv, fig. 2	78
Syphonia piriformis,? Golfuss, pl. xiv, fig. 4	78
Nucula minuta, Owen, pl. xiv, fig. 5	78
Cyathophyllum turbinatum? Goldfuss, pl. xiv, fig. 6	78
Cyathophyllum vesiculosum? Goldfuss, pl. xiv, fig. 8	78
Orthis, pl. xiv, fig. 9	78
Pentamerus oblongus? pl. xiv, fig. 10	78
Astrea fungiformis, pl. xiv, fig. 11	78
Sarcinula costata, Goldfuss, pl. xiv, fig. 12	78
Tubipora lamellosa, n. s., Owen, 1844, pl. xiv, fig. 13	78
Lingula iowensis, pl. xv, fig. 1	80

	Page.
Bellerophon (cast), pl. xv, fig. 2	80
Orthis (cast), pl. xv, fig. 3	80
Pleurotomaria? (cast), pl. xv, fig. 4	80
Trochus lenticularis? (cast), pl. xv, fig. 5	80
Pleurotomaria (cast), pl. xv, fig. 6	80
Delthyris cast), pl. xv, fig. 7	80
Euomphalus (cast), pl. xv, fig. 8	80
Atrypa orbicularis (cast), pl. xv, fig. 9	80
Pleurotomaria (cast), pl. xv, fig. 10	80
Orthis testudinaria? (cast), pl. xv, fig. 11	80
Bellerophon (cast), pl. xv, fig. 12	80
Illænus-Trentonensis? pl. xvi, fig. 1	82
Casts of *Strophomena*, pl. xvi, fig. 2	82
Casts of *Strophomena, Orthis?* and *Atrypa*, pl. xvi, fig. 4	82
Cyathophyllum profundum, Cou., pl. xvi, fig. 5	82
Pleurotomaria? pl. xvi, fig. 7	82
Siliceous casts of *Strophomena deltoidea*, pl. xvi, fig. 8	82
Orthis, pl. xvii, fig. 1	84
Pleurotomaria? pl. xvii, fig. 3	84
Atrypa, pl. xvii, fig. 4	84
Strophomena serica? pl. xvii, fig. 5	84
Strophomena deltoidea, pl. xvii, fig. 6	84
Paradoxides? pl. xvii, fig. 7	84
Cardium iowensis, n. s., Owen, 1844, pl. xvii, fig. 8	84
Thaleops? pl. xvii, fig. 9	84
Strophomena nasuta? pl. xvii, fig. 10	84
Ceraurus, pl. xvii. fig. 11	84
Cypricardites, pl. xvii, fig. 12	84
Pleurotomaria, pl. xvii, fig. 13	84
Delthyris expansus, pl. xvii, fig. 14	84
Bellerophon bilobatus, pl. xvii, fig. 15	84
Strophomena, pl. xviii, fig. 1	86
Asaphus, pl. xviii, fig. 2	86
Strophomena angulata? pl. xviii, fig. 3	86
Pleurotomaria? pl. xviii, fig. 4	86
Pleurotomaria angulata? pl. xviii, fig. 5	86
Pleurotomaria lenticularis, pl. xviii, fig. 6	86
Trochus? pl. xviii, fig. 7	86
Orthoceras, pl. xviii, fig. 9	86
Phragmoceras? pl. xviii, fig. 11	86

2.

OWEN, D. D. On the Geology of the Western States of North America. <Quart. Journ. Geol. Soc., London. Vol. ii, pp. 433–447, pl. xix. London, 1846.

This article, of which an abstract was given in the Proc. Geol. Soc., London., vol. iv, p. 1, contains many paleontological notes.

	Page.
Pentremites pyriformis, Say	437
Archimedes, Lesueur	437
Retepora archimedea	437
Aulopora tubæformis	442
Retepora indianensis, n. s., Owen, 1846	442

3.

OWEN, D. D. [Lists of fossils found.] <Rep. of a Geol. Recon. of the Chippewa land district of Wisconsin. 30th Congress, 1st session, Senate, Executive Doc. No. 57. Appendix, pp. 131–133. Washington, 1848.

	Page.
List of fossil genera found in the lower sandstones of Wisconsin, fig 1, species undetermined	131
List of fossil genera found in the lower magnesian limestone, fig. 2, species undetermined	131

	Page.
List of fossil genera and species found in the lower fossiliferous limestone at St. Peter's and Fort Snelling, which are identical with those occurring in the blue limestone of the Ohio Valley	131–132
List of organic remains found near the "Big Spring," on the Upper Iowa River	132
List of organic remains found in the limestones (fig. 3) of Turkey River, near the agency and the vicinity	133

4.

OWEN, D. D., *and* SHUMARD, B. F. Descriptions of fifteen new species of Crinoidea from the subcarboniferous limestone of Iowa, collected during the U. S. Geological Survey of Iowa, Wisconsin, and Minnesota, in the years 1848–'49. <Journ. Acad. Nat. Sci., Philadelphia. Second series, vol. ii, part i, pp. 57–70, pl., vii. Philadelphia, 1850.

This article was afterward republished in Owen's U. S. Geol. Rep. of Iowa, Wisconsin, and Minnesota.

	Page.
Platycrinus, Miller	57
P. planus, n. s., Owen & Shumard, 1850, pl. vii, figs. 4 *a–c*	57–58
P. yandellii, n. s., Owen & Shumard, 1850, pl. vii, figs. 6 *a, b*	58
P. discoideus, n. s., Owen & Shumard, 1850, pl. vii, figs. 1 *a, b*	58–59
P. corrugatus, n. s., Owen & Shumard, 1850, pl. vii, figs. 2 *a–e*	59–60
P. burlingtonensis, n. s., Owen & Shumard, 1850, pl. vii, fig. 5	60–61
Dichocrinus, Munster	61
D. ovatus, n. s., Owen & Shumard, 1850, pl. vii, figs. 9 *a, b*	61–62
D. striatus, n. s., Owen & Shumard, 1850, pl. vii, figs. 10 *a, b*	62–63
Cyathocrinus, Miller	63
C. iowensis, n. s., Owen & Shumard, 1850, pl. vii, figs. 11 *a–c*	63
C. cornutus, n. s., Owen & Shumard, 1850, pl. vii, figs. 8 *a, b*	63–64
Pentremites, Say	64
P. norwoodii, n. s., Owen & Shumard, 1850, pl. vii, figs. 13 *a–c*	64–65
P. melo, n. s., Owen & Shumard, 1850, pl. vii, figs. 14 *a–c*	65–66
P. laterniformis, n. s., Owen & Shumard, 1850, pl. vii, fig. 15	66
P. stelliformis, n. s., Owen & Shumard, 1850, pl. vii, fig. 16 *a, b*	67
Actinocrinus unicornus, n. s., Owen & Shumard, 1850, pl. vii, figs. 12 *a, b*	67–68
Actinocrinus evansii, n. s., Owen & Shumard, 1850, pl. vii, figs. 3 *a, b*	68–69

5.

OWEN, D. D., *and* SHUMARD, B. F. On the Number and Distribution of Fossil Species in the Palæozoic Rocks of Iowa, Wisconsin, and Minnesota. <Proc. Amer. Ass. Adv. Sci., fifth meeting, 1851. pp. 235–239. Washington and Cincinnati, 1851.

6.

OWEN, D. D., *and* SHUMARD, B. F. Description of seven new species of Crinoidea from the subcarboniferous limestone of Iowa and Illinois. < Journ. Acad. Nat. Sci., Philadelphia. Second series, vol. ii, part ii, pp. 89–94, pl. xi. Philadelphia, 1852.

This article was afterward republished in Owen's U. S. Geol. Rep. of Wisconsin, Iowa, and Minnesota.

	Page.
Platycrinus, Miller	89
P. americanus, n. s., Owen & Shumard, 1852, pl. xi, figs. 1 *a, b*	89
Poteriocrinus, Miller	89
P. rhombiferus, n. s., Owen & Shumard, 1852, pl. xi, figs. 2 *a–c*	89–90
P. tumidus, n. s., Owen & Shumard, 1852, pl. xi, figs. 3 *a, b*	90–91
P. spinosus, n. s., Owen & Shumard, 1852, pl. xi, fig. 4	91–92
P. occidentalis, n. s., Owen & Shumard, 1852, pl. xi. figs. 5 *a, b*	92–93
Agassizocrinus, Troost in MSS	93
A. conicus, n. s., Owen & Shumard, 1852, pl. xi, fig. 6	93
Symbathocrinus dentatus, n. s., Owen & Shumard, 1852, pl. xi, figs. 7 *a, b*	93–94

7.

OWEN, D. D. Description of new and imperfectly known Genera and Species of Organic Remains, collected during the geological surveys of Wisconsin, Iowa and Minnesota. < Rep. Geol. Surv. of Wisconsin, Iowa, and Minnesota, and incidentally of a part of Nebraska Territory, by David Dale Owen. Appendix, Article I, pp. 573-587, pls. i-id, ii-iib, iii-iiia, iv-v, vii-viiia. Philadelphia, 1852.

	Page.
Crustacea (Trilobites).	
Dike[l]locephalus, n. g., D. D. Owen, 1852	573-574
D. minnesotensis, n. s., D. D. Owen, 1852, tab. i, figs. 1, 2, 10; tab. i A, figs. 3, 6...	574
D. pepinensis, n. s., D. D. Owen, 1852, tab. i, fig. 9 *a*, *b*, fig. 13 (?) and tab. i, A, fig. 17 (?)	574
D. miniscoensis, n. s., D. D. Owen, 1852, tab. i, figs. 3, 12, and tab. i A, figs. 4, 5..	574-575
D. iowensis, n. s., D. D. Owen, 1852, tab. i, fig. 4, and tab. i A, fig. 13	575
D. granulosus, n. s., D. D. Owen, 1852, tab. i, figs. 7 (and 5 ?)	575
Lonchocephalus, n. g., D. D. Owen, 1852, tab. i A, fig. 12	575
L. chippewäensis, n. s., D. D. Owen, 1852, tab. i, figs. 6, 14; tab. i A, fig. 9	576
L. hamulus, n. s., D. D. Owen, 1852, tab. i A, figs. 8-12	576
Crepicephalus, n. g., D. D. Owen, 1852 ..	576-577
Menocephalus, n. g., D. D. Owen, 1852, tab. i, fig. 11	577
Asaphus (Isotelus) iowensis, D. D. Owen, 1852, tab. ii A, figs. 1-7	577
Cephalopoda	577
Ammonites nebrascensis, n. s., D. D. Owen, 1852, tab. viii, fig. 3, and tab. viii A, fig. 2.	577-578
Ammonites nebrascensis (?) D. D. Owen, 1852, tab viii, fig. 2	578
Ammonites cheyennensis, n. s., D. D. Owen, 1852, tab. vii, fig. 2	578-579
Ammonites opalus, n. s., D. D. Owen, 1852, tab. viii, fig. 6	579
Ammonites moreauensis, n. s., D. D. Owen, 1852, tab. viii, fig. 7	579
Ammonites lenticularis, D. D. Owen, tab. viii, fig. 5	579
Scaphites of Parkinson	579-580
S. (Ammonites ?) compriumus, n. s., D. D. Owen, 1852, tab. vii, fig. 4	580
S. (Ammonites ?) nodosus, n. s., D. D. Owen, 1852, tab. viii, fig. 4	581
Gyroceras burlingtonensis, n. s., D. D. Owen, 1852, tab. v, fig. 10	581
Discites tuberculatus, n. s., D. D. Owen, 1852, tab. v, fig. 14	581
Gasteropoda	581
Pleurotomaria muralis, n. s., D. D. Owen, 1852, tab. ii, fig. 6	581
Straparollus (Euomphalus) minnesotensis, n. s., D. D. Owen, 1852, tab. ii, figs. 12, 13.	581
Conchifera	582
Inoceramus sagensis n. s., D. D. Owen, 1852, tab. vii, fig. 3	582
Inoceramus nebrascensis, n. s., D. D. Owen, 1852, tab. viii, fig. 1	582
Cucullea nebrascensis, n. s., D. D. Owen, 1852, tab. vii, fig. 1, 1 *a*	582
Brachiopoda	583
Lingula pinnaformis, n. s., D. D. Owen, 1852, tab. i B, figs. 4, 6, 8	583
Lingula ampla, n. s., D. D. Owen, 1852, tab. i B, fig. 5	583
Orbicula prima, n. s., D. D. Owen, 1852, tab. i B, figs. 17, 19, and top figures on tab. iv .	583
Atrypa comis, n. s., D. D. Owen, 1852, tab. iii A, fig. 4	583
Chonetes granulifera, n. s., D. D. Owen, 1852, tab. v, fig. 12	583
Chonetes (?) iowensis, n. s., D. D. Owen, 1852, tab. iii A, fig. 7	584
Productus nebrascensis, n. s., D. D. Owen, 1852, tab. v, fig. 3	584
Leptaena trilobata, n. s., D. D. Owen, 1852, tab. ii, figs. 17, 18	584
Strophodonta, Hall	584
S. parva, n. s., D. D. Owen, 1852, tab. iii A. fig. 9	584
S. (?) costata, n. s., D. D. Owen, 1852, tab. iii A, fig. 5	585
S. iowensis, n. s., D. D. Owen, 1852	585
Orthis cuneata, n. s., D. D. Owen, 1852, tab. iii A, fig. 10	585
Spirifer iowensis, n. s., D. D. Owen, 1852, tab. iii, fig. 1	585
Spirifer peanatus, n. s., D. D. Owen, 1852, tab. iii, figs. 3 and 8 (?)	585
Spirifer ligus, n. s., D. D. Owen, 1852, tab. iii, fig 4	585
Spirifer eucuternus, D. D. Owen, tab. iii, figs. 2, 2 *a* and 6, 6 *a*	586
Spirifer cedarensis, D. D. Owen, 1852, tab. iii, fig. 5	586
Spirifer inequicostatus, n. s., D. D. Owen, 1852, tab. v, fig. 6	586
Foraminifera	586
Selenoides, n. g., D. D. Owen, 1852, fig. 3, *a* (and *b* ?)	586
S. iowensis, n. s., D. D. Owen, 1852, tab. ii B, fig. 13	587

8.

OWEN, D. D., and SHUMARD, B. F. Descriptions of one New Genus and twenty-two New Species of Crinoidea, from the subcarboniferous limestone of Iowa. < Rep. Geol. Surv. of Wisconsin, Iowa, and Minnesota, and incidentally of a part of Nebraska territory, by David Dale Owen. Appendix. Article II, pp. 587-598, pls. V*a*-V*b*. Philadelphia, 1852.

	Page.
Platycrinus, Miller	587
P. planus, Owen & Shumard, 1850, tab. v A, figs. 4 *a-c*	587
P. yandellii, Owen & Shumard, 1850, tab. v A, figs. 6 *a, b*	587-588
P. discoideus, Owen & Shumard, 1850, tab. v A, figs. 1 *a, b*	588
P. corrugatus, Owen & Shumard, 1850, tab. v A, figs. 2 *a-e*	589
P. burlingtonensis, Owen & Shumard, 1850, tab. v A, fig. 5	589
Dichocrinus, Munster	589-590
D. oratus, Owen & Shumard, 1850, tab. v A, figs. 9 *a, b*	590
D. striatus, Owen & Shumard, 1850, tab. v A, figs. 10 *a, b*	590
Cyathocrinus, Miller	591
C. iowensis, Owen & Shumard, 1850, tab. v A, figs. 11 *a-c*	591
C. cornutus, Owen & Shumard, 1850, tab. v A, figs. 8 *a, b*	591
Pentremites, Say	591
P. norwoodii, Owen & Shumard, 1850, tab. v A, figs. 13 *a-c*	591-592
P. melo, Owen & Shumard, 1850, tab. v A, figs. 14 *a-c*	592
P. lateriformis, Owen & Shumard, 1850, tab. v A, fig. 15	592-593
P. stelliformis, Owen & Shumard, 1850, tab. v A, figs. 16 *a, b*	593
Actinocrinus unicornus, Owen & Shumard, 1850, tab. v A, figs. 12 *a, b*	593-594
Megistocrinus, n. g., Owen & Shumard, 1852	594
M. evansii, n. s., Owen & Shumard, 1852, tab. v A, figs. 3 *a, b*	594
Platycrinus, Miller	594
P. americanus, Owen & Shumard, 1852, tab. v B, figs. 1 *a, b*	594-595
Poteriocrinus, Miller	595
P. rhombiferus, Owen & Shumard, 1852, tab. v B, figs. 2 *a-c*	595
P. tumidus, Owen & Shumard, 1852, tab. v B, figs. 3 *a, b*	595-596
P. spinosus, Owen & Shumard, 1852, tab. v B, fig. 4	596
P. occidentalis, Owen & Shumard, 1852, tab. v B, figs. 5 *a, b*	596-597
Agassizocrinus (Troost in MSS)	597
A. conicus, Owen & Shumard, 1852, tab. v B, fig. 6	597
Synbathocrinus dentatus, Owen & Shumard, 1852, tab. v B, figs. 7 *a, b*	597-598

9.

OWEN, D. D. Shumard, B. F. Summary of the distribution of Orders, Genera, and Species, in the Northwest. < Rep. Geol. Surv. of Wisconsin, Iowa, and Minnesota, and incidentally of a part of Nebraska Territory, by David Dale Owen. Appendix. Article III, pp. 598-600. Philadelphia, 1852.

Modified by additions and researches since the publication in the Proc. of the American Association, 1851.

XI.—THE WRITINGS OF JAMES SCHIEL.

SCHIEL, JAMES. List and Description of Organic Remains collected during the Exploration of the Central Pacific Railroad line, 1853–'54. <Rep. Expls. and Survs. from the Mississippi River to the Pacific Ocean. Report of Expls. for a route for the Pacific Railroad of the line of the forty-first parallel of north latitude, by Lieut. E. G. Beckwith, 1854. Vol. ii, chap. x, pp. 108–109, pls. i–iv. Washington. 1855.

	Page.
Fenestella, pl. i, fig. 1	108
Brachiopoda.	
Terebratula subtilita, Hall, pl. i. figs. 2 a, b	108
Productus splendens, Norwood & Prather, pl. i, fig. 3	108
Productus æquicostatus, Shumard, pl. ii, figs. 4, 5	108
Spirifer, pl. i, fig. 5	108
Inoceramus, pl. ii, fig. 6	108
Inoceramus confertim-annulatus, Rœmer, pl. ii, fig. 7	108
Inoceramus pseudo-mytiloides, pl. iii, fig. 8	108
Gryphæa pitcheri, Morton, pl. iii, fig. 9	108
Cardium multistriatum, Shumard, pl. i, fig. 10	109
Phillipsia ——— ?, pl. i, figs. 11–14	109
Encrinites and *bryozoa*, pl. i, fig. 12	109
Ammonites	109
Gryphæa pitcheri	109

XII.—THE WRITINGS OF BENJAMIN F. SHUMARD.

1.

SHUMARD, B. F. Description of the species of Carboniferous and Cretaceous fossils collected. <Expl. of the Red River of Louisiana, by Randolph B. Marcy, assisted by George B. McClellan. Appendix E, Paleontology, pp. 197–211, pls. i–vi. Washington, 1853.

FOSSILS OF THE CARBONIFEROUS SYSTEM.

Crinoidea. Page.
Cyathocrinus granuliferus, Yandell & Shumard, MSS., pl. —, fig. — 199
Agassizocrinus dactyliformis, Troost, MSS., pl. i, fig. 7 199–209
Pentremites florealis, Say, 1820 200
Pentremites sulcatus, F. Roemer, 1852 200

Bryozoa.
Archimedipora archimedes, Lesueur, 1842, pl. i, fig. 6 201

Brachiopoda.
Productus punctatus, Martin, pl. i, fig. 5 and pl. ii, fig. 1 201
Productus cora, D'Orbigny, 1842 202
Productus costatus, Sowerby, pl. i, fig. 2 202
Terebratula sub'ilita, Hall, pl. iv, fig. 8 202–203
Terebratula marcyi, Shumard, n. s., 1853, pl. i, figs. 4 a, b 203
Spirifer ——? Shumard, 1853, pl i, fig. 3 203

FOSSILS OF THE CRETACEOUS PERIOD.

Mollusca.
Pecten quadricostatus, Sowerby, pl. iii, fig. 6; pl. —, fig. — 204
Exogyra ponderosa, Roemer, 1849 205
Gryphæa pitcheri, Morton, pl. vi, fig. 5 205
Exogyra texana, Roemer, pl. v, figs. 1 a, b and fig. 5 205
Ostrea subovata, n. s., Shumard, 1853, pl. v, fig. 2 .... 205, 206
Inoceramus confertim-annulatus, Roemer, pl. vi, fig. 2 206
Trigonia crenulata, Lamarck, pl iv, fig. 1 206
Astarte washitensis, n. s., Shumard, 1853, pl iii, fig. 3 .... 206, 207
Cardium multistriatum, n. s., Shumard, 1853, pl. iv, fig. 2 207
Panopæa texana, n. s., Shumard, 1853, pl. vi, fig. 1 207
Terebratula choctawensis, n. s., Shumard, 1853, pl. ii, figs, a, b 207–208
Globiconcha (Tylostoma) tumida, n. s., Shumard, 1853, pl. v, fig. 3 208
Globiconcha (?) elevata, n. s., Shumard, 1853, pl. iv, fig. 4 208
Eulima (?) subfusiformis, n. s., Shumard, 1853, pl. iv, fig. 3 208
Ammonites vespertinus, Morton 209
Ammonites marcianus, n. s., Shumard, 1853, pl. iv, fig. 5 209
A. acuto carinatus, n. s., Shumard, 1853, pl. i, fig. 3 209–210
Ammonites ——? Shumard, 1853, pl. —, fig. — 210

Echinodermata.
Hemiaster eleganus, n. s., Shumard, 1853, pl. ii, figs. 4 a–c 210
Holaster simplex, n. s., Shumard, 1853, pl. iii, fig. 2 210–211
Holectypus planatus, F. Roemer 211

2.

EVANS, JOHN, *and* SHUMARD, B. F. On some New Species of Fossils from the Cretaceous formation of Nebraska territory. <Trans. Saint Louis Acad. Sci., vol. i, pp. 38–42. 1857. Saint Louis, 1856–1860.

Page.
Acephala.
Avicula nebrascana, n. s., E. & S., 1857 38
Limopsis striato-punctatus, n. s., E. & S., 1857 38

	Page.
Cardium subquadratum, n. s., E. & S., 1857	38–39
Cardium rarum, n. s., E. & S., 1857	39
Arca sulcatina, n. s., E. & S., 1857	39
Leda fibrosa n. s., E. & S., 1857	39–40
Mytilus meekii, n. s., E. & S., 1857	40
Ostrea subtrigonalis, n. s., E. & S., 1857	40
Gasteropoda.	
Pleurotoma minor, n. s., E. & S., 1857	40–41
Fusus haydeni, n. s., E. & S., 1857	41
Fusus nebrascensis n. s., E. & S., 1857	41
Turritella multilineata, n. s., E. & S., 1857	41–42
Rostellaria americana, n. s., E. & S., 1857	42
Cephalopoda.	
Ammonites galpini, n. s., E. & S., 1857	42

4.

SHUMARD, B. F. Descriptions of New Fossils from the Tertiary Formation of Oregon and Washington Territories and the Cretaceous of Vancouver's Island, collected by Dr. Jno. Evans, U. S. Geologist, under instructions from the Department of the Interior. <Trans. Saint Louis Acad. Sci., vol. i, pp. 120–125. 1858. Saint Louis, 1856–1860.

TERTIARY SPECIES.

	Page.
Lucina fibrosa, n. s., Shumard, 1858	120
Corbula evansana, n. s., Shumard, 1858	120–121
Leda willamettensis, n. s., Shumard, 1858	121
Leda oregona, n. s., Shumard, 1858	121–122
Pecten cooseensis, n. s., Shumard, 1858	122
Venus securis, n. s., Shumard, 1858	122–123

CRETACEOUS SPECIES.

Inoceramus vancouverensis, n. s., Shumard, 1858	123–124
Pinna calamitoides, n. s., Shumard, 1858	124
Pyrula glabra, n. s., Shumard, 1858	125

5.

SHUMARD, B. F. Notice of New Fossils from the Permian Strata of New Mexico and Texas, collected by Dr. George G. Shumard, Geologist of the United States Government Expedition for obtaining Water by means of Artesian Wells along the 32d Parallel, under the direction of Capt. John Pope, U. S. Corps Top. Eng. <Trans. Saint Louis Acad. Sci., vol. i, pp. 290–297. 1858. Saint Louis, 1856–1860.

	Page.
Productus popei, n. s., Shumard, 1858	290
P. mexicanus, n. s., Shumard, 1858	291
P. pileolus, n. s., Shumard, 1858	291
P. semireticulatus, Mart., sp	292
Aulosteges guadalupensis, n. s., Shumard, 1858	292
Spirifer mexicanus, n. s., Shumard, 1858	292–293
Spirifer sulcifera, n. s., Shumard, 1858	293
Spiriferina billingsii, n. s., Shumard, 1858	294
Retzia papillata, n. s., Shumard, 1858	294–295
Retzia ? meekana, n. s., Shumard, 1858	295
Rhynchonella guadalupae, n. s., Shumard, 1858	295–296
Camarophoria ? bisulcata, n. s., Shumard, 1858	296
Phillipsia peranaulata, n. s., Shumard, 1858	296–297
Fusulina elongata, n. s., Shumard, 1858	297

6.

SHUMARD, B. F. Notice of Fossils from the Permian Strata of Texas and New Mexico, obtained by the United States Expedition, under Capt. John Pope, for boring Artesian Wells along the 32d Paral., with Descriptions of New Species from these Strata and the Coal Measures of that region. <Trans. Saint Louis Acad. Sci., vol. i, pp. 387–403, 1859, pl. xi. Saint Louis, 1856–1860.

PERMIAN FOSSILS.

	Page.
Zoophyta.	
Chœtetes mackrothii, Geinitz	387
Chœtetes. sp. (?) Shumard, 1859	388
Campophyllum (?) *texanum*, n. s., Shumard, 1859	388
Polycœlia ?, Shumard, 1859	388
Crustacea.	
Phillipsia perannulata, Shumard, pl. xi, fig. 10	388
Bairdia, sp. (?), Shumard, 1859	388
Bryozoa	888
Fenestella popeana, Prout	388
Acanthocladia americana, Swallow	388
Foraminifera.	
Fusulina elongata, Shumard	388–389
Brachiopoda.	
Productus	389
P. calhounianus, Swallow	389
P. mexicanus, Shumard	389
P. pileolus, Shumard	389
P. semireticulatus var. *antiquatus*, Martin	389
P. popei, Shumard, pl. xi, figs. 8a, b	389
P. norwoodii, Swallow	389–390
P. leplayi (?), Verneuil	390
Strophalosia	390
S. (Aulosteges) guadalupensis, Shumard, pl. xi, figs. 5a, b	390
Chonetes	390
C. permiana, n. s., Shumard, 1859	390
C. flemingi (?), Norwood & Pratten	390
Spirifer	390
S. mexicanus, Shumard, pl. xi, figs. 4a, b	390–391
S. guadalupensis, n. s., Shumard, 1859	391
S. sulciferus, Shumard, pl. xi, figs. 3a–c	391
S. cameratus, Morton	391
Spiriferina	391
S. billingsii, Shumard	391–392
Terebratula	392
T. elongata, Schlotheim	392
T. perinflata, n. s., Shumard, 1859	392
Rhynchonella	392
R. guadalupæ, Shumard, pl. xi, figs. 6a–c	392
R. indentata, n. s., Shumard, 1859	393
R. texana, n. s., Shumard, 1859	393
Rhynchonella, sp. (?), Shumard, 1859	393–394
Camerophoria, Swallow, pl. xi, figs. 1a–e	394
C. bisulcata, Shumard, pl. xi, figs. 2a–d	394
C. swalloviana, n. s., Shumard, 1859, pl. xi, fig. 1 a–e	394–395
C. schlotheimi (?), Buch	395
Retzia	395
R. papillata, Shumard, pl. xi, figs. 9 a–c	395
R. meekiana, Shumard, pl. xi, figs. 7 a, b	395
Streptorhynchus	395
S. (Orthisina) shumardianus, Swallow	395
Orthisina	395
Orthisina sp. (?), Shumard, 1859	395
Crania	395
C. permiana, n. s., Shumard, 1859	395–396

	Page.
Acephala.	
Myalina	396
M. squamosa, Sow	396
M. recta, Shumard	396
Pleurophorus	396
P. occidentalis, M. & H	396
Monotis	396
M. speluncaria, Schlotheim	396–397
Monotis, sp. (?), Shumard, 1859	397
Axinus	397
A. securis, n. s., Shumard, 1859	397
Edmondia	398
E. suborbiculata, Swallow	398
Cardiomorpha	398
Cardiomorpha, sp. (?) Shumard, 1859	398
Gasteropoda.	
Turbo	398
T. guadalupensis, n. s., Shumard, 1859	398
T. helicinus (?), Schlotheim	398
Straparollus	399
Straparollus, sp. (?), Shumard, 1859	399
Bellerophon	399
Bellerophon, sp. (?), Shumard, 1859	399
Pleurotomaria	399
P. halliana, n. s., Shumard, 1859	399
Chemnitzia	399
C. swalloviana, n. s., Shumard, 1859	399
Cephalopoda.	
Nautilus, sp. (?), Shumard, 1859	399–400
Orthoceras, sp. (?), Shumard, 1859	400
CARBONIFEROUS FOSSILS.	
Gasteropoda.	
Turbo texanus, n. s., Shumard, 1859	400
Straparollus cornudanus, n. s., Shumard, 1859	400–401
Pleurotomaria proutiana, n. s., Shumard, 1859	401
Pleurotomaria obtusispira, n. s., Shumard, 1859	401
Pleurotomaria perornata, n. s., Shumard, 1859	401–402
Macrocheilus texanus, n. s., Shumard, 1859	402

7.

SHUMARD, B. F., *and* OWEN, D. D.   (See Owen, D. D., and Shumard, B. F.).

XIII.—THE WRITINGS OF ROBERT P. WHITFIELD.

1.

WHITFIELD, R. P. Descriptions of New Fossils. <Rep. of a reconnaissance of the Black Hills of Dakota, by William Ludlow, pp. 103-104, pl. i. Washington, 1875.

	Page.
Obolus	103
O. pectenoides, n. s., Whitfield, 1875, pl. —, figs. 1-3	103
Lingulepis primæformis ? pl. —, fig. 4	103
Terebratula	103
T. helena, n. s., Whitfield, 1875, pl. —, figs. 5-10	103-104

2.

WHITFIELD, R. P. Descriptions of New Species of Fossils. <Rep. of a reconnaissance from Carroll, Montana Territory, on the Upper Missouri, to the Yellowstone National Park, and return, by William Ludlow. pp. 139-145, pls. i-ii. Washington, 1876.

	Page.
Crepicephalus, Owen	141
C. (Loganellus) montanensis, n. s., Whitfield, 1876, pl. i, figs. 1, 2	141
Arionellus, Barrande	141
A. tripunctatus, n. s., Whitfield, 1876, pl. i, figs. 3-5	141-142
Gryphæa, Lam	142
G. planoconvexa, n. s., Whitfield, 1876, pl. ii, figs. 9, 10	142
Gervillia, Defrance	142
G. sparsalirata, n. s., Whitfield, 1876, pl. ii, fig. 8	142
Myalina, De Koninck	143
M. ? (Gervillea) perplana, n. s., Whitfield, 1876, pl. i, fig. 8	143
Pinna, Linn	143
P. ludlovi, n. s., Whitfield, 1876, pl. i, figs. 6, 7	143
Tapes, Mühlf	143
T. montanensis, n. s., Whitfield, 1876, pl. ii, figs. 1, 2	143-144
Mactra, Linn	144
M. maia, n. s., Whitfield, 1876, pl. ii, fig. 5	144
Sanguinolaria, Lam	144
S. oblata, n. s., Whitfield, 1876, pl. ii, figs. 3, 4	144
Thracia, Leach	144
T. (Corimya) grinnelli, n. s., Whitfield, 1876, pl. ii, figs. 6, 7	144-145
Vanikoropsis, Meek	145
V. toumeyana, M. & H., sp., 1856, pl. ii, figs. 11-13	145

3.

WHITFIELD, R. P. Preliminary report on the Paleontology of the Black Hills, containing descriptions of new species of fossils from the Potsdam, Jurassic, and Cretaceous formations of the Black Hills of Dakota. <U. S. Geographical and Geological Survey of the Rocky Mountain region, J. W. Powell in charge. pp. 1-49. Washington, July, 1877.

FOSSILS FROM THE PRIMORDIAL ROCKS.

	Page.
Plantæ.	
Palæochorda	7
P. prima, n. s., Whitfield, 1877, pl. i, fig. 2	7

	Page.
Palæophycus, Hall	7
P. occidentalis, n. s., Whitfield, 1877, pl. i, fig. 3	7–8

Molluscoida.
Brachiopoda.

Lingulepis cuneolus, n. s., Whitfield, 1877, pl. ii, figs. 5, 6	8–9
Lingulepis perattenuatus, n. s., Whitfield, 1877, pl. ii, figs. 7–9	9

Articulata.
Trilobita.
Calymenidæ.

Crepicephalus, Owen, (? *Loganellus*, Devine.)	10
C. (Loganellus) centralis, n. s., Whitfield, 1877, pl. ii, figs. 21–24	10–11
C. (Loganellus) planus, n. s., Whitfield, 1877, pl. ii, fig. 20	11

FOSSILS FROM THE JURASSIC ROCKS.

Radiata.
Echinodermata.
Asteroidea.

Asterias, Linnæus	15
A. ? dubium, n. s., Whitfield, 1877, pl. iii, fig. 3	15

Pectenidæ.

Pecten, Brug	16
P. newberryi, n. s., Whitfield, 1877, pl. iv, figs. 12–15	16–17
Pseudomonotis (Eumicrotis) orbiculata, n. s., Whitfield, 1877, pl. iii, figs. 17–19	17

Mytilidæ.

Mytilus, Linn	18
M. whitei, n. s., Whitfield, 1877, pl. v, figs. 9–12	18
Trapezium, Humph. (= *Cypricardia*, Lam.)	18
T. bellefourchensis, n. s., Whitfield, 1877, pl. v, figs. 1–4	18–19
T. subequalis, n. s., Whitfield, 1877, pl. v, figs. 5–8	19–20
Pleuromya, Agassiz (family uncertain)	20
P. newtoni, n. s., Whitfield, 1877, pl. v, figs. 19–20	20–21
Tancredia corbuliformis, n. s., Whitfield, 1877, pl. vi, figs. 5–8	21–22
Tancredia bulbosa, n. s., Whitfield, 1877, pl. vi, figs. 1–3	22
Tancredia postica, n. s., Whitfield, 1877, pl. vi, fig. 14	22–23

Veneridæ.

Dosinia, Scopoli	23
D. jurassica, n. s., Whitfield, 1877, pl. v, figs. 21–24	23–24

Psammobiidæ.

Psammobia, Lam	24
P. ? prematura, n. s., Whitfield, 1877, pl. v, fig. 31	24
Neæra, Gray	24
N. longirostra, n. s., Whitfield, 1877, pl. v, fig. 35	24–25

Gastrochænidæ.

Saxicava, Bellevue	25
S. jurassica, n. s., Whitfield, 1877, pl. v, figs. 25–30	25–26

FOSSILS OF THE CRETACEOUS.

Mollusca.
Lamellibranchiata.
Pteriidæ., Meek.

Pteria, Scop	29
P. (Pseudopteria) sublevis, n. s., Whitfield, 1877, pl. vii, fig. 6	29–31
Inoceramus perplexus, n. s., Whitfield, 1877, pl. viii, fig. 3, and pl. x, figs. 4, 5	31
Endocostea, n. g., Whitfield, 1877	31–32
E. typica, n. s., Whitfield, 1877, pl. ix, figs. 1–7	32–33

Nuculanidæ.

Nuculana	33
N. subequilatera, n. s., Whitfield, 1877, pl. xi, figs. 3, 4	33

Crassatellidæ.

Crassatella, Lam	34
C. subquadrata, n. s., Whitfield, 1877, pl. xi, fig. 12	34

Cyprinidæ.

Sphæriola	34
S. transversa, n. s., Whitfield, 1877, pl. x, figs. 14–16	34–35

	Page.
Zellinidæ.	
Leiopistha, Meek	35
Subgenus *Cymella*, Meek	35
Leiopistha (Cymella) meeki, n. s., Whitfield, 1877, pl. xi, figs. 27, 28	35–36
Anatinidæ.	
Thracia, Leach	36
T. subgracilis, n. s., Whitfield, 1877, pl. xi, figs. 29–30	36
Gasteropoda.	
Fusus cheyennensis, n. s., Whitfield, 1877, pl. xii, fig. 9	37
Aporrhais meeki, n. s., Whitfield, 1877, pl. xii, fig. 5	37
Aporrhais (Goniocheila) castorensis, n. s., Whitfield, 1877, pl. xii, fig. 1	38
Akera, O. F. Muller	38
A. glans-oryza, n. s., Whitfield, 1877, pl. xii, fig. 25	38–39
Cephalopoda.	
Helicoceras, D'Orb	39
H. stevensoni, n. s., Whitfield, 1877, pl. xiv, figs. 4–7	39–40
Heteroceras, D'Orb	40
H. newtoni, n. s., Whitfield, 1877, pl. xv, figs. 1–4	40–41
Ancyloceras, D'Orb	42
A. jenneyi, n. s., Whitfield, 1877, pl. xvi, figs. 6–8	42–43
A. tricostatus, n. s., Whitfield, 1877, pl. xv, figs. 7–8	43–44
Ptychoceras meekanum, n. s., Whitfield, 1877, pl. xvi, figs. 1–2	44–45
Ptychoceras crassum, n. s., Whitfield, 1877, pl. xvi, figs. 3–5	45–46
List of fossils described in the report of the Paleontology of the Black Hills	46–49

4.

WHITFIELD, R. P. Paleontology of the Black Hills of Dakota. < Report on the Geology and Resources of the Black Hills of Dakota, by Henry Newton, E. M., and Walter P. Jenney, E. M. < U. S. Geographical and Geological Survey of the Rocky Mountain region, J. W. Powell in charge, pp. 325–468, pls. i–xvi. Washington, 1880.

FOSSILS FROM THE PRIMORDIAL ROCKS.

	Page.
Plantæ.	
Palæochorda	331
P. prima, Whitfield, 1877, pl. i, fig. 2	331–332
Palæophycus, Hall	332
P. occidentalis, Whitfield, 1877, pl. i, fig. 3	332
Palæophycus, sp.? Whitfield, 1880, pl. i, fig. 1	333
Incertæsedes.	
Arenicolites, sp.? Whitfield, 1880, pl. ii, fig. 25	333–334
Brachiopoda.	
Lingulidæ.	
Lingulepis, Hall	335
L. pinnaformis, Owen, pl. ii, figs. 1–4	335
L. cuneolus, Whitfield, 1877, pl. ii, figs. 5–6	336
L. perattenuatus, Whitfield, 1877, pl. ii, figs. 7–9	337
L. dakotensis, M. & H., pl. ii, figs. 10–11	337–338
Obolidæ.	
Obolus, Eichwald	
O.? pectenoides, Whitfield, 1875, pl. ii, figs. 18–19	338–339
Obolella, Billings	339
O. polita, Hall, 1860, pl. ii, figs. 12, 13	339–340
O. nana, M. & H., 1861, pl. ii, figs. 14–17	340–341
Articulata.	
Trilobita.	
Calymenidæ.	
Crepicephalus, Owen (? *Loganellus*, Devine)	341
C. centralis, Whitfield, 1877, pl. ii, figs. 21–24	341–343
C. planus, Whitfield, 1877, pl. ii, fig. 20	343–344

FOSSILS FROM THE JURASSIC ROCKS.

Radiata.	
Echinodermata.	
Asteroidea.	
Asterias, Linn	344
A.? dubium, Whitfield, 1877, pl. iii, fig. 3	344–345

Crinoidea.
  Pentacrinidæ.
    Pentacrinites, Miller 345
    P. asteriscus, M. & H., 1858, pl. 3, figs. 1-2 345
Brachiopoda.
  Lingulidæ.
    Lingula, Brug. 346
    L. brevirostris, M. & H., 1858, pl. iii, figs. 4-5 346-347
  Rhynchonellidæ.
    Rhynchonella, Fischer 347
    R. myrina, pl. iii, figs. 6-7 347
Lamellibranchiata.
  Ostreidæ.
    Ostrea, Linn 348
    O. strigilecula, White, pl. iii, figs. 8-12 348-349
    Gryphæa, Lam. 349
    G. calceola var. nebrascensis, M. & H., 1861, pl. iii, figs. 13-16 349-350
  Pectenidæ.
    Pecten, Bruguières 350
    P. newberryi, Whitfield, 1877, pl. iv, figs. 12-15 350-351
    Camptonectes, Agassiz 351
    C. bellistriatus, Meek, 1860, pl. iv, figs. 6-11 351-353
    C. extenuatus, M. & H., 1860, pl. iv, figs. 4, 5 353-354
  Aviculidæ.
    Pseudomonotis, Bronn 354
    P. (Eumicrotis) curta, Hall, 1852, pl. iii, figs. 20-25 354-356
    P. (Eumicrotis) orbiculata, Whitfield, 1877, pl. iii, figs. 17-19 356-357
    Avicula, Lam. 357
    Subgenus Oxytoma, Meek 357
    A. (Oxytoma) mucronata, M. & H., pl. iv, figs. 1, 2 357-358
    Gervillia, DeFrance 358
    G. recta, Meek, pl. iv, fig. 3 358-359
  Arcidæ.
    Grammatodon, M. & H 359
    G. inornatus, M. & H., 1858, pl. v, figs. 16-18 359-360
  Mytilidæ.
    Mytilus, Linn 360
    M. whitei, Whitfield, 1877, pl. v, figs. 9-12 360-361
    Volsella, Scopoli 361
    V. (modiola) formosa, M. & H., 1861, pl. v, fig. 15 361-362
    V. pertenuis, M. & H., 1858, pl. v, figs. 13, 14 362-363
  Crassatellidæ.
    Astarte, Sowerby 363
    A. ? fragilis, M. & H., 1860, pl. v, figs. 32, 33 363-364
    Trapezium, Humph. (= Cypricardia, Lam.) 364
    T. bellefourchensis, Whitfield, 1877, pl. v, figs. 1-4 364-365
    T. subequalis, Whitfield, 1877, pl. v, figs. 5-8 365-366
    Pleuromya, Agassiz (family uncertain) 367
    P. newtoni, Whitfield, 1877, pl. v, figs. 19, 20 367-368
  Tancrediidæ.
    Tancredia, Lycett (= Hettangia, Terquem) 368
    T. ? inornata, M. & H., 1860, pl. vi, figs. 9-13 368-369
    T. corbuliformis, Whitfield, 1877, pl. vi, figs. 5-8 370
    T. bulbosa, Whitfield, 1877, pl. vi, figs. 1-3 370-371
    T. postica, Whitfield, 1877, pl. vi, fig. 14 371-372
    T. warrenana, M. & H., 1860, pl. vi, fig. 4 372
  Veneridæ.
    Dosinia, Scopoli 373
    D. jurassica, Whitfield, 1877, pl. v, figs. 21-24 373
  Psammobiidæ.
    Psammobia, Lam. 374
    P. ? prematura, Whitfield, 1877, pl. v, fig. 31 374
  Anatinidæ.
    Thracia, Leach 375
    T. ? sublævis, M. & H., 1860, pl. v, fig. 34 375

	Page.
Neæra, Gray	376
N. longirostra, Whitfield, 1877, pl. v, fig. 25	376

Gastrochænidæ.

Saxicava, Belle[v]ue	376
S. jurassica, Whitfield, 1877, pl. v, figs. 25–30	376–377

Cephalopoda.
Tetrabranchiata.
Ammonitidæ.

Ammonites, Bruguière	378
A. cordiformis, M. & H., 1858, pl. vi, figs. 20–24	378–380
A. cordiformis var. *distans*, Whitfield, 1880, pl. vi, fig. 25	380–381

Dibranchiata.
Belemnitidæ.

Belemnites, Agricola	381
B. densus, M. & H., 1858, pl. vi, figs. 15–19	381–382

FOSSILS FROM THE CRETACEOUS ROCKS.

Lamellibranchiata.
Monomyaria.
Pectinidæ.

Syncyclonema, Meek	383
S. rigida, H. & M., pl. vii, fig. 1	383–384

Heteromyaria.
Aviculidæ.

Pteria, Scopoli	384
P. linguiformis, E. & S., sp., pl. vii, figs. 2, 3	384–385
P. (Oxytoma) nebrascana, E. & S., pl. vii, fig. 4	385–386
P. (Pseudopteria) fibrosa, M. & H., sp., 1856, pl. vii, fig. 5	386
P. (Pseudopteria) sublevis, Whitfield, 1877, pl. vii, fig. 6	387
Inoceramus, Sowerby	389
I. problematicus? Schlot., pl. vii, fig. 11	389–390
I. fragilis, H. & M., pl. ix, fig. 10	390–391
I. altus, Meek, 1871, pl. ix, fig. 11	391
I. perplexus, Whitfield, 1877, pl. viii, fig. 3, and pl. x, figs. 4, 5	392
I. sublævis, H. & M., pl. x, figs. 1–3	393
I. sagensis, Owen, 1852, pl. vii, fig. 12, and pl. viii, fig. 2	393–395
I. simpsoni, Meek, 1860, pl. viii, fig. 1	395–396
I. vanuxemi, M. & H., 1860, pl. vii, figs. 8, 9, and pl. viii, figs. 4, 5	396–398
I. vanuxemi var.? Whitfield, 1880, pl. vii, fig. 10	398
I. barabini, Morton, 1834, pl. vii, fig. 7, and pl. ix, fig. 8	398–400
I. tenuilineatus, H. &. M., pl. ix, figs. 12, 13	400, 402
Endocostea, Whitfield, 1877	402–403
E. typica, Whitfield, 1877, pl. ix, figs. 1–7	403–404
E. sulcata, Roemer, sp., pl. x, fig. 6	404–405

Integropallia.
Arcidæ.

Subgenus *Idonearca*, Conrad	405
Idonearca shumardi, M. & H., pl. xi, figs. 8–11	405–406
Nucula, Lam.	406
N. planimarginata, M. & H., 1856, pl. xi, figs. 5, 6	406–407
Nuculana, Link	407
N. bisulcata, M. & H., 1861, pl. xi, fig. 7	407–408
N. subequilatera, Whitfield, 1877, pl. xi, figs. 3, 4	408
Yoldia, Möller	409
Y. evansi, M. & H., 1866, pl. xi, figs. 1, 2	409

Lucinidæ.

Lucina, Brugnière	409
L. occidentalis, Morton, pl. xi, figs. 19–21	409–410
L. ventricosa, H. & M., pl. xi, figs. 14–16	410–411
L. (Diplodonta?) subundata, H. & M., pl. xi, figs. 17, 18	411–412

Crassatellidæ.

Crassatella, Lam.	412
C. subquadrata, Whitfield, 1877, pl. xi, fig. 12	412–413
Astarte, Sowerby	413
A. evansi, H. & M., pl. xi, fig. 13	413

Venielliidæ.   Page.
   Veniella, Stoliczka 414
     V. humilis, M. & H., 1860, pl. x, figs. 7–13 414
   Sphæriola 415
     S. transversa, Whitfield, 1877, pl. x, figs. 14–16 415
Sinuopallia.
  Veneridæ.
   Dosinia, Scopoli 416
   D. missouriana? Morton, pl. xi, figs. 25, 26 416–417
   Thetis, Linn 417
   T. circularis, M. & H., 1856, pl. xi, figs. 21–24 417
  Family ?
   Leiopistha, Meek 418
   Subgenus *Cymella*, Meek 418
   L. (Cymella) meeki, Whitfield, 1877, pl. xi, figs. 27, 28 418–419
  Anatinidæ.
   Thracia, Leach 419
   T. subgracilis, Whitfield, 1877, pl. xi, figs. 29, 30 419–420
   Neæra, Gray 420
   N. moreauensis, M. & H., pl. xi, fig. 31 420
Gasteropoda.
 Siphonostomata.
  Fasciolariidæ.
   Fasciolaria, Lam 421
   F. (Cryptorhytis) fusiformis, H. & M., pl. xii, fig. 12 421–422
   F. (Cryptorhytis) contorta, M. & H., pl. xii, fig. 10 422–423
   F. (Piestocheilus) culbertsoni, M. & H., pl. xii, fig. 11 423–424
   Fusus, Lam 424
   F. shumardi, H. & M., pl. xii, figs. 7, 8 424
   F. cheyennensis, Whitfield, 1877, pl. xii, fig. 9 424–425
  Aporrhaidæ.
   Aporrhais, Dillwyn 425
   A. newberryi, Meek, pl. xii, fig. 4 425–426
   A. meeki, Whitfield, 1877, pl. xii, fig. 5 426
   A. (Goniocheila) casturensis, Whitfield, 1877, pl. xii, fig. 1 427
   Anchura, Conrad 428
   A. ? sublevis, M. & H., pl. xii, fig. 6 428
   A. (Drepanocheilus) nebrascensis, E. & S., pl. xii, figs. 2, 3 429
 Holostomata.
  Naticidæ.
   Lunatia, Gray 430
   L. concinna, H. & M., pl. xii, fig. 13 430
   Vanikora, Quoy and Gaimard 430
   V. ambigua, M. & H., pl. xii, fig. 14 430–431
   Amauropsis, Morch 431
   A. paludinæformis, H. & M., pl. xii, fig. 16 431–432
  Trochidæ.
   Margarita, Leach 432
   M. nebrascensis, M. & H., pl. xii, fig. 15 432
Pulmonifera?
 Siphonariidæ?
   Anisomyon, M. & H 433
   A. alveolus, M. & H., pl. xii, fig. 20 433–434
   A. subovatus, M. & H., pl. xii, fig. 19 434–435
   A. patelliformis, M. & H., pl. xii, figs. 17, 18 435
   A. borealis, Morton, 1842, pl. xii, figs. 21–23 436
Tectibranchiata.
 Bullidæ.
   Haminea, Leach 437
   H. subcylindrica, M. & H., pl. xii, fig. 24 437
   Akera, O. F. Muller 437
   A. glans-oryza, Whitfield, 1877, pl. xii, fig. 25 437–438
Prosopocephala.
 Solenoconchæ.
  Dentaliidæ.
   Dentalium, Linn 438
   D. gracile, H. & M., pl. xii, fig. 20 438–439

Cephalopoda.
  Tetrabranchiata.
    Nautilidæ.
      Nautilus, Breynius.
        N. dekayi var. *montanaensis,* Meek, pl. xvi, figs. 10, 11 439–440
    Ammonitidæ.
      Prionocyclus, Meek 440
      P. wyomingensis, Meek, 1870, pl. xiv, figs. 1–3 440–441
    Scaphitidæ.
      Scaphites, Parkinson 441
      S. nodosus, Owen, 1852, pl. xiii, fig. 12 441–443
      S. nodosus var. *brevis,* Meek, pl. xiii, figs. 8, 9 443
      S. nodosus var. *quadrangularis,* Meek, pl. xiii, figs. 10, 11 443–444
      S. warreni, M. & H., pl. xiii, figs. 1–4 444–446
      S. wyomingensis, M. & H., pl. xiii, figs. 5–7 446–447
      Helicoceras, D'Orb 447
      H. stevensoni, Whitfield, 1877, pl. xiv, figs. 5–8 447–449
      Heteroceras, D'Orb 449
      H. newtoni, Whitfield, 1877, pl. xv, figs. 1–4 449–451
      H.? nebrascense, M. & H., pl. xv, fig. 6, and pl. xiv, fig. 9 451–452
      Ancyloceras, D'Orb 452
      A. jenneyi, Whitfield, 1877, pl. xvi, figs. 7–9 452–454
      A. tricostatus, Whitfield, 1877, pl. xv, figs. 7, 8 454–455
      Ptychoceras, D'Orb 455–457
      P. meekanum, Whitfield, 1877, pl. xvi, figs. 1, 2 457–458
      P. crassum, Whitfield, 1877, pl. xvi, figs. 3–5 459
  Synopsis of species from the Black Hills, noticed in other works, not described in this report 460–464
  List of fossils described in this report 465–468

5.

WHITFIELD, R. P. Brachiopoda and Lamellibranchiata of the Raritan Clays and Green sand Marls of New Jersey. < Monographs of the United States Geological Survey, Vol. ix, pp. i–xx and 1–264, pls. i–xxxv. Washington, 1885.

CONTENTS.

Letter of transmittal from Prof. George H. Cook ix
Sketch of the Geology of Cretaceous and Tertiary formations of New Jersey ix
Letter of transmittal from Prof. Robert P. Whitfield xv
Preliminary remarks xvii
Brachiopoda 3
  Section I.—Brachiopoda of the Marl Beds 5
Lamellibranchiata 17
  Section II.—Lamellibranchiata from the Raritan Clays 22
  Section III.—Lamellibranchiata from the Lower Marl Beds 29
  Section IV.—Lamellibranchiata from the Middle Marl Beds 194
  Section V.—Lamellibranchiata from the base of the Upper Marls 205
  Section VI.—Lamellibranchiata from the Eocene Marls 222
  Section VII.—Unionidæ from the Camden Clays 243
  Section VIII.—Classified list of the species 253

BRACHIOPODA.

Section I.—Brachiopoda from the several Marl Beds of the State.
Brachiopoda.
  Terebratulidæ.
    Terebratula, Llhwyd 6
    T. harlani, Morton, pl. i, figs. 15–23 6–9
    Terebratulina atlantica, Morton, pl. i, figs. 10–13 9–11
    Terebratulina floridana, Morton 11–12
    Terebratulina lachryma, Morton, pl. i, fig. 14 12
    Terebratella, D'Orb 12
    T. plicata, Say, 1829, pl. i, figs. 5–9 12–14
    T. vanuxemi, Lyell & Forbes, 1845, pl. i, figs. 1–4 14–15

LAMELLIBRANCHIATA.

Section II.—Lamellibranchiate Shells from the Plastic Clay.

Astartidæ.
   Astarte, Sowerby.. 23
    A. veta, Conrad, pl. ii, fig. 1 23–24

Cyprinidæ.
   Ambonicardia, n. g., Whitfield, 1885 24–25
    A. cookii, n. s., Whitfield, 1885, pl. ii, figs. 11–14 25
   Corbicula, Megerle.. 26
    C.? emacerata, n. s., Whitfield, 1885, pl. ii, figs. 5, 6 26
    C. annosa, Conrad, pl. ii, figs. 2–4 26–27
   Gnathodon 27
    G.? tenuidens, n. s., Whitfield, 1885, pl. ii, figs. 7–10 27–28

Section III.—Lamellibranchiata from the Lower Marl Beds.

Integripalliata.
Asiphonida.
Monomyaria.
  Ostreidæ.
   Ostrea, Linn. 29
    O. denticulifera, Conrad, pl. iii, figs. 8, 9 29
    O. crenulimarginata, Gabb, pl. iii, figs. 10, 11 30
    O. panda, Morton 30
    O. plumosa, Morton, pl. iii, figs. 12, 13 31–32
    O. subspatulata, L. & Sow., 1845, pl. iii, fig. 14 32–33
    O. tecticosta, Gabb, pl. iii, figs. 1, 2 33–34
    O. larva, Lamarck, pl. iii, figs. 3, 7 34–36
   Gryphæa, Lam. 36
    G. vesicularis, Lam.? (1806), pl. iii, figs. 15, 16; pl. iv, figs. 1–3; and pl. v... 36–39
   Exogyra, Say 39
    E. costata, Say, pl. vi, figs. 1, 2 39–41

Anomiidæ.
   Anomia, Linn. 42
    A. argentaria, Morton, pl. iv, figs. 9–11 42
    A. tellinoides, Morton, pl. iv, figs. 12, 13 43
   Diploschiza, Conrad, 1866 43
    D. cretacea, Conrad, pl. iv, figs. 4–8 43–44
   Paranomia, Conrad, 1860 44
    P. scabra, Morton, pl. x, fig. 10 44
    P. lineata, Conrad, pl. ix, fig. 10 45
  Pecten, Klein 45
    P. venustus, Mort., pl. vii, figs. 1–4 45–46
    P. quinquenarius, Conrad, 1854, pl. vii, figs. 13–16 47
    P. tenuitestus, Gabb, 1861, pl. vii, figs. 5, 6 47–48
    P. planicostatus, n. s., Whitfield, 1885, pl. viii, figs. 10, 11 48–49
    P. (Chlamys) craticulus, Morton, pl. vii, figs. 17, 18 49–50
    P. (Syncyclonema ?) perlamellosus, pl. vii, fig. 7 50–51
   Amusium, Klein 51
    A. simplicum, Conrad, pl. vii, figs. 11, 12 51–52
    A. conradi, n. s., Whitfield, 1885, pl. vii, figs. 8–10 52–53
   Camptonectes (Amusium) burlingtonensis, Gabb, pl. viii, figs. 3–9 53–55
   Camptonectes parvus, n. s., Whitfield, 1885, pl. viii, figs. 1, 2 55
   Neithea, Drouet 56
    N. quinquecostata (Sowerby), pl. viii, figs. 12–14 56–57

Spondylidæ.
   Spondylus, Lam. 57
    S. gregalis, Morton, pl. ix, figs. 11, 12, and pl. x, figs. 1, 2 57–58
   Dianchora, Sowerby 58–59
    D. echinata, Morton, pl. x, figs. 3–9 59–60
   Plicatula, Lam. 61
    P. urticosa, Morton, pl. ix, figs. 1, 2 61
   Radula, Klein 61
    R. pelagica, Morton, pl. ix, figs. 3–5 61–62
    R. acutilineata, Conrad, pl. ix, figs. 6, 7 62–63
    R. reticulata, Lyell & Forbes, 1845, pl. ix, figs. 8, 9 63–64

WRITINGS OF R. P. WHITFIELD. 267

	Page.
Heteromyaria.	
Mytilidæ.	
Mytilus, Linn.	64
M. oblivius, n. s., Whitfield, 1885, pl. xvii, fig. 1.	64
Modiola, Lam.	64
M. julia, Lea, 1861, pl. xvii, figs. 6, 7?	64–65
M. burlingtonensis, n. s., Whitfield, 1885, pl. xvii, figs. 8, 9	65–66
Lithodomus, Cuvier.	66
L. affinis, Gabb, 1861, pl. xvii, figs. 2, 3.	66–67
L. ripleyana, Gabb, 1861, pl. xvii, figs. 4, 5.	67–68
Pteriidæ, Meek (=*Aviculidæ* of Authors).	68
Pteria, Scopoli.	68
P. petrosa, Conrad, pl. xiv, fig. 10.	68–69
P. laripes, Morton, pl. xiv, fig. 9.	69–70
P. navicula, n. s., Whitfield, 1885, pl. xiv, fig. 8.	70–71
Meleagrinella, n. g., Whitfield, 1885	71–72
M. abrupta, Conrad, pl. xiv, figs. 11–14.	72–73
Gervilliopsis, n. g., Whitfield, 1885	73
G. ensiformis, Conrad, pl. xv, figs. 8–11, and pl. xvi, fig. 5.	73–74
G. minima, n. s., Whitfield, 1885, pl. xv, fig. 7.	74–75
Inoceramus, Sowerby	75
I. barabini, Morton? pl. xv, figs. 3, 5.	75–76
I. sagensis, Owen, pl. xiv, fig. 15, and pl. xv, figs. 1, 2.	76–78
I. sagensis var. *quadrans*, Whitfield, 1885, pl. xiv, fig. 16.	79
I. perovalis, Conrad, pl. xv, fig. 6.	80
I. pro-obliquus, n. s., Whitfield, 1885, pl. xiv, fig. 17.	80–81
Pinnidæ.	
Pinna, Linn.	81
P. laqueata, Conrad, pl. xvi, figs. 1, 2	81–82
Dimyaria.	
Arcidæ.	
Arca, Linn.	82
A. altirostris, Gabb, 1861, pl. xii, figs. 22, 23	82–83
Nemodon, Conrad	83
N. eufaulensis, Gabb, pl. xii, figs. 3–5	83–84
N. angulatum, Gabb, 1860, pl. xii, figs. 6, 7	84–85
N. brevifrons, Conrad, pl. xii, figs. 1, 2	85–86
Nemoarca, Conrad, 1870	86
N. cretacea, Conrad, 1870, pl. xii, figs. 8–10	86–87
Breviarca, Conrad	87
B. saffordi, Gabb, pl. xii, figs. 11, 12	87–88
Trigonarca, Conrad, 1862	88
T. cuneiformis, Conrad, pl. xii, figs. 17, 18.	88–89
T. transversa, Gabb, 1861, pl. xii, figs. 13–16	89–91
Cibota, Browne.	91
C. rostellata, Morton, pl. xi, figs. 34–36	91–92
C. uniopsis, Conrad, pl. xi, figs. 32, 33.	92–93
C. obesa, n. s., Whitfield, 1885, pl. xi, figs. 30, 31	93–94
C. multiradiata, Gabb, 1860, pl. xi, figs. 21, 22	94
Idonearca, Conrad, 1872	95
I. tippana, Conrad, pl. xii, figs. 19–21.	95–96
I. antrosa, Morton, pl. xiii, figs. 6–11	96–98
I. vulgaris, Morton, pl. xiii, figs. 1–5.	98–99
Axinea, Poli	99
A. mortoni, Conrad, pl. xi, figs. 23–25.	99–101
A. alta, n. s., Whitfield, 1885, pl. xi, figs. 26–29	101
Nuculidæ.	
Nucula, Lam	102
N. percrassa, Conrad, pl. xi, figs. 4–6	102
N. monmouthensis, n. s., Whitfield, 1885, pl. xi, fig. 1.	102–103
N. slackiana, Gabb, pl. xi, figs. 2, 3	103–104
N. perequalis, Conrad	104–105
Nuculana, Link	105
N. protexta, Gabb, pl. xi, fig. 10.	105–106
N. gabbana, n. s., Whitfield, 1885, pl. xi, figs. 11–13	106–107
N. longifrons, Conrad, pl. xi, figs. 16, 17.	107–108

	Page.
N. pinnaformis, Gabb, pl. xi, figs. 7, 8	108-109
N. compressifrons, Conrad, pl. xi, fig. 9	109
Perrisonota, Conrad	110
P. protexta, Conrad, pl. xl, figs. 14, 15	110
Nucularia, Conrad	111
N. papyria, Conrad, pl. xi, figs. 18-20	111-112
Trigonidæ.	
Trigonia, Brug.	112
T. mortoni, n. s., Whitfield, 1885, pl. xiv, figs. 5, 6	112-113
T. eufaulensis, Gabb, pl. xiv, figs. 1-4	113-114
T. cerulea, n. s., Whitfield, 1885, pl. xiv, fig. 7	114-115
Siphonida.	
Integripalliata.	
Crassatellidæ.	
Crassatella, Lam.	115-116
C. vadosa, Morton ?, pl. xvii, figs. 12-15	116-117
C. cuneata, Gabb, pl. xvii, figs. 18-20	118-119
C. delawarensis, Gabb	119
C. monmouthensis, Gabb, pl. xvii, figs. 21, 22	119-120
C. prora, Conrad, pl. xvii, figs. 10, 11	120-121
C. subplana, Conrad, pl. xviii, figs. 14-16	121-122
C. transversa, Gabb, 1861, pl. xvii, figs. 16, 17	122-123
Scambula, Conrad, 1869	123
S. perplana, Conrad, pl. xviii, figs. 8-10	123-124
Astartidæ.	
Gouldia, Ad	124
G. decemnaria, Conrad, pl. xviii, fig. 4	124-125
G. conradi, n. s., Whitfield, 1885, pl. xviii, figs. 1-3	125-126
G. declivis, Conrad, pl. xviii, fig. 11	126
G. paralis, Conrad, pl. xviii, figs. 12, 13	126-127
Vetericardia, Conrad, 1872	127
V. octolirata, Gabb	127-128
V. crenulirata, Lea, 1861, pl. xviii, figs. 5-7	128-129
Lucinidæ.	
Lucina, Brug.	129
L. cretacea, Conrad, pl. xviii, figs. 23-25	129-130
L. smockana, n. s., Whitfield, 1885, pl. xviii, figs. 21, 22	130-131
Chamidæ.	
Diceras, Lam	131
D. dactyloides, n. s., Whitfield, 1885, pl. xviii, figs. 26, 27	131
Cardiidæ.	
Cardium, Linn	132
C. eufaulensis, Conrad, pl. xx, figs. 17-19	132
C. ripleyanum, Conrad, pl. xx, fig. 14	132-133
C. ripleyense, Conrad	133
Oriocardium, Conrad, 1870	133
Cardium (Oriocardium) dumosum, Conrad, pl. xx, figs. 9-13	133-135
Cardium (Oriocardium) multiradiatum, Gabb, pl. xxi, figs. 1-3	135-136
Protocardium, Beyrich	136
C. (Protocardium) perelongatum, n. s., Whitfield, 1885, pl. xx, figs. 20-22; pl. xxi, figs. 4, 5	136-138
Pachycardium, Conrad, 1870	138
P. burlingtonense, n. s., Whitfield, 1885, pl. xxi, figs. 6, 7	138
Fulvia, Grey, 1847	139
F. tenuis, n. s., Whitfield, 1885, pl. xx, fig. 8	139
Fragum, Bolton	139
F. tenuistriatum, n. s., Whitfield, 1885, pl. xx, figs. 15, 16	139-140
Leiopistha, Meek	140
L. protexta, Conrad, pl. xx, figs. 1-3	140-141
L. elegantula, Roemer	141-142
L. inflata, n. s., Whitfield, 1885, pl. xx, figs. 4, 5	142
Cymella, Meek	142
C. meeki, Whitfield, pl. xx, figs. 6, 7	142-143
Cyprinidæ.	
Veniella, Stoliczka	144
V. conradi, Morton, pl. xix, figs. 8-10	144-145

	Page
V. decisa, Morton, pl. xix, figs. 15, 16	145–147
V. inflata, Conrad, pl. xix, figs 4, 5	147–148
V. elevata, Conrad, pl. xix, figs. 6, 7	148–149
V. trigona, Gabb, 1861, pl. xix, figs. 11–14	149–150
V. subovalis, Conrad, pl. xix, figs. 1, 2	150–151
V. trapezoidea, Conrad, pl. xix, fig. 3	151–152
Sphaeriola, Stoliczka	152
S. umbonata, n. s., Whitfield, 1885, pl. xix, figs. 17, 18	152
Sinuopalliata.	
Veneridæ.	
Callista, Poli	153
C. delawarensis, Gabb, pl. xxii, figs. 8–10	153–154
Aphrodina, Conrad, 1868	154
A. tippana, Conrad, pl. xxii, figs. 6, 7	154–155
Cyprimeria, Conrad, 1864	156
C. depressa, Conrad, pl. xxii, figs. 11–13	156–157
C. densata, Conrad, pl. xxii, figs. 19–21	157–158
C. excavata, Morton, pl. xxii, figs. 16, 17	159–160
C. heilprini, n. s., Whitfield, 1885, pl. xxii, figs. 14, 15	160
C. spissa, Conrad, pl. xxii, fig. 18	160–161
Dosinia, Scopoli	161
D. gabbi, n. s., Whitfield, 1885, pl. xxii, figs. 4, 5	161–162
D.? erecta, n. s., Whitfield, 1885, pl. xviii, figs. 17–20	162–163
Tenea, Conrad, 1871	163
T. pinguis, Conrad, pl. xxii, figs. 1–3	163–164
Tellinidæ.	
Tellimera, Conrad, 1871	164
T. eborea, Conrad, pl. xxiii, figs. 12, 13	164–165
Linearia, Conrad, 1871	165
L. metastriata, Conrad, pl. xxiii, figs. 6–8	165–166
L. contracta, n. s., Whitfield, 1885, pl. xxiii, fig. 5	167
Æora, Conrad, 1871	167
Æ. cretacea, Conrad, pl. xxiii, figs. 16, 17	167–168
Aenona, Conrad, 1871	168
Æ. eufaulensis, Conrad, pl. xxiii, figs. 2, 3	168–169
Æ. papyria, Conrad, pl. xxiii, fig. 4	169–170
Oorimya, Agassiz	170
O. tennis, n. s., Whitfield, 1885, pl. xxiii, figs. 9–11	170–171
Donacinidæ.	
Donax, Linn.	171
D. fordii, Conrad, pl. xxiii, fig. 1	171–172
Mactridæ.	
Veleda, Conrad, 1871	172
V. lintea, Conrad, pl. xxiii, figs. 18–21	172–173
V. tellinoides, n. s., Whitfield, 1885, pl. xxiii, fig. 23	173–174
V. transversa, n. s., Whitfield, 1885, pl. xxiii, fig. 22	174
Anatinidæ.	
Pholadomya, Sowerby	175
P. occidentalis, Morton, pl. xxiv, figs. 1–3	175–176
P. roemeri, n. s., Whitfield, 1885, pl. xxiv, fig. 4	176–177
Periplomya, Conrad	177
P. elliptica, Gabb, 1861, pl. xxiii, figs. 14, 15	177–178
Cercomya, Agassiz	178
C. peculiaris, Conrad, pl. xxiii, figs. 24, 25	178
Corbulidæ.	
Corbula, Bruguière	178
C. crassiplica, Gabb, pl. xxiii, fig. 30	178–179
C. foulkei, Lea, 1861, pl. xxiii, figs. 27–29	180
C. subcompressa, Gabb, pl. xxiii, fig. 26	180–181
Saxicavidæ.	
Panopea, Ménard	181
P. decisa, Conrad, pl. xxiv, figs. 5–8	181–182
Solenidæ.	
Solyma, Conrad, 1871	182
S. lineolata, Conrad, pl. xxv, figs. 11–13	182–183

	Page.
Leptosolen, Conrad, pl. xxv, figs. 1-2	183-184
Legumen, Conrad, 1858	184
L. planulatum, Conrad, xxv, figs. 3-4	184-185
L. appressum, Conrad, pl. xxv, figs. 6-8	185-186
Siliqua, Muhlfeld	186
S. cretacea, Gabb, pl. xxv, figs. 9, 10	186-187
Pholadidæ.	
Pholas, Linn.	187
P. cithara, Morton, pl. xxv, 14-16	187-188
P. ? lata, n. s., Whitfield, 1885, pl. xxv, fig. 17	189-190
Martesia, Leach	190
M. (Pholas) cretacea, Gabb, pl. xxv, figs. 20-23	190
Teredidæ.	
Teredo, Linn	191
T. irregularis, Gabb, pl. xxv, figs. 18, 19	191-192
Gastrochænidæ.	
Clavagella, Lam	192
C. armata, Morton, pl. xxv, fig. 24	192-193

Section iv.—Lamellibranchiata from the Middle Marl Beds.

Ostreidæ.

Gryphæa, Lam.	194
G. vesicularis, Lam., pl. xxvi, figs. 9, 10	194
G. bryani, var. *precedens*, n. var., Whitfield, 1885, pl. xxvi, figs. 7-8	104-195
Gryphæostrea, Conrad	195
G. vomer, Morton, pl. xxvi, figs. 11-12	195-196
Mytilidæ.	
Modiola, Lam.	197
M. ovata, Gabb, pl. xxvi, figs. 13-14	197
M. (Lithodomus?) inflata, n. s., Whitfield, 1885, pl. xxvi, figs. 1-2	197-198
Pteriidæ, Meek.	
Pinna, Linn.	198
P. rostriformis, Morton, pl. xvi, figs. 3-4	198
Arcidæ.	
Idonearca, Conrad	199
I. medians, n. s., Whitfield, 1885, pl. xxvi, figs. 5-6	199
I. compressirostra, n. s., Whitfield, 1885, pl. xxvi, figs. 15-16	199-200
Isocardiidæ.	
Isocardia, Lamarck	200
I. conradi, Gabb, pl. xxvi, figs. 3-4	200-201
Teredidæ.	
Teredo, Linnæus	201
T. tibialis, Morton, pl. xxvi, figs. 19-22	201-203
Gastrochænidæ.	
Gastrochæna, Spengl	203
G. americana, Gabb, pl. xxvi, figs. 17-18	203-204

Section V.—Lamellibranchiata from the lower layers of the Upper Marl Beds of New Jersey.

Ostreidæ.

Ostrea, Linnæus	205
O. glandiformis, n. s., Whitfield, 1885, pl. xxvii, figs. 1-5	205-206
Gryphæa, Sow	206
G. bryani, Gabb, 1876, pl. xxvii, figs. 6-9	206-207
Mytilidæ.	
Modiola, Lam.	207
M. johnsoni, n. s., Whitfield, 1885, pl. xxviii, figs. 8, 9	207
Arcidæ.	
Arca, Linn	208
A. quindecemradiata, Gabb, 1860, pl. xxvii, figs. 10-13	208
Astartidæ.	
Cardita, Brug.	209
C. intermedia, n. s., Whitfield, 1885, pl. xxviii, figs. 14, 15	209
Crassatella, Lamarck	209
C. conradi, n. s., Whitfield, 1885, pl. xxviii, figs. 1-5	209-210
C. delawarensis, Gabb, pl. xxvii, figs. 14, 15	210-211

WRITINGS OF R. P. WHITFIELD. 271

	Page.
C. littoralis, Conrad, pl. xxviii, figs. 6, 7	212–213
C. rhombea, n. s., Whitfield, 1885, pl. xxvii, figs. 16–19	213–214
Cardiidæ.	
Criocardium, Conrad	214
C. nucleolus, n. s., Whitfield, 1885, pl. xxviii, figs. 10, 11	214–215
Cyprinidæ.	
Veniella, Stoliczka	215
V. rhomboidea, Conrad, pl. xxviii, figs. 12, 13	215–216
Petricolidæ.	
Petricola, Lam.	216
P. nova-ægyptica, n. s., Whitfield, 1885, pl. xxviii, fig. 22	216–217
Mactridæ.	
Veleda, Conrad, 1871	217
V. nasuta, n. s., Whitfield, 1885, pl. xxviii, fig. 23	217
Veneridæ.	
Caryatis, Roemer	218
C. ? reta, n. s., Whitfield, 1885, pl. xxviii, figs. 16–19	218–219
Saxicavidæ.	
Panopea, Ménard	219
P. elliptica, n. s., Whitfield, 1885, pl. xxviii, figs. 24, 25	219–220
Anatinidæ.	
Periplomya, Conrad	220
P. truncata, n. s., Whitfield, 1885, pl. xxviii, figs. 20, 21	220–221

Section VI.—Lamellibranchiata from the Eocene marls of New Jersey.

Ostreidæ.	
Ostrea, Linnæus	222
O. glauconoides, n. s., Whitfield, 1885, pl. xxix, fig. 2	222–223
O. (Alectrionia?) linguafelis, n. s., Whitfield, 1885, pl. xxix, fig. 1	223
Gryphæa, Lam.	
G. vesicularis, Lam., pl. xxix, figs. 7, 8	224
Pectenidæ.	
Pecten, Klein	224
P. kneiskerni, Conrad, pl. xxix, figs. 3–5	224–226
P. rigbyi, n. s., Whitfield, 1885, pl. xxix, fig. 6	226
Avicula annosa, Conrad, pl. xxix, fig. 9	226–227
Nuculidæ.	
Nucula, Lam.	227
N. circe, n. s., Whitfield, 1885, pl. xxix, fig. 12	227–228
Nuculana, Mörch	228
N. albaria, Conrad, pl. xxix, figs. 15, 16	228–229
Nucularia, Conrad	229
N. secunda, n. s., Whitfield, 1885, pl. xxix, figs. 13, 14	229–230
Axinea, Poli	230
A. conradi, n. s., Whitfield, 1885, pl. xxix, figs. 10, 11	230
Astartidæ.	
Astarte, Sowerby	231
A. castanella, n. s., Whitfield, 1885, pl. xxx, figs. 1, 2	231
A. planimarginata, n. s., Whitfield, 1885, pl. xxx, figs. 3, 4	232
Cardita, Brug.	232
C. perantiqua, Conrad, pl. xxx, figs. 8–10	232–233
C. brittoni, n. s., Whitfield, 1885, pl. xxx, figs. 11, 12	233–234
Crassatella, Lam.	234
C. alta, Conrad, pl. xxix, fig. 17	234–235
C. obliquata, n. s., Whitfield, 1885, pl. xxix, fig. 18, and pl. 30, figs. 13, 14	235–236
Cardiidæ.	
Protocardium, Beyr.	236
P. curtum, Conrad, 1870, pl. xxx, figs. 5–7	236–237
Veneridæ.	
Caryatis, Roemer	237
C. ovalis, n. s., Whitfield, 1885, pl. xxx, figs. 15, 16	237–238
Mactridæ.	
Veleda, Conrad, 1871	238
V. equilatera, n. s., Whitfield, 1885, pl. xxx, fig. 17	238–239

	Page
Corbulidæ.	
Corbula, Brug.	239
C. (Neæra) nasutoides, n. s., Whitfield, 1885, pl. xxx, figs. 18, 19	239-240
Neæra, Grey	240
N. æquivalvis, n. s., Whitfield, 1885, pl. xxx, figs. 20, 21	240-241
Pholadidæ.	
Parapholas. Conrad, 1848	241
P. kneiskerni, n. s., Whitfield, 1885, pl. xxx, figs. 22-24	241-242
Teridiæ.	
Teredo, Linn.	242
T. emacerata, n. s., Whitfield, 1885, pl. xxx, fig. 25	242
Section VII.—Unionidæ, from the clays at Fish House, Camden County.	
Unionidæ.	
Unio, Retzius	244
U. nasutoides, Lea, 1868, pl. xxxiv, figs. 4, 5	244-245
U. radiatoides, Lea, 1868, pl. xxxiv, figs. 1-3	245-246
U. subrotundoides, Lea, 1868, pl. xxxii, fig. 5	246-247
U. cariosoides, Lea, 1868, pl. xxxii, fig. 3	247
U. humerosoides, Lea, 1868, pl. xxxi, fig. 4	248
U. roanokoides, Lea, 1868, pl. xxxi, fig. 3, and pl. xxxiii, figs. 1, 2, and pl. xxxiv, fig. 7.	248-249
U. ligamentinoides, Lea, 1868, pl. xxxi, fig. 1, and pl. xxxii, fig. 4, and pl. xxxiv, fig. 8	249
U. alatoides, Lea, 1868, pl. xxxiii, figs. 3, 4, and pl. xxxiv, fig. 6	249-250
U. præanodontoides, n. s., Whitfield, 1885, pl. xxxi, fig. 2	250
U. rectoides, n. s., Whitfield, 1885, pl. xxxii, figs. 1, 2	250-251
Anodonta, Cuvier	251
A. grandoides, Lea, 1868, pl. xxxv, figs. 2, 3	251-252
A. corpulentoides, Lea, 1868, pl. xxxv, fig. 1	252
Section VIII. Appendix.	
Classified list of the species described in this volume	253-264

6.

WHITFIELD, R. P., *and* HALL, JAMES. (See Hall, James, and Whitfield, R. P.)
WHITFIELD, R. P., *and* WHITE, C. A. (See White, C. A., and Whitfield, R. P.)

SUPPLEMENT.

THE WRITINGS OF J. W. BAILEY.

BAILEY, J. W. Notes concerning the minerals and fossils collected by Lieutenant J. W. Abert, while engaged in the geographical examination of New Mexico. <Rep. of Lieut. J. W. Abert of his Examination of New Mexico in the years 1846–'47. Ex. Doc. No. 41, pp. 547–548, and 3 plates. Washington [1849].

>Plate [I] faces p. 522, and contains fossil leaves from the coal beds of the Raton.
>Plate [II] faces p. 546, and contains sharks' teeth and some *Gasteropoda* from Poblazon and an [*Athyris*] from Tuerto.
>Plate [III] faces p. 547, and contains an *Inoceramus* [*problematicus*] and a fossil leaf from the coal bed at Raton.

THE WRITINGS OF I. N. NICOLLET.

NICOLLET, I. N. List of Fossils belonging to the several formations alluded to in the Report; arranged according to localities. <Rep. intended to illustrate a Map of the Hydrographical basin of the Upper Mississippi river made by I. N. Nicollet. 26th Congress, 2d session, Senate Ex. Doc. No. 237. Appendix C, pp. 167–170. Washington, 1843.

	Page.
Atrypa lingulata, n. s., Nicollet, 1843	167
Ostrea congesta, Conrad	169

A large number of fossils are mentioned by their generic names and said to be new species, but no specific name is given and they are not described.

THE WRITINGS OF HIRAM A. PROUT.

1.

PROUT, H. A. Description of New Species of Bryozoa from Texas and New Mexico, collected by Dr. George G. Shumard, Geologist of the U. S. Expedition for Boring Artesian Wells along the 32d Parallel, under the direction of Capt. John Pope, U. S. Corps Top. Eng. <Trans. St. Louis Acad. Sci., vol. i, pp. 228–235. 1858. St. Louis, 1856–'60.

	Page.
Fenestella trituberculata, n. s., Prout, 1858	228–229
Fenestella popeana, n. s., Prout, 1858	229–230
Fenestella corticata, n. s., Prout, 1858	230
Fenestella intermedia, n. s., Prout, 1858	231
Fenestella variabilis, n. s., Prout, 1858	231–232
Fenestella shumardii, n. s., Prout, 1858	232
Fenestella norwoodiana, n. s., Prout, 1858	233
Fenestella subretiformis, n. s., Prout, 1858	233–234
Eschara ? concentrica, n. s., Prout, 1858	234
Eschara ? tuberculata, n. s., Prout, 1858	234–235

PROUT, H. A. Second Series of Descriptions of Bryozoa from the Palæozoic Rocks of the Western States and Territories. <Trans. St. Louis Acad. Sci., vol. i, pp. 266-273, pl. xvi, figs. 2 a. b. 1858. St. Louis, 1856–'60.

PERMIAN SPECIES.

	Page.
Polypora mexicana, n. s., Prout, 1858, pl. xvi, figs. 2a, b	270–271

THE WRITINGS OF BENJAMIN F. SHUMARD.

SHUMARD, B. F. Paleontology. <Rep. of a geological exploration from Fort Leavenworth to Bryan's Pass, made in connection with the survey of a road from Fort Riley to Bridger's Pass, under command of Lieutenant F. T. Bryan, topographical engineer, 1856, by H. Engelmann, geologist and mining engineer. <Rep. of the Secretary of War. Message from the President of the U. S. to the 35th Congress, 1st session, Ex. Doc. No. 2, vol. ii, pp. 517–520. Washington, 1857.

Gives two lists and notes of fossils, and describes one new species *Mytilus engelmanni*.

FOSSILS OF THE COAL MEASURES.

	Page.
Brachiopoda.	
Productus splendens, Norwood & Pratten	518
Productus rilliersi, D'Orbigny	518
Productus nebrascensis, Owen, D. D	518
Productus æquicostatus, Shumard	518
Productus semireticulatus, Mart., sp	518
Chonetes smithii, Norwood & Pratten	518
Terebratula ? subtilita, Hall	518
Spirifer plano-convexa, Shumard	518
Spirifer meusebachanus, Roemer, F	518–519
Spirifer kentuckensis, Shumard	519
Rhynconella ——— ?	519
Atrypa ——— ?	519
Orthisina umbraculum ? Buch, sp	519
Acephala.	
Myalina subquadrata, Shumard	519
Mytilus engelmanni, n. s., Shumard, 1857	519
Arca ——— ?	519
Pecten occidentalis, Shumard	519
Avicula ——— ?	519

FOSSILS OF THE CRETACEOUS FORMATION.

Ammonites peracultus ? H. & M	520
Scaphites mandenensis, Morton, sp	520
Scaphites nicolleti, Morton, sp	520
Rostellaria nebrascensis, Evans & Shumard	520
Ostrea congesta, Conrad	520
Inoceramus cripsii, Mantell	520
Inoceramus barabini, Morton	520
Inoceramus sagensis ? Owen, D. D	520
Inoceramus fragilis, H. & M	520
Inoceramus ten[u]ilineatus, Hall & Meek	520
Arca shumardi, M. & H	520
Vertebral scales and fin bones of fishes	520

INDEX OF GENERA AND SPECIES.

[The page numbers in heavy type indicate references to the original descriptions.]

A.

	Page.
Abra? formosa	48
Acambona	**119**
pinna	129
Acanthocardium	99
Acanthocladia americana	19, 257
Acanthotelson	**36**, 46, 53, 61, 179
eveni	**53**, 61, 179
inæqualis	**36**, 46
stimpsoni	**36**, 46, 61, 179
Acanthotelsonidæ	36
Acar	98
Acella	106, 141, 161
haldemani	**141**, 161
Acephala	225, 232, 255, 258, 274
Acervularia	108, 150, 155, 163, 223
adjunctiva	**150**, 163
davidsoni	155
pentagona	108, 192
Acidaspis	57, 79, 186
ceralepta	79
cincinnatiensis	79
crosotus	79
humata	57
parvula	**186**
Acila	98
Aclis	74, 86, 152
robusta	86
? stevensoni	**152**
swalloviana	74
Acmæa occidentalis	102
papillata	102
parva	102
Acmæidæ	102
Acœli	35
Acrochordiceras	**110, 239**
hyatti	**110, 239**
Acroloxus	106, 161
minutus	106, 161, 169
Acrothele	189, 198
? dichotoma	**189**
matthewi	198
Acrotreta	126, 129, 189
gemma	189
pyxidicula	**126**, 129
subsidua	**126**, 129
Actæon	94, 101, 143
attenuatus	101
concinnus	**13**
ellipticus	50
impressa	**220**

	Page.
Actæon intercalaris	**94**
melanellus	221
pygmæus	221
(Solidula) attenuuta	**18**
striatus	221
subellipticus	**14**, 101
woosteri	**143**
Actæonella	29, 210
syriaca	**210**
Actæoncma striata	220
sulcata	220
Actæonidæ	101
Actæonina	29, 86, 143
biplicata	48
minuta	86
naticoides	48
prosocheila	**143**
Actæoninæ	29
Actinaria	96, 142, 163
Actinoceramus	47, 97
Actinoceras	27
Actinocrinidæ	53, 81
Actinocrinites	81, 87
delicatus	81
longus	81
penicillus	81
sculptilis	87
sculptus	81
Actinocrinus	23, 26, 37, 38, 41, 42, 59, 118, 127, 130, 140, 163
(Alloprosallocrinus) cuconus	**38**
(Amphoracrinus?) concavus	**26**
subtuibinatus	**23**
araneolus	**23**
asteriscus	**23**
(Batocrinus) asteriscus	42
dodecadactylus	42
pistilliformis	41
pistillus	59
calyculus var. pardinensis	40
concinnus	42
delicatus	87
dodecadactylus	**26**
evansii	**249**
nashvillæ var. subtractus	118
pistillus	**37**
(Pradocrinus?) amplus	**26**
pyriformis var. rudis	26
quadrispinus	**118**

INDEX OF GENERA AND SPECIES.

Actinocrinus (Saccocrinus?) amplus 59
    scitulus **23**, 42
    sillimani **26**, 42
    speciosus **23**
    (Sphærocrinus) concavus 37
    unicornicus **249**, 251
    validus **23**
    viaticus **127**, 130
    wachsmuthi **118**, 149, 165
Actinopteria 194
    boydi 194
Actinozoa 129, 130, 131, 136, 138, 144, 147, 158,
    163, 192
Adelopthalmus mazonensis? 53
Admete (Admetopsis) gregaria 134
    ? gregaria **76**
    ? subfusiformis **76**
    ? rhomboides **76**
Admetidæ 135
Admetopsis 134, 144
    rhomboides 144
    subfusiformis 144
Ænona 269
    eufaulensis 269
    papyria 269
Æora 269
    cretacea 269
Aganides 104
Agaricocrinus 26, 82, 83, 154
    gracilis **26**
    nodosus 82
    springeri **154**
    whitfieldi 83
Agaronia punctulifera 222
Agassizocrinus 85, 87, 249, 251
    carbonaria 85
    chesterensis 85
    conicus 85, **249**, 251
    dactyliformis 255
    gibbosus 85
    pentagonus 85
Agelacrinites 54, 78, 83
    (Lepidodiscus) cincinnatien-
      sis 78
      squamosus **54**, 83
    pileus 78
    vorticellata 78
Agnostus 126, 129, 189, 198, 234
    acadicus 198
    bidens 189
    communis 189, **234**
    interstrictus **126**, 129
    neon 189, **234**
    prolongus 189, **234**
    richmondensis **189**
    seclusus **189**
    tumidosus **234**
Agraulos 31, 190, 199
Agraulos ——? 31
Agraulos ? globosus **190**
    oweni 31
    quadrangularis 199
Akera 261, 264
    glans-oryza **261**, 264
Alaba 102

Alasmodonta 98
Alcyonaria 96
Alectryonia 97, 136, 151, 158
Alethopteris whitneyi **246**
Alipes 102
Allopoxallocrinus 38, 54
Allorisma ...32, 45, 61, 73, 86, 89, 90, 132, 138, 147, 149
    163, 165, 178, 245
    ? altirostrata **19**, 20
    audax **223**
    capax **245**
    (Cercomyopsis) pleuropistha ?... 89
    (Chænomya) hybrida 61
    ? cooperi 19, 20
    costata 86, 90
    elegans 51, 52
    geinitzii 86
    ? gilberti 147, 163
    ? leavenworthensis **19**, 20
    marionensis **138**, 165
    nucula arata **232**
    (Sedgwickia) geinitzii 73
      granosa 73
      pleuropistha **66**
      reflexa **73**
      subelegans **73**
    subcuneata **19**, 20, 32, 73, 149, 178
    var 132
    terminalis **232**
    ventricosa **67**, 89
    winchelli **67**, 89
Alveolites 62, 108, 156
    goldfussi 156
    multilamella **108**
    rockfordensis 102
    vallorum **62**
Amauropsis 102, 264
    paludinæformis 102, 264
Ambocœlia 31, 119
    (Spirifer?) minuta **119**
Ambonicardia **266**
    cookii **266**
Ambonychia 55, 56, 79
    acutirostris 56
    carinata 232
    costata 79
    intermedia **55**
    (Megaptera) alata **68**, 79
    casei **41**, 79, 148
Ammonites 28, 35, 93, 104, 135, 298, 209, 210, 219,
    227, 246, 253, 255, 263
    acuto-carinatus **255**
    belknapii **243**
    cheyennensis **250**
    complexus **14**, 104
    var. sucinensis **27**, 93
    cordiformis **19**, 35, 263
    var. distans 263
    galpini **256**
    geniculatus **249**
    gibbonianus 243
    graysonensis 159
    halli **15**
    henryi **19**, 35
    lævianus **135**

INDEX OF GENERA AND SPECIES.

	Page.
Ammonites lenticularis	250
leoncnsis	**219**
libanensis	**210**
marciana	**255**
moreauensis	**250**
mullananus	**29**
? mullananus	107
nebrascensis	**250**
newberryanus	**17**, 93
novi-mexicana	**243**
opalus	**250**
peracultus	274
percarinatus	**14**, 245
peruvianus	243
pickeringi	227
placenta, var. intercalaris	21
dekayi, var. intercalaris	135
pleurisepta	**219**
safedensis	**209**
(Scaphites?) ramosus	**17**
serrato-carinatus	**66**
shumardi	**243**
swallovii	159
syriacus	**208**
texanus	219
vancouverensis	**27**
vermilionensis	**21**
vespertinus	**255**
Ammonitidæ	35, 104, 135, 263, 265
Amorphozoa	41
Ampelita	107
Amphicœlia	56
neglecta	56
Amphidiscus armacus	229
Amphion	191
nevadensis	**191**
Amphistegina	120
Amphora libyca	229
Amphoracrinus	42, 54, 82
divergens	54, 82
? spinobranchiata	82
subturbinatus	42
Amplexus	121, 136, 155, 163, 223
coralloides?	243
fragilis	**121**
yandelli	156
zaphrentiformis	**136**, 160
Ampullaria	197
powelli	180, **188**, 197
Ampullina alveata	221
Amusium	**266**
conradi	**266**
simplicum	266
Amussium	35
aurarium	**35**
propatulum	49
Anadara? canalis	50
? congesta	50
incile	50
microdonta	50
protracta	50
trigintinaria	50
trilineata	50
Anarthrocanna australis	**227**

	Page.
Anatina claibornensis	220
Anatinidæ	30, 32, 34, 100, 109, 111, 134, 261, 262, 264, 271
Anchura	94, 102, 111, 127, 134, 137, 143, 264
biangulata	48
(Drepanoch[e]ilus) americana	48, 102
decemlirata	48
mudgeana	**143**
nebrascensis	48, 102, 264
prolabiata	143
rostrata	48
ruida	143
? ? fusiformis	**111**, 134
haydeni	**143**
newberryi	**94**
nuptialis	125, **127**
parva	48, 102
prolabiata	**137**
ruida	**137**
? sublevis	48, 102, 264
Ancillaria elongata	222
Ancillopsis altile	222
Ancylidæ	106
Ancyloceras	104, 209, 261, 265
annulatum	159
? cheyenensis	**15**
(Hamites) uncus	**19**
jenneyi	**261**, 265
? nebrascensis	**15**
nicolletii	**14**
safedensis	**209**
tricostatus	**261**, 265
? uncum	104
Ancyloceratidæ	104
Ancylus	112
undulatus	**65**, 112, 166, 170
Angulus	100
Animalia	129
Anisomyon	**20**, 102, 112, 134, 143, 264
alveolus	21, 102, 264
borealis	21, 102, 134, 264
centrale	**70**, 134, 143
inæquicostatus	48
patelliformis	20, 21, 102, 264
sexsulcatus	21, 102, 112
shumardi	102
subovatus	21, 102, 264
Anisopoda	46
Anisorhynchus	101
Anisothyris	101
Anisus	34, 106, 108
Annularia	149
longifolia	149
Annulata	35
Anodonta	140, 160, 272
? angustata	167
? cattskillensis	167
corpulentoides	272
decurtata	170
grandoides	272
parallela	**140**, 160, 168,
propatoris	**139**, 160, 168
Anodontopsis	79, 194
amygdalæformis	**194**

INDEX OF GENERA AND SPECIES.

	Page.
Anodontopsis ? milleri	**66**, 79
(Modiolopsis ?) unionoides	79
unionoides	**66**
Anolax gigantea	222
plicata	222
Anomalocardia trigintinaria	50
Anomalocrinus	37, 55, 77
incurvus	77
Anomalocystites	77
(Atelocystites) balanoides	**69**, 77
Anomia	93, 95, 97, 158, 160, 211, 213, 219, 266
argentaria	266
concentrica	**22**, 95
gryphorhynchus	**71**, 160, 168
micronema	**88**, 160, 168
nitida	**93**
? obliqua	**21**, 97
(Placunopsis ?) gryphorhyncus	77
propatoris	**158**, 167
rætiformis	**111**
subcostata	211, 213, 219
subtrigonalis	**21**, 97
tellinoides	266
anomiidæ	97, 111, 266
Anomocare	190
? parvum	**190**
Anomalocardia rhomboidella	220
Anomphalus	**40**, 87
rotulus	**40**, 87
Anthophyllum cuneiforme	220
expansum	**247**
Anthracerpes	**37**, 46, 53
typus	**37**, 46
Anthracoptera	61, 165
? ? fragilis	**41**, 61
polita	**165**
Anthracopupa ohioensis	167
Anthrapalæmon	37, 46, 61, 179
gracilis	**37**, 46, 61, 179
Aphrodina	100, 269
tippana	269
Aploceras	27
Aporrhaidæ	102, 111, 134, 264
Aporrhais	102, 264
biangulata	102
decemlirata	48
(Goniocheila) castorensis	**261**, 264
meeki	**261**, 264
newberryi	264
parva	**21**, 48
sublævis	**21**, 48
Aptycha	29
Arachnida	61
Arachnophyllum	**223**
Arca	92, 137, 207, 208, 209, 211, 213, 215, 216, 217, 218, 267, 270
——?	266, 274
acclivis	**207**
altirostris	267
bhamdunensis	**209**
brevifrons	**207**
canalis	50, **215**, 216
carbonaria	20
? coalvillensis	**137**

	Page.
Arca congesta	50, **215**, 216
(Cucullæa) cordata	**15**
equilateralis	**17**
inornata	**18**, 180
shumardi	**15**
cuneus	**209**
declivis	**207**
devincta	**206**
equilateralis	92
fabiformis	**208**
incile	50
indurata	**207**
longa	**209**
microdonta	50, 211, 213
obispoana	**217**
orientalis	**207**
parallella	**70**
protracta	50
quindecemradiata	270
rhomboidella	220
shumardi	274
striata	51
subelongata	**218**
subrotundata	**207**
sulcatina	**256**
syriaca	**207**
trilineata	50, **215**, 216
vancouverensis	**17**
Arcadæ	28
Arcestes	110, 146, 163, 239
? —— ?	146, 163
? cirratus	**146**, 163, **239**
gabbi	**110**, 239
? perplanus	**110**, 239
Arcestidæ	110, 239
Archæocnaris	90
vermiformis	89
Archæocidaridæ	43
Archæocidaris	23, 44, 72, 127, 131, 136, 150, 151, 163, 190, 245
——?	19, 87
cratis	**136**, 163
dininnii	**150**, 163
gracilis	**245**
longispinus	**245**
mucronata	**23**, 44
ornatus	131, **245**
triplex	**151**
triserrata	**72**
trudifer	**127**, 131
Archæocaris vermiformis	**68**
Archimedes	147, 154, 248
laxa	154
Archimedipora	94
——?	94
archimedes	255
Architectonica ahotti	48
henrici	221
ornata	221
plana	221
pseudogranulata	221
Arcidæ	33, 98, 111, 134, 202, 203, 267, 270
Arcinæ	33
Arcopagella	**70**, 100
? macrodonta	**100**

INDEX OF GENERA AND SPECIES. 279

Arcopagella mactroides **70**, 100
Arcopagia 215, 216, 217, 218
    medialis **215**, 216
    texana 218
    unda **217**
Arenicolites ? 261
Arethusiana 190
    americana **190**
Arionellus 28, 186, 250
    (Crepicephalus) oweni **28**
    pustulatus **186**
    tripunctatus **259**
Arrhoges 102
Articulata 31, 35, 46, 55, 56, 58, 59, 74, 79, 80, 81,
                84, 91, 105, 108, 109, 135, 136,
                138, 144, 153, 159, 165, 260, 261
Asaphiscus 129
    wheeleri 129
Asaphus 79, 91, 185, 186, 191, 248
    barrandi **232**
    caribouensis **191**
    ? curiosa 191
    emoryi **233**
    homalonotoides **186**
    (Isotelus) iowensis 250
            megistos 79
            vigilans **65**, 91
    (Megalaspis ?) goniocercus **76**
    romingeri **185**
    wisconsensis **185**
Asiphonida 266
Astarte 30, 34, 164, 207, 208, 209, 218, 236, 262,
                      263, 266, 271
    arctata **207**
    ? arenosa **236**
    castanella **271**
    eugonata **207**
    ovansi 263
    ? fragilis **22**, 34, 262
    gemma **224**, 225
    gibbosa 51
    grogaria 15
    inornata **22**, 34
    lintea **208**
    lucinoides **209**
    minutissima 220
    mortonensis 51
    mucronata **208**
    nebrascensis 51
    orientalis **207**
    packardi **164**, 180
    parva 220
    pervetus **207**
    planimarginata **271**
    subcordata **209**
    sublineolata **208**
    syriaca **207**
    texana **218**
    undulosa **208**
    vallisnerianus 51
    ventricosa **30**
    veta 266
    washitensis **255**
Astartella 90, 144, 165
    ——— ? 90

Astartella gurleyi **144**, 165
    newberryi **90**
    varica 90
    astartidæ 266, 268, 270, 271
Astartila **224**, 225
    ? corpulenta **225**
    cyclas **224**, 225
    cyprina **224**, 225
    cytherea **224**, 225
    intrepida **224**, 225
    polita **224**, 225
    transversa **224**, 225
Astereidea 26
Asterias 260, 261
    dubium **260**, 261
Asteridæ 24
Asteroidea 40, 44, 60, 61, 78, 260, 261
Astræospongia 58
    hamiltonensis **41**, 58
Astrea fungiformis 247
    ? gigas **247**
    mamillaris **247**
Astrodapsis **215**, 217
    antiselli **216**, 217
Astylospongia 56, 192
    ? carbonaria **41**
    ? ? christiani **56**
    præmorsa 91
Athyris 24, 43, 72, 85, 89, 92, 94, 109, 148, 178, 193,
                    196, 235, 245, 273
    angelica 193
    claytoni **235**
    crassicardinalis **116**
    hirsuta 196
    lamellosa 89
    parvirostra **24**
    persinuata **109**
    planosulcata 43, **235**
    roissyi 109
    subquadrata ? 235
    subtilita ... 52, 72, 85, 92, 94, 169, 178, 245, 246
    vittata 148
Atlantidæ 121
Atrypa 58, 62, 94, 108, 148, 193, 247, 248
    ——— ? 274
    aspera 58, 62, 94
    comis **250**
    desquamata 193
    limitaris ? 247
    lingulata **273**
    orbicularis 248
    prisca 247
    reticularis 58, 62, 94, 108, 148, 193
Aturia angustatus 50
    orbiculata 49
    vanuxemi 222
Aucella 35, 176
    concentrica var. 176
    erringtoni 35
    var. linguiformis 35
    hausmanni 51, 52
Aulacomya 98
Aulophyllum 62
    ? richardsoni **62**
Aulopora serpens 192, 247

INDEX OF GENERA AND SPECIES.

Aulopora tubæformis 248
Aulostegos 121
   guadalupensis 256
   spondyliformis 120,121
Auriculidæ 112
Austrella rigida 227
Avalana subglobosa 14, 29, 102
Avicula 24, 73, 86, 136, 208, 262
   ———? 224, 274
   abrupta 47
   annosa 271
   convexo-plano 47
   cretacea 47
   curta 49
   ? custa 232
   ? fibrosa 15
   haydeni 13,47
   iridescens 47
   laripes 47
   linguiformis 47
   longa 73, 86
   (Monotis) tenuicostata 18
   morganensis 86
   multangula 50
   munsteri 49
   nebrascana 47, 255
   oblonga 24
   (Oxytoma) gastrodes 76
      mucronata 262
   parkensis 136
   pedernalis 47
   petrosa 47
   pinnæformis 51
   planisulca 47
   (Pseudoptera) propleura 76
      rhytophora 76
   samariensis 208
   speluncaria 51
   subglibbosa 21, 47
   ? sulcata 73
   triangularis 47
   volgensis? 226
Aviculidæ 29, 32, 262, 263, 267
Aviculinæ 29, 32
Aviculopecten ... 24, 32, 43, 45, 61, 73, 80, 89, 90, 94,
      110, 116, 119, 127, 132, 146,
      162, 164, 178, 196, 235, 236
   ———? 32
   affinis 196
   altus 146, 162
   amplus 24, 32, 43
   burlingtonensis 24, 43
   carbonarius 87
   carboniferus 73, 178
   catactus 110
   coreyana 127, 132
   coxanus 24, 45, 73
   crenistriatus 66, 89
   curto-cardinalis 236
   (Eumicrotis?) augustensis .. 236
   eurekensis 196
   fimbriatus 41
   grandocostus 119
   haguei 196
   hawni 51

Aviculopecten idahoensis 146, 162
   indianensis 41, 61
   interlineatus 24, 45, 132, 178
   koninckii 24, 45
   limaformis 116
   macoyi 32, 132
   neglectus 73, 86, 87
   nodocostatus 116
   oblongus 43
   occidaneus 110
   occidentalis 45, 73, 132, 178
   oweni 24, 43
   paralis 80
   parvulus 236
   pealei 146
   ? pealei 162
   pellucidus 24, 45
   peroccidens 196
   pintoensis 196
   (Pseudomonotis) idahoensis ... 71
   randolphensis 41
   sanduskyensis 67
   spinuliferus 64
   (Streblopteria?) hertzeri .. 66, 90
   superstrictus 164
   utahensis 94, 110
   weberensis 235
   whitei 73
   williamsi 68
   winchelli 89
Aviculo-pectininæ 32
Aviculopinna 73
   americana 52, 73, 90
   pinnæformis 52
Axinæa 98, 102, 111, 160, 215, 216
   barbarensis 215, 216, 217
   subimbricata 98
   wyomingensis 111
Axinea 217, 267, 271
   alta 267
   couradi 271
   holmesiana 160
   mortoni 267
Axininæ 34
Axinus 258
   rotundatus 20
   (Schizodus) ovatus 19, 20
   securis 258
Axophyllum 92, 121, 177
   rudis 92, 121, 177
Aztbemis 217

B.

Baculites 28, 93, 94, 103, 135, 209
   ———? 209
   anceps var. obtusus 94, 103
   asper 103
   baculus 28
   chicoensis 93
   compressus 14, 103
   grandis 14, 103
   inornatus 27
   occidentalis 27, 93
   ovatus 14, 17, 103, 135
   syriacus 209

INDEX OF GENERA AND SPECIES. 281

	Page.
Baculitidæ	103, 135
Bailiella	198
Bairdia ? ——	257
Bakevellia	32, 132
parva	19, 20, 32, 132
Balanus	227
estrellanus	217
Barbatia	98, 143, 158
barbulata	158
coalvillensis	143
(Polynema?) parallela	98
Bariosta	105
Baroda subelliptica	143
wyomingensis	143
Barrandia	191
? ——?	191
mccoyi	191
Barycrinus	54, 83, 84
geometricus	83
hoveyi	83
var. herculeus	54, 83
magnificus	54, 83
mammatus	83
pentagonus	83
spectabilis	64, 84
subtumidus	83
Baraphyllum	57, 144
?? arenarium	58
fungulus	144
Basommatophora	135
Bathyomphalus	34, 106, 137, 161
Bathyurellus (Asaphiscus) bradleyi	76
Bathyurus	185, 191, 234
armatus	186
? congeneris	191
? haydeni	76
longispinus	185
pogonipensis	234
serratus	76
? simillimus	191
? tuberculatus	191
Batocrinus	42, 54, 81
cassedayanus	54, 81
christyi	81
(Eretmocrinus) neglectus	54, 81
remibrachiatus	81
pyriformis	81, 87
quasilus	54, 81
trochiscus	54, 81
verneuillianus	82
Beaumontia	147, 158
? solitaria	147, 158
Belemnitella	105
? bulbosa	15, 105
paxillosa	49
Belemnites	31, 35, 95, 111, 176, 263
densus	19, 35, 95, 179, 263
mucritatis	176
nevadensis	111
pacificus	35
paxillosa	49
Belemnitidæ	31, 35, 105, 111, 263
Belemnocrinus	59, 118, 120
typus	118
whitii	40, 59

	Page.
Bellerophon	24, 42, 46, 55, 74, 81, 117, 118, 126, 129, 132, 138, 147, 151, 154, 164, 178, 195, 197, 235, 245, 247, 248, 258
allegoricus	126, 129
bilabiatus	117
bilobatus	248
bowmani	138
(Bucania) platystoma	55
carbonarius	74, 178
combsi	195
crassus	24, 46, 132, 178
cyrtolites	42
gibsoni	154
inspeciosus	151
interlineatus	51
leda	195
lyra	195
mera	195
majusculus	197
marcouanus	51
marcouianus	74
micromphalus	226
montfortianus	74
neleus	235
newberryi	67, 81
nodocarinatus	178
pannens	118
patulus	81
pelops	195
percarinatus	74, 178
perelegans	117
perplexa	195
propinquus	67, 81
scriptiferus	118
strictus	223, 226
sublævis	154
subpapillosus	147, 164
textilis	197
undulatus	223, 226
vinculatus	117
Bellerophontidæ	39, 132
Bellinurus	36, 46
danæ	36, 46
Beyrichia	121, 188, 191, 195
bella	188
cincinnatiensis	187
fœtoidea	121
lithofactor	120
var. velata	120
petrifactor	121
var. velata	121
(Primitia) occidentalis	195
Blastoidea	26, 44, 126
Brachiopoda	20, 22, 24, 26, 27, 30, 31, 33, 41, 43, 44, 45, 55, 56, 57, 58, 59, 60, 61, 62, 64, 68, 72, 78, 80, 83, 89, 90, 91, 92, 94, 96, 98, 108, 109, 110, 116, 118, 121, 126, 127, 129, 130, 131, 133, 136, 146, 147, 148, 149, 151, 154, 155, 162, 163, 165, 177, 189, 190, 192, 196, 223, 232, 233, 234, 235, 236, 250, 253, 255, 257, 260, 261, 262, 265, 274
Brachydontes	140, 158, 160
Brachyspira	159
Breviarca	98, 267

INDEX OF GENERA AND SPECIES.

Breviarca saffordi 267
Bryozoa 19, 154, 253, 255, 257, 274
Bucanella nana 66
Bucardinæ 34
Buccinidæ 103
Buccinofusus diegoensis 222
Buccinopsis 219
    parryi **219**
Buccinum 205
    ? devinctum **206**
    integrum **205**
    ? nebrascensis 14
    sowerbii 221
    ? vinculum **13**
Buchicoras 135
    swallovi 135
Bulimnea 106
Bulimus limnæformis **16**, 221
    nebrascensis **16**
    perversus 221
    ? teres **15**, 221
    ? vermiculus **15**, 221
Bulinus 106, 161
    atavus **139**, 161, 169
    disjunctus **145**, 161, 169
    floridanus 170
    longiusculus 106, 161, 169
    rhomboideus 106, 161, 169
    subelongatus 106, 161, 169
Bulla 212, 214
    dekayi 220
    jugularis **212**, 214
    minor **14**
    occidentalis **14**
    petrosa 50, **206**
    subcylindrica **16**
    volvaria **14**
Bullidæ 101, 264
Bursaerinus **26**, 59, 118
    confirmatus **118**
    wachsmuthi **26**, 59
Busycon ? 211, 213
    bairdi 16
    ? blakei **211**, 213
    ? oregonensis 50
Buthus ? ? carbonarius **53**
Bythinella 159
    gregaria **70**, 159, 170

C.

Cælenterata 177
Calceocrinus 63, 82, 83, 188
    barrandii **188**
    ? bradleyi **63**, 83
    ? wachsmuthi **63**, 82
Calceola 99
Callianassa 153
    danai **13**
    oregonensis **227**
    ulrichi **153**, 157
Callipteris 149
    sullivanti 149
Callista 28, 100, 209
    (Aphrodina ?) tenuis 100

Callista delawarensis **269**
    deweyi 28
    (Dosiniopsis) deweyi 100
    nebrascensis 100
    orbiculata 100
    owenana 100
    eufalensis 48
    ? pellucida 100
Callonea 98
Callonema 195
    occidentalis **195**
Calobates 101
Calophyllum **223**
Calymene 79, 148
    bufo 247
    sonaria 79, 148
Calymenidæ 260, 261
Calyptræa trochiformis 221
Camæna 106
Camarophoria 43, 196, 257
    ? bisulcata **256**, 257
    cooperensis 196
    globulina 52
    schlotheimi ? 257
    subtrigonia 43
    swalloviana **257**
Camerinidæ 31
Cameroceras 129
Campeloma 107, 112, 162, 166
    macrospira 112, 162, 168
    (Melantho) macrospira **68**
    multilineata 107, 162, 169
    multilineatum 221
    multistriata 107, 162, 169, 221
    producta **166**, 169
    vetula 107, 162, 169
    vetulum 221
Campophyllum 72, 109, 177
    ? texanum **257**
    torquium 72, 177
Camptonectes 33, 49, 95, 127, 133, 164, 236, 263
    (Amusium) burlingtonensis . 266
    bellistriata 95
    bellistriatus 33, 49, 133, 230, 262
    calvatus 220
    extenuatus 33, 49, 180, 236, 262
    parvus **266**
    pertennistriatus **236**
    platessa **127**, 133
    platessiformis 164
    stygius **127**, 133
Camptopteris remondi **246**
Campylodiscus americ ? 229
Cancellaria 210
    petrosa **210**
Caninia 223
Cantharis vaughni 49
Cantharulus 103
Cantharus 103
    (Cantharulus) vaughani 103
    julesburgensis **157**
Canthyria 105
Caprina 214, 218
    occidentalis **215**, 218

INDEX OF GENERA AND SPECIES. 283

	Page.
Capriua planata	**215**, 218
Caprinella coraloidea	**13**
Caprotina	93
(Requienia?) bicornis	93
texana	243
Capsa	220
texana	**220**
Capularia discoidea	220
Capulidæ	132
Capulus fragilis	**14**
occidentalis	**13**, 48
Carbonarea	**64**, 92
gibbosa	**64**, 92
Cardiidæ	34, 99, 111, 268, 271
Cardinia	164, 224
? costata	225
cuneata	**224**
? cuneata	225
? exilis	224
præcisa	**164**
recta	**224**, 225
Cardiola	197
? filicostata	**197**
Cardiomorpha	24, 79, 86, 89, 119, 236, 258
(Cardiopsis?) parvirostris	**119**
missouriensis	86, 236
?? obliquata	**68**, 79
radiata	**24**
subglobosa	**89**
Cardiopsis	**26**, 42, 119
radiata	42
Cardissoides	34
Cardita	211, 213, 216, 218, 219, 270, 271
abbreviata	50
brittoni	**271**
carinata	50
eminula	218
intermedia	**270**
littoralis	271
monilicosta	50
occidentalis	50, 216
perantiqua	271
planicosta	**211**, 213, 219
radiaus	50
rhombea	**271**
subtenta	50, **206**
subtetrica	**220**
Cardium	28, 93, 99, 111, 143, 207, 208, 209, 210, 211, 212, 213, 218, 268
australe	225
bellulum	**93**
bellum	**208**
biseriatum	**207**, 210
(Cerastoderma) modestum	50
choctawense	159
congestum	**218**
crebrilechuatum	**207**, 209
(Criocardium) dumosum	268
multiradiatum	268
speciosum	90
curtum	111
elegantulum	47
eufaulensis	268
ferox	**225**
(Hemicardium?) curtum	**28**
hermonense	**207**

	Page.
Cardium iowensis	**248**
kansasense	**70**, 99
lintenm	**210**, 212
mediale	**218**
modestum	50, **211**, 213
multistriatum	253, **255**
ovulum	**208**
pauperculum	**70**, 143
pertenue	**28**
(Protocardia) filosum	**218**
multistriatum	218
salinænse	**70**
texanum	218
(Protocardium) perelongatum	**268**
rarum	**256**
ripleyanum	268
ripleyense	268
sancti-sabæ	47
scitulum	**17**
shumardi	**21**
speciosum	**16**
subcurtum	111
subquadratum	**256**
syriacum	**207**
trite	**143**
Cariclla flemingii	221
Carinifex	112
binneyi	**65**
tryoni var. concava	**65**
(Vorticifex) binneyi	112, 166, 170
tryoni	**65**, 112, 166, 170
Caryatis	100, 271
ovalis	**271**
? veta	**271**
Caryophyllia	142
egeria	142
johannis	**142**
Cassidula	219
(Lacinia alveata)	219
Cassidulus patelliformis	222
Cassiope	127, 134
whitfieldi	**127**, 134
Cassiopella	**139**, 162
turricula	162, 169
Catenipora escharoides	247
gracilis	**233**
Catillocrinus	54, 59, 83
bradleyi	**54**, 83
wachsmuthi	59
Catillus	97
obliquus	221
Catopygus patelleformis	222
Cavolinidæ	31
Cellepora tubulata	222
Cemoria crucibuliformis	50
Centrocrinus	38, 42
Centronella	56
billingsiana	**56**
Cephalopoda	20, 21, 22, 25, 27, 28, 31, 33, 35, 38, 39, 40, 42, 44, 46, 55, 59, 63, 65, 67, 68, 74, 79, 80, 81, 84, 87, 91, 93, 94, 95, 103, 109, 110, 111, 117, 121, 126, 129, 132, 135, 138, 144, 146, 149, 154, 159, 162, 165, 179, 191, 195, 198, 226, 236, 250, 256, 258, 261, 262, 265
Cerastoderma	99

INDEX OF GENERA AND SPECIES.

Ceratiocaris.. 61, 89
   (Colpocaris) bradleyi............ **68**, 89
      elytroides............ **68**, 90
     sinuatus............................ **53**, 61
   (Solenocaris) strigata............ **68**, 90
Ceraurus................................79, 186, 191, 248
   ————?.. 191
   icarus... 79
   rarus... **186**
   pleurexanthemus................................. 185
Cercomya.. 269
   peculiaris... 269
Ceriphasiidæ.......................... 106, 112, 139
Cerithidea.. 161, 180
   ? nebrascensis......................... 169, 180
   (Pirenella?) nebrascensis....106, 161
Cerithiidæ... 106
Cerithiopsidæ... 102
Cerithiopsis.. 102
   moreauensis..................................... 102
Cerithium..180, 210
   bilineatum....................................... **210**
   fremonti.................................221, **231**
   mediale.. **206**
   nebrascensis.................................... **16**
   nodulosum............................221, **231**
   pillingi....................................... **180**
   tenerum...............................221, **231**
   totium-sanctorum............................ **180**
Chænocardia.. 86
   ovata... 86
Chænomya.......................... **32**, 38, 73, 84, 86
   cooperi... 32
   hybrida.. **38**
   leavenworthensis............................. 32, 73
   minehaha....................................... 73, 86
   rhomboidea................................ **38**, 84
Chætetes............. 55, 131, 138, 147, 158, 257
   ? ? dimissus.............................**147**, 158
   lycoperdon....................................... 231
   mackrothii...................................... 257
   milleporaceus.................................. 131
   muscatinensis................................ **138**
   petropolitanus................................. 55
Chamidæ..180, 208
Chariocephalus.................................190, 234
   ? tumifrons........................190, **234**
Chemnitzia........................48, 103, 134, 258
   cerithiformis..................................... 103
   coroua... 48
   meekiana....................................... 48
   swalloviana.................................... **258**
   ? texana... 48
Chenopus..207, 209, 210
   ————?.. 210
   induratus...................................... **207**
   syriacus....................................... **208**
   turriculoides................................ **207**
Chetetes crinita... 226
   gracilis... **226**
   ovata... 226
   tasmaniensis.................................. 226
Chione... 100
   (Liophora) alveatus......................... 50
       athleta......................... 50

Chione (Liophora) latilirata................ 50
Chiton.. 87
   carbonarius..................................... 87
Chladocrinus.. 33
Chlamys.. 97
   nebrascensis................................... 97
Chonetes..... 24, 31, 43, 60, 62, 72, 85, 94, 119, 127, 131,
                    177, 192, 235, 245, 257
   deflecta.. 102
   filistriata..................................... **193**
   flemingi ?.................................... 257
   geniculata.................................... **119**
   glabra....................................... 52, 72
   granulifera.....................72, 131, **250**
   hemispherica................................. 192
   illinoisensis.................................... 60
   ? iowensis.................................. **250**
   loganensis................................... **235**
   macrostriata................................ **193**
   mesoloba..................................... 131
   ? ? millepunctata.................**64**, 85
   mucronata............ **19**, 20, 31, 52, 193
   permiana................................... **257**
   planumbona............................**24**, 43
   platynota............................**127**, 131
   pusilla.. 62
   setigera..................................... 193
   smithii.................................. 85, 274
   variolata................................... 232
   verneuliana................20, 72, 177, 245
        var. utahensis.................. 94
Chonophyllum... 165
   sedaliense................................... **165**
Cibota.. 267
   multiradiata.................................. 267
   obesa.. **267**
   rostellata................................... 267
   uniopsis..................................... 267
Cidaris... 200
   hemigranosus............................... 150
Cinulia... 29, 101
   (Avellana) pulchella......................... 48
        naticoides.......................... 48
   (Oligoptycha) concinna.................... 102
Cladocora [?] lineata................................ 49
Cladocrinus.. 33
Cladodus occidentalis............................. 20
Cladopora..155, 192
   pulchra....................................... 192
   reticulata.................................... 155
Clasteria... **227**
   australis.................................. **227**
Clathropora... 155
   flabellata................................. **231**
   frondosa.................................... 155
Clathropteris.. 246
Clausilia contraria................................ 221
   teres... 221
   vermicula................................... 221
Clavagella.. 270
   armata....................................... 270
Clavatula ?.......................................211, 213
   ? californica........................**211**, 213
Clavelites (Peistochilus) scarboroughi....... 49
Clavella vicksburgensis........................... 222

INDEX OF GENERA AND SPECIES. 285

	Page.
Clavifusus altile	222
cooperi	222
Cleidophorus	32
Cleobis	**223**, 225
gracilis	**224**
grandis	**223**
recta	**224**
Clidophorus	79
(Nuculites ?) fabula	79
pallasi	51
(Pleurophorus) simplus	51
solenoides	51
Climacograptus	129
Clinopistha	**64**, 80, 86, 178
antiqua	**66**, 80
radiata	178
var. levis	64, 86
Clisiophyllum	30, **223**
gabbi	**30**
Closteriscus	**102**
tenuilineatus	102
Clydonites	110, 239
lævidorsatus	110, 239
Clydonitidæ	**110**, 239
Clypeaster joncsii	222
rogersi	222
tumidus	222
Cocconeis concentrica	229
finnica	229
gemmata	229
lineata	229
oblonga	229
prætexta	203, 229
punctata	229
Cocconema asperum	203, 229
cistula	229
cymbiforme	203
gibbum	229
gracile	229
lanceolata	229
lunula	229
Codonites	**64**, 83, 87
gracilis	**64**, 83
stelliformis	83, 87
Cœlocrinus	42, **117**
concavus	42
subspinosus	**117**
Cœlopleurus infulatus	222
Cœlospira concava	36
Colcolus	195
lævis	**195**
Colcoprion	191
minuta	**191**
Colpocaris	89
Colnmna	106, 161
teres	106, 161, 169
vermicula	106, 161, 169
var. contraria	106
Columnaria	223
? sexradiata	49
thomii	**233**
Colus	212, 214
arctatus	50, **212**, 214
Comarocystites	37, 54
shumardi	**37**, 54

	Page.
Comarocystites shumardi var. obconicus	37, 54
Combophyllum	62
multiradiatum	**62**
Complanaria	99
Conactæon	20
Conchifera	21, 22, 116, 119, 127, 130, 132, 133, 135, 136, 137, 138, 140, 142, 144, 146, 147, 151, 153, 155, 158, 159, 160, 162, 163, 165, 178, 250
Concholepas pygmæa	221
Conchopeltis	**185**
alternata	**185**
minnesotensis	**185**
Confervites ? tenella	**227**
Coniferæ	226
Conocardium	38, 80, 92, 117, 130, 194
nevadensis	**194**
obliquum	**38**, 92
ohioense	**66**, 80
pulcellum	**117**
trigonale	80
Conocephalites	186, 234
calciferus	**186**
hartii	**186**
(Pterocephalus) laticeps	**234**
subcoronatus	**234**
Conocoryphe	108, 129, 198, 234
(Bailiella) baileyi	198
(Conocephalites) kingii	**65**
elegans	198
matthewi	198
(Ptychoparia) gallatinensis	**76**
kingii	108, 129
walcotti	198
Conodictyum	40
radiatum	40
Conomitra fusoides	221
Constellaria	155
antheloidea	155
Conularia	39, 81, 84, 89, 118, 138, 149, 151, 165, 185, 195, 198, 226
byblis	**118**
crustula	151, **165**
elegantula	**67**, 81
inornata	**226**
levigata	226
micronema	**67**, 89
missouriensis	84, 149, 198
molaris	**138**
multicostata	**39**
newberryi	89
quadrata	**185**
subcarbonaria	**39**, 84
tenuistriata ?	226
victa	**118**
whitei	**39**
Coralliochama	**180**
orcutti	**180**
Corbicula	99, 105, 112, 137, 140, 160, 166, 266
æquilateralis	**76**
annosa	266
aughcyi	**166**, 168
berthoudi	**166**, 168
cardiniæformis	**140**, 160
cleburni	**140**, 160, 168
crassatelliformis	**70**

INDEX OF GENERA AND SPECIES.

	Page.
Corbicula (Cyrena?) securis	76
cytheriformis	105, 160, 169
lemacerata	266
? fracta	70
? var. crassiuscula	77
(Leptesthes) cardiniæformis	165
fracta	160, 168
macropistha	141, 160, 168
planumbona	88, 160, 169
subelliptica	105, 160
subelliptica var. moreauensis	105
nebrascensis	105, 160, 168
nucalis	70, 99
obesa	140, 160, 169
occidentalis	105, 160, 168
powelli	137
pyriformis	168
subelliptica	168
? subtrigonalis	70, 99
umbonella	168
(Veloritina) bannisteri	77
cytheriformis	77
durkeei	112, 160, 168
inflexa	76
Corbis lamellosa	220
Corbula	47, 95, 101, 105, 111, 112, 134, 137, 143, 160, 207, 208, 210, 211, 213, 218, 219, 269, 272
aleiheusis	210
(Anisorhynchus?) engelmanni	112
pyriformis	95, 112
congesta	207
crassatelliformis	160
crassimarginata	101
crassiplica	269
diegoana	211, 213
dubiosa	145
engelmanni	95
evansana	256
foulkei	269
? gregaria	15
inornata	18, 101
mactriformis	15, 161, 168
moreauensis	15
nasuta	219
nematopora	76, 134, 143
(Neæra) nasutoides	272
occidentalis	218
(Pachydon) mactriformis	105
perundata	105
subtrigonalis	105
perundata	15, 160
(Potamomya?) concentrica	22
engelmanni	22
pyriformis	22
pyriformis	161
subcompressa	269
sublineolata	208
subtrigonalis	15, 160, 168
subundifera	137
syriaca	208
tropidophora	77, 160
undifera	77, 161, 168
var. subundifera	161, 168
ventricosa	15

	Page
Corbulamella	18, 101
gregaria	18, 101
Corbulidæ	101, 105, 111, 112, 134, 269, 272
Cordieria moorei	221
Corimya	269
tenuis	269
Cornulina armigera	222
Coroceras	110, 239
Coscencis	203
Costella	107
Crania	119, 121, 177, 257
modesta	121, 177
permiana	257
reposita	119
sheldoni	119
Craniidæ	121
Crassatella	93, 99, 143, 208, 211, 213, 217, 237, 260, 263, 268, 270, 271
alta	211, 213, 271
cimarronensis	143
collina	217
conradi	270
cuneata	268
declivis	237
delawarensis	268, 270
evansii	13
monmouthensis	268
obliquata	271
(Pachythærus) evansi	99
prora	268
shumardi	93
subplana	268
subquadrata	260, 263
syriaca	208
transversa	268
uvasana	211, 213
vadosa	268
Crassatellidæ	30, 32, 33, 34, 99, 260, 262, 263, 268
Crassatellina	70, 99
oblonga	70, 99
Crenella	28, 98
concentrica	220
elegantula	28, 98
granulata cancellata	47
Crenipecten	196
hallanus	196
Crepicephalus	234, 250, 259, 260, 261
(Bathyurus?) angulatus	234
centralis	261
(Loganellus) anytus	234
centralis	260
granulosus	234
haguei	234
maculosus	234
montanensis	259
nitidus	234
planus	260
? quadrans	234
simulator	234
unisulcatus	234
planus	261
Crepidula	212, 214
? ——	206
prærupta	50, 206
princeps	212, 214

INDEX OF GENERA AND SPECIES.

	Page.
Crinoidea	26, 33, 37, 40, 41, 42, 43, 44, 45, 55, 118, 121, 127, 133, 255, 262
Crinoideæ	232
Criocardium	99, 268, 271
nucleolus	**271**
Criptogamia	128
Crucibulum	212, 214
spiuosum	**212**, 214
Crustacea	31, 36, 39, 46, 55, 56, 57, 58, 59, 61, 65, 67, 68, 74, 79, 80, 81, 89, 91, 108, 109, 121, 126, 129, 136, 148, 149, 153, 159, 179, 191, 195, 227, 234, 235, 250, 257
Cruziana	126, 128
linnarssoni	**126**, 128
rustica	**126**, 128
Crypta præerupta	50
Cryptoceras	39, 101
Cryptogramma floridana	220
? penita	220
Cryptomya	215, 216
ovalis	**215**, 216
Cryptonella	193
? circula	**193**
pinouensis	**193**
Cryptorhytis	103
Ctenobranchiata	34
Ctenoides acutilineata	47
denticulata	47
squarrosa	47
Cucullæa	98, 111, 208, 209, 218
exigua	**16**
haguei	**111**
(Idonearca ?) cordata	98
nebrascensis	98
shumardi	98
lintea	**208**
nebrascensis	**250**
ononchella	220
opiformis	**209**
paralella	**208**
subrotunda	**208**
terminalis	**218**
transversa	220
(Trigonarca ?) obliqua	**111**
? Cucullæarca	98
Cucullifera	**222**
Cupellæocrinus	38
Cyathaxonia	52
Cyathocrinites	54, 82, 87
fragilis	**54**, 82
? poterium	83
sculptilis	82
tenuidactylus	**54**, 82
Cyathocrinus	23, 26, 37, 38, 42, 43, 45, 53, 60, 81, 117, 118, 150, 154, 163, 249, 251
?	19
angulatus	**23**, 43
arboreus	**38**, 60
cornutus	**249**, 251
? crassus	**23**
enormis	37, 60
farleyi	**40**, 60
granuliferus	235
inflexus	52
iowensis	**249**, 251

	Page.
Cyathocrinus kelloggi	**118**
lamellosus	**117**
multibrachiatus	154
? poterium	**64**
quinquelobus	**37**, 60
ramosus	52
rigidus	**118**
saffordi	**23**, 43
sangamonensis	**23**, 45
scitulus	**23**, 27
sculptilus	27, 43
stillativus	150, 163
subtumidus	**37**
wachsmuthi	**26**, 60
Cyathophycus	**186**
reticulatus	**186**
subsphericus	**186**
Cyathophyllidæ	62, 108, 109, 121, 223
Cyathophyllideæ	232
Cyathophyllum	62, 108, 109, 2.3
—— ?	**192**
articum	**62**
caliculare	**247**
(Campophyllum ?) nevadense	**109**
corinthium	**247**
corniculum	192
davidsoni	192
helianthoides	247
palmeria	**108**
profundum	248
rugosum	192
subcæspitosum	**109**
turbinatum ?	247
undulatum et multiplicatum	**247**
vesiculosum ?	247
Cyathopora iowensis	247
Cyclas	99, 215, 217
estrellana	**217**
formosa	**15**
fragilis	**15**
permacra	50, **217**
subellipticus	**15**
tetrica	**215**
Cyclina ? circularis	48
Cyclobranchiata	34
Cyclonema	79, 80, 148
bilix	79, 148
crenulata	**67**, 80
Cyclopteris	246
moquensis	**246**
Cyclora	79
minuta	79
? parvula	79
Cylichna	101
dekayi	220
petrosa	50
scitula	**21**, 101
? volvaria	101
Cylichnidæ	101
Cylindrites	29
Cymatopleura ? campylodiscus	**203**
Cymbella gibba	**203**
Cymbopora	100
Cymella	48, 101, 134, 261, 264, 268

INDEX OF GENERA AND SPECIES.

Cymella bella 101
    meeki... 268
Cymomia lamarckii ... 222
Cyphaspis ... 148, 191
    ? brevimarginatus... 191
    christyi... 148
Cyphosoma texanum... 233
Cypricardella ... 117, 194, 197
    quadrata... 117
Cypricardia ... 225, 260, 262
    acutifrons... 226
    arcodes... 226
    (Avicula?) veneris... 226
    imbricata... 226
    occidentalis... 232
    prærupta... 226
    ? rigida... 117
    rugulosa... 224
    siliqua... 226
    simplex... 226
    sinuosa... 224
Cypricardinia ... 90, 194
    carbonaria... 67, 90
    indenta... 194
Cypricarditos ... 54, 55, 79, 248
    ———?... 55
    ? carinata... 79
    obliquus... 55
    sterlingensis... 79
Cypridæ... 121
Cypridinidæ ... 121, 136
Cyprimeria ... 92, 93, 111, 269
    ? crassa... 93
    densata... 269
    depressa... 269
    excavata... 269
    heilprini... 269
    spissa... 269
    subalata... 111
    ? tenuis... 92
Cyprina ... 99, 176
    arenaria... 18, 47
    compressa... 18
    cordata... 18
    ? dallii... 176
    humilis... 21, 47
    laphami... 48
    ovata... 18, 99
        var. compressa... 99
    subtumida... 18, 48
Cyprinidæ ... 260, 266, 268, 271
Cypris ... 136
    ———?... 136
Cyrena ... 99, 135, 137, 143, 158
    arenaria... 47
    carletoni... 76, 158, 167
    (Corbicula) cytheriformis... 21
        durkeei... 66
    dakotensis... 99, 167
    holmesi... 88
    inflexa... 143
    intermedia... 15
    moreanensis... 15
    occidentalis... 15

Cyrena socnris... 143
    (Veloritina) durkeei... 135
        erecta... 137
Cyrenidæ ... 99, 105, 112, 135
Cyrtia... 119
    curvilineata... 119
Cyrtina ... 57, 58, 62, 193
    billingsi... 62
    dalmani... 57
    davidsoni... 193
    hamiltonensis... 62, 193
    panda... 62
    triquetra... 58
Cyrtoceras ... 25, 46, 59, 91, 129, 138, 195, 236
    (Aploceras) curtum... 46
    cessator... 236
    conicum... 247
    curtum... 25, 27
    dardanus... 91
    dictyum... 138
    ? dilatatum... 25, 46
    nevadense... 195
    ohioense... 67
    ohioensis... 81
    sacculum... 50
Cyrtoceratites... 81
Cyrtochilu... 103
Cyrtolites ... 56, 79, 121, 191, 234
    ? costatus... 79
    dyeri... 79
    ? gillianus... 121
    imbricatus... 56
    (Microceras) inornatus... 79
    ornatus... 79
    sinuatus... 191, 234
Cysteophyllum ... 62, 155, 192, 223
    americanum... 192
        var. articum... 62
    vesiculosum... 155
Cystidea... 37
Cystoidea... 54, 77
Cystoseirites ?... 227
Cythere ... 74, 79, 121
    cincinnatiensis... 68, 79
    nebrascensis... 74
    simplex... 121
Cytherea ... 207, 211, 213, 216, 218, 219
    deweyi... 15, 48
    lamarensis... 48, 150
    lenticularis... 220
    leonensis... 48, 218
    ?(Meretrix) dariena?... 216
    missouriana... 48, 243
    nebrascensis... 15, 48
    nuttall... 219
    orbiculata... 13, 48
    oregonensis... 50
    ovata... 220
    owenana... 16, 48
    parvula... 231
    pellucida... 16, 48
    syrinea... 207
    tenuis... 13, 48
    texana... 48, 218

INDEX OF GENERA AND SPECIES.

	Page.
Cytherea tippana	48
vespertina	50
Cytherodon	194

D.

	Page.
Dalmania	39
danæ	**39**
Dalmanites	57, 58, 79, 81, 109, 186, 195
carleyi	**69,** 79
danæ	56
intermedius	**186**
meeki	**195**
(Odontocephalus) ægeria	58
ohioensis	**67,** 81
tridentiferus	57
Dawsonella meeki	167
Decapoda	35, 37, 46, 61
Delphinula depressa	221
plana	221
Dolthyris	247, 248
curnteines	**247**
expansus	248
Deltodus mercurei	**246**
Dendrocrinus	55, 77, 188
caseyi	**66**
oswegöensis	**55**
retractilis	**188**
Dendrograptus	187
compactus	**187**
simplex	**187**
tenuiramosus	**187**
Dentalidæ	34, 101, 132, 264
Dentalium	26, 34, 44, 74, 86, 93, 95, 101, 127, 132, 209, 264
annulostriatum	**65,** 86
canna	**127,** 132
cretaceum	209
(Ditrupa?) pusillum	221
fragilis	**14**
gracilis	**13,** 101, 264
komookseuse	93
meekianum	74, 86
nanaimoensis	**17**
pauperculum	**21**
? subquadratum	**22,** 34, 95
thallus	50
venustum	**26,** 43
Diadora	216
crucibuliformis	216
Diæra	132, 133, 134, 135
Dianchora	266
echinata	266
Diastropha	107
Dibranchiata	35, 105, 263
Dicellocephalus	126, 129, 189, 191
? augustifrons	**189**
bilobatus	189
? expansus	**190**
finalis	**191**
flagricaudus	**126,** 129
inexpectans	**191**
marica	**189**
nasutus	**189**
osceola	189
? quadriceps	189

	Page.
Dicellocephalus richmondensis	**189**
iole	**189**
Diceras	268
dactyloides	268
Dichocrinus	23, 42, 43, 54, 63, 82, 83, 84, 91, 118, 249, 251
angustus	**118**
constrictus	**23,** 43
conus	**23,** 42
cornigerus	84
crassitestus	**118**
expansus	**54,** 83
ficus	83, 91
lineatus	**63,** 82
ovatus	**249,** 251
pisum	**63,** 82
(Pterotocrinus) chestorensis	**23**
crassus	**23**
striatus	**249,** 251
Dicraniscus	**70**
ortoni	**70**
Dicranobranchia	132
Dictyonema fenestrata	232
Dielasma	127, 130, 132
? bovidens	127
Dikelocephalus	231, 234, **250**
bilobatus	**234**
flabellifer	**234**
gothicus	**234**
granulosus	**250**
iowensis	249
miniscænsis	**250**
minnesotensis	**250**
multicinctus	**234**
pepinensis	**250**
quadriceps	**234**
washsatchensis	**234**
Dimyaria	98, 105, 133, 267
Diodora crucibuliformis	50
Dione	100
augustifrons	50
? brevilineata	50
decisa	50
eufalensis	48
lamarensis	48
leonensis	48
[?] meekiana	48
missonriana	48
nebrascensis	48
orbiculata	48
oregonensis	50
ovata	220
owenana	48
? pellucida	48
[?] ripleyana	48
[?] tenuis	48
texana	48
tippana	48
uniomeris	50
vespertina	50
Diphyphyllum	108, 155
archiaci	155
arundinaceum	155
fasciculum	**108**
simcöense	192

	Page.
Diphyphyllum stramineum	155
Diplodon	105
Diplograptus	129
Diploschiza	266
cretacea	266
Discinidæ	110
Discina 85, 89, 110, 119, 177, 192, 196	
? 110, 192	
capax	**119**
connata	**196**
convexa	177
lodensis	192
lugubris	49
manhattanensis	**20**
minuta	192
multilineata	49
newberryi	196
nitida........85, 177, 196	
(Orbiculoidea) newberryi	89
? pleurites	**89**
tenuilineata	**20**
Discites.........................39, 104	
tuberculatus	**250**
Discoflustrellaria bouei	220
Discophycus	**186**
typicalis	**186**
Discoplea oregonica	229
atmosphærica	203
Discoscaphites	104
Discosorus conoideus	232
Dispotæa	205
constricta	**205**
Dithyrocaris 87, 179	
carbonarius **65**, 87, 179	
Ditrupa subcoarctata	221
Docoglossa	102
Dolabra	56
? carinata	**68**
sterlingensis **40**, 56	
Dolatocrinus ornatus	66
Doliopsis	**222**
Dolium petrosum	**206**
Donacilla	99
Donacinidæ	269
Donax	269
fordii	269
? protexta	50, 206
Doryerinus 54, 82	
canaliculatus	82
quenquelobus var. intermedius	54, 82
roemeri **54**, 82	
unicornis	82
Dosinia 211, 212, 215, 216, 217, 260, 262, 264, 269	
alta **211**, 213, 215, 216, 217	
erecta	**269**
gabbi	**269**
gyrata	220
jurassica **260**, 262	
longua **213**, 216, 217	
missouriana?	264
montana	**217**
subobliqua	**217**
Dosinia ? tenuis	**27**
? Dosiniopsis	100
lenticularis	220

	Page.
Drepanocheilus 48, 102	
Dreissena leucophæata	180
Drillia lonsdalii	222
texana	222
Dymyaria 132, 135	
Dysnomya	105
Dystactella	194
insularis	**194**

E.

Ecculiomphalus	195
devonicus	**195**
Echinianthus mortonis	222
Echinodermata 19, 24, 26, 33, 37, 40, 41,	
42, 43, 44, 54, 55, 56, 58, 59, 60, 64,	
66, 68, 72, 77, 81, 84, 91, 94, 96, 118,	
127, 130, 131, 133, 136, 138, 144, 147,	
148, 149, 150, 151, 154, 158, 163, 165,	
196, 206, 209, 210, 236, 255, 260, 261	
Echinoidea 23, 42, 43, 44, 54, 60, 96	
Echinosphærites ? **232**	
Echinus 206, 209, 210	
bullatus	**210**
infulatus	222
kerakensis	**209**
libanensis	**210**
syriacus	**206**
Edmondia 38, 45, 73, 86, 89, 109, 117, 178, 197, 236, 258	
aspinwallensis **70**, 73, 178	
? calhouni **19**, 20	
burlingensis	**117**
? circularis	**197**
? glabra	**73**
medon	**197**
myrina	**236**
? nebrascensis	73
ovata	87
peroblonga **38**, 86	
? pinonensis	**109**
reflexa	**73**
suborbiculata	258
subtruncata	**73**
tapesiformis	**89**
unioniformis	45
Edriocrinus	57
pocilliformis	57
Egeta	99
Egeria inflata	220
plana	220
Electroma	97
Endolobus 39, 45	
Enclimatoceras **176, 239**	
(Nautilus) ulrichi	**176**
Encrinites	253
Encrinurus	186
trentonensis	**186**
varicostatus	**186**
Endoceras multitubulatum	191
proteiforme	191
Endocostea **260**, 263	
sulcata	263
typica **260**, 263	
Endopachys maclurii	220
Entalis	101

INDEX OF GENERA AND SPECIES.

	Page.
Entalis paupercula	101
Entolium	35, 73, 86, 89, 178
aviculatum	73, 86
shumardianum	89
Ensis curtus	50
Entomostraca	36, 46, 61, 79, 89
Eocidaris	72, 83
hallianus	52, 72
squamosa	**63**
? squamosus	83
Eocystites	198
primævus	198
Eoscorpius	**53**, 61
carbonarius	61
Ephithemia	203
Eretmocrinus	41, 54
Eridophyllum	155
strictum	155
Eriphyla	99
gregaria	99
Erisocrinus	**35**, 36, 37, 45, 63, 72, 82, 85, 150, 163
antiquus	**63**, 82
(Ceriocrinus) planus	163
inflexus	163
conoideus	**37**, 45
nebrascensis	**35**, 36
planus	**150**
tuberculatus	**37**, 45
typus	**35**, 36, 45, 72, 85, 163
whitei	82, **63**
tubulata	222
Eschara concentrica	**273**
tuberculata	**273**
Ethmophyllum	**53**
whitneyi	**53**
gracilis	**53**
Etonia	57
peculiaris	57
Eudiscoceras	**110, 239**
gabbi	**110**
Eucalyptocrinus	148
crassus	148
Euchcilodon	237
creno-carinata	**237**
Euconactæon	29
Eulima	25
aciculata	221
chrysalis	**77**
funicula	**77**
? inconspicua	**77**
lugubris	221
notata	221
peracuta	**25**, 27
scalæ	221
? subfusiformis	**255**
Eulimella	134, 144, 158
funicula	134, 144
? chrysallis	158
? inconspicua	158
Eumicrotis	**29**, 32, 33, 45, 178, 236
curta	32, 33, 49, 236
hawni	32, 178
var. ovata	32
sinuata	45
Eunema	**25**

	Page.
Eunema ? salteri	**25**
Eunotia amphioxys	229
argus	229
gibba	203, 229
gibberula	229
granulata	229
librile	203, 229
subulata	229
textricula	229
uncinata	229
webstermanni	229
zebra	229
zebrina	229
Euomphalidæ	132
Euomphalus	25, 30, 80, 117, 118, 130, 132, 138, 165, 179, 195, 197, 235, 248
ammon	117
decewi	80
eurekensis	**195**
laxus	**130**, 235
michleranus	**233**
(Omphalotrochus) whitneyi	**30**
(Phanerotinus) laxus	195
planodorsatus	**25**
pernodosus	132
(Raphistoma?) rotuliformis	**65**
trochiscus	**65**
roberti	**118**
rugosus	51, 87, 179
springvalensis	**138**, 165
(Straparollus) ophirensis	**235**
subrugosus	197
utahensis	**235**
subplanus	**232**
umbilicatus	**25**
Eupachycrinus	72, 84, 85, 136, 163
boydii	**64**, 85
fayettensis	85
platybasis	**136**, 163
tuberculatus	85
verrucosus	72
Euphoberia	**53**, 61
armigera	**53**, 61
major	**53**
?? ――	
Euproops	61, 179
danæ	61, 179
colletti	**179**
Euptycha	29
Eurydesma	225
cordata ?	224, 225
elliptica	**224**, 225
gibbosa	225
globosa	**224**
sacculus	225
Eurypteridæ	61, 187
Eurypterus	61, 179
(Anthraconectes) mazonensis	**53**, 61, 179
Enspira	143, 158
coalvillensis	143
utahensis	158
Eutomoceras	**110, 239**
laubei	**110**
Eutropia haleana	48

INDEX OF GENERA AND SPECIES.

	Page
Eutropia perovata	48
punctata	48
Evactinopora	38, 40, 60
grandis	60
radiata	38, 60
sexradiata	60
Exogyra	93, 133, 142, 151, 158, 173, 206, 208, 215, 219, 266
arietina	173, 219
aquila	173
boussingaultii	206
columbella	93, 173
costata	173, 219, 266
var.	210
fluminis	133
densata	208
var.	208
fimbriata	173, 215, 219
flabellatta	243
forniculata	151, 158, 173
fragosa	173, 215, 219
interrupta	173
læviuscula	133, 173, 219
matheroniana	173, 219
parasitica	173
plicata	173
ponderosa	133, 173, 255
texana	173, 255
valkeri	142
walkeri	173
winchelli	151, 158, 173

F.

Fabulina	100
Fasciolariidæ	103, 264
Fasciolaria	103, 159, 264
buccinoides	14, 103
cretacea	14
? (Cryptorhytis) cheyennensis	103
contorta	264
flexicostata	103
fusiformis	264
(Mesorhytis) gracilenta	103
moorei	221
(Piestocheilus) alleni	159
cretacea	103
culbertsoni	103, 264
galpiniana	103
scarboroughi	103
plicata	221
Faviphyllum ? rugosum	232
Favistella	130, 155
stellata	130, 155
Favositidæ	62, 108, 109
Favosites	30, 62, 108, 117, 126, 130, 148, 155, 156
——(?)	156, 192
basaltica	156, 192
divergens	130
favosus	155
hemispherica	192
maxima ?	247
polymorpha	62, 156, 247
var.	108
dubia	156
ramosa	247

	Page
Favosites [whitfieldi]	117
whitfieldi	126
Fenestella	60, 72, 88, 226, 253
——?	72
ampla	226
corticata	273
delicata	67, 88
fossula	226
gracilis	226
intermedia	273
internata	226
media	226
(Multiporata?) var. lodiensis	88
norwoodiana	273
popeana	257, 273
shumardii	72, 273
subretiformis	273
(Lyropora) retrorsa	60
trituberculata	273
variabilis	273
Ficopsis cooperi	221
mammillatus	221
penitus	221
remondii	221
Ficus mammillatus	221
modestus	50
[? ?] ocoyanus	50
Fishes	20
Fissurella	39, 212, 214
crenulata	212, 214
Fistulipora	71, 156, 192
canadensis	156
nodulifera	72
Flabellatæ	34
cuneiforme	220
Foraminifera	19, 29, 31, 64, 71, 85, 91, 177, 250, 257
Forbesiocrinus	23, 26, 37, 44, 60
agassizi var. giganteus	26, 60
? norwoodi	23
monroensis	26
wortheni	83
? semiovatus	23
Fragillaria	203
acuta	229
amphicephala	229
rhabdosoma ?	203, 229
Fragum	268
tenuistriatum	268
Fulvia	268
tenuis	268
Fungidæ	62, 96
Fusilina	29, 31, 71, 85, 131, 177, 245
cylindrica	19, 29, 31, 71, 131, 177, 245
var. ventricosa	19
elongata	256, 257
gracilis	29, 85, 87
robusta	29
ventricosa	85, 87
Fusimitra ? lineata	221
? minima	221
Fusispira	234
compacta	234
Fusus	103, 144, 159, 209, 237, 264
arctatus	50
cheyennensis	261, 264

INDEX OF GENERA AND SPECIES. 293

	Page.
Fusus cooperi	221, 222
conybearii	222
constrictus	13
contortus	14
corpulentus	206
culbertsoni	14
dakotaensis	14, 49
ellerii	209
? flexuocostatus	14, 49
galpinianus	14
geniculus	206
haleanus	49
haydeni	256
intertextus	18, 49
mullicaensis	49
migrans	50
newberryi	14, 49
nebrascensis	256
(Neptunea?) gabbi	76, 144
utahensis	76
oregonensis	50
(Pleurotoma?) scarboroughi	18
remondii	221
scarboroughi	49
? (Serrifusus) dakotensis	103
shumardi	264
(Strepsidura) marnochi	237
subturritus	17
subturrites	49
taitii	222
tenuilineatu	13
tenuilineatus	49
? utahensis	159
vaughani	17, 49
? vinculum	49

G.

Gadus pusillus	221
subcoarctatus	221
Gafrarium liratum	220
Galaxias	106
Galeodea (Galeodaria) quinquecostata	222
Galerites oregonensis	227
Galeropsis excentricus	221
Galerus excentricus	221
Gallionella	203
crenaria	229
distans?	203, 229
granulata	229
lævis	229
punctata	229
undulata	229
varians	203
Gampsonyx	53
fimbriatus	61
Gasteropoda	20, 21, 22, 24, 26, 28, 30, 31, 32, 34, 39, 40, 42, 43, 44, 45, 56, 57, 59, 60, 63, 65, 67, 68, 74, 79, 80, 84, 86, 89, 90, 91, 93, 94, 95, 101, 105, 107, 108, 111, 112, 117, 118, 121, 126, 127, 129, 130, 132, 133, 134, 135, 136, 137, 138, 143, 147, 148, 151, 153, 154, 155, 158, 159, 161, 164, 165, 178, 191, 194, 197, 226, 232, 234, 235, 236, 250, 256, 258, 261, 264, 273
Gastrochæna	270

	Page.
Gastrochæna americana	270
Gastrochænidæ	260, 263, 270
Gastrocœli	35
Gastrosiphites	35
Genota	103
Geophila	135
Gervillia	28, 98, 119, 151, 158, 164, 259, 262
(Avicula) sulcata	51
gregaria	159
longa	51
montanaensis	164
mudgeana	151, 158
parva	51
recta	28, 98, 262
sparsilirata	259
strigosa	119
subtortuosa	16, 98
Gervilliopsis	267
ensiformis	267
minima	267
Gilbertsocrinus	42, 63, 82
bursa	42
calcaratus	42
(Goniasteroidocrinus) fiscellus	42
(Goniasteroidocrinus) obovatus	63, 82
(Goniasteroidocrinus) tuniradiatus	63, 82
Glaconema paradoxum	229
Glauconome	72, 127, 131
nereidis	127, 131
trilineata	72
Globiconcha [?] elevata	255
(Tylostoma) tumida	255
Glossidæ	99, 134
Glossopteris ampla	226
browniana	226
? cordata	226
elongata	226
linearis	226
phillipsii	231
reticulum	226
Glossus fraterna	50
markoei	50
Glycimeris	101, 143, 217
berthoudi	143
estrellanus	50, 217
occidentalis	101
Glyptocrinus	77, 187
argutus	187
baeri	69, 77
decadactylus	77
dyeri	68, 77
var. subglobosus	68, 77
nealli	77
parvus	77
? subnodosus	187
Gnathodon	266
? tenuidens	266
Gnathostomata	61, 170
Gomphoceras	39, 59, 195
(Aploceras) turbiniforme	39
sacculum	39
suboviforme	195

INDEX OF GENERA AND SPECIES.

	Page.
Gomphoceras turbiniforme	59
Gomphoncma clavatum ?	203
gracile	229
herculaneum	229
longicolle	229
mamilla	229
minutissimum	203, 229
olar	229
Gomphonema oregonicum	229
Goniasteroidocrinus	42, 53, 63, 81, 82
tuberosus	42
Goniatitidae	110, 132
Goniatites	25, 46, 110, 117, 132, 149, 195, 236
compactus	**38, 87**
desideratus	**195**
globulosus	**25,** 46
var. excelsus	92
goniolobus	**110**
iowensis	**25,** 46
kingii	**236**
lyoni	**25,** 42
oweni	149
opiauus	**117**
Goniobasis	95, 106, 112, 135, 137, 141, 161
arcta	95
? arcta	221
chrysaloidea	**137,** 161, 168
chrysallis	**70,** 161, 168
cleburni	**137,** 161, 167
convexa	106, 161, 169
var. impressa	106, 169
endlichi	**141,** 161, 168
fremontii	221
gracilenta	106, 161, 169
? insculpta	**77**
invenusta	106, 161, 169
macilenta	161, 168
nebrascensis	106, 135, 161, 169
nodulifera	**70**
? nodulosa	221
? omitta	106, 161, 169
simpsoni	95, 112
? simpsoni	221
sublaevis	106, 161, 169, 221
? subtortuosa	106, 161, 169, 221
tenera	135, 170
? tenera	221
tenuicarinata	106, 135, 162, 169
? tenuicarinata	221
Goniochasma	48, 101
stimpsoni	48
Goniocylindrites	29
Goniomya	92, 100, 164
americana	**15,** 100
borealis	92
montanaensis	164
Goniophora	194
peraugulata	194
Gonostoma yatesii	180
Gorgonidae	96
Gonidia	268
conradi	**268**
decemnaria	268
declivis	268
paralis	268

	Page.
Grammatodon	34, 92, 262
inornatus	34, 262
? vancouverensis	92
Grammostomum phyllodes	47
Grammysia	38, 59, 89, 194, 197
arcuata	197
? hannibalensis	89, 107
minor	**194**
rhomboidalis	**38,** 59
rhomboides	**67,** 89
ventricosa	**67,** 89
Granatocrinus	44, 60, 64, 83, 84, 126, 130, 147
cornutus	44
glaber	**64,** 84
granulosus	83
lotoblastus	126, 130, 147
melonoides	**64,** 83
neglectus	**64,** 83
norwoodi	40, 60, 83
pisum	**64,** 83
projectus	60
shumardi	**40,** 60
Granoarca	98
Graphiocrinus	85
dactylus	85
Graptolithus	126, 129, 187
annectans	**187**
(Climacograptus) ramulus	**126,** 129
(Diplograptus) hypniformis	126, 129
platis	129
quadrimucronatus	126, 129
Gratelupia ?	212, 214
mactropsis	**212,** 214
Griffithides	39, 198
portlocki	198
Gryphaea	28, 30, 33, 97, 133, 172, 173, 208, 219, 233, 236, 245, 259, 262, 266, 270, 271
bryani	270
var. precedens	**270**
calceola	93
var. nebrascensis	28, 33, 172, 236, 262
capuloides	**208**
dilatata	241, 242
var. tucumcarii	241, 243
mucronata	173
mutabilis	179
nebrascensis	180
navia	173
pitcheri	173, 219, 233, 241, 242, 243, 245, 253, 255
var	133
navia	233, 245
sinuata var. americana	243
thirsae	173
vesicularis	97, 173, 208, 266, 270, 271
vomer	173
Gryphaeostrea	97, 270
vomer	270
Gryphorhyncus	**29**
Gryphostrea eversa	220
Gulnaria	105
? Gymnotoceras	**110, 239**
blakei	110, 239
rotelliforme	**110, 239**

INDEX OF GENERA AND SPECIES. 295

	Page.
Gypidula	193
Gyraulus	34, 106, 153
Gyroceras	59, 63, 154
burlingtonensis	250
constrictum	59
cornutes	247
elrodi	154
logani	63
(Nantilus?) inelegans	67
rockfordensis	59
(Trochoceras?) ohioensis	67
Gyroceratites	81
(Nautilus) inelegans	81
? (Trocheras) ohioensis	81
Gyrodes	102, 111
conradi	102
depressa	111
Gyrorbis	35

H.

Hadrophyllum	165
glans	165
Haliotidæ	39
Halobia	110
(Daonella) lommeli	110
Halysites	155
catenulata	155
Haminea	101, 264
minor	101
occidentalis	101
subcylindrica	101, 264
Hamites larvatus	220
fremonti	243
leai	49
mortoni	14
verneuilii	49
Haploscapha	222
(Cucullifera) eccentrica	222
grandis	222
Harpaga tippana	48
Harpes escanabiæ	232
Harttia	198
matthewi	198
Helcion	14
alveolus	14
carinatus	14
patilliformis	14
sexsulcatus	14
subovatus	14
Helcium alveolum	21
carinatum	21
patelliforme	21
sexsulcatum	21
subovatum	21
Helicerus	224, 227
fuegiensis	224, 227
Helicidæ	106, 107, 135, 180
Helicoceras	104, 135, 261, 265
? angulatum	21, 49
tortus	18
mortoni var. tenuicostatum	104
pariense	135
tenuicostatus	19
stevensoni	261, 265
Helicotoma	191
sp?	191

	Page.
Helix	28, 106, 107, 135, 137, 141, 153, 161
(Aglaia) fidelis	180
evansi	21
evanstonensis	141, 169
kanabensis	137, 169
leidyi	14, 107, 135, 170
(Monodon?) dallii	180
occidentalis	17
(Patula) perspectiva	180
sepulta	153
peripheria	137, 170
riparia	137, 170
? sepulta	169
spatiosa	28, 221
veterna	28, 170
? veterna	107
vetusta	106, 161, 169
vitrinoides	17
(Zonites) marginicola	170
Helicoceras tortum	18, 49
Heliolites	155
elegans	155
Helisoma	34, 106
Helouix thallus	50
Hemiaster	96
elegans	255
? humphreysanus	18, 96
Hemicystites	78
(Cystaster) granulatus	78
stellatus	78
Hemifusus horni	221
Hemipronites	31, 56, 72, 78, 85, 89, 108, 109, 121, 177
chemungensis var. arctostriata	108
crassus	31, 72, 85, 177
crenistria	109, 131
subplanus	56
Hemitrypa?	226
Hercoglossa	104
Heteractis duclosii	220
Heteroceras	93, 104, 261, 265
? angulatum	49, 104
cheyennensis	49, 104
? cochleatum	104
cooperi	93
? nebrascense	104, 265
newtoni	261, 265
oweni	49
tortum	49, 104
? umbilicatum	104
Heterocrinus	37, 55, 77
(Anomalocrinus) incurvus	37
crassus	91
constrictus	77
exilis	77
exiguus	68
heterodactylus	77
(Iocrinus) subcrassus	77
juvenis	77
laxus	77
simplex	77
subcrassus	37, 55, 68
Heteromyaria	97, 263, 267
Hettangia	262
americana	76

INDEX OF GENERA AND SPECIES.

	Page.
Himantidium arcus	229
Hinnites	216
crassa	**216**
crassis	49
giganteus	49
Hipponyx borealis	21
pygmæa	221
Hippurites	209, 210
liratus	**210**
plicatus	**210**
syriacus	**209**
Holaster	206
comanchesi	**243**
simplex	**255**
syriacus	**206**
Holectypus planatus	**233**, 255
Holopea	39
(Cyclora) nana	**67**
(Isonema) depressa	**39**
Holostomata	264
Homala	100
Homalina	100
Homocrinus	91
angustatus	**64**, 91
Hormomya	98
Huronia annulata	**232**
vertebralis	232
Hyalina	106, 161
? evansi	106, 161, 169
? occidentalis	106, 161, 169
Hybocrinus	55
? incurvus	55
Hydreionocrinus	121
verrucosus	**121**
Hydrobia	106, 137, 162
anthonii	221
anthonyi	106, 162, 170
? culimoides	107, 162, 170
recta	**137**, 170
subconica	107, 162, 170
utahensis	**137**, 169
warrenana	106, 162, 170
Hydrozoa	126, 129
Hyolithes	126, 129, 189, 191, 195, 198, 199
——?	195
acadica	**198**
(Camarotheca) emmonsi	199
carbonaria	**198**
danianus	**198**
micmac	**198**
primordialus	126, 129, 189
shaleri	**199**
vanuxemi	**191**
Hyridella	105

I.

Idonearca	98, 134, 263, 267, 270
antrosa	267
compressirostra	**270**
mediana	**270**
shumardi	263
tippana	267
vulgaris	267
Igocerns	40
Illa-nurus	191

	Page.
Illænurus curekensis	**191**
Illænus	36, 55, 80, 91, 186
(Bumastus) graftonensis	**65**, 91
insignis	80
crassicauda	55
indeterminatus	**186**
milleri	**186**
taurus	55
trentonensis?	248
Incertæ sedes	261
Inoceramus	29, 30, 93, 95, 97, 111, 127, 133, 136, 143, 207, 208, 209, 218, 245, 253, 263, 267
——?	92, 231
(Actinoceramus) costellatus	47
altus	**70**, 263
aratus	**208**
aviculoides	**21**
balchii	**21**
barabini	92, 134, 263, 267, 274
(Catillus) balchii	97
convexus	97
cripsii?, var. subcompressus	97
barabini	97
incurvatus	97
proximus	97
?, var. subcircularis	97
sagensis, var. nebrascensis	97
subloevis	97
tenuilineatus	97
tenuirostris	97
undabundus	97
vanuxemi	97
conradi	**13**
confertim-annulatus	218, 253, 255
convexus	**13**
crassalatus	**133**
cripsii	218, 274
?	245
var. subundatus	92
cuneatus	**21**
deformis	111, 134
depressus	**134**
dimidius	**127**, 134
elevatus	**209**
erectus	**111**
exogyroides	**29**
flaccidus	**134**
fragilis	**13**, 93, 133, 263, 274
gilberti	**136**, 143
howelli	**136**, 143
(Inoceramus) altus	97
fragilis	97
incurvus	**16**
lerouxi	**243**
lynchii	**207**
(Mytiloides) problematicus	97
var.	
aviculoides	98
mytilopsis	**218**
nebrascensis	**250**
? obliqua	**30**
oblongus	143

INDEX OF GENERA AND SPECIES. 297

	Page.
Inoceramus perovalis	267
perplexus	260, 263
pertenuis	16
problematicus	95, 111, 133, 245, 273
problematicus?	263
pro-obliquus	267
pseudo-mytiloides	253
? rectangulus	30
sagensis	250, 263, 267, 274
var. quadrans	267
simpsoni	22, 95, 111, 263
subcompressus	21
sublævis	13, 263
subundatus	27
sulcatus	47
syriacus	209
tenuilineatus	13, 263, 274
tenuirostriatus	29
texanus	218
umbonatus	18
undabundus	29
vancouverensis	256
vanuxemi	21, 263
ventricosus	15
(Volviceramus) exogyroides	97
umbonatus	97
Inoperculata	34
Insecta	37, 47
Integripalliata	266, 268
Integropallia	263
Iocrinus	188
trentonensis	188
Iphidea (??) sculptilis	76
Iridea	105
Isocardia	205, 207, 270
conradi	270
crenulata	207
fraterna	50
? hodgei	68
markoi	50, 205
washita	243
Isocardiidæ	270
Isochilina	188
Isodora	107
Isognomon	29, 32
Isonema	30, 159
depressa	59
humilis	67
Isopleurus curviliratus	48
meekianus	48
Isopoda	36, 46, 61, 170

J.

Janira	209, 215, 216
affinis	49
bella	49, 215, 216
syriaca	209
Jeanpaulia radiata	246

K.

Kutorgina	189, 234
minutissima	234
prospectensis	189
sculptilis	189
whitfieldi	189

L.

	Page.
Lacinia	219
Lacunaria alabamiensis	221
erecta	221
Lamellibranchiata	26, 26, 27, 28, 29, 30, 32, 33, 38, 40, 41, 42, 43, 44, 54, 55, 56, 58, 59, 61, 64, 66, 68, 73, 79, 80, 84, 86, 89, 90, 91, 92, 93, 94, 95, 97, 105, 108, 110, 112, 148, 149, 190, 194, 196, 235, 236, 260, 262, 263, 271
Lamellopora	247
infundibularia	247
Lamna texana	246
Lamprodoma elongata	222
phillipsii	222
Lampsilis	105
Lapparia mooreana	221
Latia dallii	166, 170
Latiarca	98
otoucheila	220
transversa	220
Latirus (Persisternia) plicatus	221
Leaia	61, 179
tricarinata	61, 179
Lecythiocrinus	150, 163
ollicæformis	150, 163
Leda	24, 26, 28, 116
barrisii	116
bellistriata	51
bisulcata	28, 47
compsa	220
curta	26
fibrosa	48, 256
longifrons	47
(Nucula) subcituta	19
oregona	49, 220, 256
pinnæformis	47
protexta	47
slackiana	47
subangulata	47
subscituta	20
willamettensis	49, 256
(Yoldia) levistriata	24
Legumen	270
appressum	270
planulatum	270
Leiopistha	127, 134, 264, 268
(Cymella) meeki	264
undata	101, 134
elegantula	268
inflata	268
protexta	268
(Psilomya) meekii	127, 134
Leioplax	138
? turricula	138
Leiopteria	194
rafinesquii	194
Leiorhynchus	109, 193
quadricostatus	109
Lepacrinites	77
moorei	77
Leperditia	80, 126, 129, 188, 191, 195
alta	80
bivia	126, 129, 191

INDEX OF GENERA AND SPECIES.

	Page.
Leperditia (Isochilina) armeta	**188**
rotundata	**195**
Lepidesthes	**60**, 144, 154, 165
colletti	**144**, 154, 165
Lepidesthes coreyi	**60**
Lepidocentrus	**63**
irregularis	**63**
Lepidoptera	37, 47
Lepocrinites moorei	**66**
Leptæna	57, 78, 130, 189, 234
melita	189, **234**
? nucleata	57
sericea	78, 130
trilobata	**250**
Leptesthes	99, 105, 141, 160
Leptocœlia	57
flabellites	57
Leptocardia	99, 100, **236**
carditoidea	**236**
typica	**236**
Leptodesma	194, 196
———— ?	196
transversa	**194**
Leptodomus granosus	20
Leptolimnea	106, 153
Leptopora	147, 163
winchelli	**147**,163
Leptosolen	101, 270
conradi	**70**,101
Leucochcila	159
Lichas	39, 55, 91
boltoni	91
cucullus	**39**,55
Lichenocrinus	66, 69, 77
crateriformis	78
dyeri	77
Lima	39, 73, 86, 110, 133, 178, 218, 236
acutilineata	47
? cuneata	**30**
denticulata	47
leonensis	**218**
(Limatula) erecta	**110**
pelagica	47
(Plagiostoma) occidentalis	236
recticostata	**30**
retifera	52, 73, 86, 178
? sinuata	**30**
squamosa	47
wacoensis	133, 218
Limidæ	30, 110, 133
Limnæa	105, 107, 112, 153, 161
(Acella) haldemani	168
? compactilis	169
(Leptolimnea) minuscula	**153**
(Limnophysa?) compactilis	**77**
nitidula	112, 168
vetusta	112
meekiana	107
meekii	170
miniuscula	170
nitidula	95, 161, 221
(Pleurolimnæa) tenuicostata	106, 161, 169, 221
(Polyrhytis) kingii	**112**,170
shumardi	107, 170

	Page.
Limnæa (Polyrhytis) similis	**22**,95, 112, 170
tenuicosta	**16**, 221
vetusta	**22**, 95, 170
Limnæidæ	34, 105, 107, 112, 135
Limnæinæ	34
Limnophila	34
Limnophysa	106
Limopsis	98
ellipsis	220
nitens	50
parvula	98
pectunculariѕ	220
striato-punctatus	**255**
Limoptera	194
sarmenticia	**194**
Linoaria	100, 269
cancellata-sculpta	48
contracta	**269**
? formosa	100
irradians	48
nuctastriata	269
Lingula	28, 33, 55, 58, 62, 72, 85, 89, 96, 119, 126, 129, 133, 177, 189, 192, 198, 262
alba-pinensis	**192**
ampla	**250**
antiqua	231
brevirostris	**18**,33,262
? dawsoni	**198**
halli	119
iowensis	**247**
licna	192
ligea	192
var. nevadensis	**192**
(Lingulella) membranacea	89
louensis	**192**
? manticula	**126**, 129, 189
melio	89
minuta	**62**
mytiloides	85
nitida	28, 96
ovata	**223**, 225
pinnaformis	**250**
prima	231
quadrata	55
scotica	89
var. nebrascensis	72
subspatulata	**13**, 53
umbonata	177
whitei	**192**
Lingulella davisii	68
lamborni	**68**
Lingulepis	31, 189, 233, 234, 2
cuneolus	**260**, 2
dakotensis	261
ella	**234**
mœra	189, **233**
? minuta	189, **234**
perattenuatus	**260**, 261
pinnaformis	261
pinniformis	31, 68
prima	31
primæformis	259
Lingulidæ	31, 33, 62, 96, 133, 261, 262
Linnarssonia	**199**
sagittatis	199

INDEX OF GENERA AND SPECIES. 299

	Page.
Linnarssonia transversa	199
Liopistha	47, 101, 261
(Cymella) meeki	261
protexta	101
Lioplacodes	35, 49
veterna	49
voternus	35, 167
Liopodesthes	134, 158
lingulifera	134
nuptialis	134
?obscurata	158
Lithasia antiqua	166, 170
Lithodendron lineata	49
Lithodomus	38, 208, 210, 267
affinis	267
cretaceus	208
ripleyana	267
stamineus	210
Lithodontium furcatum	229
nasutum	229
scorpius	229
Lithophaga	38, 61
lingualis	38, 61
?portennis	38, 84
Lithostrotion	30, 109, 131, 149, 156, 165
——?	30, 232
?californiense	30
canadense	149
mamillare	30, 149, 156, 165
microstylum	165
whitneyi	109, 131
Lithostylidium	229
amphiodon	229
crenulatum	229
læve	229
quadratum	229
rude	229
trabecula	229
Littorina	212, 214
pedroana	212, 214
Littorinidæ	103
Lituites	80, 91
graftonensis	65, 91
?ortoni	80
Loganellus	234, 260, 261
Lonchocephalus	250
chippewaensis	250
hamulus	250
Lophophyllum	72, 85, 117, 131, 138, 165, 177
calceola	117
expansum	138, 165
proliferum	72, 85, 131, 177
Loxonema	25, 27, 46, 86, 152, 195, 197
——?	195
approximatum	195
attenuata var. semicostata	67
bella	197
cerithiformis	25, 46
eurekensis	195
inornata	25
kanei	36
multicostata	27, 46
nitidula	25
nobile	195
rugosa	25, 46, 152

	Page
Loxonema scitula	25, 46
semicostata	86
?subattenuatum	195
Lucina	80, 99, 131, 153, 207, 208, 263, 268
acutilineata	266
cleburni	157
cretacea	268
(Diplodonta?) subundata	263
fibrosa	256
gyrata	220
lirata	80
occidentalis	16, 50, 99, 263
var. ventricosa	99
(Paracyclas) ohioensis	66, 80
pornaera	50
profunda	153, 157
safedensis	269
smockana	268
subtruncata	207
subundata	13, 99, 134
syriaca	207
ventricosa	263
Lucinidæ	30, 99, 110, 134, 263, 268
Lunatia	102, 137, 264
?acutispira	48
concinna	102, 264
minima	221
occidentalis	102
subcrassa	102
utahensis	137
Lunulicardium	235
fragosum	235
Lunulites	220
bonei	220
?dactyloides	247
duclosii	220
interstitia	220
Lutraria?	211, 213, 215, 217
transmontana	215
traskei	213
Lyellia	155
americana	155
Lyonsia (Panopæa) concava	19
Lyopomata	96, 133
Lyosoma	164
powelli	164, 179
Lyropora	60

M.

Maclurea	130, 191, 234
—— (?)	130, 191
annulata	191
carinata	191
minima	234
subannulata	191
Macrocallista	100
Macrocheilidæ	132
Macrocheilus	25, 46, 74, 86, 90, 127, 132, 197
——?	46, 197
altonensis	86
auguliferus	127, 132
intercalaris	25, 46
var. pulchellus	74
klipparti	68, 90

300   INDEX OF GENERA AND SPECIES.

	Page		Page
Macrocheilus medialis	25, 16	Margarita abyssinus	48
newberryi	86	mudgeana	102
pallianus	51	nobrascensis	102, 264
pulchellus	25	Margaritana	98
texanus	258	nobrascensis	99, 167
Macrocyclis	107	Margaritella	102
spatiosa	107, 170, 221	abotti	48
Macrodon	73, 86, 90, 117, 197	flexistriata	102
?	85	Marginella biplicata	221
delicatus	64	Marsupiocriuites	42
hamiltonæ	197	Martesia	101, 270
micronema	40	cuneata	101
obsoletus	69, 90	(Pholas) cretacea	270
parvus	117	? rœssleri	69
tenuistriata	51, 73	Martinia	30, 31, 41, 44, 63, 130, 131, 155, 193
tenuistriatus	41, 85	Matthevia	199
truncatus	197	variabilis	199
Macrodontinæ	34	Mazonia	61
Macroscaphites	104	woodiana	61
Macrura	37, 46, 61, 179	Meekella	72, 85, 120, 121, 131, 177
Mactra	93, 100, 111, 207, 211, 213, 215, 218, 250	striato-costata	72, 85, 121, 131, 177
?	217	Meekoceras	146, 162, 239
alta	16	aplanatum	146, 162, 239
arciformis	207	gracilitatis	146, 163, 239
? canonensis	70, 143	var	146, 163
(Cymbophora?) formosa	100	mushbachanus	146
gracilis	100	mushbachianum	162, 239
nitidula	100	Megalaspis	126, 129
? siouxensis	100	belemnurus	126
? utahensis	111	belemnura	129
warrenana	100	Megambonia	79, 194
diegoana	211, 213	jamesi	68, 79
emmonsi	111	occidentalis	194
formosa	16	Megaptera	56, 148
gabiotensis	217	casei	56
gibbsana	27, 93	Megistocrinus	82, 118, 138, 251
gracilis	21	crassus	118
? holmesi	143	evansii	251
? incompta	134	farnsworthi	138
maia	259	parvirostris	82
potrosa	207	plenus	118
pervetus	207	(Saccocrinus) whitei	82
siouxensis	21	Melampus	158
syrinca	207	?	158, 167
texana	215, 218	americana	166, 169
(Trigonella) ? arenaria	111	antiquus	77, 158, 167
warrenana	16	? priscus	22
Mactridæ	100, 111, 134, 269, 271	Melania	28, 112, 137, 162
Mæonia	225	anthonyi	16, 221
axinia	225	areta	22, 221
? carinata	225	claiborneusis	170
elliptica	225	convexa	16, 17
elongata	225	(Goniobasis?) sculptilis	65
fragilis	225	subsculptilis	65
gigas	225	? wyomingensis	77
gracilis	225	humerosa	22, 221
grandis	225	? insculpta	162, 169
myiformis	225	invenusta	17
? recta	225	larunda	137
valida	225	minutula	16, 221
Mæra	100	multistriata	16, 221
Malea	216	nobrascensis	16
ringens	216	? nitidula	22
Malletinæ	32	omitta	17
Margarita	102, 264	(Potodoma) veterna	28, 49

INDEX OF GENERA AND SPECIES. 301

	Page.
Melania ? scalptilis	112, 166, 170
simpsoni	22, 221
sublaevis	17, 221
? subsculptilis	163, 166, 170
subtortuosa	17, 221
taylori	166, 170
tenuicarinata	17, 221
warrenana	17
wyomingensis	162, 169
Melaniidæ	112, 135
Melanopsis	166
Meleagrinella	267
abrupta	267
Melina montana	50
torta	50
Melinia nitidula	221
Melininæ	29, 32
Mellita texana	215
Melonites	23
danæ	23
multipora	43, 60, 212
Menetus	34, 106
Menocephalus	250
Meretrix	211, 212, 213, 214
californiana	211, 213
dariena	212, 214
decisa	50, 211, 213
tularana	211, 213
uniomeris	50, 211, 213
uvasana	211, 213
Merista	57
lævis	57
Meristella	56, 80, 193
Meristella ? (Meristina) cylindrica	80
(Whitfieldia) nasuta	193
Merocrinus	187
corroboratus	188
typus	187
Merostomata	61, 179
Mesalia	102
kansasensis	102
striata	221
Mesodesma	137
bishopi	137
Mesonacis	199
Mesorhytis	103
Metaptera	105
Metis	100
Metoptoma	40, 60, 186, 188, 191, 195, 197
? analoga	191
billingsi	188
cornuta formæ	186
? devonica	195
peroccidens	197
phillipsi	191
(Platyceras) umbella	40
? umbella	60
Michilinia	165, 177, 223, 235
?	235
eugeneæ	177
expansa	165
placenta	165
Micrabacia	96
americana	96
Microcyclus	58

	Page.
Microcyclus discus	58
Microdiscus	198
dawsoni	198
punctatus	198
Microdoma	40, 87
conica	40, 87
Microdon	194, 107
(Cypricardella) connatus	197
macrostriatus	194
Micromeris minutissima	220
parva	220
Micropurgus minutulus	221
Microstizia	96
millepunctata	96
Micropyrgus	107, 162
minutulus	107, 162, 169
Milthea	99
Mitra costata	221
flemingii	221
fussoides	221
lineata	221
minima	221
mooreana	221
Mnestia	101
Modiola	28, 73, 197, 211, 213, 267, 270
attenuata	47
(Brachydontes) multilinigera	76
burlingtonensis	267
concentrica-costellata	47
contracta	50, 211, 213
cretacea	47
ducatelli	50
granulato-cancellata	47
johnsoni	270
juliæ	47
julia	267
(Lithodomus?) inflata	270
meekii	47
? nevadensis	197
ovata	270
pedernalis	47
(Perna) formosa	28, 49
pertenuis	49
saffordi	47
spiniger	50
? subelliptica	73
(Volsella) subimbricata	80
Modiolaria	98
Modiolopsis	38, 54, 58, 130, 187, 190
(?)	130
acutifrons	224
arcodes	224
cancellata	187
imbricata	224
modiolaris	232
modioliformis	54
occidens	191
orthonata	54
perovata	38, 58
pholadiformis	232
pogonipensis	191
prærupta	224
siliqua	224
simplex	224
subnasuta	64, 91

INDEX OF GENERA AND SPECIES.

	Page.
Modiomorpha	110, 194, 197
altiformo	194
ambigua	197
? desiderata	197
? lata	110
oblonga	194
obtusa	194
? ovata	110
? pintocnsis	197
Mollusca 24, 31, 32, 33, 38, 40, 41, 43, 44, 45, 54, 55, 56, 57, 58, 59, 60, 61, 68, 72, 78, 80, 84, 85, 88, 90, 91, 94, 95, 96, 105, 107, 108, 109, 110, 111, 112, 118, 120, 133, 135, 136, 137, 138, 142, 144, 148, 151, 153, 154, 155, 176, 225, 255, 260	
vera	132, 133, 164
Molluscoidæ	133, 165, 260
Monoceras sulcatum	222
Monomyaria	97, 105, 132, 133, 263, 266
Monopleura	176
marcida	176
pinguiscula	176
Monopteria	41, 45, 127, 132, 178
gibbosa	178
marian	127, 132
Monotis	86, 258
——?	258
? gregaria	64, 86
lawni	19, 20, 51
speluncaria	258
Monticulipora	129, 135, 155
dalii	120
frondosa	155
monticula	138
Mortonia (Periarchus) crustuloides	222
lyelli	220
pileus-sinensis	222
tumida	222
Mortoniceras	104
shoshonense	104
? vermilionense	104
Mulinia	215, 216
densata	215, 216
Multicostatæ	34
Murchisonia	55, 74, 117, 147, 151, 164, 191
bicincta	55
copei	151
inornata	40, 87
major	232
milleri	191
nebrascensis	74
obsolete	67
? prolixa	117
subtæniata	51
terebra	147, 164
Muricidæ	103
Mya	215, 216
abrupta	205
montereyana	215, 216
? subsinuata	216
tellinoides	231
Myacites	31, 34, 111, 164
depressus	31
inconspicuus	111
nebrascensis	21, 34
(Pleuromya) subcompressa	111, 179

	Page.
Myacites (Pleuromya) weberensis	111
subcompressus	164
subellipticus	34
Myacites unionoides	49
Myalina	24, 32, 44, 45, 73, 82, 86, 132, 151, 178, 196, 236, 257, 259
——?	132
angulata	24, 32, 44
apachesi	243
avienloides	21, 32, 236
concentrica	24, 44
congeneris	196
? ((Gervillia) perplana	259
meliniformis	41, 45
(Mytelus) perattenuata	19, 20
nemesis	196
neasus	196
perattennata	32, 51, 86
permiana	32, 151, 236
recta	258
recurvirostra	24, 45
recurvirostris	178
squamosa	20, 258
st. ludovica	84
subquadrata	20, 32, 51, 73, 178, 274
swallovi	45, 51, 73, 132, 178
Myrophoria	111, 133
ambilineata	133
lineata	111
Myomia	224
elongata	224
valida	224
Myriapoda	37, 46, 61
Myrtea	99
Mytelus subarcuatus	16
Mytilarca	194
chemungensis	194
dubia	194
(Plethomytilus) oviformis	194
Mytilidæ	30, 33, 98, 110, 111, 260, 262, 267, 270
Mytiloides	97
Mytilus	30, 98, 116, 164, 211, 213, 217, 260, 262, 267
attenuatus	15
concavus ?	51
engelmanni	274
febristriatus	116
humerus	211, 213
inczensis	217
inflatus	50
meekii	256
multistriatus	30
occidentalis	116
oblivius	267
(Orthonota) ventricosa	116
pedroanus	211, 213
pertenuis	18
subarcuatus	98
whitei	164, 260, 262

N.

Naia	105
Naiadites carbonaria	167
elongata	167
lævis	167
Naidea	105

INDEX OF GENERA AND SPECIES. 303

	Page.
Narica	212, 214
diegoana	50, 212, 214, 221
Nassa	212, 214
intastriata	212
interstriata	214
pedroana	212, 214
Natica	208, 210, 211, 212, 213, 214, 217, 219, 223, 226, 236
acutispira	48
alabamiensis	221
alveata	211, 213, 221
ambigua	14, 48
collina	219
concinna	13
erecta	221
geniculata	212, 214
gibbosa	211, 213, 221
inezana	50, 217
indurata	208
? lelia	236
limula	219
minima	221
moreauensis	14
obliquata	13
occidentalis	14, 231
ocoyana	212, 214
œtites ?	211, 213
orientalis	210
paludinæformis	13
saxea	206
? scalaris	210
subcrassa	15
syriaca	208
texana	219
tuomeyana	16, 48
Naticidæ	102, 111, 132, 264
Naticina obliqua	221
Naticopsis	25, 27, 46, 80, 86, 132, 136, 152, 164, 165, 179
æquistriata	80
altonensis	86, 152
hollidayi	25
? (Idonearca) humilis	80
levis	67, 80
littonana var. genevievensis	40
monilifera	152, 165
nana	46, 132, 179
nodosus	25
(Platyostoma) acquistriata	67
remex	136, 164
subovatus	86
(Trachydomia) nodosa	46
var. hollidayii	46
ventrica	86
wheeleri	86, 179
var.	152
Nautilidæ	33, 63, 104, 121, 132, 265
Nautilina	106
Nautilus	25, 27, 29, 33, 39, 42, 45, 46, 65, 74, 84, 92, 93, 104, 132, 144, 165, 179, 245, 258, 265
——(?)	258
angustatus	50, 206
campbelli	27, 93
chesterensis	25, 45

	Page.
Nautilus (Cryptoceras) capax	39, 92
? leidyi	39
rockfordensis	40
springeri	121
danvillensis	111, 165
dekayi	104
var. montanaensis	104, 265
digonus	42
(Discites) disciformis	39, 84
ornatus var. amplus	39
(Discus) digonus	25
planorbiformis	25
sangamonensis	25
trisulcatus	25
divisus	121
elegans	104
var. nebrascensis	29
(Endolobus) peramplus	39
spectabilis	45
eccentricus	19, 20, 33
forbesianus	179
globatus	45
lamarckii	222
lasallensis	39, 87
missouriensis	179
occidentalis	74
planorbiformis	46
ponderosus	74
sangamonensis	46
(Solenocheilus) collectus	65, 84
leideyi	84
spectabilis	25
subglobosus	25
(Temnocheilus) coxanus	84
latus	87
niotensis	39, 84
winslowi	87
(Temnochilus) coxanus	65
latus	65
winslowi	65
tenui-planatus	227
(Tromatodiscus) sulcatus	40
trisulcatus	42
winslowi	179
Navicula bacilum	229
scalprum	229
semen	229
sificula	229
sigma	229
Neæra	101, 260, 263, 264, 272
acquivalvis	272
fibrosa	48
longirostra	260, 263
moreauensis	101, 264
ventricosa	101
Neithea	215, 218, 266
occidentalis	215, 218
quinquecostata	266
texana	218
Nematocrinus	40
Nemoarca	267
cretacea	267
Nemocardium	99
Nemodon	98, 267
angulatum	267

INDEX OF GENERA AND SPECIES.

	Page
Nemodon brevifrons	267
eufaulensis	267
sulcatinus	98
Nemophora floridana	220
Neptuna impressa	49
Neptunella	103
Nerinea	208, 209, 210, 219
——?	210
abbreviata	**210**
cochleæformis	**210**
cretacea	**209**
orientalis	**210**
rhamdunensis	**208**
schottii	**219**
syriaca	**208,** 210
Nerita	180
——?	180
(Nereis) donsata	48
Neritella	28, 34
nebrascensis	**28,** 34, 49
(Nereis) densata	48
Neritide	34, 133, 134, 180
Neritina	127, 133, 134, 136, 137, 141, 143, 158, 161, 166
bannisteri	158, 167
bruneri	**166,** 169
(Dostia?) bellatula	**76**
carditoides	**76**
(patelliformis)	**76**
incompta	143
naticiformis	**141,** 161, 168
nebrasconsis	49, 167
(Neritella) bannisteri	**76**
pisum	**76**
phaseolaris	**127,** 133
pisiformis	**76,** 158
pisum	143
powelli	**136**
(Velatella) baptista	**141,** 161, 169
bellatula	158, 167
carditoides	134, 158, 167
patelliformis	143
var. webe- rensis	143
volvilineata	**137,** 161, 169
Neritopsis? tuomeyana	48
Neuropteris	149, 246
angulata	**246**
hirsuta	149
rarinervis	149
Neverita gibbosa	221
Nipterocrinus	**54,** 82
arboreus	82
wachsmuthi	**54,** 82
Nodosaria	219
texana	**219**
Nœggerathia	**226**
elongata	226
media	**226**
spatulata	**226**
Notocœli	35
Notomya	224
Notosiphites	35
Nucleospira	193
barrisii	**116**
concinna	193

	Page
Nucula	73, 86, 92, 98, 116, 147, 155, 163, 178, 194, 197, 207, 208, 209, 213, 224, 247, 263, 267, 271
——?	194
abrupta, Conrad	**209**
Dana	**224,** 225
(Acila) conradi	50
? anodontoides	**69**
bellastriata	150
beyrichi	51, 73, 86
cancellata	**15,** 98
circe	**271**
concinna	**225**
crebrilineata	**208**
cultelliformis	220
decisa	213
divaricata	50, **206**
equilateralis	**15**
evansi	**15**
gabbana	**267**
glondonensis	**225**
haydeni	150
impressa	**206,** 231
insularis	**197**
iowensis	**116**
kazanensis	51
levatiforme	**197**
longifrons	267
magna	220
media	220
minuta	247
monmouthensis	**267**
myiformis	**207**
obsoletastriata	**16,** 98
?obtenta	**209**
ovula	220
paralella	**207**
parva	86, 220
pectuncularis	220
penita	49
percrassa	267
perdita	**208**
perequalis	267
perobliqua	**207**
perovata	**208**
perumbonata	147, 163
plana	220
planimarginata	98, 263
planomarginata	**15**
plicata	220
pulcherrima	220
rescuensis	**194**
scitula	**15**
semen	220
slackiana	267
submucronata	**207**
subnasuta	**13**
subplana	**15,** 98
syriaca	**207**
truskana	**17,** 92
ventricosa	**13,** 73, 155, 178
Nuculana	44, 73, 87, 98, 140, 147, 160, 163, 178, 211, 260, 263, 267, 271
albaria	271
bellistriata	178
var. attenuata	73

INDEX OF GENERA AND SPECIES.

	Page.
Nuculana bisulcata	47, 98, 263
compressifrons	268
compsa	220
cultelliformis	220
? curta	44
decisa	**244**
? equilateralis	98
inclara	**140,** 160
longifrons	47
magna	220
media	220
obesa	**147,** 163
oregona	49, 220
ovula	220
parva	220
penita	49
pinnæformis	47, 268
plana	220
plicata	220
protexta	47, 267
pulcherrima	220
semen	220
slackiana	47
subangulata	47
subequilatera	**260,** 263
subnasuta	98
willamettensis	49
Nuculanidæ	32, 260
Nuculaninæ	32, 33
Nucularia	268, 271
papyria	268
secunda	**271**
Nuculidæ	98, 267, 271
Nuculites	235
triangulus	**235**
Nullipora ? obtexta	**119**
Nummulites	209
arbiensis	**209**
floridana	220
mantelli	222
Nyassa	194
parva	**194**

O.

	Page.
Obeliscus melanellus	221
pygmæus	221
striatus	221
Obolella	27, 31, 189, 190, 198, 233, 261
?	198
? ambigua	**190**
chromatica	199
discoidea	189, **233**
nana	**27,** 31, 261
polita	261
transversa	198
?Obolidæ	261
Obolus	56, 259, 261
pectenoides	**259**
? pectenoides	261
(Trimerella) conradi	56
Obovaria	105
? Odontobasis	**103,** 137, 141, 162
buccinoides	**137**
buccinoides	162
constricta	103

	Page.
? Odontobasis formosa	**141,** 162
Odontocephalus	58
Odontopteris	149
subcuneata	149
Ozygia	190, 234
parabola	**234**
? problematia	**190**
producta	**234**
? spinosa	**190**
Olenellus	126, 129, 189
gilberti	126, 129, 189
howelli	129, 189
iddingsi	**189**
powelli	126
Olenus (Olenellus) gilberti	**88**
howelli	**88**
Oligoporus	**25,** 43, 54, 83
coreyi	**64**
danæ	43, 60
nobilus	**54,** 83
Oligoptych	101
Oliva ancillariæformis	50
phillipsii	222
Olivella ancillariformis	50
Olivula ? plicata	222
punctulifera	222
Ollacrinus	42
Omala	100
Omphiscola	106
Onychaster	**61,** 63, 83
barrisi	83
flexilis	**61**
flexilus	83
Onychocrinus	40, 43, 60, 84, 154
diversus	**40,** 60
exculptus	83, 154
monroensis	43
norwoodi	43
ramulosus	154
whitfieldi	84
Operculatum planulatum	222
Omphalotrochus	30
Ophileta	55, 108
complanata var. nana	108
owenana	**55**
Ophioderma	158
? bridgerensis	158
Ophiuroidea	78
Opis	208, 209
equalis	**209**
undatus	**208**
Orbicula	207, 209, 210
?	210
lugubris	49
multilineata	49
prima	**250**
subobliqua	**207**
? syriaca	**209**
Orbiculoidea	72, 89
?	72
Orbitolites discoidea	220
interstitia	220
(Orbitoides) mantelli	222
Orbitulites ? reticulata	**247,** 248
texanus	47

INDEX OF GENERA AND SPECIES.

	Page.
Orthis	29, 30, 57, 58, 62, 72, 78, 85, 109, 116, 129, 130, 131, 148, 151, 165, 177, 189, 190, 192, 198, 234, 235, 247, 248
—— ?	30, 58, 248
arachnoides	233
bellula	78
biforata var. acutilirata	148
lynx	130
billingsi	198
borealis	78
carbonaria	72, 85
coloradoensis	66
crenistria	243
cuneata	250
electra	129
ella	78
emacerata	78
eurekensis	189
fissicosta	78
hamburgensis	190
hibrida	57
impressa	192
insculpta	78
iowensis	62, 148
var. furnarius	58
lonensis	190
mcfarlanei	58, 62, 192
michelini	109
occidentalis	78, 130, 148
pecosii	131, 177, 243
perveta	190
plicatella	78, 130
(Platystrophia) acutilirata	78
biforata	78
dentata	78
laticosta	78
lynx	78
pogonipensis	234
resupinata ?	235
resupinoides	151
retrorsa	78
subelliptica	116
subcarinata	57
subquadrata	78, 148
testudinaria	130, 190, 248 ?
thiemei	116, 165
tricenaria	190
tulliensis	192
umbraculum ?	232
Orthisina	257
—— ?	257
crassa	19, 20
missouriana	52
missouriensis	20
shumardiana	20
umbraculum	20, 274
Orthoceras	25, 27, 36, 39, 44, 55, 74, 79, 87, 91, 94, 109, 110, 126, 129, 154, 179, 187, 195, 198, 248
—— ?	191, 198, 258
anellum	55
angulatum	91
annulato-costatum	44
annulatum	154
annulocostatum	27
baculum	22, 94

	Page.
Orthoceras blakei	110
(Cameroceras) colon	129
colon	126
cribrosum	74
crebristriatum	39, 91
eurekensis	198
expansum	25, 27, 44
fisogramma	67
jolietensis	39, 91
kingii	109
marginale	247
medullare	91
multicameratum	191
nobile	39
nova mexicana	243
oneidaense	187
(Ormoceras) beckii	55
ortoni	68, 79
randolphensis	198
rushensis	87, 170
subbaculum	39
undulatum	247
winchellii	39, 91
Orthocerata	191
Orthoceratitidæ	109, 110
Orthonema	27, 46, 74, 80, 86
conica	40, 86
newberryi	67, 80
salteri	46
subtæniata	51, 74
Orthonota	116
Orthonychia	40
Ostraca	227
Ostracoda	136
Ostrea	28, 33, 93, 94, 105, 111, 133, 136, 142, 151, 158, 160, 172, 173, 205, 206, 209, 211, 213, 216, 217, 219, 233, 236, 237, 262, 266, 270, 271
—— ?	07
alabamensis	237
(Alectryonia) bellaplicata	142
blackii	151, 158
larva	173
linguafelis	271
procumbens	172
sannionis	136, 142
americana	172
anomiæformis	172
anomioides	76, 158, 172
apressa	172
atwoodi	237, 238
barrandei	172
bella	172, 249
bellarugosa	172
belliplicata	172
blackii	172
breweri	172
borealis	237
bryani	172
bourgeoisi	238
boussingaultii	206
carinata	172, 219
carolinensis	172, 237
compressirostra	237
conchaphila	238

INDEX OF GENERA AND SPECIES. 307

Ostrea confragosa............................ 172
 congesta................97, 172, 233, 273, 274
 contracta......................**215**, 219, 237
 convexa............................... 172
 cortex........................ 133, 172, **219**
 corticosa............................ **209**
 crenulata............................. 172
 crenulimarginata.................. 172, 266
 cretacea..........................172, 237
 denticulifera.....................172, 266
 disparilis............................ 238
 diluviana............................. 172
 divaricata........................... '.37'
 elegantula........................... 172
 engelmanni................**22**, 95, 172
 oversa..........................220, 237
 exogyrella........................... 172
 falcata.............................. 172
 falciformis......................... 237
 franklini............................ 172
 fundata.............................. 238
 gabbana......................**28**, 172
 gallus............................... 238
 georgiana............................ 237
 glabra..............**18**, 160, 168, 173
 glauconoides..................... **271**
 glendiformis..................... **270**
 (Gryphæa?) patina................. 97
  uniformis............... **93**
 (Gryphæostrea?) subulata.......... 97
 heermanni.................212, 213, 238
 idriaensis........................71, 173
 inornata...................**21**, 97, 173
 insecura............................ **136**
 larva................................ 266
 lateralis............................ 173
 linguloides...................... **206**
 littlei............................... 173
 lugubris..................93, 173, **219**
 lurida.............................. 238
 lyoni............................... 173
 malleiformis....................... 173
 marshii..................241, 242, 243
 mesenterica......................... 173
 mortoni..........................173, 237
 multilirata...................173, **219**
 nasuta.............................. 173
 obrutus............................ 209
 orientalis........................... 209
 owenana............................ 173
 panda.........................173, 266
 pandæformis....................... 173
 panzana....................**217**, 238
 patercula........................... 172
 patina......................**16**, 173
 peculiaris.......................... 173
 pellucida.......................97, 173
 percrassa........................... 238
 planovata........................... 173
 plumosa.........................173, 266
 prudentia..................**133**, 173
 quadriplicata................142, 173
 robusta......................173, **219**
 sannionis........................... 173
 scapha.............................. 266

Ostrea sculpturata........................... 238
 sellæformis.....................**205**, 237
 soleniscus.................**66**, 76, 158, 173
 strigilecula................**133**, 172, 262
 subalata............................ 173
 subfalcata.......................... 238
 subjecta.....................**217**, 238
 subovata......................173, **255**
 subspatulata..............173, 219, 266
 subtrigonalis.......105, 168, 173, **236**
 syriaca........................... **206**
 tayloriana.......................... 238
 tecticostata..................173, 266
 titan..........................216, 238
 thirsæ.............................. 237
 toroso.............................. 173
 translucida...................**18**, 173
 trigonalis.......................... 237
 toomeyi.........................173, 237
 uniformis........................... 173
 veatchii............................ 238
 veleniana.....................**219**, 238
 velicata....................173, **219**
 vespertina................211, 213, 219, 238
 vicksburgensis..................... 237
 virgata.......................206, **209**
 virginiana.......................... 238
 virginica........................... 238
  var. californica............ 243
 vomer............................... 173
 wyomingensis...................... 77
Ostreidæ.....30, 33, 97, 105, 111, 133, 262, 266, 270, 271
Otozamites macombii.................. **246**
Ovales.................................. 34
Oxystele............................... 180
Oxyrhina mantelli...................... 246
Oxytoma.........**33**, 49, 97, 142, 151, 153, 158, 262

P.

Pachycardium.........................99, 268
 burlingtonense................. **268**
Pachydesma.......................215, 217
 inezana..................**215**, 217
Pachydomus............................ 225
 antiquatus..................... 225
 cuneatus....................... 225
 lævis.......................... 225
 ovalis.......................... 224
 pusillus........................ 224
 sacculus........................ 224
Pachymya...................143, 151, 158
 austinensis...................... 143
 ? compacta...................151, 158
 ? herseyi....................... 143
 ? truncata...................... **70**
Pachyodon........................101, 105
Pachyphyllum woodmani............... 192
Pachythærus........................... 99
Palæoneilo............................ 89
 bedfordensis................... **89**
Palæacis.............................. 154
 cuneatus....................... 154
 cymbia......................... 52
 obtusa.......................... 52
 umbonata....................... 52
Palæacmea........................... 193

INDEX OF GENERA AND SPECIES.

Palæaster 78
    dyeri 69, 78
    granulosus 78
    incomptus 70, 78
    ? jamesii 78
    shafferi 78
Palæchinidæ 43
Palæchinus 23, 43, 83
    burlingtonensis 23, 43
    gracilis 63, 83
Palæocampa 37, 47, 61
    anthrax 37, 47
Palæocaridæ 37
Palæocaris 37, 46, 53, 61, 179
    typus 37, 46, 61, 179
Palæochorda 259, 261
    prima 259, 261
Palæocyclus 62
    kirbyi 62
Palæomanon 193
    rœmeri 192
Palæomoera 100
Palæophycus 260, 261
    occidentalis 260, 261
Palæophyllum 155
    divaricans 155
Palasterina 24
    (Shoenaster) fimbriata 24
Paliurus 143
    pentangulatus 143
Pallium 215, 216, 217
    crassicardo 215
    estrellanum 215, 216, 217
Palmula sagittaria 47
Paludina conradi 16
    leai 16
    leidyi 16
    multilineata 16, 221
    peculiaris 16
    retusa 16
    trochiformis 16
    vetula 16, 221
Pandora 216
    bilirata 216
Panopæa 207, 210, 217, 269, 271
    cooperi 19
    decisa 269
    elliptica 271
    estrellana 50
    (myacites) subelliptica 18
    occidentalis 16
    orientalis 210
    pecterosa 207
    texana 255
Papillina altilis 222
Papyridea (Liopistha) elegantula ... 47
    rostrata 47
    sancti-sabæ 47
Paracyclas 138, 149, 194, 235
    elliptica var. occidentalis ... 149
    occidentalis 194
    peraccidens 235
    sabina 138
Paradoxidæ 31, 108
Paradoxides 108, 198, 199, 248

Paradoxides acadicus 198
    eteminicus 198
    harlani 199
    lamellatus 198
    ? nevadensis 65, 108
Paramithrax 159
    ? walkeri 159
Paranomia 266
    lineata 266
    scabra 266
Parapholas 143, 272
    kneiskerni 272
    sphenoideus 143
Pasceolus 56
    dactylioides 56
Pasithea aciculata 221
    lugubris 221
    notata 221
    scalæ 221
    sulcata 220
    striata 220
Patella 154
    levettei 154
    tenella 223
Patoceras 104
Patula 153
Pecopteris 246
    bullatus 246
    cycloloba 246
    falcatus 246
    mexicana 246
    ? odontopteroides 231
    undulata 231
    var 231
Pecten 24, 30, 205, 207, 208, 211, 212,
    213, 214, 215, 216, 217, 245, 260, 262, 266, 271
    acutiplicatus 30
    affinis 49
    altiplectus 215
    altiplicatus 217
    bella 49
    bellistriata 22
    bellistriatus 49
    broadhendii 51
    calvatus 220
    catilliformis 212, 214
    (Chlamys) craticulus 266
    comptus 224, 226
    coosensis 256
    delumbis 208
    deserti 211, 213, 217
    discus 217
    extenuatus 22, 49
    hawni 51
    humphreysii 205
    illawarrensis 226
    kneiskerni 271
    lenuisculus 226
    magnolia 217
    meekii 215, 217
    missouriensis ? 51
    mitis 226
    (Monotis ?) coloradensis 245
    nebrascensis 15
    neglectus 51

INDEX OF GENERA AND SPECIES. 309

	Page.
Pecten nevadanus	212, 214
newberryi	260, 262
obrutus	208
occidentalis	245, 274
pabloensis	216
planicostatus	266
propatulus	49, 206
quadricostatus	255
quinquenarius	266
rigbyi	241
rigida	13, 47
squamuliforus (?)	266
(Syncyclonema ?) perlamellosus	266
tenuicollis	224, 226
tenuilineatus	24
tenuisculus	224
tenuitestus	266
utahensis	22
venustus	266
Pectenidæ	30, 260, 261, 271
Pectinibranchiata	34, 102, 106, 132, 134, 135
Pectinidæ	32, 33, 97, 132, 133, 263
Pectunculus	99, 217
ellipsis	220
nitens	50, 206
patulus	206
siouxensis	13
subimbricatus	18
Pectunculina parvula	15
Pelagus vanuxemi	222
Penitella	212, 214
spelæa	214
spelæum	212
Pennularia nobilis	203
varidis	203
Pentacrinidæ	33, 133, 262
Pentacrinites	33, 94, 262
asteriscus	236, 262
Pentacrinus	133
astericus	18
asteriscus	33, 133
Pentadia	223, 226
corona	226
reniformis	223
spatangus	223
trigonia	223
Pentamerus	58, 62, 116, 193
borealis	62
comis	58, 193
huspodus	247
lenticularis	116
lotis	193
oblongus	247
subglobosus	58
Pentinidæ	133
Pentremites	26, 54, 83, 91, 118, 149, 249, 251
burlingtonensis	64, 83
conoideus	149
cornutus	26
florealis	255
godoni	149
(Granatocrinus) granulosus	38
laterniformis	249, 251
melo	249, 251
var. projectus	26

	Page.
Pentremites norwoodii	249, 251
pyriformis	149, 248
sirius	118
stelliformis	249, 251
sulcatus	255
(Tricoelocrinus) obliquatus	91
(Troostocrinus?) woodmanii	54, 83, 87
wortheni	83
Periploma claibornensis	220
Periplomya	269, 271
elliptica	269
truncata	271
Porischæchinidæ	42, 43, 44, 60
Perissoptera	102
Peronæa	100
Peronæoderma	100
Perna	29, 32, 217
montana	50, 217
torta	217
Pernopecten	59
shumardianus	59
Perisonota	268
protexta	268
Petalodus alleghaniensis	20
Petraster	26
wilberanus	26
Petricola	211, 213, 271
nova ægyptica	271
pedroana	211, 213
Petricolidæ	271
Petrospongia	43
Phacops	59, 195
anchiops	232
callicephalus	232
rana	59, 195
Phacopsidæ	109
Pharella	101, 158
dakotensis	101
? pealei	76, 158
Phasianella haleana	48
perovata	48
punctata	48
Phenopora multipora	231
Phillipsia	39, 74, 84, 87, 90, 149, 179, 196
?	74, 253
bufo	149
coronata	196
(Griffithides) bufo	65, 84
lodiensis	90
portlockii	39, 84
sangamonensis	39, 87, 179
scitula	39, 87, 179
major	74
perannulata	256, 257
scitula	74
stevensoni	69
tuberculata	65
Philocrinus	36
nebrascensis	36
pelvis	36
Pholadidæ	101, 270, 272
Pholadomya	29, 34, 35, 92, 100, 164, 205, 207, 209, 218, 269
(Cymella) undata	48
decisa	207

	Page.
Pholadomya (Goniomya) borealis	17
(Homomya) audax	225
curvata [?]	225
glendonensis	225
humilis	21, 34
kingii	164
marylandica	205
occidentalis	269
orbiculata	35
papyracea	29, 100
(Platymya) undata	225
(Procardia) hodgii	190
roemeri	269
sancti-sabæ	226
subelongata	17, 92
subventricosa	18, 100
syriaca	209
texana	218
undata M. & H	15, 48
undata Dana	223
Pholadomyidæ	100
Pholas	205, 270
cithara	270
cuncata	18
? lata	270
petrosa	205
Pholidocidaris	83
irregularis	83, 87
Pholidops	79, 192
bellula	192
cincinnatiensis	79
quadrangularis	192
Phonemus (Flabellina) cuneatus	47
sagittarius	47
Phorus	138, 208
exoneratus	138
syriacus	208
Phragmoceras	39, 91, 248
ventricosum ?	247
walshii	39, 91
Phyllites	246
coriaceous	246
venosissimus	246
Phylloceras	93, 104
? halli	104
? ramosus	93
Phylloda	100
Phyllograptus	126, 129
loringi	126, 129
Phyllopoda	61
Phyllotheca australis	227
Phyllotenthis	21, 105
subovata	105
subovatus	21
Physa	107, 135, 137, 141, 143, 161, 197
?	143, 161, 167, 168
bridgerensis	77, 135, 159, 170
carletoni	77, 143, 167
copei	139, 161, 169
felix	141, 161, 169
kanabensis	137
longiuscula	16
nebrascensis	16
pleromatis	135, 170
prisca	180, 197
rhomboidea	16

	Page.
? secalina	107, 170
subelongata	16
Physanoidæ	110, 239
Physella	107
Physetocrinus	81
Physidæ	106, 107, 135, 139
Physinæ	34
Physodon	107
Piestocheilus	49, 103, 159
vicksburgensis	222
Pileopsis tenella	226
alta	226
Pinna	73, 111, 121, 127, 132, 134, 158, 178, 196, 245, 259, 267, 270
calamitoides	256
consimilis	196
hinrichsiana	121
inexpectans	196
kingii	111
lakesi	145, 158
laqueata	267
? lingula	245
ludlovi	259
peracuta	73, 132, 178
petrina	127, 134, 152
rostriformis	270
stevensoni	152
Pinnidæ	111, 121, 132, 134, 267
Pinnularia affinis	229
amphioxys	229
digitus	229
gastrum	229
macilenta	229
mesgongyla	229
oregonica	229
pachyptera ?	203, 229
placentula	229
viridis	229
viridula	229
Pirenella	106, 161
Pisces	221
Pisidium	137
saginatum	137, 168
Pitar	100
Placenticeras	93, 104
placenta	104
var. intercalare	104
(Sphenodiscus) lenticulare	104
vancouverense	93
Placunopsis	86, 90, 142
carbonaria	41, 86, 87
hilliardensis	142
recticardinalis	90
Plagiarca	98
Plagiostoma dumosa	222
echinatum	47
pelagicum	47
Planaria nitens	221
Planella	35
Planorbella	34, 106
Planorbinæ	34
Planorbis	34, 95, 106, 107, 112, 135, 137, 153, 159, 161
?	135
æqualis	153, 170
amplexus	17

INDEX OF GENERA AND SPECIES.

Planorbis (Bathyomphalus) amplexus 106, 161, 169
  kanabensis.**137**, 169
  plano-convexus ..106,
    161, 169
 cirratus 159, 170
 convolutus............**16**, 106, 161, 169
  var................... 106
 fragilis **17**
 (Gyraulus) militaris **153**, 170
 leidyi **21**, 107, 170
 lunata............................ 170
 (Menetus) nebrascensis 167
  vetustus 167
 nebrascensis 170
 spectabilis **22**, 95, 112
  var. utahensis 95, 112
 subumbilicatus **16**
 tenuivolvis **17**
 utahensis **22**, 135, 170
  var. spectabilis......... 170
 veternus 24, 167
 retulus **21**
 vetustus 170
Plantæ.............................128, 259, 261
Planularia cuneata.................... 47
Platyceras........40, 57, 59, 60, 74, 80, 84, 86, 90, 117,
  132, 149, 165, 178, 186, 194, 197
 attenuatum...................... **67**
 biserialis 60
 bivalve **117**
 conradi......................... **194**
 dumosum var. attenuatum...... 80
 equilatera 84, 149
 fissurella 84
 haliotoides.................**40**, 59
 infundibulum 84
 lævigatum **40**
 minutissimum **186**
 multispinosum **67**, 80
 nebrascense 132, 179
 nebrascensis.................... **74**
 nodosum........................ 194
 occidens **197**
 (Orthonychia) chesterense ... **40**
  infundibulum... **40**
  lodiense **67**, 89
  pyramidatum ... 57
  quincyense....... 60
  subplicatum....**40**, 59
 paralium **117**
 piso 197
 [?] reversum 60
 spinigerum 86
 spirale.......................... 57
 subundatum... .... **57**
 thetiforme...................... **194**
 tortum **67**, 90
 tribulosum **165**
 uncum......................**40**, 84
 undulatum **194**
 ventricosum..................... 59
Platychiasma depressum 226
 oculus 226
 rotundatum 226
Platycrinites 82, 83, 85

Platycrinites æqualis 82
 burlingtonensis 82
 (Eucladocrinus) montanaensis **71**
 halli........................ 82
 hemisphericus 83
 incomptus.................. 83
 planus 82
 parvulus 85
 subspinosus................. 82
 tenuibranchiatus 82
Platycrinus .23, 26, 38, 42, 43, 59, 60, 117, 118, 127, 130,
  144, 154, 163, 165, 248, 249, 250, 251
 ——? 127, 130
 americanus **249**, 251
 boonensis................**144**, 165
 burlingtonensis **249**, 251
 corrugatus...............**249**, 251
 discoideus **249**, 251
 haydeni 163
 hemisphæricus **38**, 60, 154
 incomptus................... **117**
 multibranchiatus............... **26**
 nioteusis..................**38**, 60
 oweni **26**
 parvulus **38**
 penicillus...................**25**, 44
 planus 44, 59, **249**, 251
 plenus..................... **23**
 (Pleurocrinus) asper..........**26**, 59
  subspinosus..... 42
 pleurovimensis.............. **118**
 prattenanus................'.....**23**, 43
 quinquenodus **118**
 scobina **26**, 59
 verrucosus **117**
 yandellii **249**, 251
Platyostoma 25, 86, 148, 194, 197
 inornatum................ **197**
 lineatum 194
 nana......................**25**, 27
 niagarense 148
  var. trigonostoma.. 80
 ? trigonostoma................ **67**
 ? tumida..................... **25**
Platychisma...................... 195
 ? ambiguum **195**
 ? depressum **223**
 ? mccoyi **195**
 pelicoides 40
Platytrochus goldfussii 220
 stokesii 220
Plectosolen ? diegoensis............... 220
 parallelus 220
Plethomytilus........................ 194
Pleurocrinus 38, 42
Pleurodictyum....................... 57
 problematicum............. 57
Plenrolimnæa 106, 161
Pleuromya 260, 262
 newtoni.................**260**, 262
Pleurophorus .32, 33, 38, 45, 61, 73, 86, 90, 151, 197, 258
 92
 ? angulatus............ .... 38, 92
 calhouni 33
 ? (cardinia) subcuneata **19**

INDEX OF GENERA AND SPECIES.

Pleurophorus costatiformis..............**38**, 61
  oblongus....................**73**, 86
  occidentalis..........**19**, 32, 73, 258
  meeki......................**197**
  subcostatus............**38**, 45, 151
  ? subcuneata................ 20
  subcuneatus................ 33, 51
  tropidophorus.............. **90**
Pleurotoma 212, 214, 237
  beaumontii.................... 222
  coelata....................... 222
  childreni..................... 222
  desnoyersii.................. 222
  kellogii...................... 222
  lonsdalii..................... 222
  minor......................... **256**
  monilifera.................... 222
  nodocarinata................ 222
  obliqua....................... 222
  ocoyana....................... 212
  pagoda....................... **237**
  platysoma................... **237**
  sayi.......................... 222
  texana........................ 222
  transmontana............**212**, 214
  varicostata................. 222
  venusta...................... **237**
Pleurotomaria 24, 32, 39, 43, 44, 45, 56, 63, 74,
    79, 81, 89, 117, 147, 149, 151,
    164, 165, 179, 191, 197, 248, 258
  ——— ?......................63, 248
  angulata ?................... 248
  broadheadi................. **165**
  brazoensis................. 45
  casii....................... **56**
  chesterensis............**25**, 44
  conoides................. **40**, 87
  cornula................... **232**
  coxana.................. **40**, 87
  cyclonemoides........... **56**
  excelsa................. **246**
  granulostriata......... **24**, 45
  grayvillensis........74, 147, 164
  gurleyi................ **67**, 87
  halliana................. **258**
  haydeniana............ 74
  humerosa..........**19**, 20, 32
  inornata................ **74**
  lucina.................. 81
  lenticularis............ 248
  lonensis................ **191**
  marcoviana.............. 74
  mississippiensis........ **117**
  morrisiana..........224, 226
  multicincta............. 49
  (Murchisonia) meta...... **39**
  muralis................ **250**
  nevadensis............. 197
  newportensis........... **165**
  nodomarginata.......... 197
  nuda................**223**, 226
  obtusispira........... **258**
  perhumerosa............ **74**
  perizomata............ **151**
  peronata.............. **258**

Pleurotomaria pratteni............... **25**, 45
  prontiana..................... **258**
  (Scalites ?) tropidophora ...**70**, 79
  scitula....................**25**, 45
  shumardi..................**25**, 43
  speciosa..................**25**, 45
  sphaerulata................... 179
  spironoma................**40**, 87
  subscalaris..............**25**, 46
  subconstricta.............**24**, 45
  subdecussata................ 74
  subturbinata........**19**, 20, 32
  subsinuata................**25**, 45
  strzeleckiana............... 226
  tabulata................149, 179
  taggarti................**88**, 164
  tenuicincta...............**24**, 45
  texana....................... 40
  textiligera..............**67**, 89
  trifilata...............**223**, 224
  tumida....................... 46
  turbiniformis........**25**, 45, 179
  uniangulata............... **231**
  valvatiformis............**40**, 87
Pleurotomariidae..............32, 39, 63, 103
Plicatula 93, 136, 142, 218, 266
  arenaria.................... **237**
  hydrotheca..............**136**, 142
  incongrua.................. **219**
  striato-costata............. 52
  urticosa.................... 266
Plumulites..................... 191
Podosphenia papula............ 220
Podophthalma 32, 132, 133
Podozamites crassifolia........... **246**
Poecilopoda............189, 191, 195, 198
Polycoelia..................... 275
Polynema...................... 98
Polyphemopsis........27, 46, 86, 149, 155, 179
  ——— (?)................ 179
  chrysallis.............**40**, 86
  fusiformis.............. 149
  inornata............... 46
  nitidula...........46, 155, 179
  peracuta.............46, 179
Polypi.......71, 96, 108, 109, 121, 126, 148, 149, 177
Polypora..................72, 127, 131
  biarmica................. 52
  marginata............... 52
  mexicana................ **274**
  stragula..............**127**, 131
  submarginata........... **72**
Polyrhytes.................... 106
Polyzoa......38, 60, 68, 72, 78, 80, 88, 90, 94, 127, 131,
    144, 147, 151, 163, 178
Porambonites................. 234
  obscurus................. **234**
Porcellia..................59, 117, 118
  crassinoda............... **117**
  nodosa................... 59
  obliquinoda............. **118**
Porifera..............154, 189, 192
Porites astraformis.......... 247
Poroerinus................. 37, 55
  crassus................ **37**, 55

INDEX OF GENERA AND SPECIES.   313

Porocrinus pentagonius 37, 55
Portlandia 98
Posidonomya 90, 110, 194
   devonica 194
   fracta 90
   fragosa 110
   lævis 194
Potamida 105
Poteriocrinites 77, 82, 83, 84, 85
   biselli 84
   (Dendrocrinus) caduceus ... 77
      dyeri 68, 77
      casei 77
      cincinnatien-
        sis 68, 77
      polydactylus. 68, 77
      posticus 77
   hardinensis 84
   macoupinensis 85
   ? perplexus 82
   (Scaphiocrinus) æqualis 83
      bayensis ... 84
      carbonarius. 85
      coreyi 83
      depressus .. 83
      hemispheri-
        cus 85
      huntsvillæ . 84
      macadamsi . 83
      randolphen-
        sis 84
      unicus 83
   (Zeacrinus) arboreus 84
      armiger 84
      cariniferous ... 84
      compactilis 89
      compactilus 84
      concinnus ... 64, 83
      formosus 84
      [Hydreionocrinus]
        acanthoporus. 85
      mucrospinus ... 85
      subtumides..... 84
Poteriocrinus 23, 26, 37, 38, 42. 43, 60, 85, 91, 118
                103, 249, 251
   bursæformis 118
   carinatus 26, 60
   depressus 85
   ? enormis 26
   hemisphericus 52
   indianensis 38, 60
   montanaensis 163
   obuncus 118
   occidentalis 249, 251
   rhombiferus 249, 251
   salignoideus 118
   (Scaphiocrinus) bayensis ... 38
      carbonarius. 26
      decadacty
        lus 23, 43
      norwoodi ... 38
      solidus 26
      subtumidus. 38
      tenuidacty-
        lus 38, 60

Poteriocrinus (Scaphiocrinus) unicus..... 91
             wachsmuthi. 26
   spinosus 249, 251
   subimpressus 26, 60
   swallovi 23, 42
   tenuibrachiatus 26, 60
   tumidus 249, 251
   (Zeacrinus) carbonarius 37
Priscoficus hornii 221
Prionocyclus 94, 104, 150, 265
   macombi 94
   (Prionotropis) woolgari 104
   wyomingensis 159, 265
Prionotropis 104
Procardia 100
Productella 193
Productidæ 31, 62, 108, 109, 121
Productus .. 24, 26, 30, 44, 45, 58, 61, 62, 72, 85, 89, 92,
   94, 108, 109, 119, 130, 131, 149, 155, 163,
       177, 193, 196, 233, 235, 257
   ——? 62, 89, 109, 232
   æquicostatus 253, 274
   brachytbærus 225
   calhounianus 20, 245, 257
   cancrini 52
   cora 177, 243, 255
      var. mogayoni 243
   costatoides 245
   costatus .. 72, 109, 131. 149, 177, 243, 245, 255
   ? 232
   delawarii 243
   dissimilis 62
   elegans 235
   exanthematus 58
   flemingii 52, 243
      var. burlingtonensis ... 235
   fragilis 223, 225
   giganteus 163
   hirsutiformo 193
   horrescens 52
   horridus 52
   ivesi 245
   koninckianus 52
   lævicostatus ? 235
   lævicostus 116
   lasallensis 85
   latissimus 92
   leplayi ? 257
   longispinus 52, 72, 85, 109, 131, 177
   magnus 26, 61
   mexicanus 131, 256, 257
   multistriatus 22, 94, 109
   muricatus 131
   nanus 24, 45
   nebrascensis ... 52, 72, 85, 131, 177, 250,
                 274
   nevadensis 109
   nodosus 245, 246
   norwoodi 20, 257
   occidentalis 245
   orbinianus 52
   parvus 24, 44, 130
   pertenuis 72
   pileolus 256, 257
   popei 256, 257

INDEX OF GENERA AND SPECIES.

Productus prattenianus 20, 52, 72, 100, 131
 (Productella) hallanus **193**
  lachrymosus var.
   limus 193
  lachrymosus var.
   stigmatus 193
  lachrymosus navi-
   cella 193
  speciosus 193
  subaculeatus..... 193
  shumardianus..... 193
  truncatus 193
 punctatus....72, 85, 131, 155, 177, 243, 255
 pustulosus (?) 20, 242
 pyxidiformis 243
 rogersi 20, 52, 233, 245
 scabriculus 243, 245
 scitulus **24**, 43
 semireticulatus..30, 72, 109, 131, 177, 232,
      233, 235, 243, 245, 274
  var. antiquatus .. 257
 semistriatus **22**, 94, 109
 splendens (?) 20, 245, 253, 274
 subaculeatus 94, 108, 196
 subhorridus **109**
 symmetricus 72, 177
 villiersi 274
 viminalis **110**
Proetidæ 109
Proetus 39, 59, 79, 81, 109, 195, 232, 235
 ellipticus **39**, 59
 haldemani...................... 196
 loganensis....................... **235**
 marginalis 196
 peroccidens **235**
 (Phæton) denticulatus.............. **109**
 planimargitus.................... **67**
 planimarginatus 81
 spurlocki **69**, 79
Promacrus 68, 89
 andrewsi 89
Protarca 155
 vetusta................ 155
Protaspongia 189
 fenestrata................ 189
Protaster 78, 83
 ? granuliferus 70, 78
 ? gregarius 83
Protbyris 74, 89
 elegans **69**, 74
 mecki 76
Protista.............................. 154
Prosobranchiata 32, 34
Prosopocephala 132, 264
Protocardia 34, 92, 99, 218
 (Leptocardia?) pertenuis 100
  rara........... 100
  subquadrata ... 100
 (Protocardia) salinaensis....... 99
 scitula 92
 shumardi 34
Protocaris **199**
 marshi **199**
Protocardium 268, 271
 curtum 271

Protozoa.....31, 41, 43, 55, 56, 58, 71, 120, **121**, 131, 177
Psammobia 260, **262**
 ? cancellato-sculpta............. 46
 ? prematura................. **260**, 262
Psammobiidæ 260, 262
Pseudobuccinum.... ....**18**, 103
 nebrascensis............**18**, 103
Psendoliva sulcata 222
Psendomonotis..........................73, 262
 (Eumicrotis) curta............. 262
   orbiculata.**260**, 262
 hawni 51
 radialis 73
 sinuata 51
Pseudonautilus 194
Pseudoptera 97, 143
Psephis tantilla.......................... 50
Psilomya............................ 134
Pteria 33, 97, 142, 151, 153, 156, 260, 263, 267
 abrupta 47
 convexo plano 47
 cretacea.......... 47
 haydeni 47, 97
 iridescens........................ 47
 laripes 47, 267
 linguiformis 47, 97, 263
  var. subgibbosa........ 97
 [?] multangula 50
 navicula **267**
 nebrascana...................... 47
 (Oxytoma) erecta **153**, 157
 ? gastrodes 143
  mucronata 180
  munsteri 33, 49
  nebrascana........97, 263
  salinensis **151**, 158
 parkensis 142
 pedernalis ....?...... 47
 petrosa...............47, 267
 planisulca................ 47
 (Pseudoptera) fibrosa 97, 263
  propleura 143
  sublevis **260**, 263
 (Pterinea) morganensis **40**
 ?? stabilitatis **158**
 subgibbosa....................... 47
 triangularis 47
Pteriidæ 29, 30
 32, 33, 97, 110, 111, 132, 133, 260, 267, 270
Pteriinæ 32, 33
Pterinea 29, 45, 56, 58, 59, 194, 196
 flabella 194
 macroptera 226
 (Monoptera) gibbosa..........**41**, 45
 newarkensis **194**
 pintoensis **196**
 (Pteronites?) newarkensis **67**
 ? subpapyracea **41**, 58
 thebesensis **56**
 undulata.................... **59**
Pterininæ 29, 32, 33
Pterinopecten 196
 hoosacensis................... **196**
 spio **196**
Pterocerella....................... 48

INDEX OF GENERA AND SPECIES. 315

	Page.
Pterocerella tippana	48
Pterocephalus	190, 234
Pterophyllum robustum	**246**
fragile	**246**
Pteropoda	28, 31, 39, 56, 67, 84, 89, 118, 120, 138, 149, 151, 165, 189, 191, 195, 198
Pterotocrinus	44
chesterensis	44
crassus	44
Ptilodictya	78, 80, 90, 144, 147, 151, 163
(Stictopora) carbonaria	67, 90
gilberti	**66**, 80
sercata	90
shafferi	**68**, 78
triangulata	**144**, 147, 151, 163
Ptychaspis	186, 190, 234
minuta	190
pusulosa	**234**
speciosus	**186**
Ptychoceras	104, 265
crassum	**261**, 265
leai	49
meekanum	**261**, 265
mortoni	**17**, 104
vernenilli	49
Ptychoceratidæ	104
Ptychodus whipplei	**243**, 246
Ptychoparia	129, 190, 191, 198, 199
? annectans	**191**
anytus	190
(Daloma ?) affinis	**190**
dissimilis	**190**
granulosus	190
haguei	190
læviceps	**190**
linnarssoni	**190**
nitidus	190
occidentalis	190
orestes	199
var. thersites	199
onangondiana	198
var. aurora	199
oweni	190
permasutus	**190**
prospectensis	**190**
(Pterocephalus) laticeps	190
occidens	**190**
quadrata	199
robbi	198
rogersi	**199**
similis	**190**
var. robustus	190
(Solenopleura ?) breviceps	**190**
tener	199
unisulcatus	190
Ptychophyllum	108
? infundibulum	**108**
Ptychopteria	196
protoforme	**196**
Pulmonaria	61
Pulmonata	102, 105, 107
Pulmonifera	34, 135, 197
?	264
Psilomya	101, 127
Pupa	137, 159

	Page.
Pupa arenula	137, 159, 170
atavuncula	**159**, 170
bigsbyt	167
helicoides	**16**
incolata	**137**, 170
? leidyi	**77**
(Leucochelia) incolata	150
vermillionensis	167
vetusta	167
Purpura	212
Pyramia	225
Pyramidellidæ	103, 134
Pyramimitracostata	221
Pyramus	**224**
ellipticus	**224**
myliformis	**224**
Pyrgorhynchus mortonis	222
Pyrgulifera	**70**, 95, 162
humerosa	95, 112, 162, 168
Pyrifusus	103
? flexicostatus	49
? haleanus	49
? impressus	49
intertextus	49
(Neptunella) intertextus	103
newberryi	103
subturrites	103
newberryi	49
subturrites	49
Pyrina parryi	**233**
Pyropsis	103
bairdi	103
var. rotula	103
Pyrula bairdi	**14**, 16, 49
glabra	**256**
modesta	50

Q.

Quadrula	105

R.

Radinta	33, 37, 40, 41, 42, 43, 44, 45, 54, 55, 56, 57, 58, 60, 68, 71, 77, 85, 94, 96, 108, 109, 117, 133, 136, 138, 142, 144, 154, 165, 226, 227, 235, 236, 260, 261.
Radix	105
Radula	266
acutilineata	266
pelagica	266
reticulata	266
Raphistoma	55, 108, 130, 191, 234
acuta	**234**
lenticularis	55
nasoni	191
? rotuliformis	108
? trochiscus	108, 130
Receptaculites	55, 91, 129, 190
———?	55, 129
ellipticus	190
elongatus	**190**
formosus	**64**, 91
globularis	55
mammillaris	190
oweni	55
Remopleurides	185
striatulus	**185**

INDEX OF GENERA AND SPECIES.

Rensellæria 57, 63
    conradi 57
    lævis 63
Requienia 176
    patagiata 176
    texana 176
Retepora archimedes 248
    indianensis 248
Retzia ... 30, 72, 79, 116, 131, 147, 163, 178, 196, 257
    acumbonia 119
    (Ambona?) altirostris 119
    compressa 30
    ? meekana 256
    meekiana 257
    mormonii 20, 52, 131, 178
    papillata 256, 257
    punctulifera 73
    radialis 196
    sexplicata 116
    (Trematospira) granulifera 68, 79
    woosteri 147, 163
Rhaphoneis foliacea 229
    lanceolata 229
    oregonica 229
Rhiphidoglossa 102, 132, 133, 134
Rhiphidoglossata 32, 34
Rhizopoda 31, 129, 131, 190
Rhodea 106
Rhodocrinus 59, 150, 163
    nanus 40, 59
    vesperalis 150, 163
Rhombopora 71, 131
    lepidodendroides 71, 131
Rhynchonella 24, 30, 41, 57, 59, 72, 78, 80, 85, 116,
        118, 126, 127, 130, 131, 148, 163, 165,
        177, 193, 196, 235, 236, 245, 257, 262
    ——— ? 30, 33, 36, 62, 257, 274
    angulata 52
    argenturbica 126, 130
    capax 78, 148
    caput testudinus 118
    carolina 80
    castanea 62, 193
    dentata 78, 148
    duplicata 193
    emmonsi 193, 235
    endlichi 88, 163
    eurekensis 196
    gnathophora 30, 236
    guadalupæ 256, 257
    horsfordi 193
    indentata 257
    (Leiorhynchus) laura 193
        nevadensis ... 193
        sinuatus 193
    metallica 127, 131
    missouriensis 41, 59
    myrina 236, 262
    neglecta 80
        var. scobina 70
    nitens 49
    ? occidens 193
    opposita 116
    osagensis 72, 85
    ottumwa 118, 165

Rhynchonella pugnus 193
    pustulosa 116, 235
    rockymontana 131
    speciosa 57
    subtrigona 21
    tennesseensis 148
    tethys 193
    texana 257
    thera 196
    uta 20, 52, 131, 177, 245
    wasatchensis 127, 131
    wilmingtonensis 220
Rhynchonellidæ 30, 33, 62, 108, 109, 131, 262
Rhytophorus 95, 112, 137, 161
    meekii 137, 161, 168
    priscus 93, 112, 161, 168
Rimella curvilirata 48
Ringicula acutispira 48
    biplicata 221
    pulchella 48
    subpellucida 48
Ringiculidæ 101
Ringiculina 29
Ringinella subpellucida 48
Rissoidæ 106
Rostellaria ? 219
    americana 48, 256
    biangulata 14, 48
    ? collina 219
    (texana) 219
    fusiformis 13
    indurata 206
    nebrascensis 48, 274
    rostrata 48
Rostellites 214, 219
    bellus 48
    biplicatus 49
    conradi 49
    nasutus 49
    texana 219
    texanus 214
Rostrifera 35
Rotella 152
    nana 221
    verruculifera 152
Rotundaria 105

S.

Saccocrinus 56
    christyi 56
Sagenella 187
    ambigua 187
Sanguinolaria 259
    oblata 259
Sanguinolites 68, 80, 89, 194, 197
    aeolus 80, 197
    ? combensis 194
    ? gracilis 194
    ? mænia 197
    obliquus 66
    ? obliquus 89
    (Promacrus) missouriensis .. 69
        nasutus 69
    rigidus 194

INDEX OF GENERA AND SPECIES. 317

	Page.
Sanguinolites retusus	**197**
salteri	**197**
sanduskyensis	**66**, 80
? sanduskyensis	194
simplex	**197**
striatus	**197**
Sanguinolites ventricosus	194
Sarcinula	223
costata	247
? obsoleta	**232**
(Porites ?) glabra	**247**
Saxicava	211, 213, 260, 263
abrupta	**211**, 213
jurassica	**260**, 263
Saxicavidæ	101, 269, 271
Scala forshayii	48
Scalaria	205, 237
cerethiformis	**14**
expansa	**205**
forshayii	48
texana	48
unilineata	**237**
Scalpellum inequicostatum	48
Scambula	268
perplana	268
Scaphiocrinus	26, 43, 60, 72, 82, 83, 84, 85, 91, 117, 144, 149, 165
clio	82
delicatus	82
depressus	**64**
fiscellus	82
gibsoni	**144**, 149, 165
gurleyi	**144**, 149, 165
? hemisphærious	72
juvenis	82
macrodactylus	82
nanus	82
notabilis	82
penicillus	82
rudis	82
rusticellus	**117**
scalaris	82
striatus	82
tethys	82
(Zeacrinus) asper	82
lyra	82
scobina	82
serratus	82
Scaphites	28, 104, 135, 250, 265
(Ammonites?) comprimus	**250**
nodosus	**250**
(Discoscaphites) abyssinus	104
cheyennensis	104
conradi	104
var. gulosus	104
conradi var. intermedius	104
mandanensis	104
nicoletii	104
larvæformis	**19**, 104
mandanensis	274
nicolleti	274
nodosus	265
var. brevis	104, 265

	Page.
Scaphites nodosus var. plenus	21, 104
quadrangularis	104, 265
ventricosus	**28**, 104
vermiculus	159
vermiformis	**28**, 104
warreni	104, 135, 265
wyomingensis	265
Scaphitidæ	104, 135, 265
Scaphoides	33
Scenella	189
? conula	**189**
Schænaster	44, 60
fimbriatus	44
wachsmuthi	**40**, 60
Schizambon	**190**
typicalis	**190**
Schizocrinus nodosus ?	231
Schizodesma	100
Schizodus	24, 32, 44, 45, 73, 86, 89, 132, 178, 194, 197
——?	45
amplus	**64**, 86
chesterensis	**24**, 44
cuneatus	90, 197
curtiforme	**197**
curtus	**44**, 73, 86
(Cytherodon) orbicularis	**194**
deparcus	**197**
medinaensis	**67**, 89
obscurus	51
ovatus	32
(Prisconia) perelegans	86
rossicus	51, 52, 87
subtrigonalis	**67**
pintoensis	**197**
wheeleri	73, 132, 178
Schizopyga	215, **216**
californiana	**215, 216**
Schizothærus	211, 213
nutalli	**211**, 213
traskei	**211**
Schoenaster	24
Scoliostoma	195
americana	**195**
Scorpio carbonarius	**53**
Scutella	205
aborti	**205**
crustuloides	222
jonesii	222
lyelli	220
pileus-sinensis	222
rogersi	222
Seacrinus (Prisconaia) perelegans	**64**
Sedidæ	98
Sedgewickia	32, 38, 61, 79, 236
altirostrata	32
concava	32, 236
? compressa	**68**, 79
? fragilis	**68**, 79
(Grammysia ?) neglecta	**68**, 79
(Sanguinolites) ? subarcuata	38, 61
topekaensis	32
Semicassis sowerbii	221
Seila	102
Septopora cestriensis	87
Septastræa ? sexradiata	46

INDEX OF GENERA AND SPECIES.

	Page.
Serpula	35, 105, 135, 144, 165
insita	144, 165
lutrica	135
(Spirorbis) planorbites	51
? tenuicarinata	105
? tenuicarinatus	17
Serpulidæ	35, 105, 135
Serrifusus	163
Sigaretus scopulosum	59
scopulosus	206
Siliqua	270
cretacea	270
Sinistralia	103
Sinuopallia	264, 269
Sinum scopulosum	59
Siphonariidæ	102, 112
?	264
Siphonida	268
Siphonostomata	264
Siphonotreta ? curta	225
Skenidium	192
devonicum	192
Smithia	62, 108
hennahii	108
(Pachyphyllum) woodmani	123
verrilli	62
Solaridæ	108, 180
Solariorbis depressus	221
lineatus	221
nitens	221
Solarium	180
abyssinus	48
henrici	221
ornatum	221
pseudogranulatum	221
wallalense	180
Solecurtus ? ellipticus	225
(Psammobia ?) planulatus	225
Solemya	24, 80, 99, 153
bilix	153, 157
(Jancia) vetusta	80
protexta	59
radiata	24
suplicata	99
ventricosa	205
Solemyidæ	99
Solen curtus	50
? dakotensis	18
dieguensis	220
irradians	48
parallelus	220
(Solecurtus ?) ellipticus	223
planulatus	223
subplicatus	15
Solenidæ	101, 279
Soleniscus	25, 46, 151, 178
brevis	151
(Macrocheilus) fusiformis	175, 178
medialis	175, 178
newberryi	175, 178
paludinaformis	175, 178
ponderosus	175, 178
primigenius	175, 178
texanus	175, 178
ventricosus	175, 178

	Page.
Soleniscus planus	151, 175, 178
typicus	25, 46, 175, 178
Solenocaris	90
Solenochilus	65, 104
Solenoconcha	101, 132, 264
Solenoides	250
iowensis	250
Solenomya	45, 87, 90, 197
? ? anodontoides	90
biarmica	50
curta	197
(Jancia) vetusta	66
radiata	45
Solenopsis	74
solenoides	74
Solidula biplicata	48
Solyma	209
lineolata	260
Spatangidæ	96
Sphæra	110
whitneyi	110
Sphærella inflata	220
Sphærexochus	91
romingeri	91
Sphæriola	90, 260, 264, 269
? cordata	99
? endotrachys	90
? obliqua	88
transversa	260, 264
umbonata	269
? warrenana	99
Sphærium	105, 112, 135, 160
——?	135
formosum	105, 160, 168
idahoense	65, 112, 166, 170
planum	21, 105, 160, 168
recticardinale	21, 105, 160, 168
rugosum	65, 112, 166, 170
subellipticum	105, 160, 168
Sphærocrinus	37
Sphenodiscus	104
Sphenophyllum	155
Sphenophyllum emarginatum	155
schlotheimi	155
Sphenopteris	149, 246
acuta	149
fremonti	231
lobifolia	226
pancifolia	231
triloba	231
trifoliata	231
Sphenopterium	24, 41, 43
compressum	24, 43
cuneatum	24, 43
enorme	24, 41
var. depressum	41
obtusum	24, 43
Sphærocoryphe	185
robustus	185
Spiractæon	29
Spiraxis	221
Spirifer	30, 31, 41, 44, 57, 58, 61, 62, 73, 89, 90, 91, 92, 94, 108, 109, 116, 118, 127, 130, 131, 148, 149, 155, 163, 165, 178, 225, 233, 245, 253, 257
——?	20, 36, 255

INDEX OF GENERA AND SPECIES. 319

	Page.
Spirifer acuminata	149
agelaius	163
annectans	**196**
cameratus	20, 52, 73, 85, 94, 131, 149, 178, 233, 245, 246, 257
carteri	89
cedarensis	256
centronatus	130
compactus	**62**
cupidatus	109
darwinii	225
desiderata	**196**
duodecicostatus	225
englemanni	57, 94, 108
euritines	149, 250
extenuatus	130
fastigatus	**64**, 91
formacula	58
fultonensis	85
glaber	225
var. contracta	**26**
glans cerasus	**118**
gregaria	149
guadalupensis	257
hemicyclus	**57**
hemiplicata	20, **232**
hirtus	**116**
inequicostatis	**250**
iowensis	**250**
kentuckensis	20, 233, 274
kennicotti	**62**
keokuk	92
laminosus	52
leidyi	196
ligus	**250**
lineatus	20, 233, 243, 245
(Martinia) cooperensis	41
franklinii	**63**
glaber var. contracta	44, 127, 131
lineatus	30, 155, 178
maia	193
meristoides	**63**
peculiaris	130
plano convexa	178, 274
plano convexus	31, 73, 131
richardsoni	**63**
sublineatus	**63**
undifera	193
mensebachanus	274
mexicanus	**256**, 257
moosak paliensis	52
neglecta	196
neglectus	91
octoplicata ?	232
paradoxus	58
perextensus	**58**
perlamellosus	57
pennatus	**250**
planoconvexa	26, 274
propinquus	61
radiata	148
rocky-montani	**243**, 245
rocky montanus	131
solidirostris	**116**

	Page.
Spirifer sphalæna	**225**
(Spiriferina) cristata	196
scobina	94
pulcher	91
striatus	130, 131, 243
var. triplicatus	243
strigosus	94
subcardiiformis	163
suborbicularis	91
subundiferus	**58**
sulcifera	**256**, 257
textus	119
trigonalis	196
(Trigonotreta) argentarius	**108**
biplicatus	89
cameratus	109
opinus	90
piñonensis	**65**, 108
scobina	109
striatiformis	89
strigosus	108
texanus	**68**
? texanus	246
triplicata	**232**
utahensis	94, 108
vespertilio	225
Spirifera	193, 196, 235
?	193, 235
alba-pinensis	**235**
centronata	235
disjuncta	193
engelmanni	**22**, 193
macra	**22**
(Martinia) glabra	193
var. nevadensis	193
norwoodi	**22**
parryana	193
piñonensis	319
pulchra	**22**
raricosta	193
scobina	**22**
setigera	235
(Speriferina ?) alia	**236**
striata	235
varicosa	193
Spiriferidæ	31, 62, 108, 109
Spiriferina	30, 73, 109, 118, 127, 131, 178, 196, 236, 257
?	30, 109
billingsii	**256**, 257
houfrayi ?	236
kentuckensis	52, 73, 131, 178
(Martinia) glabra	193
octoplicata	131
pulchra	109
spinosa var. campestris	127
subtexta	**119**
Spirigera	127, 130, 132
Spirigera ?	20
monticola	**127**, 130
obmaxima	130
planosulcata	132
subtilita	20, 132
Spironema	48, 103
bella	48

INDEX OF GENERA AND SPECIES.

	Page.
Spironema tenuilineata	48, 103
Spirorbis	100, 153
? dickhauti	**153**
helix	51
Spirulæa rotula	49
Spondylidæ	266
Spondylus	217, 266
dumosus	222
echinatus	47
estrallensis	**215**
estrellanus	**217**
gregalis	266
Spongiæ	41, 43, 56, 58, 91
Spongiolites	203
Spongolithis acicularis	229
aspera	229
fustis	229
mesogonyla	229
Staguicola	100
Stralagmium concentricum	220
Stauroneis baileyi	229
semen	229
Stavelia	98
Steganocrinus	**40, 42,** 59
araneolus	42
pentagonus	**42,** 59
sculptus	42
Stephanodiscus ——?	203
Stenaster	78
grandis	**69,** 78
Stenophora columnaris	52
Stenotheca	189
acadica	198
elongata	**189**
Stramonita	212, 214
petrosa	**212,** 214
Straparollus	25, 26, 30, 42, 44, 46, 74, 87, 195, 258
——?	258
Straparollus corundanus	**258**
(Euomphalus) minnesotensis	**250**
pernodosus	**65,** 87
rugosus	74
subquadratus	**65,** 87
subrugosus	87
lens	42
newarkensis	**195**
planidorsatus	44
similis	**26,** 44
var. planus	26, 44
umbilicatus	46
Streblopteria	45, 196
similis	**196**
? tenuilineata	45
Strephona ...?	212, 214
pedroana	**212,** 214
Strepsidura	237
conybearii	222
Streptacis	86
whitfieldi	**67,** 86
Streptelasma	155
corniculum	155
Streptorhynchus	116, 119, 177, 190, 192, 235, 245, 257
chemungensis	192
equivalvis	235
inflatus	**116,** 235

	Page.
Streptorhyncus lens	119
minor	**190**
occidentalis	**245**
(Orthisimia) shumardianus	257
pyramidalis	**245**
umbraculum	245
Striarca	98
Striatopora	56, 119, 156
carbonari	**119**
limæana	**156**
missouriensis	**56**
Stricklandinia	57, 91, 135
castellana	**138**
davidsoni	152
deformis	**64,** 91
elongata var curta	57
Stricklandinia salteri	152
Strobilocystites	**138**
calvini	**138**
Stromatopora	156, 192
Strombus	208
pervetus	**208**
pustulifera	156
Stromomena rhomboidalis	130
Strophalosia	257
(Aulosteges) guadalupensis	257
Strophodonta	148, 192, 235, 250
arcuata	192
beckii	36
canace	235
calvini	192
? costata	**250**
demissa	148, 192
headleyana	36
inequiradiata	192
iowensis	**250**
parva	**250**
patersoni	192
perplana	192
punctulifera	192
Strophites grandæva	107
Strophomena	56, 57, 58, 62, 78, 89, 126, 129, 130, 148, 190, 192, 234, 235, 248
alternata	148
angulata?	248
convexa	**247**
deltoidea	248
filitexta	130
Fontinalis	**126,** 129
(Hemipronites) crenistria	89
filitexta	78
nutans	78
plano-convexa	78
planumbona	78
plicata	78
sinuata	78
sulcata	78
nemea	190, **234**
nasuta?	248
planumbona	148
rhomboidalis	36, 58, 78, 192, 235
serica?	248
(Strophodonta)	58
cavumbona	57
demissa	62

INDEX OF GENERA AND SPECIES.

	Page.
Strophomena (Strophodonta) sublemissa	62
unicostata	56
Strophomenidæ	31, 62, 108, 109
Strophostylus	57
cancellatus	57
Strotocrinus	40, 42, 81
? asperrimus	81
ectypus	81
liratus	81
perumbrosus	42, 81
(Physetocrinus ?) asper	81
dilatatus	81
regalis	42
Styliola	195
fissurella	195
fissurella var. intermittens	195
Subulites	56, 91
inflatus	65, 91
(Polyphemopsis) brevis	56
Succinea	137, 159
(Brachyspira) papillispira	159, 170
papillispira	137
Surcula	103
beaumontii	222
cœlata	222
childreni	222
desnoyersii	222
kellogii	222
monilifera	222
nodocarinata	222
obliqua	222
sayi	222
varicostata	222
Surculites	103
Surilella bifrons	229
plicata	229
Surirella	203
campylodiscus	203
splendida	203
Sycotypus	212, 214
ocohanus	214
ocoyanus	50, 212
penitus	221
Symphysurus	191
? goldfussi	191
Synbathocrinus	53, 63, 81, 82, 91
brevis	63, 82
dentatus	249, 251
robustus	91
wachsmuthi	40, 63, 82
Syncyclonema	47, 97, 263
? rigida	47, 97, 263
Synedra spendida	229
alua	229
Synocladia	72, 90, 131, 178
Syphonia biserialis	19, 52, 72, 90, 131, 178
piriformis ?	247
virgulacæ	52
var. biserialis	87
Syntrielasma	45, 72, 85, 177
hemiplicata	45, 72, 85, 177
Syringopora	56, 72, 108, 109, 130, 131, 156
?	109
harveyi	119, 130
hisingeri	192

	Page.
Syringopora lineata ?	247
maclurei	156
multattenuata	72, 131
perelegans	156, 192
Syringothyris	196
cuspidata	196

T.

	Page.
Tæniog'ossa	132, 135
Tæniopteris elegans	246
glossopteroides	246
magnifolia	246
Tanıosoma	215, 216
gregaria	215, 216
Tancredia	34, 99, 143, 164, 262
? æquilateralis	21, 34
americana	99
bulbosa	260, 262
? cælionotus	143
corbuliformis	260, 262
extensa	164
? inornata	262
postica	260, 262
warrenana	21, 34, 262
Tancrediidæ	34, 99, 262
Taonurus	155
colletti	155
Tapes	158, 211, 213, 215, 216, 217, 259
?	217
diversum	211, 213
hilgardi	158
inezensis	217
lineatum	215, 216
montana	217
montanensis	259
wyomingensis	70
Taphius	34, 106
Taxocrinus	37, 44, 58, 60, 82, 149
gracilis	37, 58
multibrachiatus var. colletti	149
semiovatus	44
thiemei	82, 87
Tectibranchiata	101, 264
Tectura ? occidentalis	48
Tecturidæ	134
Tellimera	269
chorea	269
Tellina	28, 100, 111, 207, 209, 211, 212, 213, 214
albaria	206
(Arcopagia) ? cheyennensis	107
aretata	206
bitruncata	206
? cheyennensis	15
congesta	211, 213
dartena	212, 214
diegoana	211, 213
emacerata	206
equilateralis	15
? formosa	21, 48
gracilis	15
? isonema	111
modesta	111
nasuta	206
nitidula	28
obruta	207

	Page
Tellina occidentalis	243
ocoyana	**212,**214
(Œne?) subscitula	100
pedro	**211**
pedroana	213
(Peronæa?) equilateralis	100
scitula	100
plana	220
pronti	**15**
scitula	**15**
subelliptica	**15**
subscitula	**70**
subtortuosa	**16**
syriaca	**207,**269
Tellinella	100
Tellinidæ	100, 111, 269
Tellinides	100
Tellinomya	55, 79, 190
alta	55
contracta	190
? hamburgensis	**190**
? obliqua	79
protensa	**232**
ventricosa	55
Tellinula	100
Temnochilus	65, 104
Tenea	269
pinguis	269
Tentaculites	39, 56, 138, 195
attenuatus	195
bellulus	195
gracilistriatus	195
boyti	**138**
oswegoensis	**39,** 56
scalariformis	195
sterlingensis	**39,** 56
tenuistriatus	**39,** 56
Terebra	237
plicifera	**237**
Terebratella	265
plicata	265
vanuxemi	265
Terebratula	30, 73, 85, 130, 132, 146, 154, 162, 178, 196, 209, 218, 233, 235, 236, 257, 259, 265
——?	30, 206, 225
amygdala	**223,** 225
augusta	146, 162, **236**
bovens	85
bovidens	73, 178
burlingtonensis	**116**
choctawensis	216, **255**
(Dielasma) bovidens	132
burlingtonensis	130
elongata	**223,** 225, 257
formosa	154
harlani	265
hastata	196
helena	**259**
hermonensis	**209**
humboldtensis	236
lachryma	222
leonensis	**220**
marcyi	255
mexicana	**233**
millepunctata	20, **233**

	Page
Terebratula mormonii	**243**
nitens	49, **206**
perinflata	**257**
plano-sulcata	243
rocky montana	**243**
roysii	243
semisimplex	**146,** 162
subtilita	**232,** 233, 243, 253, 255, 274
uta	**243**
utah	**235**
wacoensis	218
wilmingtonensis	220
Terebratulidæ	30, 121, 132, 265
Terebratulina atlantica	265
floridana	265
lachryma	265
Terebrispira	103
Teredidæ	101, 270, 272
Teredo	101, 270, 272
emacerata	**272**
globosa	**18,** 101
irregularis	270
selliformis	**21,** 101
substriata	206
tibialis	270
Tessarolax	158
hitzii	**158**
Testacea	206
Tetrabranchiata	33, 35, 103, 132, 135, 263, 265
Tetradecapoda	36, 46, 61 90
Teuthidæ	105
Textularia phyllodes	47
Thaleops?	248
Thallogenes	126
Thaumastus	106, 161
limnæformis	106, 161, 169
Theca	28, 31
gregaria	31
lanceolata	226
(Pugiunculus) gregarea	**28**
Thecia ramosa	192
Thecosmata	31
Thetis	100, 264
? circularis	100, 264
Thracia	34, 92, 100, 151, 158, 215, 216, 259, 261, 262, 264
? arcuata	**21,** 34
(Corimya) grinnelli	**259**
gracilis	100
mactropis	**215,** 216
mayæformis	**151,** 158
? occidentalis	**17,** 92
? prouti	101
subgracilis	**261,** 264
? sublevis	**21,** 34, 262
? subtortuosa	100
subtruncata	**17,** 92
trapezoides	**205**
Thyatira (?) bisecta	50
Tiara humerosa	221
Tinoporus (Orbitolina) texanus	47
Tornatella elliptica	50
Tornatellæa impressa	220
Toxaster elegans	233
texanus	233
Toxoglossa	134

INDEX OF GENERA AND SPECIES. 323

	Page.
Trachycardium	99
Trachyceras	110, 239
judicaricum	110, 239
var. subasperum.	110
whitneyi	110, 239
Trachyceratidæ	110, **239**
Trachydomia	46
Trachytriton	49, 102
vinculum	49, 102
Trapezium	260, 262
bellefourchensis	**260**, 262
micronema	**76**, 143
subequalis	**260**, 262
truncatum	143
Trematis	126, 129
pannulus	**126**, 129
Trematocrinus	23, 42
fiscellus	**23**
Trematodiscus	**27**, 42, 104
Trematospira	57, 193
? imbricata	57
infrequens	**193**
Triartbrus becki	187
Trichopteris	**231**
filamentosa	**231**
gracilis	**231**
Trigonarca	98, 267
(Breviarca) exigua	98
? salinænsis	98
? siouxensis	98
cuneiformis	267
transversa	267
Trigonatæ	34
Trigonella	100
Trigonellites	35
Trigonia ... 30, 33, 92, 133, 164, 207, 209, 218, 236, 268
alta	**207**
americana	164
cerulea	**268**
conradi	**22**, 23
crenulata	255
cuneiformis	**207**
distans	**210**
emoryi	218
eufanlensis	268
evausana	17
evansi	92
lorentii	**227**
montanaensis	164
mortoni	**268**
paudicosta	**30**
quadrangularis	**236**
texana	**218**
syriaca	**207**, 209
Trigoniidæ	30, 32, 33, 111, 132, 133, 268
Trigonotreta	51, 57, 58, 89
Trilobita	31, 39, 90, 179, 260, 261
Trilobites	28, 148, 149, 249
Trimerella	80
grandis	80
ohioensis	80
Triplesia	80, 190
calcifera	190
ortoni	80
Triton	216

	Page.
Tritonidæ	102
Tritonidea	103
Tritonifusus migrans	50
?tennilineatus	49
Tritonium diegoensis	222
Trochactæon	29
Trochactæonina	29
Trochidæ	102, 180, 264
Trochita	212, 214, 217
antiqua	**67**
carbonaria	**40**
costellata	**217**
diegoana	**212**, 214
trochiformis	221
Trochoceras	39, 79
baeri	**39**, 79
Trochonema	55, 80
tricarinata	**67**, 80
umbilicata	55
Trochus	180
lenticularis ?	248
(Oxystele) euryostomus	**180**
Tropidina	35, 107
Tropidiscus	106
Tropidocardium	99
Tropidoleptus	58
carinatus	58
Tuba ? bella	48
Tubicola	35, 105, 135
Tubipora lamellosa	**247**
Tubulostium dickhauti	157
Tudicola ? dakotensis	49
(Pyropsis) bairdi	49
Tulotoma	138, 162
thompsoni	**138**, 162, 169
Turbinolia	218
goldfussii	220
stokesii	220
texana	**218**
Turbo	227, 258
glabra	30
guadalupensis	**258**
helicinus ?	258
lineatus	221
mudgeanus	**70**
nebrascensis	**14**
paludinæformis	**231**
tenuilineata	48
tenuilineatus	**14**
texanus	**258**
Turbonilla	134, 143
(Chemnitzia?) coalvillensis	**77**, 143
melanopsis	134
maclurii	220
Turrus	101, 137
(Goniochasma) stimpsoni	101
sphenoideus	**137**
(Xylophagella) elegantulus	101
Turricula	103
Turrilites	**49**
cheyennensis	49
(Helicoceras) cochleatus	**18**
? umbilicatus	**18**
Turris	103
minor	103

INDEX OF GENERA AND SPECIES.

Turbis (Sutenla) ? contortus...... 163
    hitzla 163
    texanus 40
Turritella .. 46, 134, 143, 158, 208, 210, 211, 212, 213, 214, 216, 217, 219
    (Aclis?) micronema **76**, 143
    altilira **216**
    bilineata **331**
    coalvillensis........ .... **76**, 143
    convexa **14**
    gatunensis... .... **216**
    inezana **217**
    irrorata 214
    kansasensis.............. **70**
    leonensis............. **220**
    magnicostata **208**
    marnochi **143**
    moreauensis............. **14**
    multilineata.............. **256**
    ocoyana............... **212**, 214
    peralveata. **208**
    planilateris............... **219**
    spironema **76**, 158
    stevensana 46
    striata 221
    syriaca **208**, 210
    uvasana............134, **211**, 213
    variata **217**
Turritellidæ 102, 104, 134, 135
Tylostoma 153
    princeps............. **153**

U.

Uintacrinus............... **93**
    socialis **93**
Umbonium nana 221
Umbrella planulata 222
Unicardium............... 30
    gibbosum............. **30**
Unio, 34, 95, 105, 112, 135, 136, 137, 140, 159, 160, 166, 272
    alatoides 272
    aldrichi **140**, 160, 168
    (Baphia ?) nebrascensis............. **70**
    belliplicatus **66**, 112, 160, 168
    brachyopisthus **137**, 160, 168
    cariosoides 272
    clinopisthus... .... **166**, 170
    condoni **180**
    conesi **139**, 160, 168
    cristonensis **88**, 167
    cryptorhynchus **139**, 160, 168
    danai **18**, 105, 160, 168
    deweyanus **18**, 105, 160, 168
    endlichi **139**, 160, 168
    gallinensis............... **88**
    goniambonatus **140**, 160, 168
    gonionotus............**137**, 160, 168
    haydeni **22**, 95, 112, 170
    holmesianus.............**139**, 160, 168
    hubbardii 167
    humerosoides 272
    leai 170
    leanus **70**
    ligamentinoides 272
    martini 180

Unio meeki................... **139**, 159, 170
    mendax **139**, 168
    nasutoides 272
    nucalis **18**, 54, 167
    penultimus................ 167
    petrinus................ **137**
    prænodontoides **272**
    primævus............**139**, 160, 168
    priscus 15, 105, **160**, 168
    proavitus **139**, 160, 168
    propheticus............**137**, 160, 168
    radiatoides 272
    rectoides **272**
    roanokoides............. 272
    senectus **139**, 160, 168
    shoshonensis............**137**, 159, 170
    stewardi **136**, 167
    subrotundoides 272
    subspatulatus **18**, 105, 160, 168
    tellinoides 170
    terra-rubræ **88**
    vetustus **22**, 95, 112, 135, 160, 168
    washakiensis **70**, 159, 170
Unionidæ 34, 98, 105, 112, 135, 139, 272
Unionopsis 99
Uperocrinus 41
Urosthenes............... **224**
    australis... .... **224**

V.

Valvata 35, 107, 158, 162
    ? montanensis 107, 162, 169
    nana............**77**, 158, 167
    parvula **16**, 107, 162
    scabrida 167
    subumbilicata 107, 162, 169
    ? (Tropidina) scabrida............. 35
Valvatidæ 35, 107
Vanikora 102, 264
    ambigua 48, 102, 264
    diegoana 50, 221
Vanikoridæ 102
Vanikoropsis **102**, 259
    tuomeyana............102, 259
Vanuxemia 54
    dixonensis **41**
    ? dixonensis 54
Velatella 134, 141, 158, 161
Veleda 209, 271
    equilatera **271**
    lintea 269
    nasuta **271**
    tellinoides **269**
    transversa **269**
Veloritina 99, 135, 137, 160
Venericardia (Cardiocardites) carinata 70
        monilicosta 50
        subtenta 50
        occidentalis . 50
        (Pteromeris) abbreviata .. 50
        radians 50
Veneridæ 100, 111, 260, 262, 264, 269, 271
Veniella 99, 134, 264, 268, 271
    conradi 99, 268
    decisa 269

INDEX OF GENERA AND SPECIES. 325

	Page
Veniella elevata	269
goniophora	**99**, 134
humilis	264
inflata	269
mortoni	99
rhomboidea	271
subovalis	268
subtumida	99
trapezoidea	269
trigona	269
(Venilicardia ?) humilis	99
Venilliidæ	264
Venilia	29, 99
gabbana	47
humilis	47
laphami	48
mortoni	**29**
quadrata	47
subtumida	48
Venilicardia	99
Venus	207, 208, 217
——?	206
alveata	50
althleta	50
angustifrons	50, **206**
bisecta	50, **206**
brevilineata	50, 206
? circularis	**16**, 48
floridana	220
indurata	**207**
latilirata	50
lamellifera	**206**
meekiana	48
ripleyana	48
pajaroana	**217**
penita	220
perovalis	**208**
securis	**256**
(Trigonia) tantilla	50
unionides	49
vespertina	**219**
syriaca	**207**
Vermes	135, 138, 144, 153, 165
Vermetus rotula	49
Vestigia	129
Vetelletia (Ancylus) minuta	**16**
Vetericardia	268
crenulirata	268
octolirata	268
Vitrina	106, 161
obliqua	**17**, 169
? obliqua	106, 161
Vitrinidæ	106, 107
Vivipara	28
Viviparidæ	35, 107, 112, 135
Viviparus	35, 107, 112, 135, 137, 141, 162
?——?	136
conradi	107, 112, 162, 169
couesi	**141**, 162, 168
gilli	35
gillianna	167
glaber	50
ionicus	**136**, 170
leai	107, 162, 169

	Page.
Viviparus leidyi	107, 162, 169
var. formosus	107, 169
paludinæformis	170
panguitchensis	**137**, 169
peculiaris	107, 162, 169
plicapressus	**138**, 162, 169
prudentius	162, 169
prudentia	141
retusus	107, 162, 169
reynoldsana	**28**
reynoldsianus	107, 162, 169
trochiformis	107, 135, 162, 169
var	135
wyomingensis	**68**, 170
Volsella	33, 98, 111, 140, 158, 160, 164, 262
attenuata	47, 98
(Brachydontes) laticostata	**140**, 160, 168
multilinigera	158
regularis	**140**, 160, 168
concentrico-costellata	47
conradi	49
contracta	50
cretacea	47
ducatelli	50
formosa	33, 49
galpiniana	98
inflata	50
julia	47
meekii	47, 98
(Modiola) formosa	262
(Modiolina) platynota	**164**
pedernalis	47
pertennis	33, 49, 262
saffordi	47
scalprum var. isonema	111
[?] spinigera	50
subimbricata	164
Volutalithes	211, 219
Voluta[l]ithes	213
californiana	**211**, 213
sayana	219
Volutilithes bella	48
biplicata	49
nasuto	49
Volviceramus	97
Vorticifex	112

W.

Waldheimia	121
compacta	**120**, 121
Websteria	96
cretacea	96
Whitfieldia	193

X.

Xenophora	81
(Pseudophorus) antiqua	80
Xiphosura	46
Xylophaga elegantula	**18**, 48
stimpsoni	18, 48
Xylophagella	48, 101
elegantula	48
Xyphosura	36
Xystracanthus arcuatus	20

INDEX OF GENERA AND SPECIES.

Y.

Yoldia 33, 44, 73, 90, 98, 263
   evansi 98, 263
   impressa 49
   levistriata 44
   microdonta 70, 98
   (Palæoneio ?) carbonaria 69, 90
   scitula 98
   stevensoni 69, 90
   subscitula 33, 73
   ventricosa 98

Z.

Zamites occidentalis 246
Zaphrentis 58, 62, 109, 117, 119, 130, 131,
                   148, 155, 165, 177
   ——? 58, 130
   acutus 117
   calceola 165
   cylindrica 243
   elliptica 119, 165
   excentrica 109, 131
   gibsoni 177
   glans 119
   haysii 36
   mcfarlanei 62
   ? multilamella 109, 232
   Zaphrentis priscus 167
   rafinesquii 155
   recta 62
   spinulifera 125
   stansburyi 109, 232, 243
Zaptychius 188, 197
   carbonaria 180, 188, 197
Zeacrinus 23, 42, 43, 45, 72, 82, 84, 85, 118
   ? armiger 64
   crassus 45
   discus 23, 45
   (Hydreionocrinus?) acantho-
      porus 64, 87
   ? mucrospinus 72
   perangulatus 118
   planobrachiatus 23, 43
   sacculus 118
      var. concinnus 118
   troostanus 23, 42
Zellinidæ 261
Ziphosura 61
Zoophyta 55, 56, 57, 58, 119, 257
Zygospira 57, 78
   cincinnatiensis 78
   headi 79
   modesta 78
   subconcava 57

GENERAL INDEX.

A.

	Page.
Abert, J. W.	273
Africa	156
Agua doce molluscos cretaceos	181
Alabama	17, 96
Alaska fossils	176
Albino flowers	124
Albuquerque	241
Allen, J. A.	123
Alluvium	74, 75
America (North)	49, 139, 141, 147, 166, 167, 179, 198, 242
America (Northwestern)	144, 180
fossils of	217
Amérique	125
Andes fossil ammonite	227
Archimedes limestone	75
Arizona	128, 146, 187, 188, 241, 244
Arkansas	153, 157, 241, 242
River	241, 242
Artesian wells	156, 157, 256, 257, 274
Arthur, Mr	146
Ashley's Fork, Utah	145
Aspinwall	71
Astoria fossils	205
Oregon	205, 206
Atchison Landing	71
Atlantic supracretaceous	205
Tertiary fossils	205
Australia	223, 224
Azoic	123

B.

Bailey, J. W., writings of	203, 273
Bear Creek	146
River	76, 145
fossils	95, 112
Laramie fossils	168
Beckett's	205
Beckwith, E. S.	203
Bell County, Texas	146
Bellevue	71
Benner's fossil shells	205
Bennett's mill	71
Berthoud, E. S.	146
Bexar County, Texas	146
Bibliography	141, 147
Big Spring, Iowa	248
Bijou Creek, Colorado	145
Biographic sketch of F. B. Meek	11
Birds	124
Bird's eye fossils	231
Birds of Iowa	125
Bismarck	176
Bitter Creek	145
fossils	77
group	141
Black Buttes, Wyoming	145
Black Hills	18, 157, 261
fossils	259
Black River fossils	231
Black slate	25
Blake, W. P.	210, 242
Blood, circulation of	140
Braintree argillites fauna	199
Branchial appendages of Trilobites	185
Brazil	181
Brazilian National Museum	181
Bridger group	145
British America	171
North America	147
British Provinces	242
Brownville	71
Bryan's Pass	256
Bull-snake, (Pityophis)	175
Burlington	116, 117
beds	116
group fossils	42, 59, 81

C.

Cache à la poudre	145
Calciferous fossils	166, 231
California	29, 35, 152, 180, 241, 242, 244
desert	244
fossils	210, 212, 216
(North)	17, 231
(Southern)	244
Tertiary fossils	215
Cambrian	188
faunas	198
fossils	189
middle	199
Trilobites	199
Camden clays	265
company, New Jersey	272
Canada	188
de las Uvas	244
(Geol. Surv.)	179
Canadian period	126, 129
Cañon Park, Colorado	145
Carboniferous	13, 18, 21, 23, 31, 32, 35, 38, 41, 59, 63, 64, 69, 74, 75, 76, 81, 92, 108, 116, 123, 126, 127, 128, 130, 131, 136, 148, 150
crinoids	54

327

GENERAL INDEX.

	Page.
Carboniferous fossils	19, 22, 24, 29, 40, 59, 66, 70, 71, 92, 94, 109, 117, 118, 136, 114, 146, 147, 150, 151, 163, 164, 167, 175, 189, 196, 233, 241, 246, 255, 258
(lower)	147
fossils	81, 91, 116, 235, 243
Ostreidae	172
shells	223
(upper)	147
Caroll, Montana	256
Catalogue	210
Cat, domestic instinct in	175
Cedar Bluff	71
Cenozoic	135, 140, 153
Unionidæ	140
Central Pacific Railroad line fossils	253
Cerebro-spinal meningitis	119
Cephalopodes cretaceos	181
Chalk organic remains of	208
Chazy fossils	231
limestone fossils	186
Check list of fossils, Eocene and Oligocene	220
Chemung	13, 116
Chester group fossils	44, 84
Chippewa land district fossils	248
Chouteau limestone	75
Chutes River, Oregon	229
Cincinnati group	37, 68, 69
fossils	55, 77
Claiborne	205
Clinton fossils	80
group fossils	232
Coal beds	273
measures	17, 36, 50, 53, 61, 75, 88, 121, 149, 155
of Europe	128
fossils	45, 85, 90, 92, 121, 177, 235, 256
(upper)	19
United States	128
Coalville, Utah	76, 145
Coast Range	244
Collett, John	148, 154, 174, 177
Colorado	128, 136, 141, 145, 116, 153, 157, 159, 244
River	188
(Eastern)	156
group	145
(Northwestern)	142
of the West	246
Commissioner of Agriculture	156
Common schools	124
Conchiferos cretaceos	181
Cone in cone	122
Conrad, T. A., writings of	205
Contents, table of	5
Cook, G. H.	265
Cooper marble	75
Cope, E. D.	139, 171, 222
Corniferous fossils	80
group fossils	57
Cretaceous	13, 14, 15, 16, 17, 20, 35, 47, 76, 96, 108, 123, 125, 127, 128, 133, 150, 151, 152, 156, 176, 259, 265
corals	146
eastern limit in Iowa	125
(European)	27, 96

	Pag
Cretaceous fossils	21, 22, 23, 27, 28, 66, 70, 71, 76, 9 93, 95, 96, 111, 136, 142, 143, 145, 14 157, 158, 167, 180, 214, 218, 220, 222, 233, 243, 255, 256, 269, 26
invertebrates	181
Iowa	120
(lower)	19
ostreidæ	172
Crow Creek, Colorado	145
Croxton, Mr.	71
Cyathophyeus, nature of	187

D.

Dakota	123, 157, 176, 259, 26!
group	96, 120
(Northwestern)	171
Dall, W. H.	176
Dana, J. D.	242
writings of	223
Danforth Hills, Colorado	145
Dead Sea	206
Deer Creek coal field	188
Denison, Texas	146
Devonian	13, 28, 41, 54, 66, 77, 116, 123, 155, 174, 176
fossils	22, 57, 65, 94, 108, 118, 148, 167, 189, 192, 235
Dodd's ranch, Utah	145
Drift	75, 120, 122
(glacial)	171, 176
Dry Creek (Mexico), Cretaceous fossils	218
Dutton, C. E.	174

E.

Echinodermes cretaceos	181
Echo Cañon	76
Economic geology	75
Economic geology of Moniteau County Missouri	13
Ehrenberg, C. G., writings of	229
El Paso	241, 242
Cretaceous fossils	218
Emory, Major	214
Encrinital limestone	13, 75
Engelman, H.	256
Eocene fossils	170, 205, 210, 212, 220
of Southern States	237
marls	265
fossils	271
ostreidæ	237
Etats-Unis	241
Eureka district, Palæontology of	188
Europe	141, 147
Evans, Colonel	88
Evans, John, and Shumard, B. F.	255

F.

Fair Haven section	205
Far West	128
Fish house, New Jersey	272
Fœtal hydrocephalus	119
Forests	142
Forestry	166, 176
Fort Benton group	96
Clark	17
Collins, Colorado	145

GENERAL INDEX.

	Page.
Fort Leavenworth	256
Pierre group	96
Riley	256
Smith	241
Snelling	248
Tejon	244
Union	141
group	96
Fortieth parallel, palæontology of	108
Fossil corals	223
leaves	273
mollusca	180
plants	149, 155, 226, 246
ridge, Colorado	145
Fox Hill beds	27
Hills fossils	145
group	96, 145
Fremont, J. C	203, 231
French	242
Fresh and brackish water deposits fossils	139
water infusoria, Oregon	203
and land mollusca	150
miocene fossils	180
mollusks	125, 156
paleozoic fossils of Nevada	188
Frontera cretaceous fossils	218

G.

Galena beds fossils	55
Garter snake	142
Gasteropódes cretaceos	181
Gatun, Isthmus of Darien fossils	216
Geinitz, H. B	51
Genus pyrgulifera	180
Geological survey of Iowa, Wisconsin, and Minnesota	249
Geology of North America	233, 242
Georgia	152
Glacial rocks of Colorado	244
Pike's Peak	244
Golden City, Colorado	145
Goniatite limestone	25
Goode, G. Brown	175
Grand Cañon of the Colorado	188
tertiary history of	174
River	246
Great Plains	156, 157
Prairie region	166
Red Pipestone quarry	122
Salt Lake	232
Greeley, Colorado	88, 145, 146
Green River	246
group	145, 171
fossils	166
region	140
sand marls	265
Greenland	141
Grinnell	93
Grupo da Bahia	181
Guthrie County, Iowa	123

H.

Hague, Arnold	188
Haime	52
Hall James	203

	Page.
Hall, James, writings of	231
and Meek, F. B	13
and Whitfield, R. P	233
Hamilton group fossils	58
Harper's Hill Australia	223
Hartt, C. F	181
collection	198
Hawn, F	17
Hayden, F. V	66, 71, 96
Hays, Dr	36, 52
Heilprin, A	172
Angelo, writings of	237
Helderberg	52
(lower) fossils	56
upper	13
fossils	232
Helotes, Texas	146
Hersey, J. C	146
High Point mine	13
Hilgard, E. W	124
Hilliard, Wyoming	145
Hudson River fossils	231
group	157, 175
Hunter, valley of, Australia	223
Hyatt, A	146
Alpheus, writings of	239

I.

Idaho	76, 146
(Southeastern)	146, 162, 239
Illawara, Australia	223
Illinois	24, 26, 36, 37, 38, 40, 41, 50, 53, 54, 67, 81, 87, 144, 154, 247, 249
palæontology	90
survey	179
Indiana	25, 69, 144
fossils	147, 154, 177
Indianapolis	124
Infusorial deposits, Oregon	229
fossils	103
Interior states	164
Introductory note	7
Iowa	26, 116, 117, 121, 123, 138, 142, 247, 249, 250, 251
first annual report	120
and second annual reports	122
geography of	124
lakes of	122
(Northwestern)	123
River	249
soils of	119
(Southern)	120
(Southwestern)	120, 123
geology of	120
(Western)	123
Irish, C. W	123
Isthmus of Darien, Tertiary fossils	212, 214
Ives, J. C	245

J.

Jacun cretaceous fossils	213
Jenney, Walter P	157, 261
John Day group	180
Jordan River	206
Judith River	17, 141
group	96, 139

GENERAL INDEX.

Jurassic 33, 47, 70, 76, 108, 125,
    127, 128, 133, 136, 179, 241, 259
  fossils 16, 21, 22, 23, 27, 28, 29, 30, 66, 71,
    94, 110, 136, 164, 167, 209, 235, 243, 260, 261
  ostreidæ 172
Jura-Trias 144
  fossils 146
Kanab Valley 187
Kansas 19, 71, 151
  (Eastern) 35, 92
  (Northeastern) 17, 19
  River 19
Kennedy Channel 36, 52
Kennicott, Robert 62
Kentucky 124, 153
Keokuk group fossils 43, 60, 83
Kinderhook group fossils 41, 59
King 64
  Clarence 65, 233, 239
Kjærkonmæddings 124, 125

L.

Laguna, New Mexico 241
Lake Superior 232
Laramie fossils 140, 145, 166, 168
  group 136, 141, 145, 156, 171
  fossils 159, 160
  ostreidæ 173
  sea 156
Laredo 218
Leon Springs Cretaceous fossils 218
Lignite 171
  beds fossils 105
Lignitic 141
Little Thompson Creek 145
Locos 75
Loriol, P. de 174, 181
Los Angeles 241, 242, 244
Louisiana 255
Ludlow, William 259

M.

Mackensie River 62
Macomb, J. N. 93, 246
Magnesian limestone, first 75
  second 13, 75
  third 75
  fourth 75
  (lower) fossils 248
Mammoth Cave 124
Marcou, J. B 171
Marcou, Jules 17, 35, 233, 242
  writings of 241
Marcy, R. B 255
Marine Eocene fossils 180
Marl beds (Lower) 265, 266
  (Middle) 265
  fossils 270
  of New Jersey, fossils of 265
Marls (Upper) 265
Marl beds (Upper) fossils 270
Marnoch, G. W 146
Matthew 198
Mauvaises Terres 14
McClellan, G. B 255
Meek, F. B 125, 128

Meek, F. B., writings of 13
  and Haydon, F. V 14, 17, 23, 27
  and Worthen, A. H 23, 35, 40, 53
    63, 81, 90
Mesozoic 127, 133, 140, 153
  fossils 176, 239
  unionidæ 140
Metamorphoses of Triarthrus becki 186
Mexican boundary survey 218
Mexico 171, 218
  Gulf of 124
Mickleborough, John 188
Miller 125
Miller County, Missouri 74
Miller, S. A. 154, 157, 174, 175
Mine la Motte, Missouri 68
Minnesota 123, 249, 250, 251
Miocene 49, 171, 214
  fossils 170, 210, 211, 213
  ? fossils 212
  ostreidæ 237
Mississippi 96
  River 96, 156, 241, 242
  Upper, fossils of 273
  Valley 116, 118
Missouri 13, 68, 74
  River 14, 15, 28, 71, 96
  great falls of 176
  Upper 31, 95, 175, 259
Montagnes Rocheuses 241
Moniteau County, Missouri 13
Montana 139, 176, 244, 259
  (Northeastern) 171
Montreal 179
Morgan County, Missouri 74, 75
Morrison, Colorado 145, 146
Morse, E. S 124
Mountain limestone fossils 243
Mudge, B. F 146
Mullan, John 28

N.

Napoleon 241, 242
Natatory appendages of trilobites 185
National Museum 175
  Park 259
Nebraska 13, 14, 15, 16, 17, 18, 19,
    20, 21, 22, 23, 27, 35, 36, 51, 96, 122, 251, 255
  City 71
  (Eastern) 71
Nevada 53, 128, 188, 244
  paleozoic, section of 189
  (Trias) 110
Newberry, J. S., writings of 245
New genus of Eurypteridæ 187
Jersey 14, 17, 27, 96, 265, 271
  Mexican cretaceous 96
  Mexico 17, 128,
    151, 152, 241, 242, 244, 256, 257, 273, 274
  South Wales fossils 224
  York 116, 232
Newton, Henry 157, 261
Niagara fossils 80
  group 154
  fossils 56, 231

GENERAL INDEX.

N

	Page.
Nicollet, I. N., writings of	273
Niobrara group	96
Nishnabolany	120
Non-marine fossil mollusca, supplement to	180
mollusca	167
North American continent biologically considered	177
Northwest	17
Northwestern Boundary Commission	27
localities	92
Nouveau Mexique	241
Nuevo Leon	171

O.

Oak Creek, Texas, Cretaceous fossils	218
Océan, Pacifique	241
Ocoya Creek (Miocene fossils)	212, 214
Ohio	66, 67, 68, 69, 70, 77, 88, 153
valley	248
Oligocene fossils	220
Ostreidæ	237
Omaha	71
Onondaga limestone	13
Oregon	17, 208, 229, 231, 256
(Eastern)	180
(Western)	180
Organic remains in Oregon	231
Oriskany group fossils	57
Ostreidæ, fossil N. A.	172
Otœ City	71
Owen, D. D., and Shumard, B. F.	249, 251
writings of	247
Owl, great horned	144
Oyster, enemies of	176

P.

Pacific Ocean	28
Railroad	65
R. R. R. Exploration	233
Pacoña or Pacorina Cañon	244
Palæontologic notes	199
Palæontology of Black Hills	261
Colorado River of the West	245
Exploration of Iowa, Wisconsin, and Illinois	247
Fortieth Parallel Survey	233
P. R. R. Surv., vol. vii	217
United States and Mexican Boundary Survey	233
Paleozoic	37, 41, 64, 68
crinoidea	53, 63, 64
fossils	26, 138
groups	187
pteropods	199
rocks	249
of Texas	188
(Upper) fossils	51
Palestine expedition	206
Park range	142
Part I	9
II	113
III	183
IV	201
Parvin, T. S	123
Patuxent River section	205

	Page.
Permian	19, 32, 187
fossils	256, 257, 274
Permo-carboniferous fossils	235
Petermann, A	242
Pike's Peak	244
Plastic clays fossils	266
Plateau Province	136
Platte River	71
Plattsmouth	71
Pliocene	244
fossils	170
ostreidæ	238
Poblazou	273
Pogonip group fossils	190
Point of Rocks, Wyoming	145
Polar Ocean	52
Pope, John	241, 242, 256, 257, 274
Portugueso	181
Post Pliocene	205
ostreidæ	238
Potomac River, section near mouth	205
Potsdam	18, 31, 259
fossils	231, 233
Powell, J. W	126, 157
Prairie fires	124
Pre-carboniferous of Grand Cañon	188
Preservation of fossils	146
Preston	241, 242
Primordial	31, 126, 128
fossils	27, 259, 261
Progress of Invertebrate Palæontology U. S	150, 152, 156, 171
Prospect Mountain group fossils	189
Prout	64
Hiram A, writings of	274
Pueblo, Mexico	153
Pyramid Mount	17

Q.

Quaternary	13, 74, 75

R.

Raritan clays	265
Raton	273
Raynolds, W. F	27
Recent fossils	211, 213
Red quartzite	123
River	241, 242, 255
Report Geology of Iowa, Wisconsin and Minnesota	250
Survey, Iowa	123
Palæontologic field work, 1877	114
Reptilian age	33
Rio Grande del Norte	241, 242
Puercos cretaceous fossils	218
San Pedro cretaceous fossils	218
Riverside	71
Rock Bluff	71
Rockford	25
Rock Springs, Wyoming	1 5
Rocky Mountains	88, 142, 145, 157, 242
Rouge (rivière)	241
Rulo	71
Ryder, John A	172

GENERAL INDEX.

S.

	Page.
Saccharoidal sandstone	13, 74, 75
Sage Creek, Colorado	145
Salado, Texas	146
Saline Co., Missouri	74, 75
Saint John formation	108
St. John, O. H	123
Saint Joseph	71
St. Louis group fossils	43, 84
Saint Mary's river section	205
St. Peter's limestone (lower)	248
Saint Vrain's river, Colorado	145
San Fernando Mission	244
Sierra	244
Francisco Mountains	241
Francisquito ranch	244
Juan	93
Expedition	246
Lorenzo (Peru) fossils	227
Pedro Geological section	241
Santa Fe	246
Saskatchawan Valley	171
Schiel, Dr	203
James, writings of	253
Schlumberger	174
Shell heaps	125
structure	122
Shumard	256
B. F	157
and Owen, D. D	258
writings of	255
G. G	274
Sierra Liebre	244
Madre	244
Nevada	242
of Santa Monica	244
Silliman, B	242
Silurian	31, 38, 41, 53, 54, 64, 70, 76, 77, 128, 174
fossils	66, 70, 71, 76, 108
(Lower)	13, 18, 68, 74, 123, 126, 128, 153, 155
fossils	27, 65, 91, 148, 189, 190, 234
system	73
(Upper)	156, 123, 154, 155
fossils	91, 144, 148
Simpson, J. H	94
Sparrows, English	152
Spergen Hill	76
Spontaneous fission	125
Stansbury, Howard	232
Subcarboniferous	126, 130, 149, 153, 154, 156
fossils	121, 249, 251
Sucia Island	27, 92
Sulphur Creek	76
Summary of distribution of fossils in the Northwest	251
Sunflowers	166
Supplement	273
Susswasserformen, Oregon	229
Swallow, G. C	13
Syria fossils	206

T.

Tanganyika shells	156
Terra del Fuego	224

	Page.
Tertiary	15, 16, 17, 76, 92, 108, 128, 135, 130, 137, 139, 171, 265
classification of	205
fossils	21, 22, 23, 27, 28, 65, 66, 70, 71, 77, 93, 95, 112, 138, 159, 214, 215, 216, 218, 219, 243
Oregon	256
Washington Territory	256
(Lower)	205
(Medial)	205
fossils	215
of California	244
Ostroidæ	237
shells	212
upper	205
Texas	67, 68, 151, 218, 242, 256, 257, 274
cretaceous	176
fossils	214
Paleozoic	188
Tertiary fossils	215
Tierra del Fuego fossils	227
Toads	141
Triassic	88, 108
fossils	110, 144, 162, 167, 235, 239, 246
Trenton Falls, N. Y	185
fossils	187, 231
group fossils	54
limestone	75, 185
fossils	185, 186
Trilobites	185
period	126, 129
Trilobite	185
appendages of	186
eggs of	186
eye of	187
legs of	185
molting of shell	187
new genus of Cambrian	199
organization of	187
Truckee group fossils	166
Tucumcari	241
Tuerto	273
Turkey River fossils	248

U.

Uinta Mountains	142, 145
Ulrich, E. O	174
United States	147, 188, 242
Expl. Exp. fossils	224
National Museum	181
Utah	22, 71, 94, 128, 139, 145, 146, 153, 159, 232, 244
Southern	152
Utica slate	187
fossils	186

V.

Vancouver's Island	17, 92, 256

W.

Walcott, C. D., writings of	185
Walker, D. H	146
Warren, G. K	17, 18, 23
Wasatch group	145
Washington Biological Society	177
Territory	256

	Page.
Waverly group fossils	88, 235
Western States	38, 40, 41, 63, 64, 67, 157, 163, 248, 274
West Indies	141, 147
Virginia	69
Wetherby, A. G.	153
Wheeler, G. M	125, 128, 151, 244
Whipple, A. W.	211, 242
White, C. A.	237, 259
Biography of	115
and Nicholson, H. A.	141, 147
and St. John, O. H	120, 121
writings of	116
Mountain Indian reservation	188
River, Colorado	145
group	96
Indian agency, Colorado	145
Tertiary fossils	96, 107
Whiteaves, J. F.	179
Whitfield, R. P., and Hall, James	272
and White, C. A.	272

	Page.
Whitfield, R. P., writings of	259
Whitney, J. D.	20, 53
Wilkes, Charles	205, 223, 224
Williamson, R. S	210
Winchell	17
Wind River group	96
Tertiary fossils	96, 107
Wisconsin	174, 247, 248, 249, 250, 251
Woodpeckers	125
Wooster, L. C	146
Worthen, A. H	90, 179
Wyoming	68, 71, 76, 139, 145, 146, 150, 153, 159, 244
(Southwestern)	156

Y.

Yampa River, Colorado	145
Yellowstone River	171

Z.

Zoophytes	223

www.ingramcontent.com/pod-product-compliance
Lightning Source LLC
Chambersburg PA
CBHW021154230426
43667CB00006B/396